Studies in Theoretical and Applied Statistics

Selected Papers of the Statistical Societies

For further volumes:
http://www.springer.com/series/10104

Agostino Di Ciaccio • Mauro Coli
Jose Miguel Angulo Ibañez
Editors

Advanced Statistical Methods for the Analysis of Large Data-Sets

Springer

Editors
Agostino Di Ciaccio
University of Roma "La Sapienza"
Dept. of Statistics
P.le Aldo Moro 5
00185 Roma
Italy
agostino.diciaccio@uniroma1.it

Mauro Coli
Dept. of Economics
University "G. d'Annunzio", Chieti-Pescara
V.le Pindaro 42
Pescara
Italy
coli@unich.it

Jose Miguel Angulo Ibañez
Departamento de Estadística e Investigación
Operativa, Universidad de Granada
Campus de Fuentenueva s/n
18071 Granada
Spain
jmangulo@ugr.es

This volume has been published thanks to the contribution of ISTAT - Istituto Nazionale di Statistica

ISBN 978-3-642-21036-5 e-ISBN 978-3-642-21037-2
DOI 10.1007/978-3-642-21037-2
Springer Heidelberg Dordrecht London New York

Library of Congress Control Number: 2012932299

Printed on acid-free paper

Springer is part of Springer Science+Business Media (www.springer.com)

Editorial

Dear reader, on behalf of the four Scientific Statistical Societies: *SEIO, Sociedad de Estadística e Investigación Operativa* (Spanish Statistical Society and Operation Research); *SFC, Société Française de Statistique* (French Statistical Society); *SIS, Società Italiana di Statistica* (Italian Statistical Society); *SPE, Sociedade Portuguesa de Estatística* (Portuguese Statistical Society), we inform you that this is a new book series of Springer entitled *Studies in Theoretical and Applied Statistics*, with two lines of books published in the series "Advanced Studies"; "Selected Papers of the Statistical Societies." The first line of books offers constant up-to-date information on the most recent developments and methods in the fields of Theoretical Statistics, Applied Statistics, and Demography. Books in this series are solicited in constant cooperation among Statistical Societies and need to show a high-level authorship formed by a team preferably from different groups to integrate different research points of view.

The second line of books proposes a fully peer-reviewed selection of papers on specific relevant topics organized by editors, also in occasion of conferences, to show their research directions and developments in important topics, quickly and informally, but with a high quality. The explicit aim is to summarize and communicate current knowledge in an accessible way. This line of books will not include proceedings of conferences and wishes to become a premier communication medium in the scientific statistical community by obtaining the impact factor, as it is the case of other book series such as, for example, "lecture notes in mathematics."

The volumes of *Selected Papers of the Statistical Societies* will cover a broad scope of theoretical, methodological as well as application-oriented articles, surveys, and discussions. A major purpose is to show the intimate interplay between various, seemingly unrelated domains and to foster the cooperation among scientists in different fields by offering well-based and innovative solutions to urgent problems of practice.

On behalf of the founding statistical societies, I wish to thank Springer, Heidelberg and in particular Dr. Martina Bihn for the help and constant cooperation in the organization of this new and innovative book series.

Maurizio Vichi

Preface

Many research studies in the social and economic fields regard the collection and analysis of large amounts of data. These data sets vary in their nature and complexity, they may be one-off or repeated, and they may be hierarchical, spatial, or temporal. Examples include textual data, transaction-based data, medical data, and financial time series.

Today most companies use IT to support all business automatic function; so thousands of billions of digital interactions and transactions are created and carried out by various networks daily. Some of these data are stored in databases; most ends up in log files discarded on a regular basis, losing valuable information that is potentially important, but often hard to analyze. The difficulties could be due to the data size, for example thousands of variables and millions of units, but also to the assumptions about the generation process of the data, the randomness of sampling plan, the data quality, and so on. Such studies are subject to the problem of missing data when enrolled subjects do not have data recorded for all variables of interest. More specific problems may relate, for example, to the merging of administrative data or the analysis of a large number of textual documents.

Standard statistical techniques are usually not well suited to manage this type of data, and many authors have proposed extensions of classical techniques or completely new methods. The huge size of these data sets and their complexity require new strategies of analysis sometimes subsumed under the terms "data mining" or "predictive analytics." The inference uses frequentist, likelihood, or Bayesian paradigms and may utilize shrinkage and other forms of regularization. The statistical models are multivariate and are mainly evaluated by their capability to predict future outcomes.

This volume contains a peer review selection of papers, whose preliminary version was presented at the meeting of the Italian Statistical Society (SIS), held 23–25 September 2009 in Pescara, Italy.

The theme of the meeting was "Statistical Methods for the analysis of large data-sets," a topic that is gaining an increasing interest from the scientific community.

The meeting was the occasion that brought together a large number of scientists and experts, especially from Italy and European countries, with 156 papers and a large number of participants. It was a highly appreciated opportunity of discussion and mutual knowledge exchange.

This volume is structured in 11 chapters according to the following macro topics:

- Clustering large data sets
- Statistics in medicine
- Integrating administrative data
- Outliers and missing data
- Time series analysis
- Environmental statistics
- Probability and density estimation
- Application in economics
- WEB and text mining
- Advances on surveys
- Multivariate analysis

In each chapter, we included only three to four papers, selected after a careful review process carried out after the conference, thanks to the valuable work of a good number of referees. Selecting only a few representative papers from the interesting program proved to be a particularly daunting task.

We wish to thank the referees who carefully reviewed the papers.

Finally, we would like to thank Dr. M. Bihn and A. Blanck from Springer-Verlag for the excellent cooperation in publishing this volume.

It is worthy to note the wide range of different topics included in the selected papers, which underlines the large impact of the theme "statistical methods for the analysis of large data sets" on the scientific community. This book wishes to give new ideas, methods, and original applications to deal with the complexity and high dimensionality of data.

Sapienza Università di Roma, Italy — *Agostino Di Ciaccio*
Università G. d'Annunzio, Pescara, Italy — *Mauro Coli*
Universidad de Granada, Spain — *José Miguel Angulo Ibañez*

Contents

Part III Integrating Administrative Data

Part IV Outliers and Missing Data

Part V Time Series Analysis

Part I
Clustering Large Data-Sets

Clustering Large Data Set: An Applied Comparative Study

Laura Bocci and Isabella Mingo

Abstract The aim of this paper is to analyze different strategies to cluster large data sets derived from social context. For the purpose of clustering, trials on effective and efficient methods for large databases have only been carried out in recent years due to the emergence of the field of data mining. In this paper a sequential approach based on multiobjective genetic algorithm as clustering technique is proposed. The proposed strategy is applied to a real-life data set consisting of approximately 1.5 million workers and the results are compared with those obtained by other methods to find out an unambiguous partitioning of data.

1 Introduction

There are several applications where it is necessary to cluster a large collection of objects. In particular, in social sciences where millions of objects of high dimensionality are observed, clustering is often used for analyzing and summarizing information within these large data sets. The growing size of data sets and databases has led to increase demand for good clustering methods for analysis and compression, while at the same time constraints in terms of memory usage and computation time have been introduced. A majority of approaches and algorithms proposed in literature cannot handle such large data sets. Direct application of classical clustering technique to large data sets is often prohibitively expensive in terms of computer time and memory.

Clustering can be performed either referring to hierarchical procedures or to non hierarchical ones. When the number of objects to be clustered is very large, hierarchical procedures are not efficient due to either their time and space complexities

L. Bocci (✉) · I. Mingo
Department of Communication and Social Research,
Sapienza University of Rome, Via Salaria 113, Rome, Italy
e-mail: laura.bocci@uniroma1.it

A. Di Ciaccio et al. (eds.), *Advanced Statistical Methods for the Analysis of Large Data-Sets*, Studies in Theoretical and Applied Statistics,
DOI 10.1007/978-3-642-21037-2_1,

which are $O(n^2 \log n)$ and $O(n^2)$, respectively, where n is the number of objects to be grouped. Conversely, in these cases non hierarchical procedures are preferred, such as, for example, the well known K-means algorithm (MacQueen 1967). It is efficient in processing large data sets given that both time and space complexities are linear in the size of the data set when the number of clusters is fixed in advance. Although the K-means algorithm has been applied to many practical clustering problems successfully, it may fail to converge to a local minimum depending on the choice of the initial cluster centers and, even in the best case, it can produce only hyperspherical clusters.

An obvious way of clustering large datasets is to extend existing methods so that they can cope with a larger number of objects. Extensions usually rely on analyzing one or more samples of the data, and vary in how the sample-based results are used to derive a partition for the overall data. Kaufman and Rousseeuw (1990) suggested the CLARA (Clustering LARge Applications) algorithm for tackling large applications. CLARA extends their K-medoids approach called PAM (Partitioning Around Medoids) (Kaufman and Rousseeuw 1990) for a large number of objects. To find K clusters, PAM determines, for each cluster, a medoid which is the most centrally located object within the cluster. Once the medoids have been selected, each non-selected object is grouped with the medoid to which it is the most similar. CLARA draws multiple samples from the data set, applies PAM on each sample to find medoids and returns its best clustering as the output. However, the effective of CLARA depends on the samples: if samples are selected in a fairly random manner, they should closely represent the original data set.

A K-medoids type algorithm called CLARANS (Clustering Large Applications based upon RANdomized Search) was proposed by Ng and Han (1994) as a way of improving CLARA. It combines the sampling technique with PAM. However, different from CLARA, CLARANS draws a sample with some randomness in each stage of the clustering process, while CLARA has a fixed sample at each stage. Instead of exhaustively searching a random subset of objects, CLARANS proceeds by searching a random subset of the neighbours of a particular solution. Thus the search for the best representation is not confined to a local area of the data. CLARANS has been shown to out-perform the traditional K-medoids algorithms, but its complexity is about $O(n^2)$ and its clustering quality depends on the sampling method used.

The BIRCH (Balanced Iterative Reducing using Cluster Hierarchies) algorithm proposed by Zhang et al. (1996) was suggested as a way of adapting any hierarchical clustering method so that it could tackle large datasets. Objects in the dataset are arranged into sub-clusters, known as *cluster-features*, which are then clustered into K groups using a traditional hierarchical clustering procedure. BIRCH suffers from the possible "contamination" of cluster-features, i.e., cluster-features that are comprised of objects from different groups.

For the classification of very large data sets with a mixture model approach, Steiner and Hudec (2007) proposed a two-step strategy for the estimation of the mixture. In the first step data are scaled down using compression techniques which consist of clustering the single observations into a medium number of groups. Each group is represented by a prototype, i.e., a triple of sufficient statistics. In the second step the mixture is estimated by applying an adapted EM algorithm to the sufficient statistics of the compressed data. The estimated mixture allows the classification of observations according to their maximum posterior probability of component membership.

To improve results obtained by extended version of "classical" clustering algorithms, it is possible to refer to modern optimization techniques, such as, for example, *genetic algorithms* (GA) (Falkenauer 1998). These techniques use a single cluster validity measure as optimization criterion to reflect the goodness of a clustering. However, a single cluster validity measure is seldom equally applicable for several kinds of data sets having different characteristics. Hence, in many applications, especially in social sciences, optimization over more than one criterion is often required (Ferligoj and Batagelj 1992). For clustering with multiple criteria, solutions optimal according to each particular criterion are not identical. The core problem is then how to find the best solution so as to satisfy as much as possible all the criteria considered. A typical approach is to combine multiple clusterings obtained via single criterion clustering algorithms based on each criterion (Day 1986). However, there are also several recent proposals on multicriteria data clustering based on multiobjective genetic algorithm (Alhajj and Kaya 2008, Bandyopadhyay et al. 2007).

In this paper an approach called *mixed clustering strategy* (Lebart et al. 2004) is considered and applied to a real data set since it is turned out to perform well in problems with high dimensionality.

Realizing the importance of simultaneously taking into account multiple criteria, we propose a clustering strategy, called *multiobjective GA based clustering strategy*, which implements the *K*-means algorithm along with a genetic algorithm that optimizes two different functions. Therefore, the proposed strategy combines the need to optimize different criteria with the capacity of genetic algorithms to perform well in clustering problems, especially when the number of groups is unknown.

The aim of this paper is to find out strong homogeneous groups in a large real-life data set derived from social context. Often, in social sciences, data sets are characterized by a fragmented and complex structure which makes it difficult to identify a structure of homogeneous groups showing substantive meaning. Extensive studies dealing with comparative analysis of different clustering methods (Dubes and Jain 1976) suggest that there is no general strategy which works equally well in different problem domains. Different clustering algorithms have different qualities and different shortcomings. Therefore, an overview of all partitionings of several clustering algorithms gives a deeper insight to the structure of the data, thus helping in choosing the final clustering. In this framework, we aim of finding strong clusters by comparing partitionings from three clustering strategies each of which searches for the optimal clustering in a different way. We consider a classical partitioning technique, as the well known *K*-means algorithm, the *mixed clustering strategy*, which implements both a partitioning technique and a hierarchical method, and the proposed *multiobjective GA based clustering strategy* which is a randomized search technique guided from the principles of evolution and natural genetics.

The paper is organized as follows. Section 2 is devoted to the description of the above mentioned clustering strategies. The results of the comparative analysis, dealing with an application to a large real-life data set, are illustrated in Sect. 3.

2 Clustering Strategies

In this section we outline the two clustering strategies used in the analysis, i.e., the multiobjective GA based clustering strategy and the mixed clustering strategy.

Multiobjective GA (MOGA) Based Clustering Strategy

This clustering strategy combines the K-means algorithm and the multiobjective genetic clustering technique, which simultaneously optimizes more than one objective function for automatically partitioning data set.

In a multiobjective (MO) clustering problem (Ferligoj and Batagelj 1992) the search of the optimal partition is performed over a number of, often conflicting, criteria (objective functions) each of which may have different individual optimal solution. Multi-criteria optimization with such conflicting objective functions gives rise to a set of optimal solutions, instead of one optimal solution, known as Pareto-optimal solution. The MO clustering problem can be formally stated as follows (Ferligoj and Batagelj 1992). Find the clustering $\mathbf{C}^* = \{C_1, C_2, \ldots, C_K\}$ in the set of feasible clusterings Ω for which $f_t(\mathbf{C}^*) = \min_{C \in \Omega} f_t(\mathbf{C}), t = 1, \ldots, T$, where $\mathbf{C}$ is a clustering of a given set of data and $\{f_t, t = 1, \ldots, T\}$ is a set of T different (single) criterion functions. Usually, no single best solution for this optimization task exists, but instead the framework of Pareto optimality is adopted. A clustering $\mathbf{C}^*$ is called Pareto-optimal if and only if there is no feasible clustering $\mathbf{C}$ that dominates $\mathbf{C}^*$, i.e., there is no $\mathbf{C}$ that causes a reduction in some criterion without simultaneously increasing in at least one another. Pareto optimality usually admits a set of solutions called *non-dominated* solutions.

In our study we apply first the K-means algorithm to the entire population to search for a large number G of small homogeneous clusters. Only the centers of those clusters resulting from the previous step undergo the multiobjective genetic algorithm. Therefore, each center represents an object to cluster and enters in the analysis along with a weight (mass) corresponding to the number of original objects belonging to the group it represents. The total mass of the subpopulation consisting of center-units is the total number of objects. In the second step, a real-coded multi-objective genetic algorithm is applied to the subpolulation of center-units in order to determine the appropriate cluster centers and the corresponding membership matrix defining a partition of the objects into K ($K < G$) clusters. Non-Dominated Sorting

Genetic Algorithm II (NSGA-II) proposed by Deb et al. (2002) has been used for developing the proposed multiobjective clustering technique. NSGA-II was also used by Bandyopadhyay et al. (2007) for pixel clustering in remote sensing satellite image data.

A key feature of genetic algorithms is the manipulation, in each generation (iteration), of a population of individuals, called chromosomes, each of which encodes a feasible solution to the problem to be solved. NSGA-II adopts a floating-point chromosome encoding approach where each individual is a sequence of real numbers representing the coordinates of the K cluster centers. The population is initialized by randomly choosing for each chromosome K distinct points from the data set. After the initialization step, the fitness (objective) functions of every individual in the population are evaluated, and a new population is formed by applying genetic operators, such as selection, crossover and mutation, to individuals. Individuals are selected applying the crowded binary tournament selection to form new offsprings. Genetic operators, such as crossover (exchanging substrings of two individuals to obtain a new offspring) and mutation (randomly mutate individual elements), are applied probabilistically to the selected offsprings to produce a new population of individuals. Moreover, the elitist strategy is implemented so that at each generation the non-dominated solutions among the parent and child populations are propagated to the next generation. The new population is then used in the next iteration of the algorithm. The genetic algorithm will run until the population stops to improve or for a fixed number of generations. For a description of the different genetic processes refer to Deb et al. (2002).

The choice of the fitness functions depends on the problem. The Xie-Beni (XB) index (Xie and Beni 1991) and FCM (Fuzzy C-Means) measure (Bezdek 1981) are taken as the two objective functions that need to be simultaneously optimized. Since NSGA-II is applied to the data set formed by the G center-units obtained from the K-means algorithm, XB and FCM indices are adapted to take into account the weight of each center-unit to cluster.

Let $\mathbf{x}_i\,(i = 1,\ldots,G)$ be the J-dimensional vector representing the i-th unit, while the center of cluster $C_k\,(k = 1,\ldots,K)$ is represented by the J-dimensional vector $\mathbf{c}_k$. For computing the measures, the centers encoded in a chromosome are first extracted. Let these be denoted as $\mathbf{c}_1, \mathbf{c}_2, \ldots, \mathbf{c}_K$. The degree u_{ik} of membership of unit $\mathbf{x}_i$ to cluster $C_k\,(i = 1, \ldots, G$ and $k = 1, \ldots, K)$, are computed as follows (Bezdek 1981):

$$u_{ik} = \left(\sum_{h=1}^{K} \left(\frac{d^2(\mathbf{x}_i, \mathbf{c}_k)}{d^2(\mathbf{x}_i, \mathbf{c}_h)} \right)^{\frac{2}{m-1}} \right)^{-1} \quad \text{for} \quad 1 \le i \le G;\ 1 \le k \le K,$$

where $d^2(\mathbf{x}_i, \mathbf{c}_k)$ denotes the squared Euclidean distance between unit $\mathbf{x}_i$ and center $\mathbf{c}_k$ and m $(m \ge 1)$ is the fuzzy exponent. Note that $u_{ik} \in [0,1]$ $(i = 1, \ldots, G$ and $k = 1, \ldots, K)$ and if $d^2(\mathbf{x}_i, \mathbf{c}_h) = 0$ for some h, then u_{ik} is set to zero for all $k = 1, \ldots, K, k \ne h$, while u_{ih} is set equal to one. Subsequently, the centers encoded in a chromosome are updated taking into account the mass p_i of each unit $\mathbf{x}_i (i = 1, \ldots, G)$ as follows:

$$\mathbf{c}_k = \frac{\sum_{i=1}^{G} u_{ik}^m p_i \mathbf{x}_i}{\sum_{i=1}^{G} u_{ik}^m p_i}, \quad k = 1, \ldots, K,$$

and the cluster membership values are recomputed.

The XB index is defined as $\text{XB} = W/n \cdot sep$ where $W = \sum_{k=1}^{K} \sum_{i=1}^{G} u_{ik}^2 p_i d^2(\mathbf{x}_i, \mathbf{c}_k)$ is the within-clusters deviance in which the squared Euclidean distance $d^2(\mathbf{x}_i, \mathbf{c}_k)$ between object $\mathbf{x}_i$ and center $\mathbf{c}_k$ is weighted by the mass p_i of $\mathbf{x}_i$, $n = \sum_{i=1}^{G} p_i$ and $sep = \min_{k \neq h}\{d^2(\mathbf{c}_k, \mathbf{c}_h)\}$ is the minimum separation of the clusters.

The FCM measure is defined as $\text{FCM} = W$, having set $m = 2$ as in Bezdek (1981).

Since we expect a compact and good partitioning showing low W together with high *sep* values, thereby yielding lower values of both the XB and FCM indices, it is evident that both FCM and XB indices are needed to be minimized. However, these two indices can be considered contradictory. *XB* index is a combination of global (numerator) and particular (denominator) situations. The numerator is equal to FCM, but the denominator has a factor that gives the separation between two minimum distant clusters. Hence, this factor only considers the worst case, i.e. which two clusters are closest to each other and forgets about other partitions. Here, greater value of the denominator (lower value of the whole index) signifies better solution. These conflicts between the two indices balance each other critically and lead to high quality solutions.

The near-Pareto-optimal chromosomes of the last generation provide the different solutions to the clustering problem for a fixed number K of groups. As the multiobjective genetic algorithm generates a set of Pareto optimal solutions, the solution producing the best PBM index (Pakhira et al. 2004) is chosen. Therefore, the centers encoded in this optimal chromosome are extracted and each original object is assigned to the group with the nearest centroid in terms of squared Euclidean distance.

Mixed Clustering Strategy

The mixed clustering strategy, proposed by Lebart et al. (2004) and implemented in the package Spad 5.6, combines the method of clustering around moving centers and an ascending hierarchical clustering.

In the first stage the procedure uses the algorithm of moving centers to perform several partitions (called base partitions) starting with several different sets of centers. The aim is to find out a partition of n objects into a large number G of stable groups by cross-tabulating the base partitions. Therefore, the stable groups are identified by the sets of objects that are always assigned to the same cluster in each of the base partitions. The second stage consists in applying to the G centers of the stable clusters, a hierarchical classification method. The dendrogram is built according to Ward's aggregation criterion which has the advantage of accounting for the size of the elements to classify. The final partition of the population is defined by cutting the dendrogram at a suitable level identifying a smaller number K $(K < G)$ of clusters. At the third stage, a so called consolidation procedure is performed to improve the partition obtained by the hierarchical procedure. It consists of applying the method of clustering around moving centers to the entire population searching for K clusters and using as starting points the centers of the partition identified by cutting the dendrogram.

Even though simulation studies aimed at comparing clustering techniques are quite common in literature, examining differences in algorithms and assessing their performance is nontrivial and also conclusions depend on the data structure and on the simulation study itself. For these reasons and in an application perspective, we only apply our method and two other techniques to the same real data set to find out strong and unambiguous clusters. However, the effectiveness of a similar clustering strategy, which implements the K-means algorithm together with a single genetic algorithm, has been illustrated by Tseng and Yang (2001). Therefore, we try to reach some insights about the characteristics of the different methods from an application perspective. Moreover, the robustness of the partitionings is assessed by cross-tabulating the partitions obtained via each method and looking at the Modified Rand (MRand) index (Hubert and Arabie 1985) for each couple of partitions.

3 Application to Real Data

The above-mentioned clustering strategies for large data set have been applied on a real-life data set concerning with labor flexibility in Italy. We have examined the INPS (Istituto Nazionale Previdenza Sociale) administrative archive related to the special fund for self-employed workers, called *para*-subordinate, where the periodical payments made from company for its employees are recorded. The dataset contains about 9 million records, each of which corresponds to a single payment recorded in 2006. Since for each worker may be more payments, the global information about each employee has been reconstructed and the database has been restored. Thus, it was obtained a new dataset of about 1.5 million records ($n = 1,528,865$) in which each record represents an individual worker and the variables, both qualitative and quantitative, are the result of specific aggregations, considered more suitable of the original ones (Mingo 2009).

A two-step sequential, tandem approach was adopted to perform the analysis. In the first step all qualitative and quantitative variables were transformed to nominal or ordinal scale. Then, a low-dimensional representation of transformed variables was obtained via Multiple Correspondence Analysis (MCA). In order to minimize the loss of information, we have chosen to perform the cluster analysis in the space of the first five factors, that explain about 38% of inertia and 99.6% of revaluated inertia (Benzécri 1979). In the second step, the three clustering strategies presented above were applied to the low-dimensional data resulting from MCA in order to identify a set of relatively homogenous workers' groups.

The parameters of MOGA based clustering strategy were fixed as follows: 1) at the first stage, K-means was applied fixing the number of clusters $G = 500$; 2) NSGA-II, which was applied at the second stage to a data set of $G = 500$ center-units, was implemented with number of generations = 150, population size = 100, crossover probability = 0.8, mutation probability = 0.01. NSGA-II was run by varying the number of clusters K to search for from 5 to 9.

For mixed clustering strategy, in order to identify stable clusters, 4 different partitions around 10 different centers were performed. In this way, 4^{10} stable groups were potentially achievable. Since many of these were empty, the stable groups that undergo the hierarchical method were 281. Then, consolidation procedures were performed using as starting points the centers of the partitions identified by cutting the dendrogram at several levels where $K = 5, \ldots, 9$.

Finally, for the K-means algorithm the maximum number of iterations was fixed to be 200. Fixed the number of clusters $K (K = 5, \ldots, 9)$, the best solution in terms of objective function in 100 different runs of K-means was retained to prevent the algorithm from falling in local optima due to the starting solutions.

Performances of the clustering strategies were evaluated using the PBM index as well as the Variance Ratio Criterion (VRC) (Calinski and Harabasz 1974) and Davies–Bouldin (DB) (Davies and Bouldin 1979) indexes (Table 1).

Both VRC and DB index values suggest the partition in six clusters as the best partitioning solution for all the strategies. Instead, PBM index suggests this solution

Table 1 Validity index values of several clustering solutions

Index	Strategy	Number of clusters				
		5	6	7	8	9
PBM	MOGA based clustering	4.3963	5.7644	5.4627	4.7711	4.5733
	Mixed clustering	4.4010	5.7886	7.0855	6.6868	6.5648
	K-means	4.3959	5.7641	7.0831	6.6677	6.5378
VRC	MOGA based clustering	6.9003	7.7390	7.3007	6.8391	6.2709
	Mixed clustering	6.9004	7.7295	7.3772	7.2465	7.2824
	K-means	6.9003	7.7390	7.3870	7.2495	7.2858
DB	MOGA based clustering	1.0257	0.9558	0.9862	1.1014	1.3375
	Mixed clustering	1.0253	0.9470	1.0451	1.0605	1.0438
	K-means	1.0257	0.9564	1.0554	1.0656	1.0495

only for MOGA based clustering strategy, since the optimal solution resulting from MOGA is chosen right on the bases of PBM index values.

MOGA based clustering strategy is found to provide values of indexes that are only slightly poorer than those attained by the other techniques mostly when a greater number of clusters is concerned.

Table 2 reports the MRand index computed for each couple of partitions. Results clearly give an insight about the characteristics of the different methods. Mixed clustering strategy leads to partitions practically similar to those obtained with K-means.

Using MOGA based clustering strategy, the obtained partitions have high degrees of similarity with the other two techniques for K ranging from 5 to 7, while it produces partitions less similar with the others when a higher number of clusters is concerned.

Chosen a partition in six clusters, as suggested by the above validity indices, the comparison of the groups obtained by each strategy points out that they achieve rather similar results – also confirmed by MRand values always greater than 0.97 (Table 2) – leading to a grouping having substantive meanings.

The cross-tabulation of the 6 clusters obtained with each of the three methods also confirms the robustness of the obtained partitioning. In particular, for each cluster resulting from MOGA strategy there is an equivalent cluster in the partitions obtained with both mixed strategy and K-means. The level of overlapping clusters is always greater than 92.3% while mismatching cases are less than 5.8%.

A brief interpretation of the six clusters identified by the mixed clustering strategy along with the related percentage of coverage of each group in every strategy is displayed in Table 3.

The experiments were executed on a personal computer equipped with a Pentium Core 2Duo 2.2 GHz processor. Despite global performances of each strategy are

Table 2 Modified Rand (MRand) index values between couples of partitions

Number of clusters	MOGA vs mixed	MOGA vs *K*-means	Mixed vs *K*-means
5	0.9990	1.0000	0.9989
6	0.9711	0.9994	0.9705
7	0.8841	0.8742	0.9638
8	0.6874	0.6859	0.9856
9	0.6461	0.6422	0.9874

Table 3 Substantive meanings of clusters and coverage in each clustering strategy

Clusters	Mixed (%)	MOGA (%)	*K*-means (%)
1: Young people with insecure employment	30.8	31.7	30.9
2: People with more than a job	12.4	11.2	12.7
3: People with permanent insecure employment	18.6	18.9	18.3
4: Qualified young adults between insecurity and flexibility	15.6	15.5	15.6
5: Strong flexible workers	15.6	15.1	15.5
6: Flexible Northern managers	7.0	7.6	7.0

found not to differ significantly, both mixed and MOGA strategies have taken between 7 and 10 minutes to attain all solutions performing equally favorably in terms of computation time than the *K*-means algorithm.

References

Alhajj, R., Kaya, M.: Multi-objective genetic algorithms based automated clustering for fuzzy association rules mining. J. Intell. Inf. Syst. **31**, 243–264 (2008).
Bandyopadhyay, S., Maulik, U., Mukhopadhyay, A.: Multiobjective genetic clustering for pixel classification in remote sensing imagery. IEEE Trans. Geosci. Remote Sens. **45** (5), 1506–1511 (2007).
Benzécri, J.P.: Sur le calcul des taux d'inertie dans l'analyse d'un questionnaire addendum et erratum à [bin.mult.] [taux quest.]. Cahiers de l'analyse des données **4**, 377–378 (1979).
Bezdek, J.C.: Pattern recognition with fuzzy objective function algorithms. NY: Plenum (1981).
Calinski, R.B., Harabasz, J.: A dendrite method for cluster analysis. Commun. Stat. **3**, 1–27 (1974).
Davies, D.L., Bouldin, D.W.: A cluster separation measure. IEEE Trans. Pattern Anal. Mach. Intell. **1**, 224–227 (1979).
Day, W.H.E.: Foreword: comparison and consensus of classifications. J. Classif. **3**, 183–185 (1986).
Deb, K., Pratap, A., Agarwal, S., Meyarivan, T.: A fast and elitist multiobjective genetic algorithm: NSGA-II. IEEE Trans. Evol. Comput. **6** (2), 182–197 (2002).
Dubes, R.C., Jain, A.K.: Clustering techniques: the user's dilemma. Pattern Recognit. **8**, 247–260 (1976).
Falkenauer, E.: Genetic algorithms and grouping problems. Wiley, NY (1998).
Ferligoj, A., Batagelj, V.: Direct multicriteria clustering algorithm. J. Classif. **9**, 43–61 (1992).
Hubert, L., Arabie, P.: Comparing partitions. J. Classif. **2**, 193–218 (1985).
Kaufman, L., Rousseeuw, P.: Finding groups in data. Wiley, New York (1990).
Lebart, L., Morineau, A., Piron, M.: Statistique exploratoire multidimensionnelle. Dunod, Paris (2004).
MacQueen, J.B.: Some methods for classification and analysis of multivariate observations. In: Proc. Symp. Math. Statist. and Prob. (5th), Univ. of California, Berkeley, Vol. I: Statistics, pp. 281–297 (1967).
Mingo, I.: Concetti e quantità, percorsi di statistica sociale. Bonanno Editore, Rome (2009).
Ng, R., Han, J.: Efficient and effective clustering methods for spatial data mining. In: Bocca, J., Jarke, M., Zaniolo, C. (eds.) Proceedings of the 20th International Conference on Very Large Data Bases, Santiago de Chile, Chile, pp. 144–155 (1994).
Pakhira, M. K., Bandyopadhyay, S., Maulik, U.: Validity index for crisp and fuzzy clusters. Pattern Recognit. **37**, 487–501 (2004).
Steiner, P.M., Hudec, M.: Classification of large data sets with mixture models via sufficient EM. Comput. Stat. Data Anal. **51**, 5416–5428 (2007).
Tseng, L.Y., Yang, S.B.: A genetic approach to the automatic clustering problem. Pattern Recognit. **34**, 415–424 (2001).
Xie, X.L., Beni, G.: A validity measure for fuzzy clustering. IEEE Trans. Pattern Anal. Mach. Intell. **13**, 841–847 (1991).
Zhang, T., Ramakrishnan, R., Livny, M.: BIRCH: an efficient data clustering method for very large databases. In: Proceedings of the Fifteenth ACM SIGACT-SIGMOD-SIGART Symposium on Principles of Database Systems, pp. 103–114 (1996).

Clustering in Feature Space for Interesting Pattern Identification of Categorical Data

Marina Marino, Francesco Palumbo and Cristina Tortora

Abstract Standard clustering methods fail when data are characterized by non-linear associations. A suitable solution consists in mapping data in a higher dimensional feature space where clusters are separable. The aim of the present contribution is to propose a new technique in this context to identify interesting patterns in large datasets.

1 Introduction

Cluster Analysis is, in a wide definition, a multivariate analysis technique that seeks to organize information about variables in order to discover homogeneous groups, or "clusters", into data. In other words, clustering algorithms aim at finding homogeneous groups with respect to their association structure among variables. Proximity measures or distances can be properly used to separate homogeneous groups. The presence of groups in data depends on the association structure over the data. Not all the association structures are of interest for the user. Interesting patterns represent association structures that permit to define groups of interest for the user. According to this point of view the interestingness of a pattern depends on its capability of identifying groups of interest according to the user's aims. It not always corresponds to optimize a statistical criterion (Silberschatz and Tuzhilin 1996).

M. Marino · C. Tortora (✉)
Dip. di Matematica e Statistica, Univ. di Napoli Federico II,
Via Cintia, Monte S. Angelo, I-80126 Napoli, Italy
e-mail: marina.marino@unina.it; cristina.tortora@unina.it

F. Palumbo
Dip. di Teorie e Metodi delle Scienze Umane e Sociali, Università di Napoli Federico II,
Via Porta di Massa 1, 80133 Napoli, Italy
e-mail: fpalumbo@unina.it

A. Di Ciaccio et al. (eds.), *Advanced Statistical Methods for the Analysis of Large Data-Sets*, Studies in Theoretical and Applied Statistics,
DOI 10.1007/978-3-642-21037-2_2, © Springer-Verlag Berlin Heidelberg 2012

For numerical variables one widely used criterion consists in minimizing the within variance; if variables are linearly independent this is equivalent minimizing the sum of the squared Euclidean distances within classes. Dealing with a large dataset it is necessary to reduce the dimensionality of the problem before applying clustering algorithms. When there is linear association between variables, suitable transformations of the original variables or proper distance measures allow to obtain satisfactory solutions (Saporta 1990). However when data are characterized by non-linear association the interesting cluster structure remains masked to these approaches.

Categorical data clustering and classification present well known issues. Categorical data can be combined forming a limited subspace of data space. This type of data is consequently characterized by non-linear association. Moreover when dealing with variables having different number of categories, the usually adopted complete binary coding leads to very sparse binary data matrices. There are two main strategies to cope with the clustering in presence of categorical data: (a) to transform categorical variables into continuous ones and then to perform clustering on the transformed variables; (b) to adopt non-metric matching measures (Lenca et al. 2008). It is worth noticing that matching measures become less effective as the number of variables increases.

This paper focuses the attention on the cluster analysis for categorical data under the following general hypotheses: there is nonlinear association between variables and the number of variables is quite large. In this framework we propose a clustering approach based on a multistep strategy: (a) Factor Analysis on the raw data matrix; (b) projection of the first factor coordinates into a higher dimensional space; (c) clusters identification in the high dimensional space; (d) clusters visualisation in the factorial space (Marino and Tortora 2009).

2 Support Vector Clustering on MCA Factors

The core of the proposed approach consists of steps (a) and (b) indicated at the end of the previous section. This section aims at motivating the synergic advantage of this mixed strategy.

When the number of variables is large, projecting data into a higher dimensional space is a self-defeating and computationally unfeasible task. In order to carry only significant association structures in the analysis, dealing with continuous variables, some authors propose to perform a Principal Component Analysis on the raw data, and then to project first components in a higher dimensional feature space (Ben-hur et al. 2001). In the case of categorical variables, the dimensionality depends on the whole number of categories, this implies an even more dramatic problem of sparseness. Moreover, as categories are a finite number, the association between variables is non-linear.

Multiple Correspondence Analysis (MCA) on raw data matrix permits to combine the categorical variables into continuous variables that preserve the non-linear association structure and to reduce the number of variables, dealing with sparseness few factorial axes can represent a great part of the variability of the data. Let us indicate with $\mathbf{Y}$ the $n \times q$ coordinates matrix of n points into the orthogonal space spanned by the first q MCA factors. For the sake of brevity we do not go into the MCA details; interested readers are referred to Greenacre book (2006). Mapping the first factorial coordinates into a feature space permits to cluster data via a Support Vector Clustering approach.

Support Vector Clustering (SVC) is a non parametric cluster method based on support vector machine that maps data points from the original variable space to a higher dimensional feature space trough a proper kernel function (Muller et al. 2001).

A feature space is an abstract t-dimensional space where each statistical unit is represented as a point. Given an units $\times$ variables data matrix $\mathbf{X}$ with general term $x_{ij}, i = 1, 2, \ldots, n$ and $j = 1, 2, \ldots, p$, any generic row or column vector of $\mathbf{X}$ can be represented into a feature space using a non linear mapping function. Formally, the generic column (row) vector $\mathbf{x}_j$ ($\mathbf{x}'_i$) of $\mathbf{X}$ is mapped into a higher dimensional space F trough a function

$$\varphi(\mathbf{x}_j) = \left(\phi_1(\mathbf{x}_j), \phi_2(\mathbf{x}_j), \ldots, \phi_t(\mathbf{x}_j)\right),$$

with $t > p$ ($t > n$ in the case of row vectors) and $t \in \mathbb{N}$.

The solution of the problem implies the identification of the minimal radius hypersphere that includes the images of all data points; points that are on the surface of the hypersphere are called *support vectors*. In the data space the support vectors divide the data in clusters. The problem consists in minimizing the radius subject to the restriction that all points belong to the hypersphere: $r^2 \geq \left\|\varphi(\mathbf{x}_j) - \mathbf{a}\right\|^2 \quad \forall j$, where $\mathbf{a}$ is the center of the hypersphere and $\|\cdot\|$ denotes the Euclidean norm.

To avoid that only the most far point determines the solution, slack variables $\xi_j \geq 0$ can be added:

$$r^2 + \xi_j \geq \left\|\varphi(\mathbf{x}_j) - \mathbf{a}\right\|^2 \quad \forall j.$$

This problem can be solved using the Lagrangian:

$$L(r, \mathbf{a}, \xi_j) = r^2 - \sum_j \left(r^2 + \xi_j - \left\|\varphi(\mathbf{x}_j) - \mathbf{a}\right\|^2\right)\beta_j - \sum_j \xi_j \mu_j + C\sum_j \xi_j, \tag{1}$$

where $\beta_j \geq 0$ and $\mu_j \geq 0$ are Lagrange multipliers, C is a constant and $C\sum_j \xi_j$ is a penalty term. To solve the minimization problem we set to zero the derivate of L with respect to r, $\mathbf{a}$ and ξ_j and we get the following solutions:

$$\sum_j \beta_j = 1$$

$$\mathbf{a} = \sum_j \beta_j \varphi(\mathbf{x}_j)$$

$$\beta_j = C - \mu_j$$

We remind that Karush–Kuhn–Tucker complementary condition implies:

$$\xi_j \mu_j = 0$$

$$\left(r^2 + \xi_j - \|\varphi(\mathbf{x}_j) - \mathbf{a}\|^2\right) \beta_j = 0$$

The Lagrangian is a function of r, $\mathbf{a}$ and μ_j. Turning the Lagrangian into the more simple Wolfe dual form, which is a function of the variables β_j, we obtain:

$$W = \sum_j \varphi(\mathbf{x}_j)^2 \beta_j - \sum_{j,j'} \beta_j \beta_{j'} \varphi(\mathbf{x}_j) \cdot \varphi(\mathbf{x}_{j'}) \quad \forall \{j, j'\}, \tag{2}$$

with the constraints $0 \leq \beta_j \leq C$.

It is worth noticing that in (2) the function $\varphi(\cdot)$ only appear in products. The dot products $\varphi(\mathbf{x}_j) \cdot \varphi(\mathbf{x}_{j'})$ can be computed using an appropriate kernel function $K(\mathbf{x}_j, \mathbf{x}_{j'})$. The Lagrangian W is now written as:

$$W = \sum_j K(\mathbf{x}_j, \mathbf{x}_j)\beta_j - \sum_{j,j'} \beta_j \beta_{j'} K(\mathbf{x}_j, \mathbf{x}_{j'}). \tag{3}$$

The SVC problem requires the choice of a kernel function. The choice of the kernel function remains a still open issue (Shawe-Taylor and Cristianini 2004). There are several proposal in the recent literature: *Linear Kernel* ($k(x_i, x_j) = \langle x_i \cdot x_j \rangle$), *Gaussian Kernel* ($k(x_i, x_j) = exp(-q\|x_i - x_j\|^2/2\sigma^2)$) and *polynomial Kernel* ($k(x_i, x_j) = (\langle x_i \cdot x_j \rangle + 1)^d$ with $d \in N$ and $d \neq 0$) are among the most largely used functions. In the present work we adopt a polynomial kernel function; the choice was based on the empirical comparison of the results (Abe 2005).

The choice of the parameter d is the most important for the final clustering result, because it affects the number of clusters.

To have a simplified notation, we indicate with $K^*(\cdot)$ the parametrised kernel function: then in our specific context the problem consists in maximising the following quantity with respect to β

$$W = \sum_m K^*(\mathbf{y}_m, \mathbf{y}_m)\beta_m - \sum_{m,m'} \beta_j \beta_{m'} K^*(\mathbf{y}_m, \mathbf{y}_{m'}), \tag{4}$$

where $\mathbf{y}_m$ represents the generic coordinate obtained via MCA, $1 \leq m \leq q$.

This involves a quadratic programming problem solution, the objective function is convex and has a globally optimal solution (Ben-hur et al. 2001).

The distance of the image of each point in the feature space and the center of the hypersphere is:

$$R^2(\mathbf{y}) = \|\varphi(\mathbf{y}) - \mathbf{a}\|^2 \tag{5}$$

Applying previous results, the distance is obtained as:

$$R^2(\mathbf{y}) = K^*(\mathbf{y},\mathbf{y}) - 2\sum_j K^*(\mathbf{y}_j,\mathbf{y})\beta_j + \sum_{j,j'} \beta_j \beta_{j'} K^*(\mathbf{y}_j,\mathbf{y}_{j'}). \tag{6}$$

Points, whose distance from the surface of the hypersphere is less than ξ, are the support vectors and they define a partition of the feature space. These points are characterized by $0 < \beta_i < C$; points with $\beta_i = C$ are called bounded support vectors and they are outside the feature-space hypersphere. If $\beta_i = 0$ the point is inside the feature-space hypersphere. The number of support vectors affects the number of clusters, as the number of support vectors increases the number of clusters increases. The numbers of support vectors depend on d and C: as d increases the number of support vectors increases because the contours of the hypersphere fit better the data; as C decreases the number of bounded support vectors increases and their influence on the shape of the cluster contour decreases.

The (squared) radius of the hypersphere is:

$$r^2 = \left\{ R(y_i)^2 | y_i \text{ is a support vector} \right\}. \tag{7}$$

The last clustering phase consists in assigning the points projected in the feature space to the classes. It is worth reminding that the analytic form of the mapping function $\varphi(\mathbf{x}) = (\phi_1(\mathbf{x}), \phi_2(\mathbf{x}), \ldots, \phi_t(\mathbf{x}))$ is unknown, so that computing points coordinates in the feature space is an unfeasible task. Alternative approaches permit to define points memberships without computing all coordinates. In this paper, in order to assign points to clusters we use the *cone cluster labeling algorithm* (Lee and Daniels 2006) adapted to the case of polynomial kernel.

The Cone Cluster Labeling (CCL) is different from other classical methods because it is not based on distances between pairs of points. This method look for a surface that cover the hypersphere, this surface consists of a union of coned-shaped regions. Each region is associated with a support vector's features space image, the phase of each cone $\Phi_i = \angle(\phi(v_i) O \mathbf{a})$ is the same, where v_i is a support vector, $\mathbf{a}$ is the center of the minimal hypersphere and O is the feature space origin. The image of each cone in the data space is an hypersphere, if two hyperspheres overlap the two support vectors belong to the same class. So the objective is to find the radius of these hyperspheres in the data space $\|v_i - g_i\|$ where g is a generic point on the surface of the hypersphere. It can be demonstrated that $\mathbf{K}(v_i, g_i) = \sqrt{1 - r^2}$ (Lee and Daniels 2006), so in case of polynomial kernel we obtain:

$$\mathbf{K}(v_i, g_i) = ((v_i g_i') + 1)^d,$$
$$\sqrt{1 - r^2} = ((v_i g_i') + 1)^d. \tag{8}$$

Starting from (8) we can compute the coordinate of g_i: $g_i' = \left[(1 - r^2)^{\frac{1}{2d}} - 1\right] v_i'$ and consequently the value of $\|v_i - g_i\|$. If distances between two generic support vectors is less than the sum of the two radii they belong to the same cluster.

Defined by N the number of units and by N_{SV} the number of support vectors, computational cost of the CCL method is $O(N_{SV}^2)$, while computational cost of the classical method, complete graph (CG), is $O(N^2)$. When the number of support vectors is small, CCL is faster then CG.

3 Empirical Evidence

The method has been applied to the 1984 United States Congressional Voting Records Database. The access information is available at the UCI Machine Learning Repository home page[1]. The dataset includes votes for each of the U.S. House of Representatives Congressmen on the 16 key votes identified in the Congressional Quarterly Almanac (CQA). The data matrix we use in this paper represents a simplified version of the original dataset. It consists of 435 rows, one for each congressman, and 16 columns referring to the 16 key votes of 1984. Each cell can assume 3 different categories: *in favor, against* and *unknown*. An additional column indicates the political party of each congressman (democrat or republican). We assume it represents the "true" classification.

In the first step we used an MCA algorithm in order to reduce the number of variables. Looking at the eigenvalues scree plot in Fig. 1, we observe that the growth of the explained inertia is minimal starting from the third factor.

So we computed the coordinates of the units on the first two MCA factors that explain 81% of the inertia. The items represented in this new space are characterized by non linear relations.

In the second step we use SVC algorithm. The method identifies nested clusters and clusters of arbitrary form. We chose a gaussian kernel because it gives a better performance with this dataset. The number of classes (Fig. 4) depends on kernels parameters: with parameters $d = 3$ and $C = 0.001$ we obtained 3 classes.

Applying a Cone cluster labeling algorithm, we obtained the solution in Fig. 2.

In order to appreciate the quality of our results we propose a comparison with k-means method. We are aware that the two method optimize different criteria however the k-means algorithm popularity makes it a fundamental reference method. We used the k-means method to cluster the coordinates of the items on

[1] http://archive.ics.uci.edu/ml/

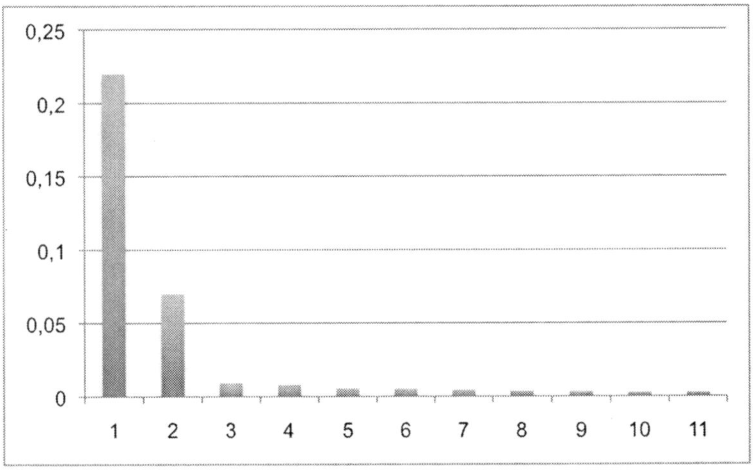

Fig. 1 Eigenvalues scree plot (first eleven eigenvalues)

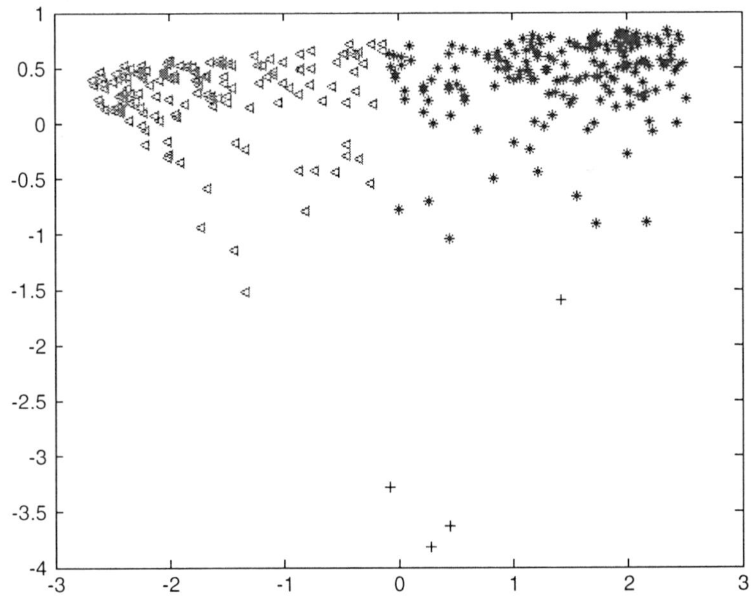

Fig. 2 Clusters obtained by SVC algorithm

factorial axes Fig. 3. We reiterated the algorithm 1,000 times because, as it is well known, k-means can converge to local minima while SVC find the optimal solution if any. With this dataset in 99% of cases the results converge always to the same minimum. In 1% we find not satisfactory solutions because the presence of a singleton.

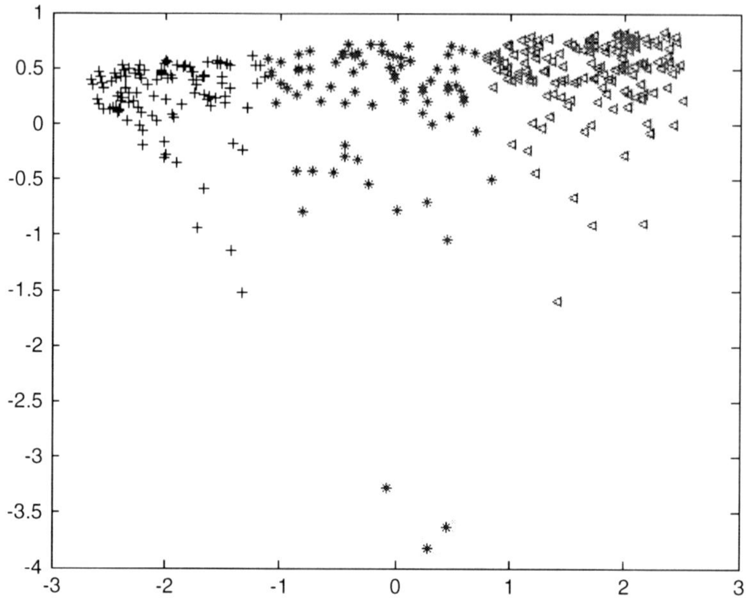

Fig. 3 Clustering obtained with k-means

The results obtained with two classes are not presented because solutions are unstable and group separations are not satisfactory.

It can be reasonably assumed that the "true" classification were defined by the variable *political party*, not involved in the analysis. SVC algorithm results (Fig. 2) can be summarized as follow:

- (blue $*$) 223 republicans, 9 democrats
- (red Δ) 157 democrats, 42 republicans
- (black $+$) 2 republicans, 2 democrats.

Figure 3 shows that using k-means the clusterings structures changes a lot with respect to the "true classification". The results can be summarized as follow:

- (black $+$) 137 democrats, 23 republicans
- (red Δ) 194 republicans, 4 democrats
- (blue $*$) 50 republicans, 27 democrats.

In order to appreciate the procedure performing, we also use the CATANOVA method (Singh 1993). This method is analogous to the ANOVA method for the case of categorical data. The CATANOVA method tests the null hypothesis that all the k classes have the same probability structure q_i: $H_0 : q_{ij} = q_i$ for all $i = 1, \ldots, p$ and $j = 1, \ldots, k$ where p is the number of variables and k the number of clusters. The null hypothesis is rejected using both k-means and SVC methods, in both cases there are significative differences between clusters. The statistic CA is distributed as a χ^2 with $(n-1)(k-1)$ degrees of freedom, where n is the number of observations. The

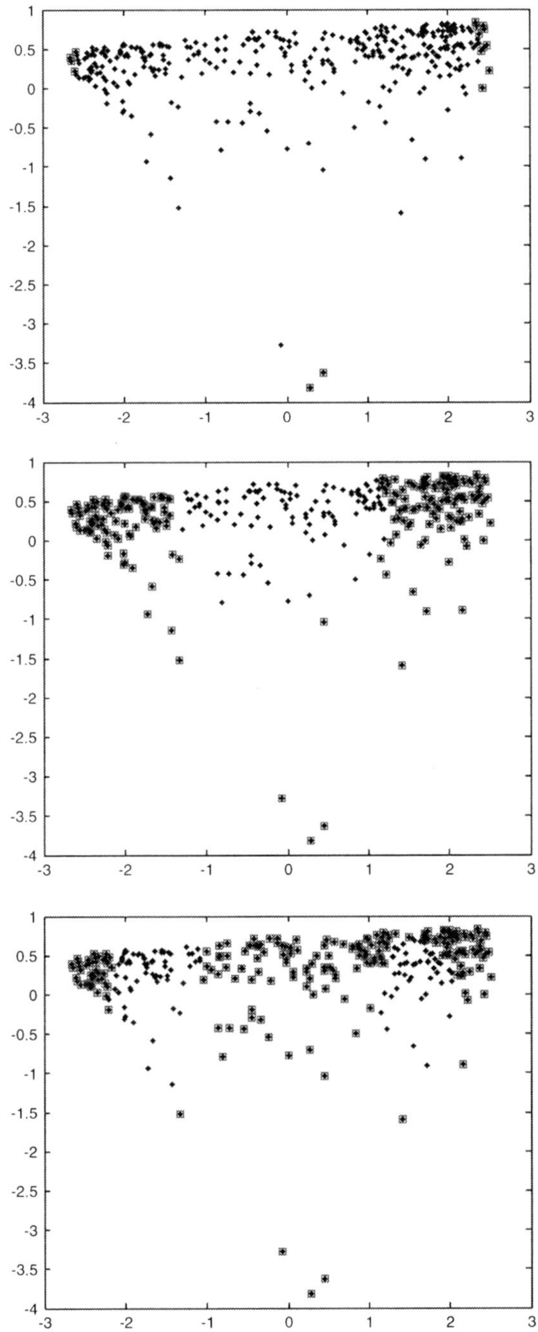

Fig. 4 Number of support vector changing the value of the kernel parameter

value of *CA* for k-means on this dataset is 1.8275×10^4. The value of *CA* applying to the SVC algorithm is 1.8834×10^4; using SVC we obtain an higher value of the CATANOVA index, we can conclude that, with this dataset, SVC performs better than k-means.

4 Conclusion

This method can be an alternative to traditional clustering methods when dealing with large data-sets of categorical data. The first issue solved is the quantification of categorical data with the MCA that reduces the dimensionality without losing nonlinear relations between variables. The second is the adaptation of the cone cluster labeling method, used in the case of Gaussian kernel, to the case of polynomial kernel. One of the advantages is that the classes can be seen on factorial axes and this can help in the interpretation of the results; moreover, the method proposed gives stable results. There are still some open issues: the choice of the kernel function is made empirically and there is no analytic way to choose it; the number of classes depends on the kernel parameters so it can not be chosen directly.

The next task is to classify new items. To do this it can be useful to project the data into a new space that maximizes the distances between classes. The new item can be projected in this space where it can be classified.

References

Abe S. (2005) *Support vector machine for pattern classification*, Springer.

Ben-hur A., Horn D., Siegelmann H. T., Vapnik, V. (2001) Support vector clustering, *Journal of machine learning research* 2: 125–137.

Greenacre M. J., Blasius J. (2006) *Multiple correspondence analysis and related methods*, Chapman & Hall/CRC Press, Boca Raton, FL Boca-Raton.

Lenca P., Patrick M., Benot V., Stphane L. (2008) On selecting interestingness measures for association rules: User oriented description and multiple criteria decision aid, *European Journal of Operational Research* 184: 610–626.

Lee S.H., Daniels K.M. (2006) Cone cluster labeling for support vector clustering, *Proceeding of the sixth SIAM international conference on data mining*, Bethesda: 484-488.

Marino M., Tortora C. (2009) A comparison between k-means and support vector clustering of categorical data, *Statistica applicata*, Vol 21 n.1: 5–16.

Muller K. R., Mika S., Ratsch G., Tsuda K., Scholkopf B. (2001) An introduction to Kernel-based learning algorithms, *IEEE transiction on neural networks*, 12: 181–201.

Saporta G. (1990) Simultaneous analysis of qualitative and quantitative data, *Atti* 35° *Riunione Scientifica della Società italiana di Statistica*, CEDAM: 63–72.

Shawe-Taylor J., Cristianini N. (2004) *Kernel methods for pattern analysis*, Cambridge University Press. Boston.

Silberschatz A., Tuzhilin A. (1996) What makes patterns interesting in knowledge discovery systems, *IEEE Transactions on Knowledge and Data Engineering* Vol 8: 291–304.

Singh B. (1993) On the Analysis of Variance Method for Nominal Data, *The Indian Journal of Statistics*, Series B, Vol 55: 40–47.

Clustering Geostatistical Functional Data

Elvira Romano and Rosanna Verde

Abstract In this paper, we among functional data. A first strategy aims to classify curves spatially dependent and to obtain a spatio-functional model prototype for each cluster. It is based on a Dynamic Clustering Algorithm with on an optimization problem that minimizes the spatial variability among the curves in each cluster. A second one looks simultaneously for an optimal partition of spatial functional data set and a set of bivariate functional regression models associated to each cluster. These models take into account both the interactions among different functional variables and the spatial relations among the observations.

1 Introduction

There is a large number of applicative fields like metereology, geology, agronomy, ecology, where data are curves observed in spatial varying way. This is leading, in these years to the development of a new branch of statistics: Spatio-Functional Data Analysis (Ramsay and Silverman 2005). In this paper we focus on clustering methods for geostatistical functional data which are a kind of spatio-functional data (Delicado et al. 2007).

Clustering approaches in this framework can be categorized in functional methods and in spatio-temporal methods. The first ones take only into account the functional nature of data (Romano 2006) while the second ones are time-dependent clustering methods incorporating spatial dependence information between variables (Blekas et al. 2007). Existing work on clustering spatiotemporal data has been mostly studied by computer scientists most often offering a specific solution to clustering under nuisance spatial dependence. With the aim of overcoming the

E. Romano (✉) · R. Verde
Seconda Universitá degli Studi di Napoli, Via del Setificio 81100 Caserta, Italy
e-mail: elvira.romano@unina2.it; verde.rosanna@unina2.it

A. Di Ciaccio et al. (eds.), *Advanced Statistical Methods for the Analysis of Large Data-Sets*, Studies in Theoretical and Applied Statistics,
DOI 10.1007/978-3-642-21037-2_3,

restrictive independence assumption between functionals in many real applications, we evaluate the performances of two distinct clustering strategies according to a spatial functional point of view.

A first approach is a special case of Dynamic Clustering Algorithm (Diday 1971) based on an optimization criterion that minimizes the spatial variability among the curves in each cluster (Romano et al. 2009a). The centroids of the clusters, are estimated curves in sampled and unsampled area of the space that summarize spatio-functional behaviors.

The second one (Romano et al. 2009b) is a clusterwise linear regression approach that attempts to discover spatial functional linear regression models with two functional predictors, an interaction term, and with spatially correlated residuals. This approach is such to establish a spatial organization in relation to the interaction among different functional data. The algorithm is a k-means clustering with a criterion based on the minimization of the squared residuals instead of the classical within cluster dispersion.

Both the strategies have the main aim of obtaining clusters in relation to the spatial interaction among functional data.

In the next sections after a short introduction on the spatial functional data, we present the main details of the methods and their performances on a real dataset.

2 Geostatistical Functional Data

Spatially dependent functional data may be defined as the data for which the measurements on each observation, that is a curve, are part of a single underlying continuous Spatial functional process defined as: $\Xi = \left\{\chi_{\mathbf{s}} : s \in D \subseteq R^d\right\}$, where s is a generic data location in the $d-$dimensional Euclidean space (d is usually equal to 2), the set $D \subseteq R^d$ can be fixed or random and χ_s are functional random variables, defined as random elements taking values in an infinite dimensional space.

The nature of the set D allows to classify the kind of Spatial Functional Data. Following Giraldo et al. (2009) these can be distinguished in geostatistical functional data, functional marked point pattern and functional areal data.

We focus on geostatistical functional data, that appear when D is a fixed subset of R^d with positive volume, in particular we assume to observe a sample of curves $\chi_{s_i}(t)$ for $t \in T$ and $s_i \in D, i = 1, \ldots, n$. It is usually assumed that these curves belong to a separable Hilbert space $\mathbf{H}$ of square integrable functions defined in T. We assume for each $t \in T$ we have a second order stationary and isotropic random process, that is, the mean and variance functions are constant and the covariance depends only on the distance between sampling points. Formally, we have that:

- $E(\chi_{\mathbf{s}}(t)) = m(t)$, for all $t \in T,\ s \in D$.
- $V(\chi_{\mathbf{s}}(t)) = \sigma^2(t)$, for all $t \in T,\ s \in D$.
- $Cov(\chi_{\mathbf{s_i}}(t), \chi_{\mathbf{s_j}}(t)) = C(h,t)$ where $h_{ij} = \left\| s_i - s_j \right\|$ and all $s_i, s_j \in D$

- $\frac{1}{2}V(\chi_{\mathbf{s_i}}(t), \chi_{\mathbf{s_j}}(t)) = \gamma(h,t) = \gamma_{s_i s_j}(t)$ where $h_{ij} = \left\| s_i - s_j \right\|$ and all s_i, $s_j \in D$.

The function $\gamma(h,t)$ as function of h is called variogram of $\chi_{\mathbf{s}}$.

3 Dynamic Clustering for Spatio-Functional Data

Our first proposal is to partition, through a Dynamic clustering algorithm, the random field $\{\chi_s : s \in D \subset R^d\}$ into a set of C clusters such that the obtained clusters contain spatially related curves. Dynamic clustering algorithm optimizes a criterion of best fitting between a partition P_c of a set of objects in C clusters and the way to represent the clusters. Since we assumed that data are generated from a functional linear concurrent model (Ramsay and Silverman 2005) we advise to optimize the following criterion:

$$\Delta(P,G) = \sum_{c=1}^{C} \sum_{i \in P_c} \int_T V\left(\chi_{s_i}(t) - \sum_{i=1}^{n_c} \lambda_i \chi_{s_i}(t)\right) dt \qquad u.c. \sum_{i=1}^{n_c} \lambda_i = 1 \quad (1)$$

where n_c is the number of the elements in each cluster and the prototype $\chi_{s_c} = \sum_{i=1}^{n_c} \lambda_i \chi_{s_i}(t)$ is an ordinary kriging predictor for curves in the clusters. According to this criterion the kriging coefficients represent the contribute of each curve to the prototype estimate in an optimal location s_c. Where s_c is chosen among all possible locations of the space, obtained by considering a rectangular spatial grid which covers the area under the study. Among the possible locations are also included the sampled locations of the space. Thus, the parameters to estimate are: the kriging coefficients, the spatial location of the prototypes, the residuals spatial variance for each cluster.

For fixed values of the spatial locations of the prototypes s_c this is a constrained minimization problem.

The parameters λ_i $i = 1, \ldots, n_c$ are the solutions of a linear system based on the Lagrange multiplier method. In this paper we refer to the method proposed by (Delicado et al. 2007), that in matrix notation, can be seen as the minimization of trace of the mean-squared prediction error matrix in the functional setting.

According to this approach a global uncertainty measure is the prediction of trace-semivariogram $\int_T \gamma_{s_i, s_c(t)} dt$, given by:

$$\int_T V\left(\chi_{s_i}(t) - \sum_{i=1}^{n_c} \lambda_i \chi_{s_i}(t)\right) dt = \sum_{i=1}^{n_c} \lambda_i \int_T \gamma_{s_i, s_c(t)} dt - \mu \quad u.c. \sum_{i=1}^{n_c} \lambda_i = 1 \quad (2)$$

It is an integrated version of the classical pointwise prediction variance of ordinary kriging and gives indication on the goodness of fit of the predicted model.

In the ordinary kriging for functional data the problem is to obtain an estimate of a curve in an unsampled location.

In our case also the location is a parameter to estimate. We propose to solve this problem evaluating for each cluster, kriging on the locations of the grid in order to obtain the best representative kriging predictor. The prototype is the best predictor in terms of the best spatio-functional fitting (5) among the set of the estimated prototype on different spatial locations.

Once we have estimated the prototypes we allocate each new curve to the cluster according to the following *allocation* function:

$$\kappa = \chi_f \longmapsto \mathscr{P}_c \tag{3}$$

It allows to assign χ_s to cluster c of P_c $\kappa(G) = P = \{P_1, \ldots, P_C\}$, according to the minimum-spatial variability rule:

$$P_c := \{i \in \chi_s : \delta(\{i\}, \chi_{s_c}) \leq \delta(\{i\}, \chi_{s_{c*}}) \ for \ 1 \leq c^* \leq C\} \tag{4}$$

with:

$$\delta(\{i\}, \chi_{s_c}) = \frac{1}{\lambda_\alpha} \int_T V\left(\chi_{s_i}(t) - \chi_{s_c}(t)\right) dt \tag{5}$$

where λ_α is the kriging coefficient or weight such that $|s_\alpha - s_c| \cong h$ where $h = |s_i - s_c|$. Note that, it is possible however that some weights may be negative. For solving this problem we set to the latter a zero value.

Applying iteratively the assignment function followed by the allocation function under some conditions the algorithm converges to a stationary value. The convergence of the criterion is guaranteed by the consistency between the way to represent the classes and the proprieties of the allocation function.

4 Clusterwise Linear Regression for Spatially Correlated Functional Data

According to (Spaeth 1979) the clusterwise linear regression is defined as a kind of regression, such that given a data set of observations of an explanatory variable and a response variable, the aim is to find simultaneously an optimal partition of the data and the regression models associated to each cluster which maximize the overall fit.

Given a multivariate spatial functional dataset $(\chi_{s_i}(t), \theta_{s_i}(t), Y(s_i))_{i=1\ldots n}$, where $\chi_{s_1}, \ldots, \chi_{s_n}$ and $\theta_{s_1}, \ldots, \theta_{s_n}$ are samples of curves, realization of two random processes $\chi_{\mathbf{s}}(t)$ and $`_{\mathbf{s}}(t)$ with $t \in T$, $s \in D$, and $Y(s)$ is a unique observation of $\mathbf{Y(t)}$ a random function at location. We consider the following functional regression model with scalar response (Bar-Hen et al., 2008):

$$Y(s) = \mu + \langle A; \chi_s \rangle + \langle B; \theta_s \rangle + \langle C\theta_s; \chi_s \rangle + \epsilon_s \tag{6}$$

where

$$\langle A;\chi_s\rangle = \int_{[0,T]} A(t)\chi_s(t)dt \quad \langle B;\theta_s\rangle = \int_{[0,T]} B(t)\theta_s(t)dt \tag{7}$$

are the linear terms in χ_s, θ_s and

$$\langle C\theta_s;\chi_s\rangle = \int\int_{[0,T]^2} C(t,u)\chi_s(t)\theta_s(u)dtdu \tag{8}$$

is the bilinear term and ϵ_s is a spatial stationary random field with spatial correlation function ρ_s. Our aim is to get an optimal partition $P = (P_1, \ldots, P_C)$ of the spatio-functional dataset $(\chi_s(t), \theta_s(t), Y(s))$ into a set of C clusters such that curves in each cluster maximize the fit to the following functional linear model with scalar response (Bar Hen et al. 2008):

$$Y_s = f_c(\chi_s, \theta_s) + \epsilon_s \tag{9}$$

where: ϵ_s, is a spatial stationary random field with spatial correlation function ρ_s,

f_c is defined for each cluster $c \in C$ and it is assumed to be the sum of linear terms in χ_s, θ_s and a bilinear term modeling the spatial interaction between χ_s and θ_s.

Formally, f_c can be expressed as:

$$\begin{aligned} f_c(\chi_s, \theta_s) &= \mu_c + \int_{[0,T]} A_c(t)\chi_s(t)\mathrm{d}t + \int_{[0,T]} B_c(t)\theta_s(t)\mathrm{d}t \\ &+ \int\int_{[0,T]^2} C_c(t,u)\chi_s(t)\theta_s(u)dudt \\ &= \mu + \langle A_c;\chi\rangle + \langle B_c;\theta\rangle + \langle C_c\chi;\theta\rangle \end{aligned}$$

where:

$A_c(t)$ and $B_c(t)$ are the functional regression coefficients,

$V_c(t,u)$ is a coefficient to be estimated which accounts for the interaction between $\chi_s(t)$ and $\theta_s(t)$.

We assume that exists a group-variable $\mathcal{G} : \Omega \longmapsto \{1, \ldots, C\}$,(where Ω is the probability space on which $\chi_s(t)$ and $\theta_s(t)$ are defined) such that

$$E(Y_s/\chi_s = \chi_s, \theta_s = \theta_s, \mathcal{G} = c) = \mu^c + \langle A^c;\chi_s\rangle + \langle B^c;\theta_s\rangle + \langle C^c\theta_s;\chi_s\rangle \tag{10}$$

where $\{\mu^c, A^c, B^c, C^c\}_{c=1,\ldots,C}$ are the estimated regression functions given by the generalized least squares criterion.

If n data points have been recorded, the clusterwise linear regression algorithm finds simultaneously an optimal partition of the n points and the regression models $\{\mu^c, A^c, B^c, C^c\}_{c=1,\ldots,C}$ associated to each cluster, which optimize the criterion:

$$\Delta(P,G) = \sum_{c=1}^{C} \sum_{i \in P_c} [Y_i(s) - (\mu^c + \langle A^c; \chi_s \rangle + \langle B^c; \theta_s \rangle + \langle C^c \theta_s; \chi_s \rangle)]^2 \quad (11)$$

that is minimizing the sum of the squares errors SSE_c over the C clusters.

The algorithm used for finding a local minimum of the criterion Δ is a variation of the k-means.

It starts from defining an initial random partition $P^1 = (P_1, \ldots, P_C)$, then it constructs iteratively a sequence of partitions and regression coefficients as follows:

- Run until the convergence
  - For each cluster P_c
    - Estimate the coefficients $A_c(t), B_c(t), V_c(t,u)$ of f_c by Quasi Generalized Least Square by using a local spatial covariance matrices Σ_c computed on the estimated residuals of local model for $c = 1 \ldots C$
  - Allocate each $\chi_{s_i}, \theta_{s_i}, Y_{s_i}$ to the clusters such that:

$$\hat{\epsilon}_c^{+\prime} \Sigma_c^{+-1} \hat{\epsilon}_c^{+} < \hat{\epsilon}_{c'}^{+\prime} \Sigma_{c'}^{+-1} \hat{\epsilon}_{c'}^{+} \quad (12)$$

Where $\hat{\epsilon}_c = (Y_c - X_c \Phi^*)$ and $\Phi^* = (X^t X)^{-1} X^t Y$.

In order to estimate the models parameters, the functions involved in f_c are expanded on an orthonormal basis of $L^2([0,T])$ truncated at $h = l$. Where l is suitably large that it does not entail any significant loss of information. The problem becomes a linear regression with spatially correlated residuals. We use Quasi Generalized Least Square where at first, the coefficients $A_c(t), B_c(t), V_c(t,u)$ are estimated by Ordinary Least Square and the spatial correlation is estimated from the residuals. Then, the final estimate of the coefficients is performed by introducing the estimated correlation matrix in the Generalized Least Square formula.

5 Mareographic Network Analysis

We illustrate our approach on a real dataset, which contains the atmospheric pressure and air temperature of 26 sea stations of the Italian Coast (http://www.mareografico.it). For each location, we have the curve of atmospheric pressure and air temperature recorded in a period of two weeks with spatial coordinates (s_x, s_y) correspondent respectively to the latitude and longitude. The aim of the analysis is:

- To find homogeneous spatial areas and a local functional model able to summarize atmospheric behaviors. With this aim we perform Dynamic Clustering for spatio-functional data by using separately the atmospheric pressure curve and the air temperature curve.

- To investigate on the atmospheric diversity at different areas by accounting for the weight of atmospheric pressure and air temperature curves (together with their interaction) on such diversity. This is performed by Clusterwise linear regression.

As first stage of all the analysis we use Fourier basis to smooth each observed curve. Thus we performed three cluster analysis. For all the analysis, in order to choose the number of clusters, we run the algorithms with a variable number of clusters C. We look at the value of the optimized criterion as a function of the number of clusters, then we choose a number of clusters so that adding an another cluster it does not give a much better value of the criterion.

On the evaluated dataset this involves to set $C = 3$ for the first two analyses and $C = 2$ for the third one.

Moreover to initialize the clustering procedures, we run a standard k-means algorithm on the spatial locations of the observed data, such to get a partitioning of data into spatially contiguous regions.

By the results of the first two analysis, the 3 obtained clusters include quite similar stations and areas but are characterized by different prototypes. Especially, looking at the clustering structure of pressure curves, the clusters contain respectively 10, 11, 5 elements. Moreover we observe that:

- In the first cluster, according to the obtained functional parameters $\lambda_i, i = 1, \ldots, 10$, the greatest contribute to the prototype estimation corresponds to 0.59. This functional parameter corresponds to Napoli.
- In the second cluster, the greatest functional parameter that contributes to the prototype estimation corresponds to 0.46, this functional parameter corresponds to Ancona.
- In the third cluster, the greatest functional parameter that contributes to the prototype estimation corresponds to 0.66, this functional parameter corresponds to La Spezia.

At the same time looking at the clustering structure on air temperature curve, each cluster contains respectively 10, 12, 4 elements, we can observe that:

- In the first cluster, according to the obtained functional parameters $\lambda_i, i = 1, \ldots, 10$, the greatest contribute to the prototype estimation corresponds to 0.49. This functional parameter corresponds to Salerno.
- In the second cluster, the greatest functional parameter that contributes to the prototype estimation corresponds to 0.36, this functional parameter corresponds to Ortona.
- In the third cluster, the greatest functional parameter that contributes to the prototype estimation corresponds to 0.56, this functional parameter corresponds to Genova.

The third analysis, which is performed using the Clusterwise Linear regression method, takes into account both, the curves for air temperature and pressure so that it is able to provide a global view of the atmospheric diversity on the considered area

We obtain two clusters, each of them has a predicted response range respectively of $[0,28;0,43]$ and $[0,43;0,69]$. These two macroarea are respectively north and south of the Italian coast, we could conclude that this is the effect of the spatial correlation among the curve.

The location of the prototypes are Salerno and Ancona. The spatio-functional models that describe the two area are:

$$f_{c_i}(\chi_s,\theta_s) = \mu + \langle A_{c_i};\chi\rangle + \langle B_{c_i};\theta\rangle + \langle C_{c_i}\chi;\theta\rangle \;\; i = 1,2 \tag{13}$$

where A_{c_i}, B_{c_i} are respectively the coefficient functions of the atmospheric pressure and of air temperature and C_{c_i} is the interaction function among atmospheric pressure and air temperature. Thus we can observe for the first cluster that the function A_{c_1} has a crescent shape with variability in the range $[1003C;1006C]$ and of the function B_{c_1} has a decrescent shape with variability in the range $[-0.1hPa;8.8hPa]$; while for the second cluster we have more variability in the range $[1004C;1010C]$ for the function A_{c_2} and for the function B_{c_2} in the range $[4.7hPa;12.2hPa]$. In order to evaluate the effect of the interaction among the two variable, we have also performed the analysis of variance with a different fitted model for the two obtained clusters: $f_{c_i}(\chi_s,\theta_s) = \mu + \langle A_{c_i};\chi\rangle + \langle B_{c_i};\theta\rangle \;\; i = 1,2$.

The p-values($1,235e^{-12}$, $1,115e^{-7}$) respectively for the first and second cluster, show that the interaction term have a strong effect.

References

C. Abraham, P. Corillon, E. Matnzer-Lö̸ber, N. Molinari. *Unsupervised curve clustering using B-splines*. Scandinavian Journal of Statistics, **30**, 581–595, 2005.

A., Bar Hen, L., Bel, R., Cheddadi and R., Petit. *Spatio-temporal Functional Regression on Paleoecological Data*, Functional and Operatorial Statistics, 54-56. Physica-Verlag HD, 2008.

K. Blekas, C. Nikou, N. Galatsanos, N. V. Tsekos. *Curve Clustering with Spatial Constraints for Analysis of Spatiotemporal Data*. In Proceedings of the 19th IEEE international Conference on Tools with Artificial intelligence - Volume 01 (October 29 - 31, 2007). ICTAI. IEEE Computer Society, Washington, DC, 529-535, 2007.

H. Cardot, F. Ferraty, P. Sarda. *Functional linear model*. Statistics and Probability Letters, 45:11–22, 1999.

E. Diday. La Méthode des nueés dynamiques. *Rev. Stat.Appl.* **XXX**, 2, 19–34, 1971.

P. Delicado, R. Giraldo, J. Mateu. Geostatistics for functional data: An ordinary kriging approach. Technical Report, http://hdl.handle.net/2117/1099, Universitat Politecnica de Catalunya, 2007.

P. Delicado, R. Giraldo, C. Comas, J. Mateu. Statistics for Spatial Functional data. Environmetrics. Forthcoming. Technical Report, http://hdl.handle.net/2117/2446, Universitat Politecnica de Catalunya, 2009.

G. James, C. Sugar. Clustering for Sparsely Sampled Functional Data. *Journal of the American Statistical Association*, **98**, 397–408, 2005.

N. Heckman , R. Zamar Comparing the shapes of regression functions. *Biometrika*, **87**, 135–144, 2000.

C., Preda and G., Saporta. *PLS approach for clusterwise linear regression on functional data.* In Classification, Clustering, and Data Mining Applications (D. Banks, L. House, F. R. McMorris, P. Arabie and W. Gaul, eds.) 167–176. Springer, Berlin, 2004.

J.E. Ramsay, B.W. Silverman. Functional Data Analysis (Second ed.) *Springer*, 2005.

E. Romano. Dynamical curves clustering with free knots spline estimation. PHD Thesis, University of Federico II, Naples, 2006.

E. Romano, A., Balzanella, R., Verde. *Clustering Spatio-functional data: a model based approach.* Studies in Classification, Data Analysis, and Knowledge Organization Springer, Berlin-Heidelberg, New York, 2009a. ISBN: 978-3-642-10744-3.

E. Romano, A., Balzanella, R., Verde. *A clusterwise regression strategy for spatio-functional data.* In Book of Short Papers 7 Meeting of the Classification and Data Analysis Group of the Italian Statistical Society. Catania - September 9-11, 2009b Editors Salvatore Ingrassia and Roberto Rocci, p. 609–613. ISBN 978-88-6129-406-6

H. Spaeth. (1979) *Clusterwise linear regression* Computing 22, p. 367–373.

Joint Clustering and Alignment of Functional Data: An Application to Vascular Geometries

Laura M. Sangalli, Piercesare Secchi, Simone Vantini, and Valeria Vitelli

Abstract We show an application of the *k-mean alignment* method presented in Sangalli et al. (Comput. Stat. Data Anal. 54:1219–1233). This is a method for the joint clustering and alignment of functional data, that sets in a unique framework two widely used methods of functional data analysis: Procrustes continuous alignment and functional k-mean clustering. These two methods turn out to be two special cases of the new method. In detail we use this algorithm to analyze 65 Internal Carotid Arteries in relation to the presence and rupture of cerebral aneurysms. Some interesting issues pointed out by the analysis and amenable of a biological interpretation are briefly discussed.

1 Introduction

The onset and the rupture of cerebral aneurysms are still matters of research among neuro-surgeons. A cerebral aneurysm is a bulge in the wall of a brain vessel; it is generally not disrupting, and it is not rare among adult population: epidemiological studies suggest that between 1% and 6% of adults develop a cerebral aneurysm during their lives. On the contrary, the rupture of a cerebral aneurysm is quite uncommon but very severe event: about 1 event every 10,000 adults per year, with a mortality rate exceeding 50%.

L.M. Sangalli · P. Secchi · S. Vantini (✉) · V. Vitelli
MOX - Department of Mathematics "Francesco Brioschi", Politecnico di Milano, Piazza Leonardo da Vinci, 32, 20133, Milano, Italy
e-mail: simone.vantini@polimi.it

A. Di Ciaccio et al. (eds.), *Advanced Statistical Methods for the Analysis of Large Data-Sets*, Studies in Theoretical and Applied Statistics,
DOI 10.1007/978-3-642-21037-2_4,

The aim of the Aneurisk Project[1] is to provide evidence of an existing relation between this pathology and the geometry and hemodynamics of brain vessels. In particular, the present analysis considers the centerlines of 65 Internal Carotid Arteries (ICA), whose functional form is obtained from discrete observations by means of free-knot regression splines, as shown in Sangalli et al. (2009b). Details about the elicitation of discrete observations from raw data can be found in Antiga et al. (2008). Before the analysis, the 65 centerlines are jointly aligned and clustered by means of the *k-mean alignment* method proposed in Sangalli et al. (2010a,b). The aligned and clustered centerlines are then here analyzed along the paradigm of functional data analysis as advocated by Ramsay and Silverman (2005). In the end, some interesting issues amenable of a biological interpretation are discussed.

2 The k-Mean Alignment Algorithm

The k-mean alignment algorithm – whose technical details can be found in Sangalli et al. (2010a,b) – originates from the need of consistently aligning and clustering a set of functional data. This algorithm can be seen as the result of an integration of two algorithms that are currently widely used in functional data analysis: the Procrustes continuous registration algorithm (e.g., Sangalli et al., 2009a) and the functional k-mean clustering algorithm (e.g., Tarpey and Kinateder, 2003). With these two mother algorithms, the new algorithm shares both aims and basic operations. Schematic flowcharts of both Procrustes continuous registration algorithm and functional k-mean clustering algorithm are sketched in Fig. 1. Alternative approaches to the joint clustering and alignment of curves can be found for instance in Liu and Yang (2009), and Boudaoud et al. (2010).

The aim of the Procrustes continuous alignment algorithm is to align functional data by decoupling phase and amplitude variability; this task is essentially achieved by iteratively performing an *identification step* and an *alignment step*. The former step consists in the identification of a template function on the basis of the n functions as aligned at the previous iteration; the latter step consists instead in the maximization of the similarity between each function and the template, as identified at the previous identification step, by means of subject-by-subject warping of the abscissa. The problem of curve alignment is theoretically well set when a similarity index ρ between two functions and a set W of admissible warping functions of the abscissa are chosen.

[1]The project involves MOX Laboratory for Modeling and Scientific Computing (Dip. di Matematica, Politecnico di Milano), Laboratory of Biological Structure Mechanics (Dip. di Ingegneria Strutturale, Politecnico di Milano), Istituto Mario Negri (Ranica), Ospedale Niguarda Ca' Granda (Milano), and Ospedale Maggiore Policlinico (Milano), and is supported by Fondazione Politecnico di Milano and Siemens Medical Solutions Italia.

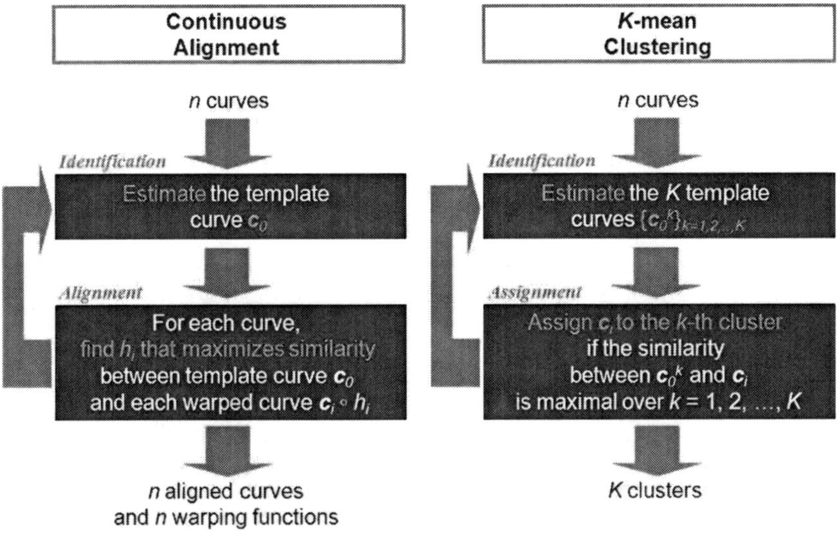

Fig. 1 Schematic flowcharts of the Procrustes continuous registration algorithm (*left*) and the functional k-mean clustering algorithm (*right*). Index i refers to the sample unit while index k to the cluster

The aim of the k-mean clustering algorithm is instead to cluster functional data by decoupling within and between-cluster variability (in this context within and between-cluster amplitude variability); this task is here achieved by iteratively performing an *identification step* and an *assignment step*. In this algorithm, the identification step consists in the identification of k cluster template functions on the basis of the k clusters detected at the previous iteration; the assignment step consists in the assignment of each function to one of the k clusters, this assignment is achieved by maximizing the similarity between each function and the k templates, as identified at the previous identification step. The problem of clustering curves is theoretically well set when a similarity index ρ between two functions and a number of cluster k to be detected are chosen.

The k-mean alignment algorithm, as a fall out of the two previous algorithms, aims at jointly aligning and clustering functional data by decoupling phase variability, within-cluster amplitude variability, and between-cluster amplitude variability. It reaches this task by putting together the basic operations of the two mother algorithms (a schematic flowchart of the k-mean alignment algorithm is sketched in Fig. 2) and thus iteratively performing an *identification step*, an *alignment step*, and an *assignment step*. Indeed, in the identification step, k template functions are identified on the basis of the k clusters and of the n aligned functions detected at the previous iteration. In the alignment step the n functions are aligned to the k templates detected at the previous iteration and k candidate aligned versions of each curve are obtained. In the assignment step, each curve is then assigned to the cluster

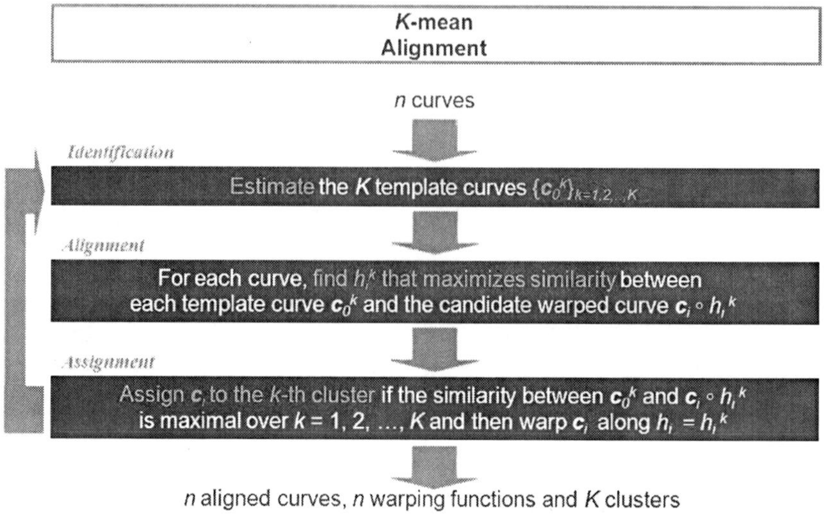

Fig. 2 Schematic flowchart of the k-mean alignment algorithm. Index i refers to the sample unit while index k to the cluster

whom the curve can be best aligned to, i.e., the cluster for which the similarity among its template and the corresponding candidate aligned curve is maximized.

On the whole, the k-mean alignment algorithm takes as input a set of n functions $\{\mathbf{c}_1, \ldots, \mathbf{c}_n\}$ (like both mother algorithms do) and gives as output k clusters (like the k-mean clustering algorithm does) and n aligned functions together with the corresponding n warping functions $\{h_1, \ldots, h_n\}$ (like the continuous alignment algorithm does).

From a theoretical point of view, the problem of jointly aligning and clustering curves is soundly posed when the number of cluster k, the similarity index ρ between two functions, and the set W of admissible warping functions are chosen. Let us mention two special choices that make the k-mean alignment algorithm degenerate to the continuous alignment algorithm and to the k-mean clustering algorithm, respectively: $k = 1$ and $W = \{\mathbf{1}\}$.

3 Analysis of Internal Carotid Artery Centerlines

In this section we discuss a real application of the k-mean alignment procedure that is also the one that urged us to develop such method: the analysis of the AneuRisk dataset. In detail, we deal with 65 three-dimensional curves representing the centerlines of 65 ICAs. Details about the elicitation of a discrete representation of the centerline from the three-dimensional angiography can be found in Sangalli et al. (2009a), while the consequent elicitation of the curve from the discrete data –

by means of three-dimensional free-knot splines – is detailed in Sangalli et al. (2009b). The outline of the analysis is to perform a k-mean alignment algorithm for different values of k, to compare the performances (measured by means of the mean similarity achieved after the k-mean alignment) in order to choose a reasonable value for k, and then to find out possible relations between both geometry and clusters membership of the aligned curves on the one hand, and presence, rupture, and location of cerebral aneurysms on the other one.

Consistently with Sangalli et al. (2009a), where another analysis of the AneuRisk dataset is presented, we use, as similarity index ρ between two curves $\mathbf{c}_1$ and $\mathbf{c}_2$, the average of the cosine of the angles between the first derivatives of homologous components of the two curves:

$$\rho(\mathbf{c}_1, \mathbf{c}_2) = \frac{1}{3} \sum_{p \in \{x,y,z\}} \frac{\int c'_{1p}(s) c'_{2p}(s) ds}{\sqrt{\int c'_{1p}(s)^2 ds} \sqrt{\int c'_{2p}(s)^2 ds}}, \tag{1}$$

and, as the set of warping functions W, we use the group of affine transformations with positive slope. This joint choice for ρ and W descends from both theoretical and medical considerations that are detailed in Sangalli et al. (2009a).

From a practical point of view, different procedures can be used for the implementation of the identification, the alignment, and the assignment steps. In particular, the procedure used within the identification step can be expected to be a very sensitive point for the good outcome of the k-mean alignment, since straightforward procedures tackling this issue cannot be easily implemented. Indeed, (a) each identification step is theoretically declined in the solution of k separate functional optimization problems; (b) each alignment step in the solution of kn separate bivariate optimization problems; and (c) each assignment step in the solution of n separate discrete optimization problems.

Effective and computationally unexpensive methods for solving optimization problems (b) and (c) – that rise in this context – exist; while, on the contrary, a numerical solution of the functional optimization problem (a) here encountered appears to be quite hard to handle. For this reason, we prefer to use a more "statistically flavored" approach to look for good estimates of the templates (i.e., local polynomial regression). This choice could of course pose some questions about the consistence among theoretical templates and their obtained estimates. This issue is extensively discussed in Sangalli et al. (2010a).

Figure 3 graphically reports the main results of the application of the k-mean alignment algorithm to the analysis of the 65 ICA centerlines. In the top-left plots, the original data are plotted. In the bottom-left and in the bottom-right plots, the output provided by the one-mean and by the two-mean alignment are respectively reported (in the latter ones the two detected clusters are identified by different colors). In the top-center plot: boxplots of similarity indexes between each curve and the corresponding template are reported for original curves and for k-mean aligned curves, $k = 1, 2, 3$. Finally, in the top-right plot, the performances of the algorithm are shown: the orange line reports, as a function of the number of clusters k,

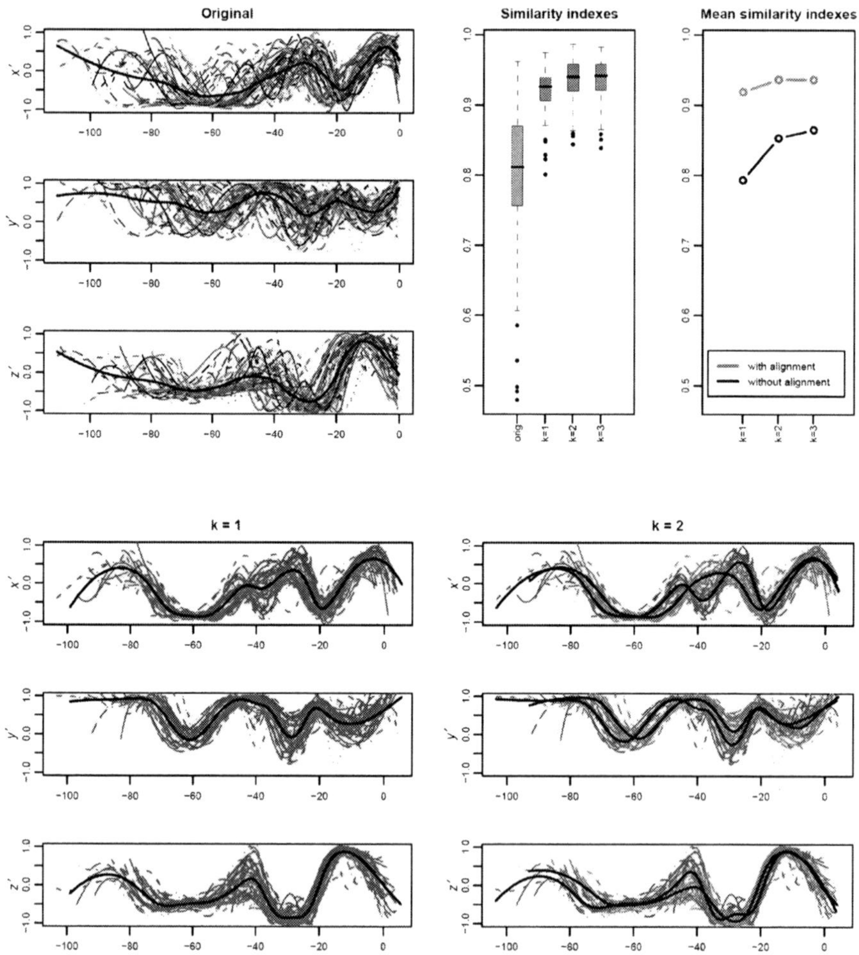

Fig. 3 *Top left*: first derivative of the three spatial coordinates x', y', z' of ICA centerlines. *Bottom*: first derivatives of one-mean and two-mean aligned curves (left and right respectively); first derivatives of templates always in black. *Top center*: boxplots of similarity indexes between each curve and the corresponding template for original curves and for k-mean aligned curves, $k = 1, 2, 3$. *Top right*: means of similarity indexes between each curve and the corresponding template obtained by k-mean alignment and by k-mean without alignment

the mean of similarity indexes (between curves and the corresponding template) obtained by k-mean alignment; the black line reports the mean of similarity indexes (between curves and the corresponding template) obtained by k-mean clustering without alignment.

Focussing on the last plot, at least two features need to be discussed. First, note the clear vertical shift between the orange and the black line: this points out the presence of a non-negligible phase variability within the original data and thus the necessity of aligning the data before undertaking any further analysis.

Second, once decided that alignment is needed, note the absence in the orange line of an evident improvement in the performance when three clusters are used in place of two: this suggests that $k = 2$ is the correct number of clusters. Consequently, the two-mean alignment algorithm will be used to jointly cluster and align the 65 ICA centerlines.

In the next two sections, we will discuss some interesting issues amenable of a biological interpretation that the two-mean alignment algorithm points out while neither the simple two-mean clustering without alignment nor the simple one-mean alignment (i.e., continuous alignment) have been able to disclose. In particular, the most interesting findings relative to the association between cluster membership and the aneurysmal pathologies are tackled in Sect. 3.1; the ones relative to the association between the shape of the aligned centerlines and the aneurysmal pathologies are instead shown in Sect. 3.2; the analysis of the warping functions is omitted since no interesting associations have been found.

3.1 Centerline Clusters vs Cerebral Aneurysms

Focussing on the two clusters detected by the two-mean alignment algorithm (bottom-right plots of Fig. 3), it is noticeable that the two clusters essentially differ within the region between 20 and 50 mm from the end of the ICA, that is also the region where the amplitude variability is maximal within the one-mean aligned data (bottom-left plots of Fig. 3). In particular (left plot of Fig. 4 where the two cluster templates are reported), we can identify a cluster associated to S-shaped ICAs (two siphons in the distal part of the ICA), i.e. the 30 green curves, and a cluster associated to Ω-shaped ICAs (just one siphon in the distal part of the ICA) i.e. the 35 orange curves. To our knowledge, it is the first time that this categorization, proposed in Krayenbuehl et al. (1982), is statistically detected. To show the primacy of the two-mean alignment, not only over the one-mean alignment but also over the simple two-mean clustering, in Fig. 4 the cluster templates detected by two-mean

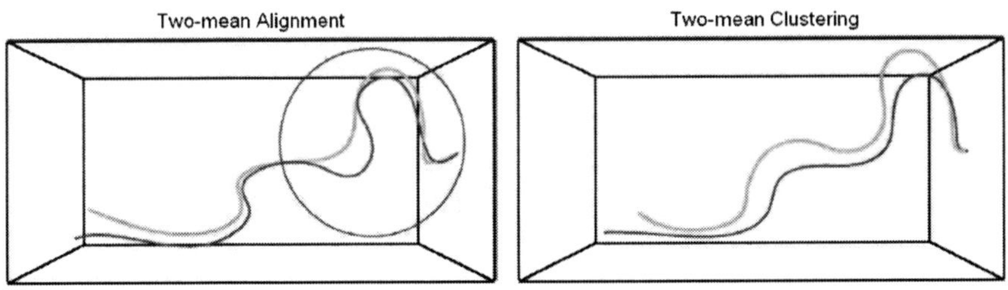

Fig. 4 *Left*: cluster template curves detected by two-mean alignment (S group in green and Ω group in orange). *Right*: cluster template curves detected by the simple two-mean clustering

alignment (left) and by the simple two-mean clustering (right) are compared. It is evident that while the former algorithm detects two morphologically different templates (the S and the Ω are clearly visible within the red circle), the latter detects two templates that are essentially equal in shape but just shifted. This is not surprising since the two-mean clustering algorithm (that is optimal if no phase variability is present within the data) is completely driven in this case by phase variability, providing fictitious and uninteresting amplitude clusters.

Moreover, the two clusters detected by the two-mean alignment turn out to be associated to the aneurysmal pathology, since there is statistical evidence of a dependence between cluster membership, and aneurysm presence and location (MC simulated p-value of Pearson's χ^2 test for independence equal to 0.0013): indeed, if we label the 65 patients according to the absence of an aneurysm (NO group), the presence of an aneurysm along the ICA (YES-ICA group), and the presence of an aneurysm downstream of the ICA (YES-DS group), we obtain the following conditional contingency table:

	NO	YES-ICA	YES-DS
S	100.0%	52.0%	30.3%
Ω	00.0%	48.0%	69.7%

A close look at the previous table makes evident that: (a) within this data set, there are no healthy subjects within the Ω cluster and all healthy subjects belong to the S cluster; (b) within the YES-DS group the number of Ω patients is more than twice the number of S patients, while within the YES-ICA group the two proportions are nearly equal. Wall shear stress is suspected to be associated to aneurysm onset and rupture and thus vessel geometry and hemodynamics could possibly explain this dependence.

Indeed, both ICA and arteries downstream of the ICA are very stressed vessels from a mechanical point of view: the former because its bends are expected to act like a fluid dynamical dissipator for the brain; the latter ones because they are floating in the brain humor without being surrounded by any muscle tissues. In particular, while S-shaped ICAs (two-bend syphon) are expected to be very effective in making the flow steadier; Ω-shaped ICAs (one-bend syphon) are instead expected to be less effective (this could be a possible explanation to (a)). Moreover for this same reason, in Ω-shaped ICAs, the blood flow downstream of the ICA is expected to be less steady, providing an overloaded mechanical stress to downstream arteries (this could be an explanation to (b)).

3.2 Centerline Shapes vs Cerebral Aneurysms

Let us now focus on the two-mean aligned curves in order to find out possible relations between centerline geometry and aneurysms. In order to reduce data dimensionality, we perform a three-dimensional functional principal component

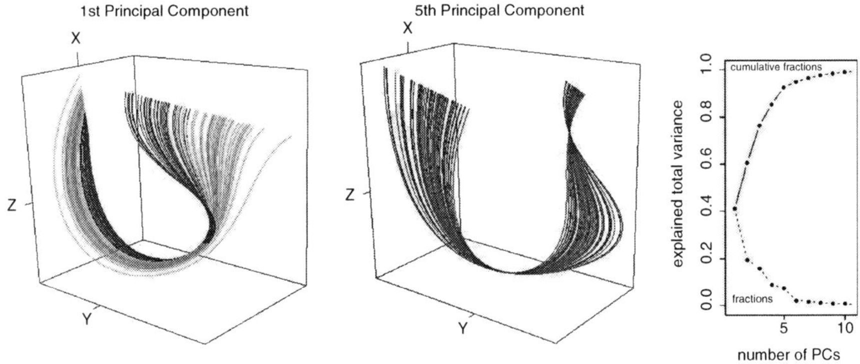

Fig. 5 *Left*: the projections of the 65 ICA centerlines along the first principal component (in orange centerlines belonging to the Ω cluster and in green centerlines belonging to the S cluster). *Center*: the projections of the 65 ICA centerlines along the fifth principal component (in red centerlines associated to patients with a ruptured aneurysm and in blue patients without aneurysm or with unruptured aneurysm). *Right*: fractions of explained total variance

analysis (e.g., Ramsay and Silverman, 2005) of the aligned centerlines for values of the registered abscissa between -34.0 and -6.9 mm, i.e., the abscissa interval where all records are available. In the right plot of Fig. 5 the fractions and the cumulative fractions of explained total variance are displayed, it is evident that one, three, or five principal components can be used to represent the centerlines. We decide to summarize the data by means of the first five principal components comforted by the fact that they provide a visually good representation of the data, by the fact that they explain more than the 90% of the total variance, and by the fact that all remaining principal components seem not to be related to any structural mode of variability but just noise.

In the left plot of Fig. 5 the projections of the 65 ICA centerlines along the first principal component are reported (orange for the Ω cluster centerlines and green for the S cluster ones). Nearly 42% of the total variability is explained by this component. It is evident that the variability associated to the first component is mostly concentrated at the top-right extreme (i.e. the proximal part of the portion of centerline under investigation), and moreover it is indicating the presence and magnitude of a second syphon before the distal one (in this picture blood flows from right to left). The Mann-Whitney test for the first principal component scores of the S and the Ω cluster centerline projections presents a p-value equal to 10^{-14}. This result strongly supports the identification of the two clusters – detected by the two-mean alignment – with the S and Ω shaped ICAs proposed by Krayenbuehl et al. (1982).

The second, third, and fourth principal components are difficult to interpret and moreover no associations have been found between these principal components and the aneurysmal pathologies. For this reason they will not be discussed in this work.

The fifth principal component (explained total variance 7%, cumulative 93%) appears instead to be surprisingly easy to interpret (in the center plot of Fig. 5 the projections of the 65 ICA centerlines along the fifth principal component are reported: in red the centerlines associated to patients with a ruptured aneurysm and in blue the ones associated to patients without aneurysm or with unruptured aneurysm). Indeed, it expresses the prominence of the distal syphon, i.e., along the fifth principal component, ICA centerlines progressively evolve from having a very sharped distal syphon (lower scores) toward smoother distal bend (higher scores). It is known that the more curved the vessel is, the higher the vorticity in the fluid and the shear stress on the wall are. Analyzing the scores relevant to the fifth components, we find that patients with a ruptured aneurysm present significant lower scores than patients with an unruptured aneurysm or without aneurysm (bends-Whitney test p-value 0.0072), i.e. the former ones present more marked bends than the latter ones. These results could support the idea of a fluid dynamical origin of the onset and/or rupture of cerebral aneurysms.

All these fluid dynamical hypotheses are going to be evaluated, in the future, by fluid dynamical simulations in order to provide a mechanical interpretation of the relation between geometry and hemodynamics on one side, and aneurysm onset and rupture on the other, that this analysis partially already highlights.

4 Conclusions

We showed in this work an application of the k-mean alignment method proposed in Sangalli et al. (2010b) that jointly clusters and aligns curves. This method puts in a unique framework two widely used methods of functional data analysis: functional k-mean clustering and Procrustes continuous alignment. Indeed, these latter two methods turn out to be two special cases of the new one.

In particular, we used this method to perform a functional data analysis of 65 three-dimensional curves representing 65 internal carotid artery centerlines. In this application the k-mean alignment algorithm outdoes both functional k-mean clustering and Procrustes continuous alignment by pointing out interesting features from a medical and fluid dynamical point of view that former methods were not able to point out.

References

Antiga, L., Piccinelli, M., Botti, L., Ene-Iordache, B., Remuzzi, A., and Steinman, D. (2008), "An image-based modeling framework for patient-specific computational hemodynamics," *Medical and Biological Engineering and Computing*, 1097–112.

Boudaoud, S., Rix, H., and Meste, O. (2010), "Core Shape modelling of a set of curves," *Computational Statistics and Data Analysis*, 308–325.

Krayenbuehl, H., Huber, P., and Yasargil, M. G. (1982), *Krayenbuhl/Yasargil Cerebral Angiography*, Thieme Medical Publishers, 2nd ed.

Liu, X. and Yang, M. C. K. (2009), "Simultaneous curve registration and clustering for functional data," *Computational Statistics and Data Analysis*, 1361–1376.

Ramsay, J. O. and Silverman, B. W. (2005), *Functional Data Analysis*, Springer New York NY, 2nd ed.

Sangalli, L. M., Secchi, P., Vantini, S., and Veneziani, A. (2009a), "A case study in exploratory functional data analysis: geometrical features of the internal carotid artery," *Journal of the American Statistical Association*, 104, 37–48.

Sangalli, L. M., Secchi, P., Vantini, S., and Veneziani, A. (2009b), "Efficient estimation of three-dimensional curves and their derivatives by free-knot regression splines applied to the analysis of inner carotid artery centrelines," *Journal of the Royal Statistical Society, Ser. C, Applied Statistics*, 58, 285–306.

Sangalli, L. M., Secchi, P., Vantini, S., and Vitelli, V. (2010a), "Functional clustering and alignment methods with applications," *Communications in Applied and Industrial Mathematics*, 1, 205–224.

Sangalli, L. M., Secchi, P., Vantini, S., and Vitelli, V. (2010b), "K-mean alignment for curve clustering," *Computational Statistics and Data Analysis*, 54, 1219–1233.

Tarpey, T. and Kinateder, K. K. J. (2003), "Clustering functional data," *Journal of Classification*, 20, 93–114.

Part II
Statistics in Medicine

Bayesian Methods for Time Course Microarray Analysis: From Genes' Detection to Clustering

Claudia Angelini, Daniela De Canditiis, and Marianna Pensky

Abstract Time-course microarray experiments are an increasingly popular approach for understanding the dynamical behavior of a wide range of biological systems. In this paper we discuss some recently developed functional Bayesian methods specifically designed for time-course microarray data. The methods allow one to identify differentially expressed genes, to rank them, to estimate their expression profiles and to cluster the genes associated with the treatment according to their behavior across time. The methods successfully deal with various technical difficulties that arise in this type of experiments such as a large number of genes, a small number of observations, non-uniform sampling intervals, missing or multiple data and temporal dependence between observations for each gene. The procedures are illustrated using both simulated and real data.

1 Introduction

Time course microarray experiments are an increasingly popular approach for understanding the dynamical behavior of a wide range of biological systems. From a biological viewpoint, experiments can be carried out both to identify the genes that are associated to a given condition (disease status) or that respond to a given treatment, as well as to determine the genes with similar responses and behaviors under a given condition, or to infer the genes' network for studying their regulation mechanisms. From the computational viewpoint, analyzing high-dimensional time

C. Angelini (✉) · D. De Canditiis
Istituto per le Applicazioni del Calcolo, CNR
e-mail: c.angelini@iac.cnr.it; d.decanditiis@iac.cnr.it

M. Pensky
Department of Mathematics, University of Central Florida
e-mail: Marianna.Pensky@ucf.edu

A. Di Ciaccio et al. (eds.), *Advanced Statistical Methods for the Analysis of Large Data-Sets*, Studies in Theoretical and Applied Statistics,
DOI 10.1007/978-3-642-21037-2_5,

course-microarray data requires very specific statistical methods which are usually not included in standard software packages, so, as a consequence, the potential of these experiments has not yet been fully exploited. In fact, most of the existing software packages essentially apply techniques designed for static data to time-course microarray data. For example, the SAM software package (see Tusher et al. (2001)) was recently adapted to handle time course data by regarding the different time points as different groups. Similar approach was also used by Kerr et al. (2000) and Wu et al. (2003) among many others.

These methods can still be very useful for analysis of very short time course experiments (up to about 4–5 time points), however, the shortcoming of these approaches is that they ignore the biological temporal structure of the data producing results that are invariant under permutation of the time points. On the other hand, most classical time series or signal processing algorithms can not be employed since they have rigid requirements on the data (high number of time-points, uniform sampling intervals, absence of replicated or missing data) which microarray experiments rarely meet. However, due to the importance of the topic, the past few years saw new developments in the area of analysis of time-course microarray data. For example, procedures for detecting differentially expressed genes were proposed in Storey et al. (2005), Conesa et al. (2006), and Tai and Speed (2006) and implemented in the software *EDGE* (Leek et al., 2006) and in the R-packages *maSigPro* and *timecourse*), respectively. Similarly, procedures for clustering time-course microarray data also appeared recently in the literature, among them Heard et al. (2006), Ray and Mallick (2006), Ma et al. (2008) and Kim et al. (2008), the latter being specifically designed for periodic gene-expression profiles.

In this paper, we first discuss the recent functional Bayesian methods developed in Angelini et al. (2007) (and their generalization Angelini et al. (2009)) for detecting differentially expressed genes in time course microarray experiments. These methods allow one to identify genes associated with the treatment or condition of interest in both the "one-sample" and the "two-sample" experimental designs, to rank them and to estimate their expression profiles. Then, we describe a novel functional approach for clustering the genes which are differentially expressed according to their temporal expression profiles. The latter approach automatically determines the number of existing groups and the "optimal" partition of the genes in these groups. Additionally, the method can also provide a set of "quasi-optimal" solutions that the experimenter can investigate.

All the above methods successfully deal with various technical difficulties that arise in microarray time-course experiments such as a large number of genes, a small number of observations, non-uniform sampling intervals, missing or multiple data and temporal dependence between observations for each gene.

The rest of the paper is organized as follow. Section 2 describes a Bayesian procedure for detecting differentially expressed genes in time course experiments. Section 3 introduces the infinite mixture model for clustering gene expression profiles. Finally, Sect. 4 illustrates these statistical procedures using both simulated and real data.

2 Detection of Differentially Expressed Genes in Time Course Microarray Experiments

In a "one sample" experimental design the data consists of the records, for N genes, of the differences in gene expression levels between the sample of interest and a reference (i.e., treated and control) in course of time. Each record is modeled as a noisy measurement of a function $s_i(t)$ at a time instant $t^{(j)} \in [0, T]$ which represents the differential gene expression profile.

The data $z_i^{j,k}$, for each gene i, each time instant j and each replicate k, are represented by the following expression:

$$z_i^{j,k} = s_i(t^{(j)}) + \zeta_i^{j,k}, \; i = 1, \ldots, N, \; j = 1, \ldots, n, \; k = 1, \ldots, k_i^{(j)}. \quad (1)$$

Here, n is the number of time points which is relatively small, $k_i^{(j)}$ is the number of replicates available at time instant $t^{(j)}$ for gene i, while the number of genes N is very large. For each gene, $M_i = \sum_{j=1}^{n} k_i^{(j)}$ observations are available. The objective is to identify the genes showing nonzero differential expression profile between treated and control samples, and then to evaluate the effect of the treatment. For each gene i, we expand its functional expression profile $s_i(t)$ into a series over some standard orthonormal basis on $[0, T]$ with coefficients $c_i^{(l)}$, $l = 0, \cdots, L_i$:

$$s_i(t) = \sum_{l=0}^{L_i} c_i^{(l)} \phi_l(t). \quad (2)$$

Following Angelini et al. (2007), genes are treated as conditionally independent and their expressions are matrix-wise modeled as $\mathbf{z}_i = \mathbf{D}_i \mathbf{c}_i + \boldsymbol{\zeta}_i$. Here, $\mathbf{D}_i$ is a block design matrix, the j-row of which is the block vector $[\phi_0(t^{(j)}) \; \phi_1(t^{(j)}) \; \ldots \; \phi_{L_i}(t^{(j)})]$ replicated $k_i^{(j)}$ times; $\mathbf{z}_i = (z_i^{1,1} \ldots z_i^{1,k_i^{(1)}}, \ldots, z_i^{n,1}, \ldots z_i^{n,k_i^{(n)}})^T$, $\mathbf{c}_i = (c_i^{(0)}, \ldots, c_i^{(L_i)})^T$ and $\boldsymbol{\zeta}_i = (\zeta_i^{1,1}, \ldots, \zeta_i^{1,k_i^{(1)}}, \ldots, \zeta_i^{n,1}, \ldots, \zeta_i^{n,k_i^{(n)}})^T$ are, respectively, the column vectors of all measurements for gene i, the coefficients of $s_i(t)$ in the chosen basis and the random errors. The following hierarchical model is imposed on the data:

$$\mathbf{z}_i \mid L_i, \mathbf{c}_i, \sigma^2 \sim \mathcal{N}(\mathbf{D}_i \mathbf{c}_i, \sigma^2 \mathbf{I}_{M_i}) \qquad L_i \sim \text{Truncated Poisson}\,(\lambda, L_{\max}) \quad (3)$$

$$\mathbf{c}_i \mid L_i, \sigma^2 \sim \pi_0 \delta(0, \ldots, 0) + (1 - \pi_0)\mathcal{N}(0, \sigma^2 \tau_i^2 \mathbf{Q}_i^{-1}) \quad (4)$$

All parameters in the model are treated either as random variables or as nuisance parameters, recovered from the data. Noise variance σ^2 is assumed to be random, $\sigma^2 \sim \rho(\sigma^2)$, and the following priors allow to account for possibly non-Gaussian errors:

Model 1: $\rho(\sigma^2) = \delta(\sigma^2 - \sigma_0^2)$, the point mass at σ_0^2.
Model 2: $\rho(\sigma^2) = IG(\gamma, b)$, the Inverse Gamma distribution.
Model 3: $\rho(\sigma^2) = c_\mu \sigma^{M_i - 1} e^{-\sigma^2 \mu/2}$.

The automatic detection of differentially expressed genes is carried out on the basis of Bayes Factors (BF), that, following Abramovich and Angelini (2006), are also used for taking into account multiplicity of errors within a Bayesian framework. Subsequently, the curves $s_i(t)$ are estimated by the posterior means. The algorithm is self-contained and the hyperparameters are estimated from the data. Explicit formulae and other details of calculations can be found in Angelini et al. (2007); the procedure is implemented in the software packages *BATS*, see Angelini et al. (2008).

The above approach has been extended to the case of the two-sample experimental design in Angelini et al. (2009). Although, mathematically, the two-sample set-up appears homogeneous, in reality it is not. In fact, it may involve comparison between the cells under different biological conditions (e.g., for the same species residing in different parts of the habitat) or estimating an effect of a treatment (e.g., effect of estrogen treatment on a breast cell). Hence, we will distinguish between "interchangeable" and "non-interchangeable" models. For a gene i in the sample $\aleph$ ($\aleph = 1, 2$), we assume that evolution in time of its expression level is governed by a function $s_{\aleph i}(t)$ as in (1) and each of the measurements $z_{\aleph i}^{j,k}$ involves some measurement error. The quantity of interest is the difference between expression levels $s_i(t) = s_{1i}(t) - s_{2i}(t)$.

Given the orthogonal basis, we expand $s_{\aleph i}(t)$ ($\aleph = 1, 2$) as in (2) and model $\mathbf{z}_{\aleph i} \mid L_i, \mathbf{c}_{\aleph i}, \sigma^2 \sim \mathcal{N}(\mathbf{D}_{\aleph i}\mathbf{c}_{\aleph i}, \sigma^2 \mathbf{I}_{M_{\aleph i}})$ as in (3).

The vectors of coefficients $\mathbf{c}_{\aleph i}$ are modeled differently in the interchangeable and non-interchangeable cases.

Interchangeable set-up: Assume that a-priori vectors of the coefficients $\mathbf{c}_{\aleph i}$ are either equal to each other or are independent and have identical distributions: $\mathbf{c}_{1i}, \mathbf{c}_{2i} \mid L_i, \sigma^2 \sim \pi_0 \mathcal{N}(\mathbf{c}_i \mid \mathbf{0}, \sigma^2 \tau_i^2 \mathbf{Q}_i^{-1})\ \delta(\mathbf{c}_{1i} = \mathbf{c}_{2i}) + (1 - \pi_0) \prod_{\aleph=1}^{2} \mathcal{N}(\mathbf{c}_{\aleph i} \mid \mathbf{0}, \sigma^2 \lambda_i^2 \mathbf{Q}_i^{-1})$.

Non-interchangeable set-up: Assume that the expression level in sample 2 is the sum of the expression level in sample 1 and the effect of the treatment $\mathbf{d}_i$, i.e. $\mathbf{c}_{2i} = \mathbf{c}_{1i} + \mathbf{d}_i$: $\mathbf{c}_{1i} \mid L_i, \sigma^2 \sim \mathcal{N}(\mathbf{c}_{1i} \mid \mathbf{0}, \sigma^2 \tau_i^2 \mathbf{Q}_i^{-1})$ and $\mathbf{d}_i \mid L_i, \sigma^2 \sim \pi_0 \delta(\mathbf{d}_i = \mathbf{0}) + (1 - \pi_0)\mathcal{N}(\mathbf{d}_i | \mathbf{0}, \sigma^2 \lambda_i^2 \mathbf{Q}_i^{-1})$.

In both set ups, parameter σ^2 is treated as a random variable with the same three choices for $\rho(\sigma^2)$ as in the "one-sample case". The inference is carried out similarly to the "one-sample case". For brevity, we do not report the formulae, see Angelini et al. (2009) for details. Note that the model allows that the two samples can be observed on different grids $t_{\aleph}^{(j)}$ taken in the same interval $[0, T]$.

A great advantage of both Bayesian models described above is that all evaluations are carried out in analytic form, with very efficient computations.

3 Clustering Time Course Profiles

Once the genes associated to the condition and treatment of interest have been selected, using, for example, the Bayesian methods described in Sect. 2, the experimenter is usually interested in investigating gene-regulation's mechanisms, i.e., to infer the genes network. For this purpose, and in order to reveal genes that show similar behavior, it is useful to cluster genes according to their temporal expression profiles. Based on the model (1), (2), we have recently proposed a novel functional Bayesain clustering procedure. In particular, we assume that the coefficients $\mathbf{c}_i$ follow the two-level hierarchical model:

Level 1: For each gene i, given the degree of the polynomial L_i, the vector of coefficients $\mathbf{c}_i$ and τ_i^2, we assume that $\mathbf{z}_i \mid L_i, \mathbf{c}_i, \tau_i^2 \sim \mathcal{N}(\mathbf{D}_i \mathbf{c}_i, \sigma^2 \tau_i^2 \mathbf{I}_{M_i})$.
Level 2: Given the degree of the polynomial L_i, the mean $\boldsymbol{\mu}_i$ and the precision τ_i^2, we assume that $\mathbf{c}_i | L_i, \boldsymbol{\mu}_i, \tau_i^2 \sim \mathcal{N}(\boldsymbol{\mu}_i, \tau_i^2 \mathbf{I}_{L_i})$.

In the above notations, τ_i^2 represents the cluster's variability, while σ^2 is a parameter that scales the variances of instrumental errors. Integrating out vectors of parameters $\mathbf{c}_i$, we obtain the following marginal distributions of the data vectors

$$\mathbf{z}_i \mid L_i, \boldsymbol{\mu}_i, \tau_i^2 \sim \mathcal{N}(\mathbf{D}_i \boldsymbol{\mu}_i, \tau_i^2 \mathbf{A}_i), \tag{5}$$

where $\mathbf{A}_i = \mathbf{D}_i \mathbf{D}_i^t + \sigma^2 \mathbf{I}_{M_i}$ is a known gene-specific design-related matrix.

Assuming that two genes, i and j, belong to the same cluster if and only if the corresponding vectors of coefficients $\mathbf{c}_i$ and $\mathbf{c}_j$ are of the same dimension $(L+1)$ and are sampled from the same $(L+1)$-variate normal distribution with the mean $\boldsymbol{\mu}$ and the scale parameter τ^2, we can characterize each cluster by the triplet of parameters $(L, \boldsymbol{\mu}, \tau^2)$. Then, introducing a latent vector $\boldsymbol{\gamma}$ of length N (number of genes to be clustered) such that $\gamma_i = \gamma_j$ if and only if genes i and j belong to the same cluster, we can rewrite expression (5) as

$$\mathbf{z}_i \mid \gamma_i, L_{\gamma_i}, \boldsymbol{\mu}_{\gamma_i}, \tau_{\gamma_i}^2 \sim \mathcal{N}(\mathbf{D}_i \boldsymbol{\mu}_{\gamma_i}, \tau_{\gamma_i}^2 A_i). \tag{6}$$

A Dirichlet process prior (see Ferguson (1973)) is elicited in (6) on the distribution of the triplet (L,μ,τ^2) of parameters that define the cluster: $(L, \mu, \tau^2) \sim DP(\alpha, G_0)$ where α is the scale parameter and G_0 the base distribution. Note that $DP(\alpha, G_0)$ is a discrete process and it has the great advantage to allow any number of clusters to be present in the dataset. We choose the following conjugate base distribution $G_0(L, \boldsymbol{\mu}, \tau^2) = g_\lambda(L)\, NIG(\boldsymbol{\mu}, \tau^2 | a, b, 0, \rho^2 Q)$, where $NIG(\boldsymbol{\mu}, \tau^2 | a, b, 0, \rho^2 \mathbf{Q}) = N(\mu | 0, \tau^2 \rho^2 \mathbf{Q}) IG(\tau^2 | a, b)$, and $g_\lambda(L)$ as in (3).

Finally, after having integrated out the parameters (L,μ,τ^2), we carry out inference on the posterior $p(\boldsymbol{\gamma}|\mathbf{z})$ using the improved MCMC Split-Merge procedure proposed by Dahl (2005).

In particular, after burn-in, we choose the MAP criterion for selecting an "optimal" allocation. The algorithm, named *FBMMC* (Functional Bayesian Mixture

Model Clustering), consists of 4 steps: – choice of an initial configuration (usually random); – estimation of the hyperparameters form the data; – evaluation of the MCMC chain; – choice of the optimal configuration. Explicit formulae and other details can be found in Angelini et al. (2011).

One of the great advantages of this Bayesian approach is that, at the price of higher computational cost, it automatically identifies the number of clusters hidden in the data-sets and the "optimal" partition. Moreover, by pooling together results of different chains or looking at the marginal posterior probability of each visited model, the algorithm can also provide a limited number of "quasi-optimal" solutions which an experimenter can easily investigate.

4 Results

In the following, we first show the performance of the above described procedures on a simulated dataset, then we use them for a case study of a human breast cancer cell line stimulated with estrogen. The synthetic and the real data-set used for illustrating the performance of our methodology and the Matlab routines used for carrying out simulations and analysis are available upon request from the first author.

4.1 Simulations

In Angelini et al. (2007) we investigated the performance of *BATS* software for the detection of time-course differentially expressed genes' profiles and compared the results with those obtained by other procedures available in the literature. In those simulation studies several data-sets were generated in which each temporal profile was independently sampled using formulae (3) and (4) (the synthetic dataset generator is contained in the software Angelini et al. (2008)).

In this paper, the objective is to study the procedure which combines the gene detection algorithm described in Sect. 2 with the clustering approach of Sect. 3. In particular, we want to investigate the precision of the following two stage procedure: at the first stage, a user selects a subset of differentially expressed genes from the whole dataset using *BATS*, then, only those genes are clustered using the *FBMMC* software. Clearly, the clustering algorithm could be also applied to the original dataset, however, limiting its application only to the genes which are associated to the treatment under consideration not only reduces the computational cost, but also allows one to obtain more robust results without any loss of biological knowledge.

For this purpose, we chose the grid of $n = 11$ non-equispaced time instants $[1, 2, 4, 6, 8, 12, 16, 20, 24, 28, 32]$, with $k_i^j = 2$ for all $j = 1, \ldots, 11$, except $k_i^{2,5,7} = 3$. This grid coincides with the grid in the real data example presented in the next section. We chose the Legendre polynomial basis in (2) and randomly partitioned a set of N_1 genes into R clusters, encoding the partition in the vector $\boldsymbol{\gamma}_{true}$. Then, we generated R triplets of cluster parameters, $(L_i, \boldsymbol{\mu}_i, \tau_i^2)$, each of them characterizing a specific cluster, sampling L_i from the discrete uniform distribution in the interval $[1, L_{\max}^{true}]$, $\boldsymbol{\mu}_i$ from $\mathcal{N}(0, \rho^2\tau_i^2\mathbf{Q})$ and τ_i^2 from the uniform distribution of width 0.1 centered at τ_0^2. After that, given allocation vector $\boldsymbol{\gamma}_{true}$ associated with the partition, we randomly sampled the gene profile coefficients $\mathbf{c}_i$ from the $(L_{\gamma_i} + 1)$-variate normal distribution $\mathcal{N}(\boldsymbol{\mu}_{\gamma_i}, \tau_{\gamma_i}^2\mathbf{Q})$ where L_{γ_i}, the vector $\boldsymbol{\mu}_{\gamma_i}$ and the scale parameter $\tau_{\gamma_i}^2$ correspond to the specific cluster where gene i belongs. Then, to study the effect of the noise we added a normal random error with zero mean and variance $\sigma^2\tau_0^2$ to each data point. For simplicity, matrix $\mathbf{Q}_i$ was set to be identity. Finally, in order to mimic a real context where only a certain (usually small) percentage of the genes are affected by the experimental condition under study, the N_1 profiles were randomly mixed with N_0 noisy profiles generated from a normal $N(0, s^2)$ distribution with standard deviation s.

For the sake of brevity, below we report only the results corresponding to the case where $\sigma = 0.9$, $\rho = 2.5$, $\tau_0 = 0.2$, $L_{\max^{true}} = 6$ and we generated $R = 9$ clusters on a set of $N_1 = 500$ truly differentially expressed genes and $N_0 = 9{,}500$ noisy profiles with standard deviations $s = 0.35$ in order to constitute a synthetic data-set of $N = 10,000$ temporal profiles in which 500 elements are differentially expressed.

First, we applied *BATS* (version 1.0) with different combinations of parameters and models to such data-set. In particular we fixed $L_{\max} = 6$, $\nu = 0$ and for each model we allowed λ's ranging between 6 and 12, corresponding to an expected prior degree of polynomials from 2.5 to 3.5. Model 2 (in Sect. 2) was applied with the two versions of the procedure for estimating the hyper-parameters of the Inverse Gamma distribution (i.e., by using the MLE or by fixing one of the two parameters and then estimating the other one by matching the mean of the IG with $\hat{\sigma}^2$ and using BATS default settings for all other parameters, see Angelini et al. (2008) for details). The resulting models are denoted as Model 2 (a) and Model 2 (b), respectively.

The number of genes detected as differentially expressed in the 28 version of the analysis (i.e., Model 1, Model 2 (a), Model 2 (b), Model 3 and $\lambda = 6, \ldots, 12$) was ranging between 491 and 500. All the genes detected by BATS as differentially expressed were correctly identified. Note that for this type of noise the results are very similar in terms of quality to those illustrated in a different simulation set-up in Angelini et al. (2007). Additionally, when varying the data-set, we found very high reproducibility of the results with limited number of false positive detections occurring only when the level of noise increases significantly. The number of genes detected by intersection of all combinations of methods and parameters was 491 out of the original 500, showing negligible loss of power, and then the *FBMMC* clustering algorithm was applied only to those profiles.

For this purpose, we ran 8 parallel MCMC chains of length $2,000,000$ in two set-ups. In the first set-up we carried out the analysis with $L_{\max} = 5$ and in the second with $L_{\max} = 6$. Each chain was initialized with a completely random configuration $\boldsymbol{\gamma}$. In each chain, we estimated hyper-parameters ρ^2, σ^2 and τ_0^2 by the off-line procedure described in Angelini et al. (2011), in which a preliminary allocation was obtained using the k-means algorithm with any number of clusters between 5 and 20. In both set-ups, we fixed $\alpha = 1$ and $a = 20$ and allowed λ to vary from 9 and 12. For each chain, after a burn-in period of about 100,000 iterations, we chose the allocation which maximizes the posterior distribution as an "optimal" one. As a measure of the precision of the algorithm we used Adjusted Rand index (ARI) (see, e.g., Yeung et al. (2001)) with respect to $\boldsymbol{\gamma}_{true}$.

For all data-sets and all MCMC chains, we observed high reproducibility of the results both with respect to the random initialization and to the choice of parameters $L_{\max}$ and λ. The number of clusters detected by the algorithm was always between 9 and 10 (9 being the true number of clusters) with average ARI index of 0.910 (std 0.015) for the first set-up and 0.903 (std 0.005) for the second set-up.

4.2 *Case Study*

In order to illustrate the performance of the proposed methodologies, we applied them to the time-course microarray study described in Cicatiello et al. (2004) (the original data are available on the GEO repository – `http://www.ncbi.nlm.nih.gov/geo/`, accession number GSE1864). The objective of the experiment was to identify genes involved in the estrogen response in a human breast cancer cell line and to understand their functional relationship. For this purpose, we applied the two-stage procedure described in Sect. 4.1.

In the original experiment, ZR-75.1 cells were stimulated with a mitogenic dose of 17β-estradiol, after 5 days of starvation on a hormone-free medium. Samples were taken after $t = 1, 2, 4, 6, 8, 12, 16, 20, 24, 28, 32$ hours, with a total of 11 time points covering the completion of a full mitotic cycle in hormone-stimulated cells. For each time point at least two replicates were available (three replicates at $t = 2, 8, 16$).

After standard normalization procedures (see, e.g., Wit and McClure (2004)) 8161 genes were selected for our analysis (among them about 350 contained at least one missing value), see Angelini et al. (2007) for details. The pre-processed data-set is freely available as an example data-set in the software *BATS* (Angelini et al., 2008).

As the first step of our procedure, we analyzed the above described data-set using *BATS* with various combination of prior models and choices of the parameter λ. Table 1 shows the number of genes identified as affected by the treatment for each of the Models 1, 2(a), 2(b) and 3, $L_{\max} = 6$, $\nu = 0$ and λ's ranging between 6 and 12, corresponding to an expected prior degree of polynomials from 2.5 to 3.5. We note that 574 genes were common to all 28 lists (combination of different methods and different parameter values) while 958 genes have been selected in at least one of the 28 lists. Note also that the list of 574 common genes includes 270 genes out of the 344 genes identified as significant in Cicatiello et al. (2004). Additionally, some of the newly identified genes are replicated spots, others are known nowadays

Table 1 Total number of genes detected as differentially expressed by the methods described in Sect. 2 (with $\nu = 0$ and $Ł_{max} = 6$) on the real dataset by Cicatiello et al. (2004). Model 2 (a) and Model 2 (b) refer to two different estimation procedures for the hyper-parameters of the IG

Model	$\lambda = 6$	$\lambda = 7$	$\lambda = 8$	$\lambda = 9$	$\lambda = 10$	$\lambda = 11$	$\lambda = 12$
Model 1	867	808	753	712	692	688	691
Model 2 (a)	893	823	765	711	679	657	650
Model 2 (b)	869	810	755	714	694	690	693
Model 3	855	786	726	676	640	617	609

Table 2 Number of clusters obtained (and occurrence of that cardinality in the 20 chains) for the 574 genes that where found differentially expressed in the real data-set Cicatiello et al. (2004)

set-up	Inferred number of clusters
FBMMC: $L_{max} = 5$	19(1); 20(3); 21(4); 22(6); 23(6)
FBMMC: $L_{max} = 6$	20(1); 21(3); 22(9); 23(2); 24(5)

to be involved in biological processes related to estrogen response. In Angelini et al. (2007), we demonstrated that *BATS* approach delivers superior performance in comparison with other techniques such as Leek et al. (2006) and Tai and Speed (2006).

At a second stage of the procedure, we applied the *FBMMC* algorithm described in Sect. 3 to the $N = 574$ genes selected as differentially expressed by intersection of all combination of methods and parameters. We ran 20 parallel MCMC chains of length $2,000,000$ in two set-ups. In the first set-up we carried out the analysis with $L_{max} = 5$ in the second with $L_{max} = 6$.

Each chain was initialized with a completely random configuration $\boldsymbol{\gamma}$. In each chain, we estimated hyper-parameters ρ^2, σ^2 and τ_0^2 by the off-line procedure described in Angelini et al. (2011), in which a preliminary allocation was obtained using the k-means algorithm with any number of clusters between 5 and 25. In both set-ups, we fixed $\alpha = 1$ and $a = 20$ and allowed λ to vary from 9 and 12. For all data-sets and all MCMC chains we observed high reproducibility of the results, both with respect to the random initialization and to the choice of parameters L_{max} and λ. Table 2 shows the numbers of clusters obtained (with the number of times it occurs in the 20 chains) for each set-up.

Acknowledgements This research was supported in part by the CNR-Bioinformatics projects, the CNR DG.RSTL.004.002 Project. Marianna Pensky was supported in part by National Science Foundation (NSF), grant DMS-1106564 and the CNR Short term mobility grant 2008.

References

Abramovich, F. and Angelini, C.: Bayesian maximum a posteriori multiple testing procedure. Sankhya, **68**, 436-460, (2006)

Angelini, C., De Canditiis, D., Mutarelli, M., and Pensky, M.: A Bayesian Approach to Estimation and Testing in Time-course Microarray Experiments. Statistical Applications in Genetics and Molecular Biology **6**, art. 24, (2007)

Angelini, C., Cutillo, L., De Canditiis, D., Mutarelli, M., and Pensky, M.: BATS: A Bayesian user friendly Software for analyzing time series microarray experiments. BMC Bioinformatics **9**, (2008)

Angelini, C., De Canditiis, D., and Pensky, M.: Bayesian models for the two-sample time-course microarray experiments. Computational Statistics & Data Analysis **53**, 1547-1565, (2009)

Angelini, C., De Canditiis, D., and Pensky, M.: Clustering time-course microarray data using functional Bayesian Infinite Mixture Model, Journal of Applied Statistics, (2011), DOI: 10.1080/02664763.2011.578620

Cicatiello, L., Scafoglio, C., Altucci, L., Cancemi, M., Natoli, G., Facchiano, A., Iazzetti, G., Calogero, R., Biglia, N., De Bortoli, M., Sfiligol, C., Sismondi, P., Bresciani, F., Weisz, A.: A genomic view of estrogen actions in human breast cancer cells by expression profiling of the hormone-responsive transcriptome. Journal of Molecular Endocrinology **32**, 719-775, (2004)

Conesa, A. Nueda, M. J., Ferrer, A., and Talon, M.: MaSigPro: a method to identify significantly differential expression profiles in time-course microarray-experiments. Bioinformatics **22**, 1096–1102, (2006)

Dahl D.B.: Sequentially-Allocated Merge-Split Sampler for Conjugate and Nonconjugate Dirichlet Process Mixture Models. Technical Report, Department of Statistics, University of Wisconsin – Madison (2005)

Ferguson T.S.: A bayesian analysis of some nonparametric problems. Annals of Statistics **1**, 209-230 (1973)

Heard, N.A., Holmes C.C., Stephens D.A.: A quantitative study of gene regulation involved in the Immune response of Anopheline Mosquitoes: An application of Bayesian hierarchical clustering of curves. Journal of the American Statistical Association, **101**, 18-29 (2006)

Leek, J. T., Monsen, E., Dabney, A. R., Storey, J. D.: EDGE: extraction and analysis of differential gene expression. Bioinformatics, **22**, 507-508, (2006)

Kerr, M.K., Martin M., and Churchill, G.A.: Analysis of variance for gene expression microarray data, Journal of Computational Biology, **7**, 819–837, (2000)

Kim B.R., Zhang,L., Berg, A., Fan J., Wu R.: A computational approach to the functional clustering of periodic gene-expression profiles. Genetics, **180**, 821-834, (2008)

Ma, P., Zhong,W., Feng,Y., Liu JS.: Bayesian functional data clustering for temporal microarray data. International journal of Plant Genomics., art. 231897, (2008)

Qin, Z. S.,: Clustering microarray gene expression data using weighted Chinese restaurant process. Bioinformatics, **22** 1988-1997, (2006)

Ray, S., Mallick B. . Functional clustering by Bayesian wavelet methods. *J. Royal Statistical Society: Series B*, **68**, 302-332 (2006)

Storey, J. D., Xiao, W., Leek, J. T., Tompkins, R. G., Davis, R. W.: Significance analysis of time course microarray experiments. PNAS **102**, 12837-12842, (2005)

Tai, Y. C., Speed, T.P.: A multivariate empirical Bayes statistic for replicated microarray time course data. Annals of Statistics, **34**, 2387-2412, (2006)

Tusher, V., Tibshirani, R., Chu, C.: Significance analysis of microarrays applied to the ionizing radiation response. PNAS, **98**, 5116-5121, (2001)

Yeung, K.Y., Fraley, C., Murua, A., Raftery, A.E., Ruzzo, W.L. Model-based clustering and data transformations for gene expression data. Bioinformatics, **17** 977-987, (2001)

Wit, E., and McClure, J. *Statistics for Microarrays: Design, Analysis and Inference*, Wiley, Chichester, West Sussex, England, (2004)

Wu, H., Kerr, M. K., Cui, X., and Churchill, G.A. MAANOVA: A software package for Analysis of spotted cDNA Microarray experiments. In *The Analysis of Gene Expression Data: Methods and Software* eds. Parmigiani, G.,Garrett, E.S., Irizarry, R. A., and Zeger, S.L. Statistics for Biology and Health. Springer, pp. 313–341, (2003)

Longitudinal Analysis of Gene Expression Profiles Using Functional Mixed-Effects Models

Maurice Berk, Cheryl Hemingway, Michael Levin, and Giovanni Montana

Abstract In many longitudinal microarray studies, the gene expression levels in a random sample are observed repeatedly over time under two or more conditions. The resulting time courses are generally very short, high-dimensional, and may have missing values. Moreover, for every gene, a certain amount of variability in the temporal profiles, among biological replicates, is generally observed. We propose a functional mixed-effects model for estimating the temporal pattern of each gene, which is assumed to be a smooth function. A statistical test based on the distance between the fitted curves is then carried out to detect differential expression. A simulation procedure for assessing the statistical power of our model is also suggested. We evaluate the model performance using both simulations and a real data set investigating the human host response to BCG exposure.

1 Introduction

In a longitudinal microarray experiment, the gene expression levels of a group of biological replicates – for example human patients – are observed repeatedly over time. A typical study might involve two or more biological groups, for instance a control group versus a drug-treated group, with the goal to identify genes whose temporal profiles differ between them. It can be challenging to model these

M. Berk · G. Montana (✉)
Department of Mathematics, Imperial College London
e-mail: maurice.berk01@imperial.ac.uk; g.montana@imperial.ac.uk

C. Hemingway
Great Ormond Street Hospital
e-mail: HeminC@gosh.nhs.uk

M. Levin
Division of Medicine, Imperial College London
e-mail: m.levin@imperial.ac.uk

A. Di Ciaccio et al. (eds.), *Advanced Statistical Methods for the Analysis of Large Data-Sets*, Studies in Theoretical and Applied Statistics,
DOI 10.1007/978-3-642-21037-2_6,

experiments in such a way that accounts for both the within-individual (temporal) and between-individual correlation – failure to do so may lead to poor parameter estimates and ultimately a loss of power and incorrect inference. Further challenges are presented by the small number of time points over which the observations are made, typically fewer than 10, the high dimensionality of the data with many thousands of genes studied simultaneously, and the presence of noise, with many missing observations.

In order to address these issues we present here a functional data analysis (FDA) approach to microarray time series analysis. In the FDA paradigm we treat observations as noisy realisations of an underlying smooth function of time which is to be estimated. These estimated functions are then treated as the fundamental unit of observation in the subsequent data analysis as in Ramsay and Silverman (2006). Similar approaches have been used for the clustering of time series gene expression data without replication (Bar-Joseph et al., 2003; Ma et al., 2006) but these cannot be applied to longitudinal designs such as the one described in Sect. 2. Our approach is much more closely related to, and can be considered a generalisation of, the EDGE model presented by Storey et al. (2005).

The rest of this paper is organised as follows. Our motivating study is introduced in Sect. 2. In Sect. 3 we present our methodology based on functional mixed-effects models. A simulation study is discussed in Sect. 4 where we compare our model to EDGE. Section 5 provides a brief summary of our experimental findings.

2 A Case Study: Tuberculosis and BCG Vaccination

Tuberculosis (TB) is the leading cause of death world-wide from a curable infectious disease. In 2006 it accounted for over 9 million new patients and over 2 million deaths; these figures are in spite of effective medication and a vaccine being available since 1921. This discrepancy is due in part to the HIV-epidemic, government cutbacks, increased immigration from high prevalence areas and the development of multi-drug resistant strains of the disease but ultimately due to our limited understanding of the complex interaction between the host and the pathogen *M. tuberculosis*. In particular, it has been a longstanding observation that the BCG vaccine conveys different levels of protection in different populations (Fine 1995). A total of 17 controlled trials of the efficacy of BCG vaccination have been carried out and efficacy has varied between 95% and 0%; some studies even show a negative effect (Colditz et al., 1994).

The purpose of this case study was to characterise the host response to BCG exposure by using microarrays to identify genes which were induced or repressed over time in the presence of BCG. Nine children with previous exposure to TB but who were then healthy, having completed TB therapy at least 6 months prior to the study were recruited from the Red Cross Children's Hospital Welcome TB research database, matched for age and ethnicity. A complete description of the experimental

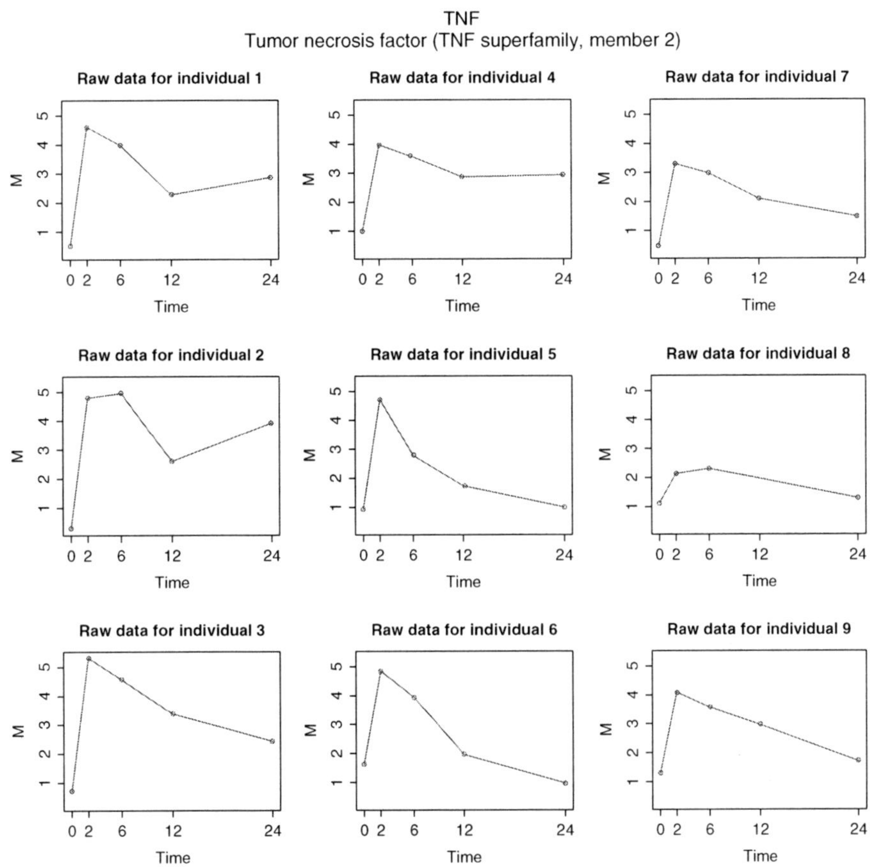

Fig. 1 An example of 9 individual gene expression profiles (biological replicates) for the TNF gene. The experimental setup is described in Sect. 2. Shown here are the raw data, before model fitting. Some of the peculiar features of the data can be observed here: (**a**) very short temporal profiles, (**b**) irregularly spaced design time points, (**c**) missing data, and (**d**) individual variability

procedures will be reported in a separate publication. In summary, each child contributed a BCG treated and a BCG negative control time series observed at $0, 2, 6, 12$ and 24 h after the addition of BCG or, in the case of the controls, 100 μl PBS. A two-colour array platform – the Stanford "lymphochip" – was used. Data preprocessing and quality control were performed using the GenePix4.0 software and in R using BioConductor (www.bioconductor.org). Figure 1 shows 9 biological replicates that have been observed for the TNF (tumor necrosis factor) gene, from which three typical features of the longitudinal data under analysis can be noted: (a) all time series are short and exhibit a clear serial correlation structure; (b) a few time points are missing (for instance, individual 8 has only 4 time points); (c) there is variability in the gene expression profiles across all individuals.

3 Mixed-Effects Smoothing Splines Models

Each observation being modelled is denoted by $y(t_{ij})$ and represents the gene expression measure observed on individual i at time t_{ij}, where $i = 1, 2, \ldots, n_k$, $j = 1, 2, \ldots, m_i$, n_k denotes the sample size in group k and m_i is the number of observations on individual i. In order to properly account for the features observed in Fig. 1, we suggest to model $y(t_{ij})$ non-parametrically:

$$y(t_{ij}) = \mu(t_{ij}) + v_i(t_{ij}) + \epsilon_{ij} \tag{1}$$

The model postulates that the observed gene expression measure $y(t_{ij})$ can be explained by the additive effect of three components: a mean response $\mu(\cdot)$, which is assumed to be a smooth, non-random curve defined over the time range of interest; an individual-specific deviation from that mean curve, $v_i(\cdot)$, which is assumed to be a smooth and random curve observed over the same time range; and an error term ϵ_{ij} which accounts for the variability not explained by the first two terms. Formally, we treat each $v_i(t_{ij})$, for $i = 1, 2, \ldots, n_k$, as independent and identically distributed realisations of an underlying stochastic process; specifically, we assume that $v_i(t_{ij})$ is a Gaussian process with zero-mean and covariance function $\gamma(s, t)$, that is $v_i(t_{ij}) \sim \text{GP}(0, \gamma)$. The errors terms ϵ_{ij} are assumed to be independent and normally distributed with zero mean and covariance matrix $\mathbf{R}_i$. We do not assume that all individual have been observed at the same design time points, and all the distinct design time points are denoted by $(\tau_1, \tau_2, \ldots, \tau_m)$.

We suggest to represent the curves using cubic smoothing splines; see, for instance, Green and Silverman (1994). The key idea of smoothing splines consists in making full use of all the design time points and then fitting the model by adding a smoothness or roughness constraint; by controlling the size of this constraint, we are then able to avoid curves that appear too wiggly. A natural way of measuring the roughness of a curve is by means of its integrated squared second derivative, assuming that the curve is twice-differentiable. We call $\boldsymbol{\eta} = (\eta(\tau_1), \ldots, \eta(\tau_m))^T$ the vector containing the values of the mean curve estimated at all design time points and, analogously, the individual-specific deviations from the mean curve, for individual i, are collected in $\mathbf{v}_i = (v_i(\tau_1), \ldots, v_i(\tau_m))^T$. The mean and individual curves featuring in model (1) can be written as, respectively, $\mu(t_{ij}) = \mathbf{x}_{ij}^T \boldsymbol{\eta}$ and $v_i(t_{ij}) = \mathbf{x}_{ij}^T \mathbf{v}_i$, with $i = 1, 2, \ldots, n$, and $\mathbf{x}_{ij} = (x_{ij1}, \ldots, x_{ijm})^T$, with $x_{ijr} = 1$ if $t_{ij} = \tau_r$, $r = 1, \ldots, m$ and $x_{ijr} = 0$ otherwise. The fact that the individual curves are assumed to be Gaussian processes is captured by assuming that the individual deviations are random and follow a zero-centred Gaussian distribution with covariance $\mathbf{D}$, where $\mathbf{D}(r, s) = \gamma(\tau_s, \tau_r)$, $r, s = 1, \ldots, m$. Finally, in matrix form, model (1) can then be rewritten as

$$\mathbf{y}_i = \mathbf{X}_i \boldsymbol{\eta} + \mathbf{X}_i \mathbf{v}_i + \boldsymbol{\epsilon}_i \tag{2}$$

$$\mathbf{v}_i \sim \text{N}(\mathbf{0}, \mathbf{D}) \qquad \boldsymbol{\epsilon}_i \sim \text{N}(\mathbf{0}, \mathbf{R}_i)$$

For simplicity, we assume that $\mathbf{R}_i = \sigma^2\mathbf{I}$. In this form, we obtain a linear mixed-effects model (Laird and Ware, 1982). Clearly, the model accounts for the fact that, for a given gene, the individual repeated measurements are correlated. Specifically, under the assumptions above, we have that $\text{cov}(\mathbf{y}_i) = \mathbf{X}_i\mathbf{D}\mathbf{X}_i^T + \mathbf{R}_i$.

3.1 Statistical Inference

A common approach to estimating the unknown parameters of a linear mixed-effects model is by maximum likelihood (ML) estimation. In our model (2), the twice negative logarithm of the (unconstrained) generalised log-likelihood for is given by

$$\sum_{i=1}^{n_k}\{(\mathbf{y}_i - \mathbf{X}_i\eta - \mathbf{X}_i\mathbf{v}_i)^T\mathbf{R}_i^{-1}(\mathbf{y}_i - \mathbf{X}_i\eta - \mathbf{X}_i\mathbf{v}_i) + \log|\mathbf{D}| + \mathbf{v}_i^T\mathbf{D}^{-1}\mathbf{v}_i + \log|\mathbf{R}_i|\}.$$

The ML estimators of the mean curve $\mu(\cdot)$ and each individual curve $v_i(\cdot)$ can be obtained by minimising a penalised version of this log-likelihood obtained by adding a term $\lambda\eta^T\mathbf{G}\eta$ and a term $\lambda_v\sum_{i=1}^{n_k}\{\mathbf{v}_i^T\mathbf{G}\mathbf{v}_i\}$, which impose a penalty on the roughness of the mean and individual curves, respectively. The matrix $\mathbf{G}$ is the roughness matrix that quantifies the smoothness of the curve (Green and Silverman, 1994) whilst the two scalars λ_v and λ are smoothing parameters controlling the size of the penalties. In principle, n_k distinct individual smoothing parameters can be introduced in the model but such a choice would incur a great computational cost during model fitting and selection. For this reason, we assume that, for each given gene, all individual curves share the same degree of smoothness and we use only one smoothing parameter λ_v.

After a rearrangement on the terms featuring in the penalised log-likelihood, the model can be re-written in terms of the regularised covariance matrices, $\mathbf{D}_v = (\mathbf{D}^{-1} + \lambda_v\mathbf{G})^{-1}$ and $\mathbf{V}_i = \mathbf{X}_i\mathbf{D}_v\mathbf{X}_i^T + \mathbf{R}_i$. When both these variance components are known, the ML estimators $\hat{\eta}$ and $\hat{\mathbf{v}}_i$, for $i = 1, 2, \ldots, n_k$, can be derived in closed-form as the minimisers of the penalised generalised log-likelihood. However, the variance components are generally unknown. All parameters can be estimated iteratively using an EM algorithm, which begins with some initial guesses of the variance components. The smoothing parameters λ and λ_v are found as those values, in the two-dimensional space $(\Lambda \times \Lambda_v)$, that optimise the *corrected* AIC, which includes a small sample size adjustment. The search for optimal smoothing values $\hat{\lambda}$ and $\hat{\lambda}_v$ is performed using a downhill simplex optimisation algorithm (Nelder and Mead, 1965).

The objective of our analysis is to compare the estimated mean curves observed under the two experimental groups and assess the null hypothesis that the curves are the same. After fitting model (2) to the data, independently for each group and each gene, we obtain the estimated mean curves $\hat{\mu}^{(1)}(t)$ and $\hat{\mu}^{(2)}(t)$. One way of measuring the dissimilarity between these two curves consists in computing the

L_2 distance between them, which can be evaluated using the smoothed curves $\hat{\mu}^{(1)}(t)$ and $\hat{\mu}^{(2)}(t)$, thus yielding the observed distance $\hat{D}$. We use this dissimilarity measure as a test statistic. Since the null distribution of this statistic is not available in closed form, we resort to a non-parametric bootstrap approach in order to approximate it.

4 Performance Assessment Using Simulated Longitudinal Data

In order to assess the performance of the proposed MESS model we compared it using a simulation study to the EDGE model developed by Storey et al. (2005). While the EDGE model takes the same form as (1), their parameterisation differs from ours in that the mean function $\mu(\cdot)$ is represented using B-splines and the individual curves $v_i(\cdot)$ are constrained to be a scalar shift. In the case of the mean curve, the B-spline representation requires specification of both the number and location of the knots which, unlike smoothing splines, offers discontinuous control over the degree of smoothing. Furthermore Storey et al. (2005) represent each gene using the same number of basis functions which, if taken to be too small, implies a poor fit to those genes with the most complex temporal profiles. Conversely, if the number of basis functions is sufficient to model these complex genes there is a danger that some genes will be overfit. In the case of the individual curves $v_i(\cdot)$, it should be clear that scalar shifts would be unable to fully capture the individual variability we observe in the raw data given in Fig. 1. This problem is compounded by the fact that Storey et al. (2005) propose an F-type test statistic for inference which makes use of the model residuals.

To determine the practical impact of these features we have set up a simulation procedure that generates individual curves that look similar to the real experimental data. Our procedure is based on a mixed-effects model with the fixed- and random-effects parameterized using B-splines, where the observations on individual i belonging to group j are given as

$$\mathbf{y}_i = \mathbf{X}_i \boldsymbol{\beta}_j + \mathbf{Z}_i \mathbf{b}_{ij} + \boldsymbol{\epsilon}_{ij} \tag{3}$$

$$\mathbf{b}_{ij} \sim \text{MVN}(\mathbf{0}, \mathbf{D}_j) \qquad \boldsymbol{\epsilon}_{ij} \sim \text{MVN}(\mathbf{0}, \sigma_j \mathbf{I}_{n_i \times n_i})$$

where $i = 1, \ldots, n$ and $j = 1, 2$. For simplicity, we use the same basis for the fixed- and random-effects so that $\mathbf{X}_i = \mathbf{Z}_i$. The parameters that need to be controlled in this setting therefore consist of the variance components $\sigma_1, \sigma_2, \mathbf{D}_1, \mathbf{D}_2$, the B-spline parameters for the group means $\boldsymbol{\beta}_1, \boldsymbol{\beta}_2$, and the B-spline basis $\mathbf{X}_i$ which is determined by the number and locations of the knots, K. Given the simple patterns often observed in real data sets, we place a single knot at the center of the time course. Wherever possible, we tune the variance components based on summary statistics computed from data produced in real studies such as the experiment described in Sect. 2. We further parameterise the covariance matrices as

$$\mathbf{D} = \tau \begin{bmatrix} \rho^0 & \rho^1 & \rho^2 & \cdots \\ \rho^1 & \rho^0 & \rho^1 & \cdots \\ \rho^2 & \rho^1 & \rho^0 & \ddots \\ \vdots & \ddots & \ddots & \ddots \end{bmatrix} \quad (4)$$

and introduce the notation $\mathbf{D}(\tau, \rho)$ for specifying these parameters. In this way we can vary the complexity of the individual curves by varying the parameter ρ and control the amount of variation between individuals by varying τ. When $\rho = 1$, the individual "curves" are scalar shifts, as in the EDGE model. As ρ tends to 0, $\mathbf{D}$ tends to $\tau\mathbf{I}$, where the B-spline parameters $\mathbf{b}_i$ are independent.

We begin simulating a given gene by first randomly generating the B-spline coefficients for the mean curve of group 1, $\boldsymbol{\beta}_1$, and for each individual belonging to this group, $\mathbf{b}_{i1}$, according to the following probability distributions $\boldsymbol{\beta}_1 \sim \text{MVN}(\mathbf{0}, \mathbf{D}_\beta)$ and $\mathbf{b}_{i1} \sim \text{MVN}(\mathbf{0}, \mathbf{D}_{b_1})$, with covariance matrices given by $\mathbf{D}_\beta = \mathbf{D}(0.25, 0.6)$ and $\mathbf{D}_{b_1} = \mathbf{D}(\tau_{b_1}, 0.6)$, where $\tau_{b_1} \sim U(0.1, 0.2)$. As in (3), the error term is normally distributed with variance σ_1. We set this variance component to be log-normally distributed with mean -2 and variance 0.35, values estimated from the real data.

Each simulated gene is differentially expressed with probability 0.1. If a gene is not differentially expressed then observations are generated for group 2 using exactly the same parameters as for group 1, i.e. $\boldsymbol{\beta}_1 = \boldsymbol{\beta}_2, \mathbf{D}_{b_1} = \mathbf{D}_{b_2}, \sigma_1 = \sigma_2$. On the other hand, if a gene is differentially expressed, then we randomly generate a vector $\boldsymbol{\beta}_\delta$ representing the difference in B-spline parameters for the group mean curves, distributed as $\boldsymbol{\beta}_\delta \sim \text{MVN}(\mathbf{0}, \mathbf{D}_{\boldsymbol{\beta}_\delta})$ and $\mathbf{D}_{\boldsymbol{\beta}_\delta} = \mathbf{D}(0.25, 0.9)$, with $\beta_{\delta 1} = 0$. We then normalise the vector $\boldsymbol{\beta}_\delta$ so that its L_2-norm is 1 before setting $\boldsymbol{\beta}_2 = \boldsymbol{\beta}_1 + \boldsymbol{\beta}_\delta$. By setting $\beta_{\delta 1} = 0$ we ensure that both mean curves began the time course at the same value, which we have observed in the real data and would similarly be the case if the data had been $t = 0$ transformed. Setting $\rho = 0.9$ for $\mathbf{D}_{\boldsymbol{\beta}_\delta}$ limits the difference between the curves in terms of complexity which, again, we observe in the real data where frequently the mean curves are simply vertically shifted versions of each other. Normalising the vector $\boldsymbol{\beta}_\delta$ enables us to control exactly how large an effect size we are trying to detect by multiplying the vector by a scaling factor.

Finally, we generate the individual curves for group 2 for a differentially expressed gene as before: $\mathbf{b}_{i2} \sim \text{MVN}(\mathbf{0}, \mathbf{D}_{b_2})$ and $\mathbf{D}_{b_2} = \mathbf{D}(\tau_{b_2}, 0.6)$, where $\tau_{b_2} \sim U(0.1, 0.2)$. The key point to note is that $\tau_{b_2} \neq \tau_{b_1}$. By doing so, a differentially expressed gene varies both in terms of the mean curve and the degree of individual variation. Similarly, σ_2 is distributed identically to yet independently of σ_1 so that the noise of the two groups is also different.

Using this simulation framework with the parameters laid out as above, we generated a data set consisting of 100,000 simulated genes observed with 9 individuals per group with 5 timepoints at 0, 2, 6, 12 and 24 h, following the same pattern of observations as the case study. 10% of genes were differentially expressed. We then used both the MESS and EDGE models to fit the data and identify differentially expressed genes. Figure 2 shows an example of a simulated gene with fitted mean and individual curves for both MESS and EDGE. In this instance EDGE's B-spline

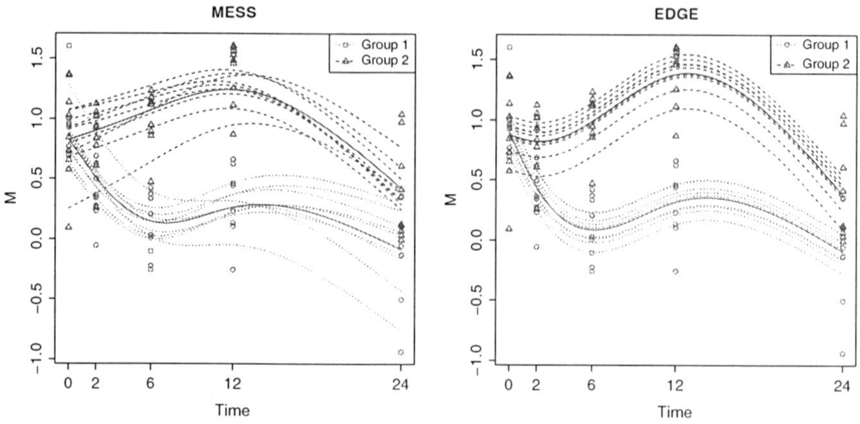

Fig. 2 An example of simulated longitudinal data and fitted curves using both MESS and EDGE. The *thick solid lines* correspond to the fitted means for each group. The *dotted lines* are the fitted individual curves for group 1 and the *dashed lines* are the fitted individual curves for group 2

parameterisation seems sufficient for representing the mean curves but the scalar shifts do not model the data as closely as MESS does. Compare this simulated gene to a real example from the experimental data shown in Fig. 3. This is the fit to the gene TNF, for which the raw data for the control group was given in Fig. 1. We can see here that EDGE has selected too few knots to adequately capture the rapid spike in gene expression levels at 2 h and that again the MESS model with individual curves provides a much closer fit to the data. Figure 4 gives the ROC curve for the simulation study based on 100,000 simulated genes. At a fixed specificity of 90%, the corresponding power for MESS is 85.1% compared to 70.4% for EDGE.

5 Experimental Results

We fit the MESS model to the BCG case study data and generated 100 bootstrap samples giving 3.2 million null genes from which to calculate empirical p-values based on the L_2 distance as a test statistic. After correcting these p-values for multiple testing, 359 probes were identified as being significantly differentially expressed, corresponding to 276 unique genes. We provide here a brief summary of the results, leaving the full biological interpretation to a dedicated forthcoming publication.

The top ten differentially expressed genes were found to be CCL20, PTGS2, SFRP1, IL1A, INHBA, FABP4, TNF, CXCL3, CCL19 and DHRS9. Many of these genes have previously been identified as being implicated in TB infection. For instance, CCL20 was found to be upregulated in human macrophages infected with *M.tuberculosis* (Ragno et al., 2001) and *in vivo* in patients suffering from

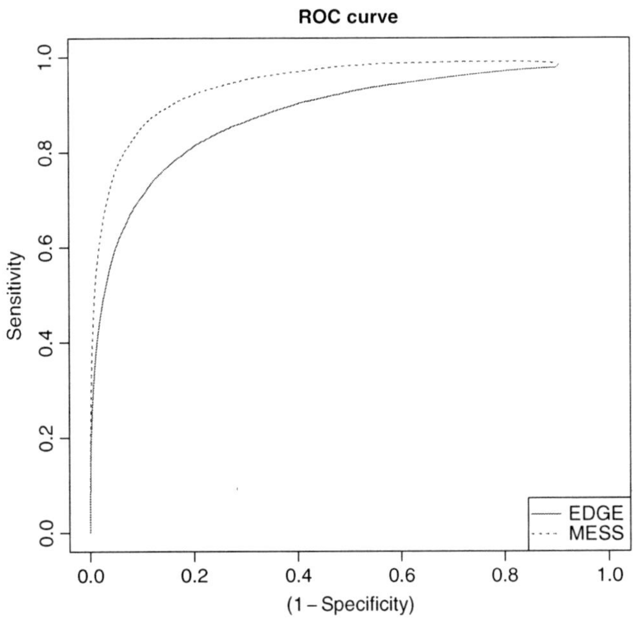

Fig. 3 BCG case study: a top scoring genes according to MESS (*left*), but not to EDGE (*right*). TNF has been strongly implicated in TB infection (Flynn et al., 1995) and we would expect it to be ranked as highly significant. EDGE's low ranking can be partly explained by poor model selection failing to accurately capture the gene expression dynamics, and the inadequacy of scalar shifts to fully explain the variation between individuals

pulmonary TB (Lee et al., 2008), while TNF-α has had a long association with the disease (Flynn et al., 1995). In total, using the GeneCards online database (www.genecards.org), 58 of the 276 significant genes were found to have existing citations in the literature corresponding to *M.tuberculosis* infection or the BCG vaccine. Those which were upregulated mainly consisted of chemokines and cytokines such as CCL1, CCL2, CCL7, CCL18, CCL20, CXCL1, CXCL2, CXCL3, CXCL9, CXCL10, TNF, CSF2, CSF3, IFNG, IL1A, IL1B, IL6 and IL8 while the downregulated genes were exemplified by transmembrane receptors CD86, CD163, TLR1, TLR4 and IL8RB. The large number of differentially expressed genes that we would have expected to identify lends credence to those genes without citations and whose role in the host response to BCG is currently unknown.

6 Conclusions

In this work we have presented a non-parametric mixed-effects model based on smoothing splines for the analysis of longitudinal gene expression profiles. Experimental results based on both simulated and real data demonstrate that the

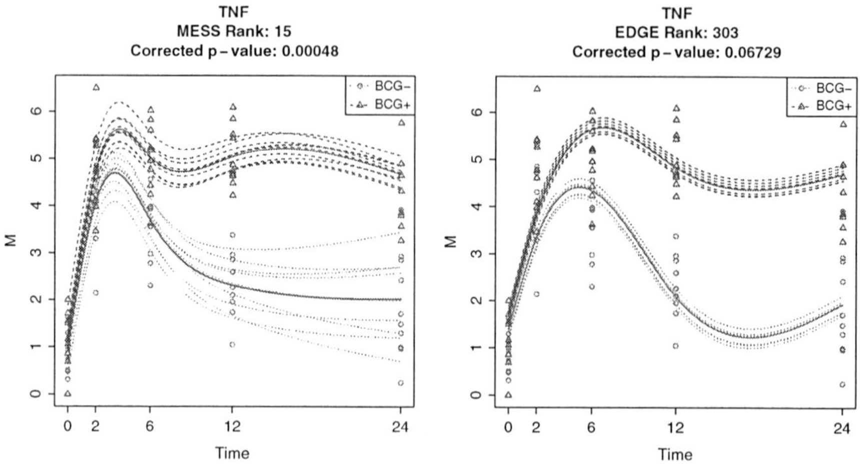

Fig. 4 ROC curve comparing the performance between the EDGE and MESS models. Results are based on 100,000 simulated genes as described in Sect. 4

use of both a flexible model that incorporates individual curves and an appropriate test-statistic yields higher statistical power than existing functional data analysis approaches.

Acknowledgements The authors thank the Welcome Trust for their financial support.

References

Z. Bar-Joseph, G. K. Gerber, D. K. Gifford, T. S. Jaakkola, and I. Simon. Continuous representations of time-series gene expression data. *Journal of Computational Biology*, 10(3-4):341–356, 2003.

G. A. Colditz, T. F. Brewer, C. S. Berkey, M. E. Wilson, E. Burdick, H. V. Fineberg, and F. Mosteller. Efficacy of BCG vaccine in the prevention of tuberculosis. Meta-analysis of the published literature. *Journal of the American Medical Association*, 271(9):698–702, 1994.

P. E. Fine. Variation in protection by BCG: implications of and for heterologous immunity. *Lancet*, 346(8986):1339–45, 1995.

J. L. Flynn, M. M. Goldstein, J. Chan, K. J. Triebold, K. Pfeffer, C. J. Lowenstein, R. Schrelber, T. W. Mak, and B. R. Bloom. Tumor necrosis factor-α is required in the protective immune response against mycobacterium tuberculosis in mice. *Immunity*, 2(6):561 – 572, 1995.

P. J. Green and B. W. Silverman. *Nonparametric Regression and Generalized Linear Models*. Chapman and Hall, 1994.

N. M. Laird and J. H. Ware. Random-effects models for longitudinal data. *Biometrics*, 38:963–974, 1982.

J. S. Lee, J. Y. Lee, J. W. Son, J. H. Oh, D. M. Shin, J. M. Yuk, C. H. Song, T. H. Paik, and E. K. Jo. Expression and regulation of the cc-chemokine ligand 20 during human tuberculosis. *Scandinavian Journal of Immunology*, 67(1):77–85, 2008.

P. Ma, C. I. Castillo-Davis, W. Zhong, and J. S. Liu. A data-driven clustering method for time course gene expression data. *Nucleic Acids Res*, 34(4):1261–1269, 2006.

J. A. Nelder and R. Mead. A simplex method for function minimiztion. *Computer Journal*, 7:308–313, 1965.

S. Ragno, M. Romano, S. Howell, D. J. C. Pappin, P. J. Jenner, and M. J. Colston. Changes in gene expression in macrophages infected with *Mycobacterium tuberculosis*: a combined transcriptomic and proteomic approach. *Immunology*, 104(1):99–108, 2001.

J. Ramsay and B. Silverman. *Functional data analysis*. Springer, 2006.

J. D. Storey, W. Xiao, J. T. Leek, R. G. Tompkins, and R. W. Davis. Significance analysis of time course microarray experiments. *Proc Natl Acad Sci U S A*, 102(36):12837–12842, Sep 2005.

A Permutation Solution to Compare Two Hepatocellular Carcinoma Markers

Agata Zirilli and Angela Alibrandi

Abstract In medical literature Alpha-fetoprotein (AFP) is the most commonly used marker for hepatocellular carcinoma (HCC) diagnosis. Some researches showed that there is over-expression of insulin-like growth factor (IGF)-II in HCC tissue, especially in small HCC. In this background our study investigates the diagnostic utility of IGF-II in HCC. Serum levels of IGF-II and AFP were determined on 96 HCC patients, 102 cirrhotic patients and 30 healthy controls. The application of NPC test, stratified for small and large tumours, allowed us to notice that IGF-II and AFP levels in HCC were significantly higher than cirrhotic patients and controls, the IGF-II levels in cirrhotic patients were significantly lower than controls. The optimal cut-off values for diagnosing HCC were determined with ROC curve. The sensitivity, specificity and diagnostic accuracy values for AFP and IGF-II have been estimated for diagnosis of HCC and, subsequently, for small or large HCC. Determination of jointly used markers significantly increases the diagnostic accuracy and sensitivity, with a high specificity. So IGF-II serum can be considered a useful tumour marker to be jointly used with AFP, especially for diagnosis of small HCC.

1 Introduction

Hepatocellular carcinoma (HCC) is among the most common fatal solid tumours world-wide. The principal causes of HCC development are the B and C virus hepatitis, particularly if they are responsible of cirrhosis presence. Alpha-Fetoprotein (AFP) is correlated to HCC presence and represents a marker of liver tumour; the

This note, though it is the result of a close collaboration, was specifically elaborated as follows: paragraphs 1, 2, 3, 7 by A. Zirilli and paragraphs 4, 5, 6 by A. Alibrandi

A. Zirilli (✉) · A. Alibrandi
Department of Economical, Financial, Social, Environmental, Statistical and Territorial Sciences, University of Messina, Viale Italia, 137 - 98122 Messina, Italy
e-mail: azirilli@unime.it; aalibrandi@unime.it

A. Di Ciaccio et al. (eds.), *Advanced Statistical Methods for the Analysis of Large Data-Sets*, Studies in Theoretical and Applied Statistics,
DOI 10.1007/978-3-642-21037-2_7, © Springer-Verlag Berlin Heidelberg 2012

test, when it is used with the conventional cut-off point of 400 ng/ml, has a sensitivity of about 48%–63% and a specificity of 100% in detecting the presence of HCC in patients with compensated cirrhosis. So, it is little sensitive and it is not a useful tumour marker for diagnosis of small hepatocellular carcinoma. In recent years various other serological markers have been developed for HCC diagnosis. However, most of these markers have been shown to be unsatisfactory in diagnosing small HCC due to low sensitivity. The Insulin-like Growth Factors (IGF-II) is a promoting growth factor, necessary during the embryonic development. Tsai et al. (2007) noticed the existence of an IGF-II over-expression in HCC tissue. In consideration of the results obtained by these authors, concerning a sensibility value for IGF-II (44%), we focalized our interest toward IGF-II as HCC marker. The present paper aims to assess the IGF-II diagnostic utility in hepatocellular carcinoma and to underline its utility as complementary tumour marker to AFP. In particular, we want:

- To compare both IGF-II and AFP serum levels among three groups: HCC patients, cirrhotic patients and healthy controls.
- To individuate an appropriate cut-off point.
- To estimate the sensibility and specificity for both markers, singly and jointly used.
- To perform a stratified analysis for small and large tumour size.

2 The Data

AFP and IGF-II serum levels were measured on 224 subjects: 96 HCC patients and 102 cirrhotic patients (hospitalized at hepatology ward of Universitary Policlinic in Messina from January 1st 2007 until December 31th 2008) and 30 healthy controls. In this context, we have to thank Prof. Maria Antonietta Freni and Prof. Aldo Spadaro because their medical and scientific support assumed an indispensable role into the realization of this paper. For each subject we collected information about sex, age, AFP serum levels, IGF-II levels and, for only HCC patients, the maximum diameter of nodules; the tumour size represents, in our case, a relevant variable to be taking in account and its measuring allows us to perform a stratified analysis. The nodules have been classified as small (if their diameter is less or equal to 3) and large (if their diameter is greater than 3), where 3 represents the median value of all measured diameters.

3 The NPC Test Methodology

In order to assess the existence of possible statistically significant differences among the three groups of subjects a non parametric inference based on permutation tests (or NPC Test) has been applied.

Permutation tests (Pesarin 2001) represent an effective solution for problems concerning the verifying of multidimensional hypotheses, because they are difficult to face in parametric context. This multivariate and multistrata procedure allows to reach effective solutions concerning problems of multidimensional hypotheses verifying within the non parametric permutation inference (Pesarin 1997); it is used in different application fields that concern verifying of multidimensional hypotheses with a complexity that can't be managed in parametric context.

In comparison to the classical approach, NPC test is characterized by several advantages:

- It doesn't request normality and homoscedasticity assumption.
- It draws any type of variable.
- It also assumes a good behavior in presence of missing data.
- It is also powerful in low sampling dimension.
- It resolves multidimensional problems, without the necessity to specify the structure of dependence among variables.
- It allows to test multivariate restricted alternative hypothesis (allowing the verifying of the directionality for a specific alternative hypothesis).
- It allows stratified analysis.
- It can be applied also when the sampling number is smaller than the number of variables.

All these properties make NPC test very flexible and widely applicable in several fields; in particular we cite recent applications in medical context (Zirilli et Alibrandi 2009; Bonnini et al. 2003; Bonnini et al. 2006; Zirilli et al. 2005; Callegaro et al. 2003; Arboretti et al. 2005; Salmaso 2005; Alibrandi and Zirilli 2007) and in genetics (Di Castelnuovo et al. 2000; Finos et al. 2004).

We supposed to notice K variables on N observations (dataset NK) and that an appropriate K-dimensional distribution P exists. The null hypothesis postulates the equality in distribution of k-dimensional distribution among all C groups $H_0 = [P_1 = \cdots = P_C] = [X_1 \overset{d}{=} \cdots \overset{d}{=} X_C]$ i.e. $H_0 = \cap_{i=1}^{k} X_{1i} \overset{d}{=} \cdots \overset{d}{=} X_{Ci}] = [\cap_{i=1}^{k} H_{0i}]$ against the alternative hypothesis $H_1 = \cup_{i=1}^{k} H_{1i}$.

Let's assume that, without loss of generality, the partial tests assume real values and they are marginally correct, consistent and significant for great values; the NPC test procedure (based on Conditional Monte Carlo resampling) develops into the following phases, such as illustrated in Fig. 1.

The null hypothesis, that postulates the indifference among the distributions, and the alternative one are expressed as follows:

$$H_0 : \left\{ X_{11} \overset{d}{=} X_{12} \right\} \cap \cdots \cap \left\{ X_{n1} \overset{d}{=} X_{n2} \right\} \tag{1}$$

$$H_1 : \left\{ X_{11} \overset{d}{\neq} X_{12} \right\} \cup \cdots \cup \left\{ X_{n1} \overset{d}{\neq} X_{n2} \right\} \tag{2}$$

In presence of a stratification variable, the hypotheses system is:

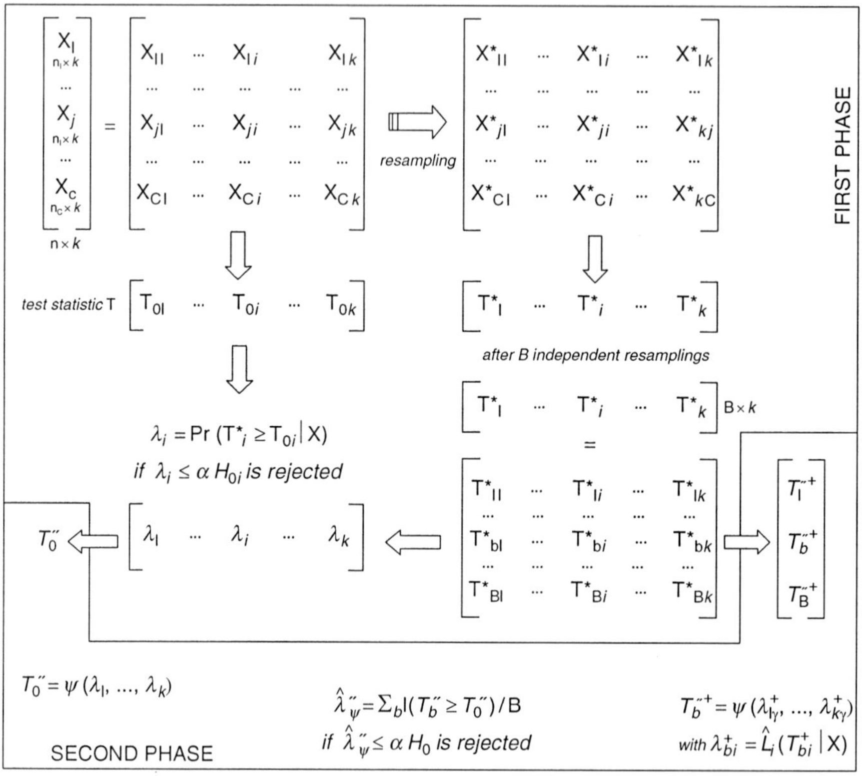

Fig. 1 NPC test procedure

$$H_{0i} : \left\{X_{11i} \stackrel{d}{=} X_{12i}\right\} \cap \cdots \cap \left\{X_{n1i} \stackrel{d}{=} X_{n2i}\right\} \tag{3}$$

$$H_{1i} : \left\{X_{11i} \stackrel{d}{\neq} X_{12i}\right\} \cup \cdots \cup \left\{X_{n1i} \stackrel{d}{\neq} X_{n2i}\right\} \tag{4}$$

The hypotheses systems are verified by the determination of partial tests (first order) that allow to evaluate the existence of statistically significant differences. By means of this methodology we can preliminarily define a set of k (k>1) unidimensional permutation tests (partial tests); they allow to examine every marginal contribution of answer variable, in the comparison among the examined groups. The partial tests are combined, in a non parametric way, in a second order test that globally verifies the existence of differences among the multivariate distributions. A procedure of conditioned resampling CMC (Conditional Monte Carlo, Pesarin 2001) allows to estimate the p-values, associated both to partial tests and to second order tests.

Under the exchangeability data among groups condition, according to null hypothesis, NPC test is characterized by two properties:

- Similarity: whatever the underlying distribution data, the probability to refute the null hypothesis is invariant to the actually observed dataset, whatever the type of data collection.
- For each α, for each distribution and for each set of observed data, if under the alternative hypothesis, the distribution dominates the null hypothesis, then an unbiased conditional test exists and, therefore, the probability of refuting the null hypothesis is always no less than the α significance level.

4 The Closed Testing Procedure

In order to check the multiplicity effects, the Closed Testing procedure (Finos et al. 2003) has been applied for correcting the p-values of the two-by-two comparisons. By Closed Testing it's possible to get the adjusted p-value for a certain hypothesis Hi, that is equal to the maximum p-values of hypotheses that implicate Hi. It uses the MinP Bonferroni-Holm Procedure that applies Bonferroni Method to derive a Stepwise procedure. It foresees the calculation of the alone p-values associated to the minimal hypotheses from which we can obtain those associated to the not minimal hypotheses. Especially in multivariate contexts we need to check, through an inferencial procedure of hypotheses verification, the Familywise Error Rate (FWE) that is the probability to commit at least an univariate first type error or probability to commit a multivariate first type error (Westfall and Wolfinger 2000). When we investigate significant differences among more groups, we need that the inferences control the FWE value, at the fixed α level. This procedure has two important properties:

- Coherence (necessary): given a couple of hypotheses (H_i,H_j), such that H_j is included in H_i, the acceptance of H_j involves the acceptance of H_i.
- Consonance (desirable): it is reached when a not minimal hypothesis H_j is rejected and there is at least a minimal hypothesis that must have rejected.
- We have a set of statistical hypotheses that are closed as regards the intersection and for which each associated test has a significance level.

5 Diagnostic Tests

Receiver Operating Characteristic (ROC) analysis was performed to establish the best discriminator limit of AFP and IGF-II in detecting the presence of HCC in patients with compensated cirrhosis and, also, in detecting the different tumour size (Zou et al. 2004). This methodology was used not only to quantify but also to compare the diagnostic performance of two examined tumour markers, singly and jointly used (Zou et al. 2007). ROC curves were constructed by calculating the sensitivity and specificity of IGF-II and AFP levels at several cut-off points.

The cut-off value with the highest accuracy was selected as the diagnostic cut-off point. Sensitivity, specificity and relative intervals confidence, positive and negative predictive value and diagnostic accuracy were calculated for AFP and IGF-II, singly and jointly used (Altman and Bland 1994).

6 The Analysis

p-values obtained by applying the NPC test and corrected by Closed Testing procedure show that, in patients affected by HCC, both AFP and IGF-II levels are significantly higher than in cirrhotic patients ($p = 0{,}0001$) and in healthy controls ($p = 0{,}0001$) (Fig. 2). The AFP serum levels are statistically higher in cirrhotic patients than healthy controls ($p = 0{,}0002$). However, IGF-II serum levels are lower in cirrhotic patients when compared to healthy controls ($p = 0{,}0001$) (Table 1 and Fig. 3).

According to the ROC curve analysis (Fig. 4), the optimal cut-off values are:

- 796 ng/ml for IGF-II with area under ROC curve of 54.7%.
- 132 ng/ml for AFP, with area under ROC curve of 68.7%.

By means of these cut-off values we estimated diagnostic tests to diagnosing HCC presence from cirrhosis and, also, to diagnosis of large or small HCC. Tables 2 and 3 show the results of diagnostic tests for both diagnoses, respectively.

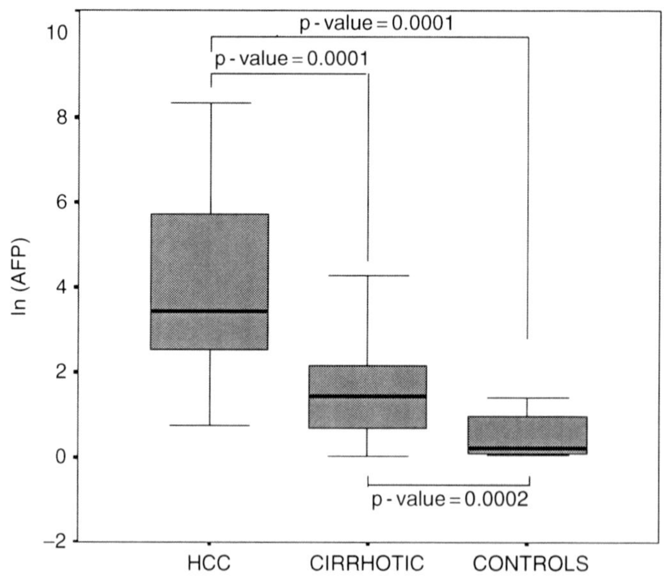

Fig. 2 Boxplot for AFP serum levels

Table 1 Means ± standard deviation, minimum and maximum value of IGF-II and AFP serum levels, in HCC patients, in cirrhotic patients and in healthy controls

Group	AFP (ng/ml)	IGF-II (ng/ml)
HCC	862.88 ± 2056.15 (2.10–9731)	515.04 ± 344.70 (26–1436.60)
HCC small	*1194.01 ± 2494.00 (2.10–9731)*	*590.9 ± 393.8 (26–1436.6)*
HCC large	*312.17 ± 701.24 (2.13–2574)*	*388.6 ± 186.4 (64.6–690.1)*
Cirrhotic	15.73 ± 34.19 (1.02–154.80)	330.17 ± 275.15 (1.83–975)
Controls	1.84 ± 1.05 (1.06–4.10)	566.16 ± 295.06 (1.10–969.70)

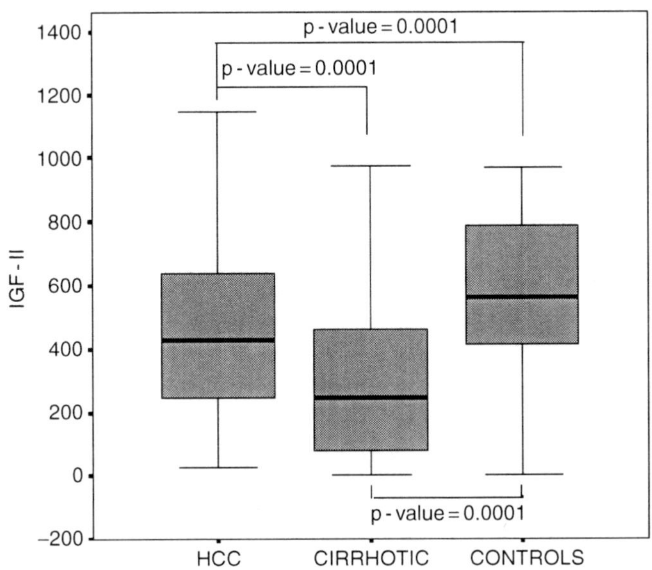

Fig. 3 Boxplot for IGF-II serum levels

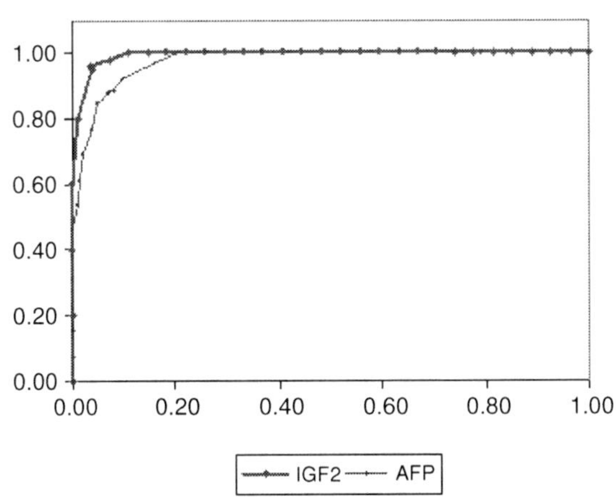

Fig. 4 ROC curves for AFP and IGF-II levels

Table 2 Diagnostic tests in discriminating HCC from cirrhosis for AFP e IGF-II

Marker	Sensibility and C.I.	Specificity and C.I.	VP+	VP−	Acc
AFP	0.406 (0.313–0.506)	0.969 (0.912–0.989)	0.929	0.620	0.688
IGF-II	0.189 (0.122–0.277)	0.906 (0.831–0.950)	0.667	0.527	0.547
AFP and IGF-II	0.469 (0.372–0.568)	0.875 (0.794–0.927)	0.789	0.622	0.672

Table 3 Diagnostic tests in discriminating small HCC from large HCC for AFP e IGF-II

Marker	Sensibility and C.I.	Specificity and C.I.	VP+	VP−	Acc
AFP	0.400 (0.286–0.526)	0.583 (0.422–0.729)	0.615	0.368	0.469
IGF-II	0.300 (0.199–0.425)	1.000 (0.904–1.000)	1.000	0.462	0.563
AFP and IGF-II	0.500 (0.377–0.623)	0.583 (0.422–0.729)	0.667	0.412	0.531

Comparing the two markers, we can notice that the AFP introduces a more elevated sensibility value than the IGF-II; with reference to the specificity, the difference between the two markers is lower, instead (Table 2). Evaluating the sensibility and specificity of AFP e IGF-II jointly used, we obtained a more elevated sensibility (in comparison to every marker singly used) even if the specificity is lower. This underlines the informative and diagnostic utility of IGF-II.

As we can see, considering the tumour size, the IGF-II seems to be the best marker because its sensibility is slightly lower than AFP but its specificity is much higher than AFP. Just for the sensitivity, the joint use of two markers assumes a considerable interest. Regarding the specificity and the accuracy, it is clear that the IGF-II, singly used, is the marker to be preferred.

7 Final Remarks

AFP serum is among the most intensively studied tumor markers for HCC. The test, when it is used with the conventional cut-off point of 400 ng/ml, has a sensitivity of about 48%–63% and a specificity of 100% in detecting the presence of HCC in patients with compensated cirrhosis.

In recent years various other serological markers have been developed for the diagnosis of HCC. However, most of these markers have been shown to be unsatisfactory in diagnosing small HCC due to low sensitivity. Starting on the results obtained by Tsai et al. (2007) concerning the sensibility value for IGF-II (44%), we focalized our interest toward IGF-II as HCC marker.

With reference to our examined casuistry, both IGF-II and AFP levels in HCC patients are significantly higher than cirrhotic patients and controls levels. The IGF-II levels in cirrhotic patients are significantly lower than healthy controls, instead.

The ROC analysis allowed us to estimate the optimal cut-off values for diagnosing HCC, for both examined markers.

Comparing the two markers, our results show that the AFP introduces a more elevated sensibility value than the IGF-II; for the specificity, however, the difference between the two markers isn't meaningful. The sensibility and specificity of both markers jointly used is higher in comparison to each marker singly used, even if the specificity is lower. This result proves the informative and diagnostic utility of the joined use of two markers.

Moreover, considering the tumour size, the IGF-II appears the best marker, since its sensibility is slightly lower than AFP and its specificity is much higher. On the bases of the obtained specificity and accuracy, the IGF-II, singly used, seems to be the marker to be preferred for diagnosing of small HCC.

References

Alibrandi A., Zirilli A.: A statistical evaluation on high seric levels of D-Dimer: a case control study to test the influence of ascites. In: Atti S.Co.2007 Conference, CLEUP, Padova, pp. 9–14, (2007)

Altman DG, Bland JM.: Diagnostic tests: sensitivity and specificity. BMJ, **308**, 1552 (1994)

Arboretti Giancristofaro R., Marozzi M., Salmaso L.: Repeated measures designs: a permutation approach for testing for active effects. Far East Journal of Theoretical Statistics, Special Volume on Biostatistics, **16**, 2, pp. 303–325 (2005)

Bonnini S., Pesarin F., Salmaso L.: Statistical Analysis in biomedical studies: an application of NPC Test to a clinical trial on a respiratory drug. In Atti 5° Congresso Nazionale della Societ Italiana di Biometria, pp. 107–110 (2003)

Bonnini S., Corain L., Munaò F., Salmaso L.: Neurocognitive Effects in Welders Exposed to Aluminium: An Application of the NPC Test and NPC Ranking Methods. Statistical Methods and Applications, Journal of the Statistical Society, **15**, 2, pp. 191–208 (2006)

Callegaro A., Pesarin F., Salmaso L.: Test di permutazione per il confronto di curve di sopravvivenza. Statistica Applicata, **15**, 2, pp. 241–261 (2003)

Di Castelnuovo A., Mazzaro D., Pesarin F., Salmaso L.: Test di permutazione multidimensionali in problemi d'inferenza isotonica: un'applicazione alla genetica. Statistica. **60**,4, pp. 691–700 (2000)

Finos L., Pesarin F., Salmaso L., Solari A.: Nonparametric iterated procedure for testing genetic differentiation. In: Atti XLIII Riunione Scientifica SIS, CLEUP, Padova (2004)

Finos L., Pesarin F., Salmaso L.: Test combinati per il controllo della molteplicit mediante procedure di Closed Testing, Statistica Applicata, **15**, 2, pp. 301–329 (2003)

Pesarin F.: Permutation testing of multidimensional Hypotheses, CLEUP, Padova (1997)

Pesarin F.: Multivariate permutation tests with applications in biostatistics. Wiley, Chichester (2001)

Salmaso L.: Permutation tests in screening two-level factorial experiments. Advances and Applications in Statistics, **5**, 1, pp. 91–110 (2005)

Tsai J. F., Jeng J. E., Chuang L. Y., You H. L., Wang L. Y., Hsieh M. Y., Chen S. C., Chuang W. L., Lin Z. Y., Yu M. L., Dai C. Y.: Serum insulin-like growth factor-II as a serologic marker of small hepatocellular carcinoma, Scandinavian Journal of Gastroenterology, **40**:1, pp. 68–75 (2007)

Westfall P. H., Wolfinger R. D.: Closed Multiple Testing Procedures and PROC MULTTEST, SAS institute Inc.(2000)

Zirilli A., Alibrandi A., Spadaro A., Freni M.A.: Prognostic factors of survival in the cirrhosis of the liver: A statistical evaluation in a multivariate approach. In: Atti S.Co.2005 Conference, CLEUP, Padova, pp. 173–178 (2005)

Zirilli A., Alibrandi A.: A permutation approach to evaluate hyperhomocysteinemia in epileptic patients. In: Supplemento ai rendiconti del circolo matematico di palermo. VII International Conference in "Stochastic Geometry, Convex Bodies, Empirical Measures and application to mechanics and Engineering train-transport", Messina, 22-24 April 2009, pp. 369–378 (2009)
Zou X.H., Castelluccio P., Zhou C.: Non parametric estimation of ROC Curves in the absence of a gold Standard. Berkeley Electronic Press UW Biostatistics Working Paper Series (2004)
Zou X.H, O'Malley AJ, Mauri L., Receiver-Operating Characteristic Analysis for Evaluating Diagnostic Tests and Predictive Models. Circulation, **115**, pp. 654–657 (2007)

Part III
Integrating Administrative Data

Statistical Perspective on Blocking Methods When Linking Large Data-sets

Nicoletta Cibella and Tiziana Tuoto

Abstract The combined use of data from different sources is largely widespread. Record linkage is a complex process aiming at recognizing the same real world entity, differently represented in data sources. Many problems arise when dealing with large data-sets, connected with both computational and statistical aspects. The well-know blocking methods can reduce the number of record comparisons to a suitable number. In this context, the research and the debate are very animated among the information technology scientists. On the contrary, the statistical implications of different blocking methods are often neglected. This work is focused on highlighting the advantages and disadvantages of the main blocking methods in carrying out successfully a probabilistic record linkage process on large data-sets, stressing the statistical point of view.

1 Introduction

The main purpose of record linkage techniques is to accurately recognize the same real world entity which can be differently stored in sources of various type. In official statistics the data integration procedures are becoming extremely important due to many reasons: some of the most crucial ones are the cut of the cost, the reduction of response burden and the statistical use of information derived from administrative data. The many possible applications of record linkage techniques and their wide use make them a powerful instrument. The most widespread utilizations of record linkage procedures are the elimination of duplicates within a data frame, the study of the relationship among variables reported in different sources, the creation of sampling lists, the check of the confidentiality of public-use

N. Cibella (✉) · T. Tuoto
Istituto Nazionale di Statistica (ISTAT), via Cesare Balbo 16, 00184 Roma, Italy
e-mail: cibella@istat.it

A. Di Ciaccio et al. (eds.), *Advanced Statistical Methods for the Analysis of Large Data-Sets*, Studies in Theoretical and Applied Statistics,
DOI 10.1007/978-3-642-21037-2_8,

micro-data, the calculation of the total amount of a population by means of capture-recapture models, etc.

Generally, the difficulties in record linkage project are related to the number of records to be linked. Actually, in a record linkage process, all candidate pairs belonging to the cross product of the two considered files, say A and B, must be classified as matches, un-matches and possible matches. This approach is computationally prohibitive when the two data frames become large: as a matter of fact, while the number of possible matches increases linearly, the computational problem raises quadratically and the complexity is $O(n^2)$ (Christen and Goiser 2005). To reduce this complexity, which is an obvious cause of problems for large data sets, it is necessary to decrease the number of comparisons. Then expensive and sophisticate record linkage decision model can be applied only within the reduced search space and computational costs are significantly saved. In order to reduce the candidate pairs space, several methods exist, i.e. techniques of sorting, filtering, clustering and indexing may all be used to reduce the search space of candidate pairs. The selection of the suitable reduction method is a delicate step for the overall linkage procedure because the same method can yield opposite results against different applications.

The debate concerning the performances of different blocking methods is very vivacious among the information technology scientists (Baxter et al. 2003, Jin et al. 2003). In this paper the focus is instead on the statistical advantages of using data reduction methods in performing a probabilistic record linkage process on large data-sets. The outline of the paper is as follow: in Sect. 2 details on the most widespread blocking methods, i.e. standard blocking and sorted neighbourhood, are given; Sect. 3 stresses the statistical point of view on the choice between the compared methods; Sect. 4 reports experimental results proving the statements given in Sect. 3 by means of real data application; finally in Sect. 5 some concluding remarks and future works are sketched.

2 Blocking Methods

Actually, two of the most challenging problems in record linkage are the computational complexity and the linkage quality. Since an exhaustive comparison between all records is unfeasible, efficient blocking methods can be applied in order to greatly reduce the number of pairs comparisons to be performed, achieving significant performance speed-ups. In fact, blocking methods, directly or indirectly, affect the linkage accuracy:

- they can cause missing true matches: when record pairs of true matches are not in the same block, they will not be compared and can never be matched
- thanks to a better reduction of the search space, more suitable, intensive and expensive models can be employed.

So, blocking procedures have two main goals that represent a trade-off. First, the number of candidate pairs generated by the procedures should be small to minimize the number of comparisons in the further record linkage steps. Second, the candidate set should not leave out any possible true matches, since only record pairs in the candidate set will be examined in detail.

The developments in the modern computer power, the machine learning, the data mining and statistical studies improve undoubtedly the performances and the accuracy of the record linkage procedure and help in finding more efficient blocking methods (e.g. the new method with the use of clustering algorithms or high-dimensional indexing). Nevertheless the potential advantages and disadvantages of the several different existing blocking methods make the choice among them a difficult task and there is not a general rule for privileging a method against the others.

In the following subsections, some details on the most widespread blocking methods, i.e. standard blocking and sorted neighbourhood, are given so as to stress the basic characteristics of the two methods compared herewith from a statistical perspective.

2.1 *The Standard Blocking Method*

The standard blocking method consists of partitioning the two datasets A and B into mutually exclusive blocks where they share the identical value of the blocking key (Jaro 1989) and of considering as candidate pairs only records within each block. A blocking key can be composed by a single record attribute, common to the data sets, or combining more than one attribute. There is a cost-benefit trade-off to be considered in choosing the blocking keys: from one hand, if the resulting blocks contain a large number of records, then more candidate pairs than necessary will be generated, with an inefficient large number of comparisons. From the other hand, if the blocks are too small then true record pairs may be lost, reducing the linkage accuracy. Moreover, to achieve good linkage quality, also the error characteristics of the blocking key is relevant, i.e. it is preferable to use the least error-prone attributes available.

In theory, when the size of the two data sets to be linked is of n records each and the blocking method creates b blocks (all of the same size with n/b records), the resulting number of record pair comparisons is $\boldsymbol{O}(n^2/b)$. This is an ideal case, hardly ever achievable with real data, where the number of record pair comparisons will be dominated by the largest block. So the selection of the blocking keys is one of the crucial point for improving the accuracy of the whole process. To mitigate also the effects of errors in blocking keys, multiple keys can be used and several passes, with different keys, can be performed (Hernandez and Stolfo 1998). Multiple passes improve linkage accuracy but the implementation is often inefficient. Roughly speaking, the multi-pass approach generates candidate pairs using different attributes and methods across independent runs. Intuitively, different runs cover different true matches, so the union should cover most of the true matches. Of course, the effectiveness of a multi-pass approach depends on which attributes are chosen and on the methods used.

2.2 The Sorted Neighbourhood Method

Another of the most well-known blocking method is the sorted neighbourhood one (Hernandez and Stolfo 1995). This method sorts together the two record sets, A and B, by the selected blocking variable. Only records within a window of a fixed dimension, w, are paired and included in the candidate record pair list. The window slides on the two ordered record sets and its use limits to $(2w - 1)$ the number of possible record pair comparisons for each record in the window. Actually, in order to identify matches, the first unit of the list is compared to all the others in the window w and then the window slides down by one unit until the end (Yan et al. 2007). Assuming two data sets of n records each, with the sorted neighbourhood method, the total number of record comparisons is $\boldsymbol{O}(wn)$.

The original sorted neighbourhood method expects a lexicographic ordering of the two data sets. Anyway, records with similar values might not appear close to each other when considering lexicographic order. In general, the effectiveness of this approach is based on the expectation that if two records are duplicates, they will appear lexicographically close to each other in the sorted list based on at least one key.

Similar to standard blocking method whereas the sliding window works as a blocking key, it is preferable to do several passes with different sorting keys and a smaller window size than only one pass with a large window size. Even if multiple keys are chosen, the effectiveness of the method is still susceptible to deterministic data-entry errors, e.g., the first character of a key attribute is always erroneous.

3 A Statistical Perspective in Comparing Blocking Methods

As stated in Sect. 1, when managing huge amount of data, the search space reduction is useful to limit the execution time and the used memory space by means of a suitable partition of the whole candidate pairs space, corresponding to the cross product of the input files. The information technologist community is really active in analyzing characteristics and performances of the most widespread blocking methods as well as in data linkage project at all: a proof of the statement is given by the proliferation of different names to refer the record linkage problem – citation matching, identity uncertainty, merge-purge, entity resolution, authority control, approximate string join, etc. A further evidence is the emergence of numerous organizations (e.g., Trillium, FirstLogic, Vality, DataFlux) that are developing specialized domain-specific record-linkage tools devoted to a variety of data-analysis applications.

In the last years, new attractive techniques for blocking have been proposed: clustering algorithms, high-dimensional indexing, by-gram indexing, canopy. Moreover machine learning methods have been developed in order to define the best blocking strategy for a given problem using training data. Generally speaking, the blocking strategy states a set of parameters for the search space reduction phase: the blocking keys, the method that combines the variables (e.g. conjunction, disjunction), the choice of the blocking algorithms, the window size, the choice of the similarity functions and so on.

In this paper we approach the problem of stating the most suitable blocking method keeping in mind also the statistical perspective on the record linkage problem. In fact, when probabilistic approach is applied, "statistical" problems arise in dealing with huge amount of data. Usually, the probabilistic model estimates the conditional probabilities of being match or un-match assuming that the whole set of candidate pairs is a mixture of the two unknown distributions: the true links and the true non-links (Armstrong and Mayda 1993, Larsen and Rubin 2001). Generally, an EM algorithm is applied in order to estimate the conditional probabilities in presence of latent classification. The statistical problem arises when the number of expected links is extremely small with respect to the whole set of candidate pairs; in other words, if one of the two unknown populations (the matches) is really too small, it is possible that the estimation mechanism is not able to correctly identify the linkage probabilities: it could happen that the EM algorithm still converges, but in fact it estimates another latent phenomenon different from the linkage one. This is why some authors suggest that, when the conditional probabilities are estimated via the EM algorithm, it is appropriate that the expected number of links is not below 5% of the overall compared pairs (Yancey 2004). A solution to this situation is the creation of suitable groups of the whole set of pairs, i.e. a blocking scheme, so that, in each sub-group, the number of expected links is suitable with respect to the number of candidate pairs.

From the statistical perspective, the choice among blocking methods depends on several characteristics, only partially connected to the computational aspects. In this work, some of the most important issues of the blocking strategy are stressed, i.e. the expected match rate and the frequency distribution of the available blocking keys dramatically influence the effectiveness of the chosen blocking method. For instance, if the standard blocking method is really useful to solve linkage problems where the overlap between files is very high, i.e. in de-duplication or post-enumeration survey context, it could be unhelpful when the expected number of matches is very small with respect to the largest file to be linked. Moreover, when most of the blocking key categories are very sparse with low frequencies (no more than five), even if identification power of the key is high, the standard blocking method can't help in defining a probabilistic linkage strategy.

4 Experimental Results

The previous aspects are highlighted in the study of the fecundity of married foreign-women with residence in Italy. This study requires the integration of two data sources: the list of the marriages and the register of births. The two data sets have a common identifier, the fiscal code of the bride/mother. Unfortunately it is affected by errors, particularly when the bride/mother is foreign. Moreover, considering a certain year, the number of births is quite small with respect to the amount of marriages of the same year, so the expected match rate is very low, below the 5% of the largest file.

The data considered in this paper referred to marriages with almost one of the married couple foreign and resident in Lombardy in 2005 and to babies born in the same Region in 2005–2006. The size of each file is about 30,000 records. The common variables are: fiscal code of the bride/mother, the three-digit-standardized name and surname of both spouses/parents, the day/month/year of birth of the bridegroom/father and of the bride/mother, the municipality of the event (marriage/birth). A probabilistic procedure based on EM solution of the Fellegi–Sunter model has been applied.

Due to the file size, a reduction method is needed, avoiding to deal with 900 millions of candidate pairs. The performances of the standard blocking method and of the sorted neighbourhood one are compared.

A previous analysis of the accuracy and of the frequency distribution of the available variables has limited the choice to the three-digit-standardized name and surname of the bride/mother as blocking keys.

We have experimented several strategies in reducing the number of the candidate pairs. The results of the different tests have been compared by means of two different groups of diagnostic for blocking methods: the first one is common in the information technology context while the second one is typical in the statistic community.

The computer scientists currently adopt the reduction ratio (RR) and the pairs completeness (PC) indexes to compare blocking techniques. The RR quantifies how well the blocking method minimizes the number of candidates: RR $= 1 - C/N$, where C is the number of candidate matches and N is the size of the cross product between data sets. The PC measures the coverage of true matches with respect to the adopted blocking method, i.e. how many of the true matches are in the candidate set versus those in the whole set: PC $=$ Cm/Nm where Cm is the number of true matches in the candidate set and Nm is the number of matches in the whole set. A blocking scheme that optimizes both PC and RR reduces the computational costs for record linkage, decreasing the candidate pairs, and, at the same time, saves the linkage accuracy by means of not loosing true matches. From the statistical perspective, the match rate (MR) is one of the measure, with the linkage error rates, to evaluate the linkage procedure. The MR represents the coverage of true matches of the overall linkage procedure, considering also the applied classification model: MR $= M/\mathrm{Nm}$ where M is the number of true matches identified at the end of the linkage procedure and Nm as already defined.

All these metrics require that the true linkage status for the record pairs is known; we consider as a benchmark the total amount of pairs with common fiscal code and we also refer to such a number when evaluating the improvements in the results of the probabilistic linkage procedures at all. In this study the pairs with common fiscal code are 517 records. As such a key is not error-free, it is possible to find a higher number of pairs that are true matches almost surely, given that they share the same values for high-identification powerful variables: standardized name and surname, day/month/year of birth. This point implies values greater than 1 for the PC and MR metrics.

The first blocking strategy consists in standard blocking method with 3-digit-standardized surname of the bride/mother as key: the categories in each file are about 4,000, resulting in 2,200 blocks and about 580,000 candidate pairs. A probabilistic procedure based on the Fellegi–Sunter model has been applied, considering as matching variables: the three-digit-standardized name of the mother and her day/month/year of birth. The results in terms of matches, possible matches, true matches and MR are shown in Table 1. The relative PC and RR are reported in Table 2.

The inefficiency of standard blocking method, compared to the benchmark, has lead us to test an alternative blocking strategy, based on sorted neighbourhood method. The six-digit-string key composed by joining standardized name and surname and a window of size 15 creates about 400,000 candidate pairs. The probabilistic procedure based on the same Fellegi–Sunter model has been applied and 567 matches and 457 possible matches have been identified. Tables 1 and 2 report the results in terms of matches, possible matches, true matches and MR and PC and RR respectively.

The standard blocking method was tested also with six-digit-string name and surname but, due to the about 24,000 categories of this key, often without any overlap between the two files, the EM algorithm for the estimation of the probabilistic

Table 1 Statistical diagnostics for blocking strategies comparison

	Blocking Surname three-digit	Sorted neighbourhood surname six-digit
Matches	439	567
Possible matches	359	457
True matches	448	592
MR	0.867	1.145

Table 2 Computer scientist diagnostics for blocking strategies comparison

	Blocking surname three-digit	Sorted neighbourhood surname six-digit
PC	1.064	1.145
RR	0.999	0.999

linkage parameters doesn't work so the MR is not evaluable. Anyway, the PC and RR are equal to 1.039 and 0.999.

As showed in the above tables, the differences between the two blocking strategies emerge basically from the statistical perspective, see the MR values, whereas the measures in Table 2 highlight smoothed differences.

5 Concluding Remarks and Future Works

The linkages presented in this paper have been performed by RELAIS, an open source software designed and implemented by ISTAT. It provides a set of standard methods and techniques in order to execute record linkage applications. In order to better face with the complexity of linkage problem, it is decomposed into its constituting phases; the software allows the dynamic selection of the most appropriate technique for each phase and the combination of the selected techniques so that the resulting workflow is actually built on the basis of application and data specific requirements. In fact, RELAIS has been designed with a modular structure: the modules implement distinct record linkage techniques and each one has a well defined interface towards other modules. In this way it is possible to have a parallel development of the different modules, and to easily include new ones in the system. Moreover, the overall record linkage process can be designed according to specific application requirements, combining the available modules. The RELAIS approach overcomes the question on which method is better compared to the others, being convinced that at the moment there is not a unique technique dominating all the others. The strength of RELAIS consists in fact of considering alternative techniques for the different phases composing the record linkage process. RELAIS wants to help and guide users in defining their specific linkage strategy, supporting the practitioner's skill, due to the fact that most of the available techniques are inherently complex, thus requiring not trivial knowledge in order to be appropriately combined. RELAIS is proposed also as a toolkit for researchers: in fact, it gives the possibility to experiment alternative criteria and parameters in the same application scenario, that's really important from the analyst's point of view. For all these reasons RELAIS is configured as an open source project, released under the EU Public License.

This paper is a first step in comparing blocking methods for record linkage, keeping in mind the statistical perspectives. Further tests are needed. It could be useful for instance to exploit data sets where the true linkage status is completely known; unfortunately these is hard to achieve without a clerical review, but manual checks are quite prohibitive for very large data-sets. A possible approach to these issues could be to replicate these experiments with synthetic data sets.

Further analyses can also be useful in comparing blocking methods in other contexts, with an expected match rate intermediate compared with the post-enumeration survey one, that is about the 99%, and with that considered in this paper, that is lower than the 5% of the largest file.

Other goals of future studies concern the examine of the statistical impact of more complicated blocking methods, such as the bigram indexing, the canopy clustering, etc. and the evaluation of the comparability of blocking choices suggested by domain experts with respect to those learned by machine learning algorithms, both supervised and unsupervised and fully automatic ones.

References

Armstrong J.B. and Mayda J.E (1993) Model-based estimation of record linkage error rates. *Survey Methodology*, 19, 137–147

Baxter R., Christen P., Churches T. (2003) A comparison of fast blocking methods for record linkage http://www.act.cmis.csiro.au/rohanb/PAPERS/kdd03clean.pdf

Christen P. and Goiser K. (2005) Assessing duplication and data linkage quality: what to measure?, *Proceedings of the fourth Australasian Data Mining Conference,* Sydney, December 2005, http://datamining.anu.edu.au/linkage.html

Hernandez M.A. and Stolfo S.J. (1995) The merge/purge problem for large databases. In M. J. Carey and D. A. Schneider, editors, *SIGMOD*, pp. 127–138

Hernandez M.A. and Stolfo S.J. (1998) Real-world data is dirty: Data cleansing and the merge/purge problem. *Data Mining and Knowledge Discovery* 2(1), 9–37

Jaro M.A. (1989) Advances in record-linkage methodology as applied to matching the 1985 Census of Tampa, Florida. *Journal of the American Statistical Association*, 84, 414–420

Jin L., Li C., Mehrotra S. (2003) Efficient Record Linkage in Large Data Sets. *Proceedings of the 8th International Conference on Database Systems for Advanced Applications* (DASFAA) 2003, Kyoto, Japan, http://flamingo.ics.uci.edu/pub/dasfaa03.pdf

Larsen M.D. and Rubin D.B. (2001). Iterative automated record linkage using mixture models. *Journal of the American Statistical Association*, 96, 32–41

Relais, Record linkage at Istat, http://www.istat.it/it/strumenti/metodi-e-software/software/relais

Yan S., Lee D., Kan M.-Y., Giles C. L. (2007) Adaptive sorted neighborhood methods for efficient record linkage, JCDL'07, Vancouver, British Columbia, Canada

Yancey W.E. (2004) A program for large-scale record linkage. In Proceedings of the Section on Survey Research Methods, *Journal of the American Statistical Association*

Integrating Households Income Microdata in the Estimate of the Italian GDP

Alessandra Coli and Francesca Tartamella

Abstract National accounts statistics are the result of the integration of several data sources. At present, in Italy, sample surveys data on households income are not included in the estimation process of national accounts aggregates. In this paper we investigate the possibility of using such data within an independent estimate of GDP, based on the income approach. The aim of this paper is to assess whether (and to what extent) sample survey microdata on household income may contribute to the estimate of GDP. To this end, surveys variables are recoded and harmonized according to the national accounting concepts and definitions in order to point out discrepancies or similarities with respect to national accounts estimates. The analysis focuses particularly on compensation of employees. Applications are based on the EU statistics on income and living conditions and on the Bank of Italy survey on income and wealth.

1 Introduction

Gross domestic product (GDP) can be measured according to three different methods: the production (or value added) approach, the expenditure approach and the income approach. The first method measures GDP as the value of goods and services produced by the nation, net of intermediate consumption. The expenditure approach estimates GDP in terms of the total amount of money spent for final uses of goods and services. The income approach focuses on GDP as the amount of income paid by the production units for the use of productive factors (work and capital).

A. Coli (✉)
Department of Statistics and Mathematics applied to Economics, University of Pisa
e-mail: a.coli@ec.unipi.it

F. Tartamella
National Accounts Office, Istat

A. Di Ciaccio et al. (eds.), *Advanced Statistical Methods for the Analysis of Large Data-Sets*, Studies in Theoretical and Applied Statistics,
DOI 10.1007/978-3-642-21037-2_9,

Due to the use of different data sources and methods, the three approaches produce different estimates of GDP. A reconciliation process is then required in order to obtain the best estimate (which often is the only one actually published). This reconciliation is usually carried out through the construction of input–output (or supply and use) tables. The simultaneous publication of different measures is certainly useful but also misleading. On one hand it helps understanding how reliable figures are, reducing the risk of excessively trusting an inaccurate estimate. On the other hand the user has to choose among three different estimates of GDP.

The three different approaches rely on the use of, as far as possible, independent sources of information. In the Italian national accounts (NA) however, only the production and expenditure methods lead to exhaustive and independent estimates of GDP, the income approach providing independent estimates only for some income components. This depends mainly on the insufficient quality of available data sources (Istat 2004).

The paper addresses the use of sample surveys on households budgets for a better application of the income approach. As of 2004, in fact Italy and other European countries can take advantage of a new and very rich survey, the European survey on income and living conditions (Eu-silc). Moreover, the Bank of Italy has been carrying out a sample survey on households income and wealth (Shiw) since 1966, at first on a yearly basis, every two years starting from 1987.

The introduction of surveys micro data on households income in the GDP estimation process would allow to analyse income by groups of individuals as well as by household typology. Moreover it would improve reconciliation of micro and macro data on income, which is an essential piece of information for sound micro-founded macroeconomic modelling.

2 The Income Approach

The income approach focuses on income paid and earned by individuals and corporations in the production of goods and services, and can be represented by the following equation:

$$GDP = WS + GOS + GMI + NIT \tag{1}$$

where:

- WS: compensation of employees (wages and salaries and employers' social contributions).
- GOS: gross operating surplus, i.e. the profit generated by corporations and quasi corporations (public and private); it also includes imputed rents of owner-occupied dwellings.
- GMI: gross mixed income, i.e. the operating surplus of unincorporated enterprises owned by households. It also includes actual rents.
- NIT: taxes on production and imports net of any subsidies on production.

Table 1 Italian GDP and its composition – years 2004–2007. Current million euro

	2004		2005		2006		2007	
	values	(%)	values	(%)	values	(%)	values	(%)
Compensation of employees	536,229	39.1%	581,995	41.0%	608,864	41.0%	631,384	40.9%
Gross mixed income	220,495	16.1%	220,495	15.6%	223,414	15.0%	227,493	14.7%
Gross operating surplus	435,764	31.7%	435,764	30.7%	447,099	30.1%	474,330	30.7%
Taxes on production and imports net of subsidies	179,787	13.1%	179,787	12.7%	206,000	13.9%	211,708	13.7%
Gross domestic product	1,372,275	100.0%	1,418,041	100.0%	1,485,377	100.0%	1,544,915	100.0%

Istat, National Accounts, July 2009 version

Table 1 shows the Italian GDP and its composition for the years 2004–2007.

The largest part of the Italian GDP consists of compensation of employees, followed by the operating surplus, the mixed income and the net indirect taxes.

We wonder whether it is possible (and to what extent) to estimate the GDP components on the basis of the households budgets surveys.

Obviously surveys could contribute to estimate only the part of GDP concerning households as payers or earners.

In principle surveys should cover 100% of compensation of employees since this kind of income is earned exclusively by households.

Households are also engaged in production as owners of unincorporated enterprises. Profits/losses from such activity are defined as mixed income in national accounts. Therefore surveys could help estimating the part of mixed income which households withdraw from the enterprise for their needs. This aggregate accounts for around 82% of gross mixed income (about 13% of GDP), according to the Italian national accounts statistics of the last five years[1].

Furthermore, households disposable income includes imputed rents of owner-occupied dwellings. In national accounting, such earning is recorded as a component (about 15%) of gross operating surplus, i.e. around 5% of GDP. Summarising, surveys might help estimate up to 60% of GDP, around 70% if we consider GDP net of taxes.

[1] In the Italian national accounts the Household sector is split into two sub-sectors, namely producer households and consumer households. Particularly producer households include non financial unincorporated enterprises with 5 or less employees and unincorporated auxiliary financial activities with no employees. In the allocation of primary income account, a quota of mixed income moves from producer households to consumer households. This income is supposed to be used by households for consumption and saving.

3 How Much do Survey Microdata fit National Accounts?

Before thinking of any method for integrating surveys microdata in national accounts (NA), it is essential to verify how much this information fits national accounts estimates.

For this purpose, we try to estimate the GDP components on the basis of survey data. Particularly, we consider the Bank of Italy survey on income and wealth (Shiw)[2] and the European statistics on income and living conditions (Eu-silc)[3]. The year 2006 is the most recent for which Shiw and Eu-silc results can be compared.

The surveys variables have been fully harmonised to national accounts with respect both to the observed population (i.e. Households) and to the content of each income component (Coli and Tartamella 2008). Results are then compared with national accounts[4] in Tables 2 and 3.

The Shiw survey collects income data at a finer detail, thus allowing more interesting comparisons. For example the coverage of the mixed income component can be evaluated only for the Shiw. In fact, in order to identify producer households, i.e. households generating mixed income (see also footnote 1), surveys are required to record both the size and legal status of the enterprise whom the self employed belongs. Up to 2008, Eu-silc does not collect this information thus not allowing to disentangle mixed income from other kind of self employed income. For this reason the NA Eu-silc comparison is made for the self-employed income component as a whole which is computed as to include income withdrawn by households both from Producer households (the mixed income share) and from the Corporation sector.

Coming to results, we notice that both surveys sensibly underrate self employed income. On the contrary both surveys overestimate the gross operating surplus (i.e. imputed rents of owner-occupied dwellings) with respect to national accounts. The reason may be an overestimated number of dwellings and/or imputed rents. Surveys probably overrate the stock of dwellings since, for fiscal reasons, owners may declare an imputed rent on a dwelling that is actually let. Moreover, surveys estimate imputed rents as the amount of money that the owners expect to pay for renting their own house, which may be biased, e.g. reflecting only current market prices, whereas most dwellings are rent according to earlier and cheaper contracts. On the contrary national accounts estimate imputed rents using the figure on actually paid rents from the household sample surveys. Fiscal reasons often probably urge people not to declare the real amount of received/paid rents. As a consequence NA imputed rents may be underestimated.

---

[2]Shiw data can be downloaded from the Bank of Italy web site.

[3]IT-SILC XUDB 2007–May 2009.

[4]To compare data correctly it is necessary to estimate national accounts income components net of taxes and social contributions. We assume a proportional taxation on all income components. Actually, a microsimulation model should be used to correctly allocate taxation among different income categories.

Table 2 Households income components[a] in NA and Shiw: a comparison - Italy, 2006.

	2006					Shiw (total)/NA
	National Accounts (households)	Shiw[b]				
		Total	Household mean	Lower 95% limit for the mean	Upper 95% limit for the mean	
Wages and salaries net of social contributions paid by employees	348,235	295,508	23,110	22,669	23,550	84.86%
Mixed income quota assigned to households, net of social contribution paid by self employed	145,903	87,263	20,544	18,857	22,231	59.81%
Gross operating surplus	86,949	140,131	7,628	7,455	7,800	161.17%

[a]Total income is in current million euros. Average income is in euros
[b]Income estimates are grossed up using the survey sampling weights

Wage and salaries is the best covered income source, especially by Eu-silc (over 95%). An in-depth analysis is shown in Tables 4 and 5 where employees and remuneration per capita values are analysed by economic activity.

We observe a general better fit of Eu-silc estimates to national accounts data. Absolute differences computed for economic sector are on average smaller with respect to Shiw, both for per capita values, number of employees and wages and salaries.

The main point about the employees distribution is that surveys record lower weights for services, in favor of industry. The difference is particularly evident for the "Other services for business activities" where NA record almost 10% of employees whereas Shiw and Eu-silc account for respectively 4% and 6%. The "Other services: public administration, health etc" category is an exception. This is not a surprise since this sector is only marginally effected by non registered employment, which on the contrary affects strongly the other services activities. Generally we notice milder differences between Eu-silc and NA employee distributions, Eu-silc providing a sort of average distribution between the Shiw and NA.

Table 3 Households income components[a] in NA and Eu-silc: a comparison - Italy, 2006

	2006					Eu-silc (total)/NA
	National Accounts (households)	Eu-silc[b]				
		Total	Household mean	Lower 95% limit for the mean	Upper 95% limit for the mean	
Wages and salaries net of social contributions paid by employees	348,235	332,585	23,493	23,199	23,787	95.51%
Self employed income	233,109	141,376	19,030	18,426	19,635	60.65%
Gross operating surplus	86,949	121,744	5,774	5,742	5,806	140.02%

[a]Total income is in current million euros. Average income is in euros
[b]Income estimates are grossed up using the survey sampling weights

Table 4 Distribution of employees by Economic activity, Italy, 2006

Economic activity (NACE Rev. 1 classification)	National accounts	Shiw	Eu-silc
Agriculture	2.90%	4.70%	2.75%
Industry	30.20%	35.40%	33.43%
– Industry (without construction)	23.5%	28.4%	26.83%
– Construction	6.7%	7.0%	6.60%
Services	66.9%	59.8%	63.83%
– Trade, hotel and restaurants	14.8%	11.7%	12.75%
– Transport storage and comm.	5.6%	4.6%	5.75%
– Financial intermediation	2.9%	3.4%	3.33%
– Other services for business activities	9.7%	3.8%	5.90%
– Other services (Public admin., education, etc.)	33.8%	36.3%	36.11%
Total economy	100.0%	100.0%	100.0%

Table 5 Wages and salaries by Economic activity, Italy, 2006[a]

Economic activity (NACE Rev. 1 classification)	National accounts	Shiw			Eu-silc[b]		
	WS (per capita)	WS (per capita)	Lower 95% limit	Upper 95% limit	WS (per capita)	Lower 95% limit	Upper 95% limit
Agriculture	11,546	10,928	10,197	11,659	10,096	9,424	10,769
Industry	17,798	15,751	15,434	16,069	18,268	17,995	18,541
Industry (without construction)	18,638	15,975	15,620	16,330	18,781	18,474	19,088
Construction	14,870	14,849	14,144	15,554	16,180	15,607	16,754
Services	19,309	16,880	16,576	17,183	19,456	19,217	19,696
Trade, hotel and restaurants	17,655	13,427	13,006	13,848	15,287	14,929	15,645
Transport storage and comm.	27,137	17,936	17,155	18,716	20,855	19,952	21,758
Financial intermediation	33,831	26,713	24,197	29,229	27,460	26,268	28,652
Other services for business activities	17,571	16,271	15,272	17,270	17,979	17,178	18,780
Other services (Public admin., education, etc.)	17,991	17,010	16,650	17,369	20,209	19,889	20,530
Total economy	18,626	16,199	15,979	16,418	18,802	18,621	18,984

[a]Current euros
[b]The economic activity refers to the current year (see text for details on the method used to calculate distributions)

Coming to wage and salaries per capita values, we notice lower values in the Shiw for every category with the only exception of "Agriculture" and "Construction". On the other hand Eu-silc reports higher remunerations even with respect to NA for some activities (industry and the "Other services" in particular).

Analysis by economic activity, though extremely interesting, is affected by the not so accurate estimate of the economic activity variable which often shows a consistent number of missing values in surveys. Moreover, as far as Eu-silc is concerned, the economic activity variable is collected with respect to the interview year whereas information on income applies to the year before. As a consequence the distribution of income by economic activity has been estimated only on a subset of sampled employees, namely the ones who declare not having changed job in the previous 12 months. This is obviously an approximation.

4 Conclusions

As it is well known national accounts statistics are the result of the integration of several data sources. At present, sample surveys data on households income are not used as an input for estimating the Italian national accounts aggregates. This is one of the reason which prevents an independent estimate of GDP on the basis of the so called income approach. Household microdata would impact on the estimate of the Italian GDP differently according to the applied methodology. On the one hand household income microdata could be embodied into the GDP estimate without impacting on its amount; the advantage would be anyhow relevant in order to analyse the distribution of income among the different sectors of population. On the other hand household income microdata could be used to provide independent estimates of some GDP components with an impact on the value of the Italian GDP itself.

In this paper we have tried to assess weather and to what extent the Bank of Italy survey on households budgets and the European Statistics on Living condition might contribute to the estimate of the GDP income components. To this purpose national accounts and surveys data on compensation of income, mixed income and operating surplus have been fully harmonized and compared. Our analysis, though preliminary, suggests that surveys data (Eu-silc in particular) would provide valuable information at least for the computation of compensation of income. The advantage would be twofold: a more accurate estimate of wages and salaries for some categories of employees, the possibility of analysing compensation of employees according to the employee's characteristics and those of her/his household.

National accounts have traditionally given relevance to the analysis of productive processes and final uses of income. On the contrary the information on institutional sectors has not been satisfying for a long time. Moreover among institutional sectors, households have been given the lowest attention. This perhaps may help understanding while in NA the need of new and rich statistics on households income has not been compelling for years. With the system of national accounts of 1993 (recently updated) national accounts increased the attention on households through the proposal of an accounting by groups of households and the introduction of the social accounting matrix. Nowadays, even in official statistics the interest is moving from production units to people in order to supply indicators of economic growth as well as of people well-being and happiness. The estimate of this new set of indicators asks for a strongest and better integration of "people" data in the national accounts framework. This is in line with the recommendations of the Commission on the Measurement of Economic Performance and Social Progress (Stiglitz et al. 2009). In fact, one of the key recommendations of the Report is "*to shift emphasis from measuring economic production to measuring people's well-being*" (p. 12, Stiglitz et al. 2009). As suggested in the Report itself, this objective may be achieved emphasising the household perspective in national accounts.

References

Brandolini, A.: The distribution of personal income, in post-war italy: source description, date quality, and the time pattern of income inequality, Temi e discussioni, Banca Italia (1999).

Coli A., Tartamella F.: Income and consumption expenditure by households groups in national accounts, Iariw Conference, Slovenia (2008).

Istat: Metodologie di stima degli aggregati di contabilità nazionale a prezzi correnti. Metodi e norme n. 21, Istat, Roma (2004).

Reid David J.: Combining three estimates of gross domestic product, economica, New Series, Vol. 35, No. 140, pp. 431–444 (1968).

Stiglitz E., Sen A., Fitoussi J.P.: Report by the commission on the measurement of economic performance and social progress (2009).

The Employment Consequences of Globalization: Linking Data on Employers and Employees in the Netherlands

Fabienne Fortanier, Marjolein Korvorst, and Martin Luppes

1 Introduction

Globalization – or the increased interconnectedness of nations, peoples and economies – is often illustrated by the strong growth of international trade, foreign direct investment (FDI) and multinational enterprises (MNEs). At the moment, more firms, in more industries and countries than ever before, are expanding abroad through direct investment and trade. The advent of globalization has been paired with intense debates among policy makers and academics about its consequences for a range of social issues related to employment, labor conditions, in-come equality and overall human wellbeing. On the one hand, the growing inter-nationalization of production may lead to economic growth, increased employment and higher wages. In setting up affiliates and hiring workers, MNEs directly and indirectly affect employment, wages and labor conditions in host countries (see e.g. Driffield 1999; Görg 2000; and Radosevic et al. 2003). On the other hand, fears are often expressed that economic growth may be decoupled from job creation, partly due to increased competition from low-wage countries, or through outsourcing and offshoring activities of enterprises (Klein 2000; Korten 1995). These concerns about the employment consequences of globalization are not entirely unwarranted, as studies by Kletzer (2005) and Barnet and Cavenagh (1994) have shown.

These contradictory findings imply that little is yet known about the net consequences of economic globalization for employment, or, more specifically, about the extent to which firm characteristics related to globalization – such as foreign ownership – affect the employment, labor conditions and careers of employees. Answering such questions requires data that includes information not only about firms and their exact features, but also details about the characteristics

F. Fortanier · M. Korvorst · M. Luppes (✉)
Statistics Netherlands
e-mail: ffor@cbs.nl; m.korvorst@cbs.nl; mlps@cbs.nl

A. Di Ciaccio et al. (eds.), *Advanced Statistical Methods for the Analysis of Large Data-Sets*, Studies in Theoretical and Applied Statistics,
DOI 10.1007/978-3-642-21037-2_10,

of the employees that work for them. By linking, at the micro level, business and social data from various surveys and registers, Statistics Netherlands is now able to shed new light on these questions for the Dutch context. This chapter documents the intricacies involved in creating such an integrated employer–employee dataset. In addition, this chapter highlights some of the novel analyses that are possible with this new dataset, and addresses some first conclusions that can be drawn from this data integration exercise regarding the impact of economic globalization for employment in the Netherlands (see also Fortanier and Korvorst 2009).

The Netherlands is not the first country to construct such a linked employer-employee dataset (LEED). Statisticians and academics have created similar datasets for e.g. Germany (Alda et al. 2005); Finland (Ilmakunnas et al. 2004); the US (Abowd and Kramarz 1999) and Denmark (Munch and Skaksen 2008), to name but a few examples. Studies based on these datasets illustrate the wealth of policy relevant research questions that can be answered, both with respect to employment and labor market issues as well as for more detailed analyses of e.g. the role of human capital in firm performance (Bryson et al. 2006).

The question regarding the employment consequences of globalization has however not yet been extensively addressed in studies based on LEED data (but exceptions include e.g. Munch and Skaksen 2008). We expect that an analysis of both firm and employee characteristics should improve our understanding of the social implication of e.g. increased international trade (exports and imports), outsourcing and offshoring, and the growing direct investment flows that imply that locally operating firms are increasingly owned, controlled and managed by foreign enterprises.

In the remainder of this chapter, we first detail the various methodological steps we took to construct the dataset, before presenting the results on globalization and employment for the Netherlands. We conclude by addressing some important methodological considerations for creating linked employer-employee datasets, which may serve as input for other National Statistical Offices, and provide suggestions for further research.

2 Methodology: Creating a Linked Employer-Employee Dataset for the Netherlands

The creation of the linked employer employee dataset (LEED) for the Netherlands primarily involved the integration of the Social Statistical Database, which includes a wide variety of variables on (the composition of) employees and the labor force, with the General Business Register and various firm-level surveys (mainly the Statistics on Finances of Large Enterprises, and the Community Innovation Survey) that provide for various firm-level variables, including foreign ownership (see also De Winden et al. 2007). All data in this chapter pertains to the 2000–2005 period, partly due to data availability and partly due to the methodological changes in both the social and firm statistics in 2006 that would have created a (potential) break in the time series.

The social statistical database (SSB) of Statistics Netherlands consists of administrative data on persons, households, jobs, benefits and pensions. It covers the entire Dutch population, including people living abroad but working in the Netherlands or receiving a benefit or pension from a Dutch institution. All persons with an official employment contract at a registered enterprise are recorded in the SSB database with a job. Since our analysis focuses on employment, we included all jobs that have existed within a year in the Netherlands. Note that this means that self-employed individuals and business owners are not included in our analysis, and neither are e.g. pensioners or unemployed students.

At the micro level, a direct relationship between employees and enterprises can be established because employees' social security numbers are available in administrative sources (e.g. insurance) together with the administrative enterprise numbers. Subsequently, the administrative units of enterprises, for example tax numbers, can be translated to statistical units of enterprises, which are recorded in the general business register (GBR). Since all employees can be assigned to an employer (enterprise), detailed information is available per enterprise on e.g. the number of jobs per year, gross job creation and destruction rates, labor force composition with respect to gender, age and ethnicity, as well as average wages and a range of other variables.

In turn, the GBR includes basic enterprise information on e.g. the industry of activity, size class, location in the Netherlands, and enterprise status (entries and exits). From this register, additional enterprise data can be derived from other administrative sources and surveys (e.g., business surveys on international orientation and foreign ownership) and be matched with data on employees from the SSB.

In our study, we were particularly interested in one of the key enterprise characteristics for globalization, i.e. the locus of control (foreign versus domestic). Using the concept of Ultimate Controlling Institute (UCI), foreign controlled enterprises are defined as those that have their centre of control or headquarters outside the Netherlands, whereas Dutch-controlled enterprises are nationally owned. The distinction enables an analysis of the consequences of inward foreign direct investments (FDI) in the Netherlands at the micro level. Information on the UCI of enterprises is primarily derived from two sources: the Financial Statistics of Large Enterprise Groups (SFGO) and the Community Innovation Survey (CIS). Unfortunately, since both of these sources are surveys, the number of enterprises for which we can positively establish their UCI is limited.

Table 1 reports the exact results of our data matching exercise in which we linked the SSB with the UCI list, resulting in a linked employer–employee dataset (LEED) for the Netherlands. The micro data integration of registered and survey data on employers and employees resulted in a sample of approximately 20 thousand enterprises each year for which the locus of control was known. Although the size of the final matched sample is quite modest, a disproportionate share of large enterprises is included, accounting for nearly 3 million jobs. This represents 40% of the total number of jobs in the Netherlands, and 55% of the jobs in the private sector. While the dataset does not form a balanced panel, the largest enterprises are

Table 1 Matching results: number of enterprises and employees

	2000	2001	2002	2003	2004	2005
No. of enterprises with employees						
In the SSB	380,339	397,726	405,544	415,172	419,263	429,440
With UCI data	26,818	25,084	26,786	24,955	24,487	23,088
In matched SSB-UCI sample	18,865	17,681	19,077	17,837	18,481	17,469
% matched/SSB	3%	3%	3%	3%	3%	3%
Number of employees						
In the SSB	7,334,333	7,504,183	7,547,416	7,492,459	7,384,313	7,441,746
In matched SSB-UCI sample	2,980,670	2,864,322	2,880,627	2,746,898	2,790,631	2,707,695
% matched/SSB	41%	38%	38%	37%	38%	37%

automatically included each year. The share of foreign controlled enterprises in the sample is about 15% of the total number of enterprises included.

Foreign controlled enterprises are however not equally represented in each size group in the sample. Data on Dutch controlled enterprises are mainly available for small to medium (<250 employees) sized enterprises, whereas foreign controlled enterprises are relatively more represented at larger size classes in our sample (see Table 2). This reflects reality, where foreign ownership is also concentrated among the largest enterprises. Yet, to prevent the risk of misrepresentation, size class is explicitly taken into account in the analysis of this LEED dataset. Unless otherwise specified, the results reported below apply to all size classes (small, medium and large).

Other methodological challenges that were tackled during the matching process included selecting the optimal matching window for business and social statistics (end-of-year date), improving business statistics concerning enterprise dynamics

Table 2 Number of enterprises in the linked employer-employee dataset by size class, 2000–2005

	2000	2001	2002	2003	2004	2005
Total	18,865	17,681	19,077	17,837	18,481	17,469
Dutch controlled	16,149	15,067	16,144	15,080	15,640	14,798
0–4 employees	2,161	2,225	1,796	1,740	1,642	1,617
5–9 employees	1,669	1,603	1,123	1,251	1,043	1,226
10–19 employees	2,997	2,734	3,368	2,711	3,455	3,204
20–49 employees	3,597	3,258	3,665	3,605	4,073	3,757
50–99 employees	3,010	2,536	3,251	2,857	2,495	2,276
100–149 employees	927	913	1,079	1,097	1,070	948
150–199 employees	451	467	466	504	536	486
200–249 employees	257	255	270	266	255	252
250–499 employees	551	547	552	524	523	520
500–999 employees	304	308	290	277	306	268
1000–1999 employees	127	121	153	148	140	138
2000 and more employees	98	100	101	100	102	106
Foreign controlled	2,716	2,614	2,933	2,757	2,841	2,671
0–4 employees	209	209	204	200	189	185
5–9 employees	165	156	158	156	131	134
10–19 employees	294	265	308	248	324	303
20–49 employees	525	471	527	509	548	494
50–99 employees	522	497	627	546	512	460
100–149 employees	273	261	296	293	319	310
150–199 employees	192	170	181	192	192	202
200–249 employees	118	128	142	126	122	112
250–499 employees	222	240	259	263	274	246
500–999 employees	123	124	141	139	141	145
1000–1999 employees	49	64	60	51	49	44
2000 and more employees	24	29	30	34	40	36

and locus of control, bridging existing time lags of data availability, and streamlining best practices across statistical divisions etcetera.

Although caution in interpreting the results is warranted – in particular with respect to the sample of enterprises – the data give a clear perspective on the consequences for employees of working for foreign versus Dutch controlled enterprises.

3 First Results

The linked employer-employee dataset that was thus constructed for the Netherlands includes a large number of employment related variables that can now be compared between foreign and domestically controlled firms: not only average employment magnitudes, but also share of high- and low-paid staff, labor force composition, worker characterics and job dynamics and labor conditions. For matters of brevity we highlight three key indicators in the remainder of this chapter that best capture the essential differences in labor market dynamics and wage distribution between foreign and domestically controlled firms in the Netherlands: (i) the total and average number of employees; (ii) wages and share of high-paid staff; and (iii) labor market dynamics (turnover rate).

3.1 Number of Employees

First of all, substantial differences with respect to the number of employees can be observed in our linked employer-employee dataset for the Netherlands (average employment was calculated as the unweighted average number of jobs per year). As shown in Fig. 1, foreign controlled enterprises have on average a larger workforce than Dutch controlled enterprises. In terms of mean number of employees foreign enterprises are 40 to 60% larger than Dutch-controlled enterprises. Furthermore, foreign enterprises in the Netherlands have shown an increase in employment from 2002 onwards, whereas Dutch controlled enterprises have shown a small decline in terms of average number of employees. This trend may be caused by foreign takeovers of (or mergers with) Dutch controlled enterprises of medium to large size, in terms of total number of employees and the creation of jobs. The sectors that showed the highest growth in employment at foreign controlled enterprises in the Netherlands were concerned with agriculture, forestry and fishing, mining, and quarrying, construction, trade and repairs, transport, storage and communication and financial inter-mediation. In contrast, at Dutch controlled enterprises small increases in average number of jobs were only realized in the food and beverages and chemicals and plastic products industries, whereas all other sectors showed a decline.

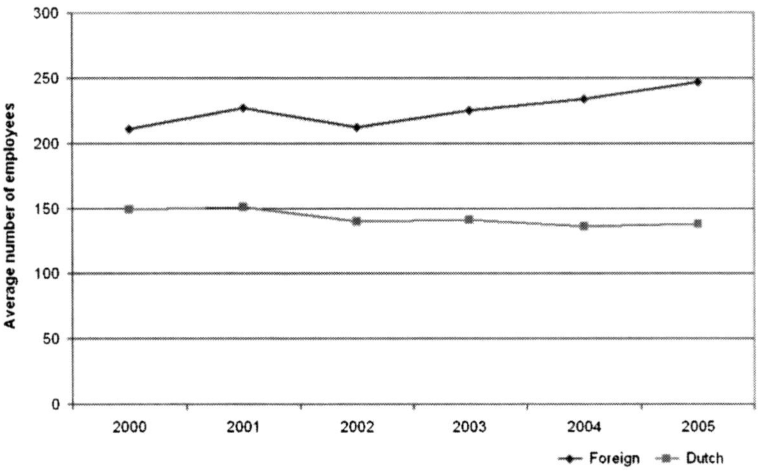

Fig. 1 Average employment at foreign and Dutch controlled enterprises in the Netherlands, 2000–2005

3.2 Wages and Pay

Several explanations have been proposed in the academic literature for the wage differential between foreign and domestically controlled firms. First of all, foreign enterprises are on average more productive than domestic enterprises – part of that productivity differential may translate into higher salaries. A second often-cited reason in the academic literature that could explain for the wage differences between foreign and domestically owned enterprises is that exactly because foreign-owned enterprises compete with local enterprises based on their technological advantages, they will pay their employees more than they would earn at local enterprises, in order to prevent labor migration (and subsequent unintentional knowledge spillovers) to domestic enterprises (Fosfuri et al. 2001).

The ratio of skilled versus non-skilled wage is called the relative wage, and may serve as a proxy for overall income inequality. Most models assume that foreign enterprises hire relatively high skilled labor, making it scarcer and thereby indirectly increase wage inequality (e.g. Wu 2000). Foreign enterprises tend to pay higher wages, to attract higher educated employees and at the same time preventing labor migration to nearby (domestic) enterprises or setting up own enterprises. Furthermore, foreign enterprises may be more productive in general, substantiating a higher wage level.

This wage differential between foreign and Dutch-controlled enterprises is also evident in the Netherlands. As Fig. 2 shows, foreign enterprises have more high- than low-paid employees, whereas Dutch-controlled enterprises have an equal share in their workforce. This difference in share of high- versus low-paid workers is stable over time (2000–2005). The difference between foreign and Dutch-controlled

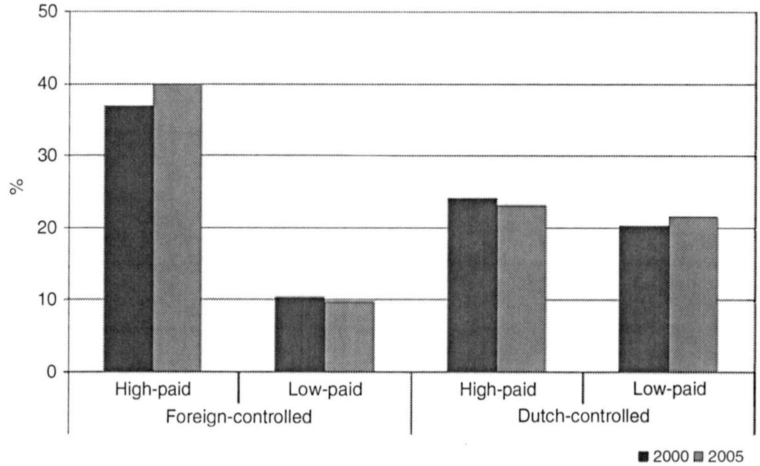

Fig. 2 Share of high- and low-paid employees at foreign and Dutch controlled enterprises in the Netherlands, 2000 and 2005

enterprises in terms of high-paid workers might be a result of foreign direct investment (FDI) demanding more managerial capacity and other high-skilled functions to coordinate the new foreign venture in the Netherlands.

The prominence of high-paid workers is negatively correlated with size class: both foreign and Dutch controlled enterprises tend to have fewer highly paid workers as they become larger. Such large enterprises often involve production plants and the like with a large share of low-skilled labor. Furthermore, both foreign and Dutch-controlled enterprises have the highest share of high-paid workers in the mining and quarrying, chemical and plastic products and financial intermediation industries.

3.3 Employment Turnover

An important indicator of labor dynamics is labor turnover, or the job separation rate per enterprise, determined by the outflow of jobs as a share of the average number of jobs per year. Information on labor turnover is valuable in the proper analysis and interpretation of labor market developments and as a complement to the unemployment rate. Job creation and job destruction play a dominant role in determining the overall labor turnover rate (see also Davis and Haltiwanger 1995). Dutch controlled enterprises show a larger labor turn-over, such as outflow of jobs, than foreign controlled enterprises, as shown in Fig. 3.

Furthermore, for the 2000–2005 period, a steady decline in labor turnover is observed in our linked employer-employee dataset for the Netherlands, both in foreign and Dutch controlled enterprises. Driven by changes in the business cycle,

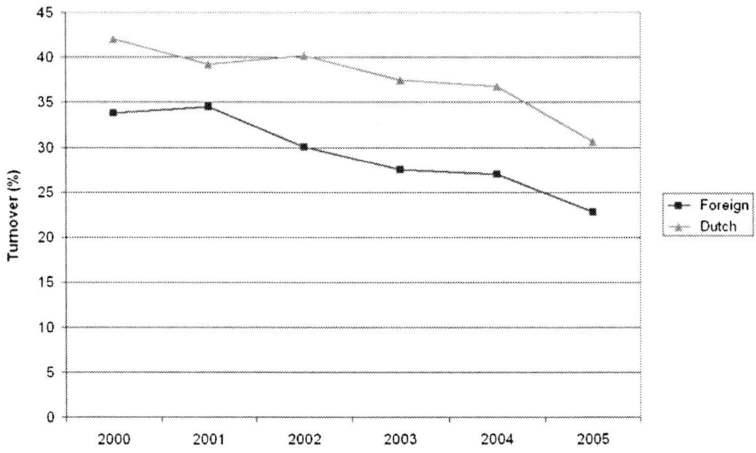

Fig. 3 Labor turnover at foreign and Dutch controlled enterprises in the Netherlands, 2000–2005

with unemployment rates increasing from 3 to almost 7%, employees were thus more willing to stay with their employer. This applies especially to employees at foreign controlled firms, resulting in an increasing retention rate. The sectors in which labor turnover is highest are the hotels and restaurants industries, real estate, renting and business sectors. This is likely due to short-term work con-tracts and seasonal employment at both foreign and Dutch controlled enterprises in these sectors, leading to a large outflow of jobs per year.

4 Conclusions and Further Research

This chapter presented the methodological considerations involved in creating a linked employer-employee dataset (LEED) for the Netherlands, in order to answer a range of questions related to the impact of firm characteristics related to globalization – i.e. foreign ownership – on employment, wages and employee characteristics. As shown above, the first outcomes based on these initial analyses generate innovative and policy relevant findings. We found for example that foreign enterprises in the Netherlands on average pay significantly higher wages than domestically owned enterprises, and that employee turnover (i.e., employees leaving an enterprise each year as a share of total employees at that enterprise) does vary substantially between foreign and domestically owned enterprises. This would mean that enterprises pay higher wages not only as a reflection of productivity differentials but also to prevent labor migration, resulting in better retention of skilled labor. However, to the extent that labor migration indeed represents the transfer of knowledge and skills across enterprises, these results bode somewhat less positive in the short run for the Dutch economy as a whole: knowledge that is embedded within foreign enterprises does not spread at the same rate as knowledge at domestic enterprises.

Methodologically, the creation of this longitudinal linked employer-employee dataset involved several challenges. The underlying datasets differed greatly, not only in terms of combining administrative records with survey information on enterprises and jobs, but also with respect to sampling frequency, units of observation and time period. For example, one problem in constructing the LEED dataset concerned the harmonization of the end-of-year date, which is default at Statistics Netherlands at the end of September for employee data and at the end of December for enterprise ownership information. By choosing the end of the year as fixed matching moment, major mismatches between employees and enterprises were avoided. At the same time, our endeavor of matching social and business statistics had an indirect positive effect of an improvement of existing business statistics within Statistics Netherlands concerning enterprise dynamics and locus of control.

Still, some quality issues remain that pertain to the integration of different sources, such as coverage errors and coherence with macro outcomes (national accounts), which imply that the results presented in this paper should be used as initial findings. Several actions are currently undertaken at Statistics Netherlands to improve the longitudinal linked employer-employee dataset and underlying statistical micro-data integration. First of all, the timeliness of data availability on jobs and enterprises will be enhanced (reducing the existing time-lag by $\sim$1 to 2 years). Furthermore, in order to enable the effective integration of data on enterprises and jobs, procedural best practices (concerning access, storage and linkage of large data sets) and operational definitions are more streamlined and documented for future reference. Secondly, more external sources will be accessed in the future to enable a higher coverage of internationalization information on enterprises in the Netherlands. Finally, when several administrative and survey data are combined, as is the case in the present LEED data set, general custom-made weighting procedures that partial out selective biases, for instance concerning size class distribution differences as noted above, are developed in order to warrant statements about the entire Dutch population of firms and employees.

In summary, without any additional data collection, Statistics 'Netherlands is thus able to relate business and social data from various surveys and registers by linking them at the micro level, creating linked employer–employee time series. In this way, a new socio-economic statistical framework is established, enabling analyses and statistical output on the relation between business information and jobs of persons and their social background.

References

Abowd, J. and Kramarz, F. (1999) The analysis of labor markets using matched employer–employee data. In: Ashenfelter, O. and Card, D. (Eds.) Handbook of Labor Economics, North-Holland: Amsterdam, pp. 2629–2710.

Alda, H., Bender, S. and Gartner, H. (2005) The linked employer-employee data-set of the IAB (LIAB), IAB Discussion paper 6/2005, Nörnberg: IAB.

Barnet and Cavenagh (1994) Global Dreams: imperial Corporations and the New World Order, New York: Simon&Schuster.

Bryson, A., Forth, J. and Barber, C. (2006) Making Employer-employee data relevant to policy, DTI Occasional Paper 4/2006, London: DTI.

Davis, S. J. and Haltiwanger, J. (1995) Measuring Gross Worker and Job Flows, NBER Working Papers 5133, National Bureau of Economic Research, Inc.

Driffield, N. (1999) Indirect employment effects of Foreign Direct Investment into the UK, Bulletin of Economic Research, 38(10):1091–1110.

Fortanier, F. and Korvorst, M. (2009) Do foreign investors pay better? Wage differences between employees at Dutch and foreign-controlled enterprises in the Netherlands: 2001–2005. In: Statistics Netherlands (ed.) *The Internationalization Monitor*, The Hague: Statistics Netherlands.

Fosfuri, A., Motta, M. and Ronde, T (2001) Foreign Direct Investment and Spill-overs through workers' mobility, Journal of International Economics, 53(1): 205–222.

Görg, H. (2000) Multinational companies and indirect employment: measurement and evidence, Applied Economics, 32(14): 1809–1818.

Ilmakunnas, P., Maliranta, M. and Vainioma, J. (2004) The roles of employer and employee characteristics for plant productivity, Journal of productivity analysis, 21(1): 249–276.

Klein, N. (2000) No Logo, London: Flamengo.

Kletzer, L. (2005) Globalization and job loss, from manufacturing to services, Economic perspectives, 29(2):38–46.

Korten, D. (1995) When corporations rule the world, London: Earthscan.

Munch, J.R. and Skaksen, J.R. (2008) Human capital and wages in exporting firms, Journal of International Economics, 75(2): 363–372.

Radosevic, S., Varblane, U. and Mickiewicz, T. (2003) Foreign Direct Investment and its effect on employment in Central Europe, Transnational Corporations, 6(1): 117–21.

De Winden, P., Arts, K. and Luppes, M. (2007) A proposed model for micro integration of economic and social data, Paper prepared for the Conference of the Federal Committee on Survey Methodology, Arlington (USA).

Wu, X. (2000) Foreign Direct Investment, intellectual property rights and wage inequality in China, China Economic Review, 11(4): 361–384.

Applications of Bayesian Networks in Official Statistics

Paola Vicard and Mauro Scanu

Abstract In this paper recent results about the application of Bayesian networks to official statistics are presented. Bayesian networks are multivariate statistical models able to represent and manage complex dependence structures. Here they are proposed as a useful and unique framework by which it is possible to deal with many problems typical of survey data analysis. In particular here we focus on categorical variables and show how to derive classes of contingency table estimators in case of stratified sampling designs. Having this technology poststratification, integration and missing data imputation become possible. Furthermore we briefly discuss how to use Bayesian networks for decision as a support system to monitor and manage the data production process.

1 Introduction

Statistical analyses can be particularly complex when are referred to surveys and databases produced by a National Institute of Statistics. The complexity is mainly due to: high number of surveys carried out by the institute, sampling design complexity, high number of variables and huge sample size. In this context it can be useful to analyse and exploit the dependence structures. Bayesian networks (Cowell et al., 1999), from now on BNs, are multivariate statistical models able to represent and manage complex dependence structures. The theoretical setting of BNs is the basis for developing methods for efficiently representing and managing

P. Vicard (✉)
Università Roma Tre, via Silvio D'Amico 77, 00145 Roma, Italy
e-mail: vicard@uniroma3.it

M. Scanu
Istat, via Cesare Balbo 16, 00184 Roma, Italy
e-mail: scanu@istat.it

A. Di Ciaccio et al. (eds.), *Advanced Statistical Methods for the Analysis of Large Data-Sets*, Studies in Theoretical and Applied Statistics,
DOI 10.1007/978-3-642-21037-2_11,

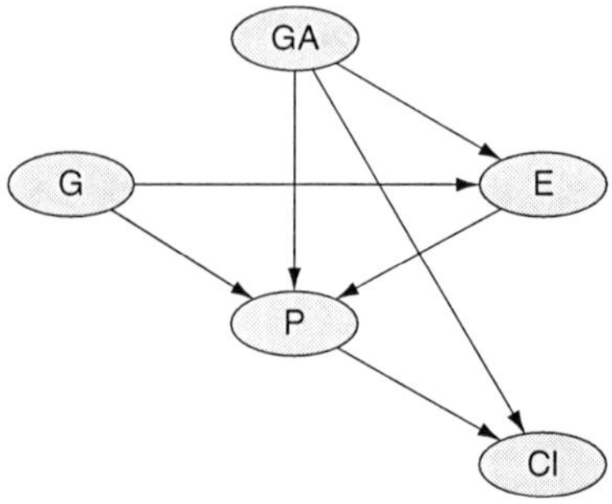

Fig. 1 Example of DAG for the five categorical variables: *Geographical area* (GA), *Gender* (G), *Education* (E), *Profession* (P), *Class of income* (CI)

survey systems. A known (or previously estimated) dependence structure can help: in computing estimators (with the sampling design either explicitly or implicitly modelled); when coherence constraints among different surveys must be fulfilled; in integrating different sample surveys (in terms of their joint distribution); in updating the estimate of a joint distribution once new knowledge on a variable marginal distribution occurs (survey weights poststratification is a special case); in missing data imputation. Furthermore, since BNs can be extended to embody decision and utility nodes giving rise to BNs for decisions (Jensen, 2001; Lauritzen and Nilsson, 2001), they can be used for data collection monitoring. Therefore BNs can be thought of as a unique framework by which it is possible to manage a survey from planning, passing through imputation and estimation to poststratification and integration with partially overlapping surveys. In the next sections we will survey recent results on the application of BNs in official statistics contexts focusing on categorical variables.

2 Background on Bayesian Networks

Bayesian networks (BN) are multivariate statistical models satisfying sets of (conditional) independence statements representable by a graph composed of nodes and directed edges (arrows) between pairs of nodes. Each node represents a variable, while missing arrows between nodes imply (conditional) independence between the corresponding variables. This graph is named directed acyclic graph (DAG); it is acyclic in the sense that it is forbidden to start from a node and, following arrows directions, go back to the starting node. Figure 1 shows an example of DAG.

Notation related to BNs describes the relationship between nodes in the following way: there are parents (e.g. P is a parent of CI) and children (e.g. CI is a child of P). Each node is associated with the distribution of the corresponding variable given its parents (if a node has no parents, as GA and G, it is associated with its marginal distribution). There are properties connecting the concept of independence between variables and absence of an arrow in the graph; these are encoded in the Markov properties (see Lauritzen (1996) Sect. 3.2.2). For example, in Fig. 1, it is possible to read that G and CI, and E and CI are independent given P, while G and GA are pairwise independent, but conditional dependent given any of these variables: E or P. Formally speaking a BN is a pair DAG/joint probability distribution satisfying the Markov condition. On the basis of these probabilistic conditional independence properties, the complex global model can be decomposed into simpler submodels. The BN definition implicitly associates a factorization of the joint distribution that highlights the dependence structure of the variables (chain rule). For k variables X_j, $j = 1, \ldots, k$, the chain rule states that the joint distribution of $(X_1, \ldots, X_k)$ can be factorized as:

$$P(X_1 = x_1, \ldots, X_k = x_k) = \prod_{j=1}^{k} P(X_j = x_j | \mathrm{pa}(X_j)), \tag{1}$$

where $\mathrm{pa}(X_j)$ is the set of parents of X_j. For instance for the BN in Fig. 1, the chain rule is

$$\begin{aligned} P(GA = x_1, G = x_2, E = x_3, P = x_4, CI = x_5) &= P(GA = x_1) \times P(G = x_2) \\ \times P(E = x_3 | GA = x_1, G = x_2) &\times P(P = x_4 | GA = x_1, G = x_2, E = x_3) \\ \times P(CI = x_5 | GA = x_1, P = x_4). \end{aligned}$$

Notice that when a BN is designed with the additional aim to make inference on its nodes (variables), it is possible to call it a *probabilistic expert system* (PES). For more details on BNs and PES see (Cowell et al., 1999).

3 Use of Bayesian Networks in Survey Sampling

In an official statistics context, PES have, among the others, two positive aspects (Ballin and Vicard, 2001): an easy-to-interpret, concise and informative way to represent both surveys and sets of surveys with their dependence structure; an inferential machine to update marginal distributions when new information arrives, i.e. to propagate information among the variables of one or more surveys. In order to exploit these properties, it is crucial to develop methods to derive PES based estimators for complex sampling designs.

In Ballin et al., 2010 contingency table estimators based on PES under a stratified sampling design have been proposed. Let $\mathbf{X} = (X_1, \ldots, X_k)$ and $\mathbf{x} = (x_1, \ldots, x_k)$ be k categorical variables and their possible values respectively. Let $\mathcal{P}$ be a population of size N generated from a superpopulation model. The parameter of interest is the contingency table of $(X_1, \ldots, X_k)$ in $\mathcal{P}$

$$\theta_{x_1,\ldots,x_k} = \sum_{i=1}^{N} \frac{I_{x_1,\ldots,x_k}(x_{1i}, \ldots, x_{ki})}{N}, \tag{2}$$

where $I_y(w)$ is the indicator function that is equal to 1 when $x = w$ and 0 otherwise.

Assume that a random sample $\mathcal{S}$ of size n is drawn from $\mathcal{P}$ according to a stratified sampling design with H strata s_h, $h = 1, \ldots, H$. Let N_h, n_h and w_h be the stratum size in $\mathcal{P}$, the stratum size in $\mathcal{S}$ and the stratum weight respectively, $h = 1, \ldots, H$. The parameter (2) can be estimated by means of the Horvitz-Thompson estimator in case no auxiliary information is available. The dependence structure of the k variables of interest can be considered as a kind of auxiliary information and estimators can be derived on its basis. The dependence structure may be known in advance otherwise it can be estimated by means of specific algorithms (for more details we refer to Ballin et al., 2010, and for a general presentation of structural learning to Neapolitan, 2004).

The relation structure among the sampling design and $X_1, \ldots, X_k$ can be taken into account in the contingency table estimation process either *explicitly*, i.e. modelling the statistical relationship between the design variable and the variables of interest, or *implicitly*, i.e. incorporating the information on sampling design via survey weights. In both cases the estimators can be derived in a likelihood-based approach, allowing also to learn the dependence structure (if not previously known).

Let us consider the first case. Let U be one design variable with as many states (H) as the strata, having frequency distribution:

$$\theta_h = \frac{N_h}{N} = \frac{n_h w_h}{\sum_{h=1}^{H} n_h w_h}, \quad h = 1, \ldots, H. \tag{3}$$

The design variable node U is represented together with the other variables, and the PES for $(U, X_1, \ldots, X_k)$ has the characteristic that U is a root, i.e. it has no parents. The estimators based on the network explicitly modelling U in the graph are named E-PES estimators and can be formally derived considering the maximum likelihood estimators under the assumed network for $(U, X_1, \ldots, X_k)$. Applying the chain rule (1) to the PES for $(U, X_1, \ldots, X_k)$, it follows that the E-PES estimator of $\theta_{\mathbf{x}}$ is

$$_{PES}\hat{\theta}_{\mathbf{x}}^{(E)} = \sum_{h=1}^{H} \theta_h \prod_{j=1}^{k} \hat{\theta}_{x_j|pa(x_j)}. \tag{4}$$

θ_h is known by design and $\hat{\theta}_{x_j|pa(x_j)}$ is the unweighted sample relative frequency of x_j given the categories of X_j parents in $\mathbf{x}$, i.e.

$$\hat{\theta}_{x_j|pa(x_j)} = \frac{\sum_{i=1}^{n} I_{(h,\mathbf{x})}\left(x_{ji}, pa\left(x_{ji}\right)\right)}{\sum_{i=1}^{n} I_{(h,\mathbf{x})}\left(pa\left(x_{ji}\right)\right)}.$$

where $I_{(h,\mathbf{x})}\left(x_{ji}, pa\left(x_{ji}\right)\right) = 1$ when $x_{ji} = x_j$ and $pa\left(x_{ji}\right) = pa\left(x_j\right)$ (categories of X_j parents in $(h, \mathbf{x})$), and zero otherwise.

It is remarkable how easily E-PES estimators can be built from the graph by simply applying the chain rule and *plugging-in* the single variable components estimates, i.e. the unweighed sample estimates of $\theta_{x_j|pa(x_j)}$.

Let us now consider the case where the design variable is not modelled together with $(X_1, \ldots, X_k)$. In this case the estimators are named I-PES estimators. As E-PES, also I-PES estimators are derived using a likelihood-based approach. By means of standard survey pseudolikelihood techniques (Rao, 2010) and the chain rule (1), given a PES for $(X_1, \ldots, X_k)$, we have that I-PES estimators are defined as follows:

$$_{PES}\tilde{\theta}_{\mathbf{x}}^{(I)} = \prod_{j=1}^{k} \tilde{\theta}_{x_j|pa(x_j)} \tag{5}$$

where each factor is a weighted estimator of the conditional distributions, i.e.

$$\tilde{\theta}_{x_j|pa(x_j)} = \sum_{i=1}^{n} \frac{w_i I_{\mathbf{x}}\left(x_{ji}, pa\left(x_{ji}\right)\right)}{\sum_{i=1}^{n} w_i I_{\mathbf{x}}\left(pa\left(x_{ji}\right)\right)}.$$

It is easy to show that when the network is complete, i.e. all the nodes in the graph are directly connected between each other, E-PES and I-PES estimators coincide with the Horvitz-Thompson estimator.

Simulation studies have been performed to analyze the sensitivity of estimators (4) and (5) to model misspecification and to compare them with the Horvitz-Thompson estimator and among themselves. Moreover approximations for the MSE of E-PES and I-PES estimators have been computed (Ballin et al., 2010).

When the true PES is not complete, the Horvitz-Thompson estimator is not efficient since it implicitly relies on a complete graph. In fact variance is highly affected by the presence in the complete graph of edges that do not exist in the true PES. Additional edges increase variability because they generate an extra number of parameters to be estimated. Therefore E-PES and I-PES estimators based on the correct incomplete graph are more efficient than the Horvitz-Thompson estimator. Moreover, if E-PES or I-PES estimators are based on a graph with additional edges compared to the true model, these estimators will have higher variability. A model can be misspecified also having less arrows than the true one. In this case relevant dependence relations are overlooked and then there is a remarkable increase in bias component. The bias increase becomes dramatic for E-PES estimators when the misspecified network has at least one arrow missing from the design variable node to a variable of interest. In this sense I-PES estimators are more robust against model misspecification. In fact they are always design consistent, i.e. able to give reasonable estimates for any descriptive population quantity without assuming any model (Pfeffermann, 1993). In general model misspecification can result both in some additional and in some missing arrows. The leading effect is that due to absence of true edges.

3.1 Use of Bayesian Networks for Poststratification

Weighting adjustment is often used in order to make survey data consistent with known population aggregates. Poststratification is used when the population distribution of a variable Z is known. It can be defined as the use of a stratified (with respect to variable Z) sample estimator when the design is not stratified on Z. We still assume that all the considered variables (the original stratification variable, the variables of interest and the poststratification variables) are categorical. When these variables are modelled by means of a PES, poststratification can be usefully reinterpreted and performed by standard propagation algorithms developed for PES.

Let z_q, $q = 1, \ldots, Q$ be the Q mutually exclusive categories of Z, and N_q, $q = 1, \ldots, Q$ the corresponding population counts. As defined in the seminal paper (Holt and Smith, 1979), poststratification modifies the original survey weights w_i, $i = 1, \ldots, N$ into:

$$w_i^* = w_i \frac{N_q}{\widehat{N}_q}, \qquad i \in s_q,\ q = 1, \ldots, Q, \tag{6}$$

where s_q is the set of sample units with $Z = z_q$, and $\widehat{N}_q$ is the estimator of N_q that uses the original weights:

$$\widehat{N}_q = \sum_{i \in s_q} w_i, \qquad q = 1, \ldots, Q.$$

The idea is to rebalance the Horvitz–Thompson estimator when some categories of Z are over- or under-sampled.

The same result can be obtained when: we consider the design variable U, the variables of interest $(X_1, \ldots, X_k)$ and the poststratification variable Z as part of a PES whose structure is complete (as for the E-PES structure of the Horvitz–Thompson estimator), and we apply the rules for dealing with an informative shock on the distribution of Z. The informative shock consists in changing the marginal distribution of Z in the PES for $(U, X_1, \ldots, X_k, Z)$ from

$$\widehat{\theta}_{z_q} = \frac{\sum_{i \in s_q} w_i}{N}, \qquad q = 1, \ldots, Q,$$

to

$$\theta^*_{z_q} = \frac{N_q}{N}, \qquad q = 1, \ldots, Q.$$

In order to illustrate this, it is convenient to define the PES so that U is a parent of Z (this can always be done for complete graphs). In this way, it is possible to consider the pair of nodes (U, Z) as a unique node, with as many categories as the Cartesian product of the categories in U and Z respectively. The updated distribution of the poststrata (U, Z) after the informative shock becomes:

$$\theta^*_{hz_q} = \theta_{h|z_q} \theta^*_{z_q} = \frac{\theta_h \theta_{z_q|h}}{\sum_{h=1}^{H} \theta_h \theta_{z_q|h}} \theta^*_{z_q} = \frac{n_h w_h}{\sum_{h=1}^{H} n_h w_h} \frac{n_{hq}}{n_h} \frac{\theta^*_{z_q}}{\widehat{\theta}_{z_q}}$$

$$= w_h \frac{n_{hq}}{N} \frac{\theta^*_{z_q}}{\widehat{\theta}_{z_q}}, \qquad q = 1, \ldots, Q; h = 1, \ldots, H. \tag{7}$$

The new weight $w^*_{(hz_q)}$ must be constant for all the units in the same (U, Z) category, of size n_{hq}. Hence:

$$w^*_{(hz_q)} = \frac{N}{n_{hq}} \theta^*_{hz_q} = w_h \frac{\theta^*_{z_q}}{\widehat{\theta}_{z_q}} = w_h \frac{N_q}{\widehat{N}_q}. \tag{8}$$

As a matter of fact, PES allow a wide range of weight adjustments by post-stratification that does not only take into account the actual distribution of the poststratification variable, but also the dependence structure of all the variables. Anyway, the graphical structure for poststratification must satisfy a fundamental rule: the stratification variable U and the poststratification one Z should be directly connected, and they should be considered as a unique node after poststratification.

3.2 *Use of Bayesian Networks for Integration*

When results of a sample survey are disseminated, it would be mandatory that the figures are consistent with the others of the same survey and with the ones of other surveys (on similar or overlapping topics). In the first case, internal coherence can be defined as the situation where all the figures of a survey can be produced marginalizing any disseminated table. In the second case, external coherence represents the situation where the figures of a variable studied in two or more different surveys (with the same reference population and time) are the same. As a matter of fact, the dependence relationship among the variables of interest and the survey design is an important aspect to be considered in order to fulfill coherence properties.

This problem can be solved by means of the updating algorithm of a PES, based on the *junction tree* (see Ballin et al. (2009)). The junction tree is a hypergraph whose nodes, named hypernodes, are complete subsets of variables (named cliques). Furthermore, a junction tree should fulfill the running intersection property that is: for any two cliques C_i and C_j in the set of cliques and any clique C' on the unique path between them in the junction tree, $C_i \bigcap C_j \subset C'$. Rules for obtaining a junction tree from a DAG are described in Cowell et al. 1999.

The idea of integration by means of PES can be illustrated by an example. Consider the case (Fig. 2) of two surveys, A and B, collecting information on respectively A_1, A_2, X_1, X_2, X_3 and B_1, B_2, B_3, X_1, X_2, X_3. This is a typical situation in many multipurpose surveys, where there is a core of common variables, i.e. X_1, X_2, X_3, and each survey investigates a particular topic. Integration of the two surveys essentially means coherence of information. Coherence can be obtained

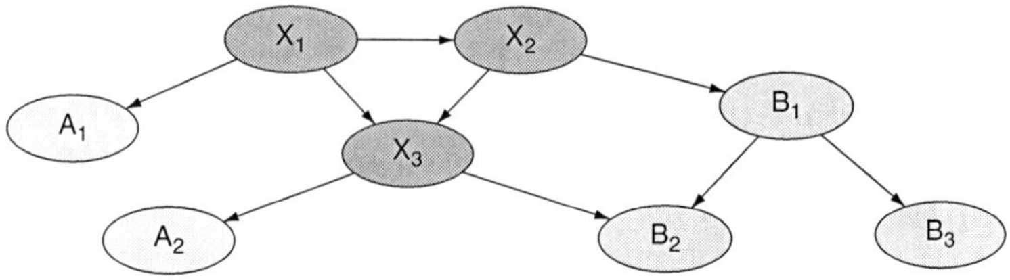

Fig. 2 Example of an integration DAG for two surveys A and B

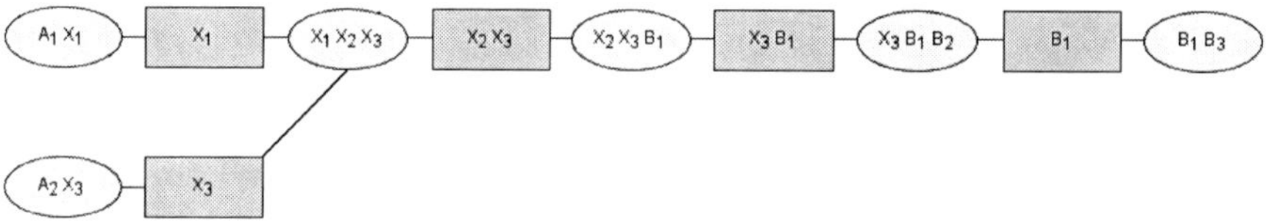

Fig. 3 Junction tree of the integration DAG in Fig. 2

when the distributions of the common variables in two surveys are the same. This rarely happens in two surveys performed in distinct times (e.g. A before B). The junction tree algorithm can be applied to update X_1, X_2, X_3, in A forcing them to have the same distribution estimated in B. The junction tree relative ti this example is shown in Fig. 3. Note that the integration network in Fig. 2 is not a real PES, unless (A_1, A_2) and (B_1, B_2, B_3) are independent given (X_1, X_2, X_3). In general the integration network is obtained overlapping the PES of different surveys with the requirement that the common variables (X_1, X_2, X_3 in Fig. 2) form a complete subgraph and separate the sets of variables observed distinctly (A_1, A_2 and B_1, B_2, B_3 in Fig. 2). In order to integrate the two surveys, let us distinguish between two types of nodes in the integration network. The first group of nodes corresponds to those that are observed in only one survey (as A_1 and A_2 in A and B_1, B_2, and B_3 in B). We suggest that every distribution to be attached to these nodes in the integration network is estimated from the corresponding survey, according to the survey weights observed in those surveys, using a Horvitz-Thompson estimator. The second group of nodes corresponds to those nodes that are observed in more than one survey. There are two possibilities:

1. Insert new information on the common variables X_1, X_2, X_3 in one survey from the other (e.g. from B to A, because B is more recent, or more accurate, etc).
2. Estimate the joint distribution of X_1, X_2, and X_3 using A and B together, as a unique sample.

4 Use of Bayesian Networks for Imputation of Missing Data

The treatment of missing values in survey data is one of the most important aspects to be taken into account in the data production process. A common approach is to impute missing items with artificial plausible values, under the assumption that the missing data mechanism is missing at random. A wide variety of imputation techniques has been developed; among them hot-deck methods are generally used in statistical institutes. Roughly speaking, hot-deck methods are based on the idea of filling in the missing items of an observation (record) with the values of a *similar* completely observed unit. When dealing with categorical variables, the concept of similarity is often accounted for by introducing a stratification that partitions the sample in clusters of similar units (i.e. showing the same categories with respect to specific variables). Hot-deck methods have desirable properties for univariate characteristics, but the preservation of relationships among variables can be a problem. Since BNs are a well-known tool to study and model the relationships among variables, they can be usefully applied for imputation. A first approach was presented in Thibaudeau and Winkler, 2002. Further research has been developed (see (Di Zio et al., 2004, 2005, 2006)). In Sect. 3 we have seen that it is possible to derive two classes of contingency table estimators based on PES and to estimate the PES structure (if it is unknown) in case the sampling design is complex. Given the dependence structure of the variables of interest it is possible to account for it while performing imputation and to easily identify those variables that are maximally informative to perform a good imputation. In fact, given a network, the information on a node (or on a set of nodes), is carried by its *Markov blanket* constituted by its parents, its children and the parents of its children (e.g. in Fig. 1 the Markov blanket of E is given by G, GA and P). Propagation algorithms developed for BNs (Cowell et al., 1999) then help to efficiently collect and spread the relevant information about the variables to be imputed. Algorithms based on the Markov blanket have been proposed in Di Zio et al., 2006. For each record, the overall vector of missing variables is imputed simultaneously by a random draw from its probability distribution conditional on its Markov blanket. This algorithm maintains a concept of similarity among units since imputation is done using a kind of stratification based on the most informative variables (pointed out by the Markov blanket). Differently from hot-deck methods, now stratification variables are found in an adaptive way, i.e. identifying them specifically for each missing pattern of each record. This dynamic *donors* selection produces improvements in terms of dependence structure preservation. Simulations and experiments have been carried out and it was shown that BN-based algorithms have good performance when compared with other imputation practices. Furthermore a software, BNimput, has been developed to perform BN based algorithms.

5 Monitoring Data Production Process Using Bayesian Networks

Data collection monitoring is carried out on the basis of various factors determining the final quality of the survey. The indices used to monitor the survey are called paradata while a survey process that can be modified on the basis of paradata is called responsive design (Groves and Heeringa, 2006). In a responsive design the survey production is treated as a decision process where every decision (with associated cost/benefit) is taken by minimizing a cost function – the expected cost is updated on the basis of paradata. In this context it is important to first formalize the decision process. In Ballin et al., 2006 it is proposed to use BNs for decisions, usually named influence diagrams (Jensen, 2001), and their improvement called LIMIDs (limited memory influence diagrams, (Lauritzen and Nilsson, 2001)) to model and solve this problem. These networks are made up of: chance nodes representing random variables; decision nodes representing decisions; utility nodes representing utility functions. When applied to monitoring a data production process, the different kinds of nodes play the following roles: chance nodes represent paradata; decision nodes represent the eventual intervention needed to improve the survey quality and utility nodes represent survey costs. In this way a monitoring and decision supporting system is set up. Once evidence on quality indicators (paradata) arrives, it is inserted and both propagation and optimization algorithms (Lauritzen and Nilsson, 2001) allow to efficiently propagate and solve a decision problem about the necessity to eventually modify the data collection process. Notice that when using LIMIDs it is not necessary to assume a unique decision maker but different decision makers can be encharged of different survey quality aspects.

Acknowledgements Thanks are due to Marco Ballin, Marco Di Zio and Giuseppe Sacco, coauthors of most of the citations in this paper. The research was supported by MIUR/PRIN 2007.

References

Ballin, M., De Francisci, S., Scanu, M., Tininini, L., Vicard, P.: Integrated statistical systems: an approach to preserve coherence between a set of surveys based on the use of probabilistic expert systems. In: NTTS (New Techniques and Technologies for Statistics) seminar (2009) Available via DIALOG http://tinyurl.com/yzh3l28.Cited20February2009

Ballin, M., Scanu, M., Vicard, P.: Paradata and Bayesian networks: a tool for monitoring and troubleshooting the data production process. In: Working Paper n 66 - 2006. Dipartimento di Economia Università Roma Tre (2006) Available via DIALOG http://host.uniroma3.it/dipartimenti/economia/pdf/wp66.pdf.

Ballin, M., Scanu, M., Vicard, P.: Estimation of contingency tables in complex survey sampling using probabilistic expert systems. J. Statist. Plann. Inference **140**, 1501–1512 (2010)

Ballin, M., Vicard, P.: A proposal for the use of graphical representation in official statistics. In: Proceeding of SCO2001 (Bressanone, 24-26 settembre 2001), pp. 487-492. CLEUP, Padova (2001)

Cowell, R.G., Dawid, A.P., Lauritzen, S.L., Spiegelhalter, D.J.: Probabilistic Networks and Expert Systems. Springer Verlag, Heidelberg (1999)

Di Zio, M., Sacco, G., Scanu, M., Vicard, P.: Multivariate techniques for imputation based on Bayesian networks. Neural Network World **4**, 303–309 (2005)

Di Zio, M., Sacco, G., Scanu, M., Vicard, P.: Metodologia e software per l'imputazione di dati mancanti tramite le reti bayesiane. In: Liseo, B., Montanari, G. E. and Torelli, N. (eds.) Software Pioneers, pp. 307-321. Franco Angeli, Milano (2006)

Di Zio, M., Scanu, M., Coppola, L., Luzi, O., Ponti, A.: Bayesian Networks for Imputation. J. Roy. Statist. Soc./A **167**, 2, 309–322 (2004)

Groves, R.M., Heeringa, S.G.: Responsive design for household surveys: tools for actively controlling survey errors and costs. J. Roy. Statist. Soc./A, **169**, 3, 439–457 (2006)
Holt, D., Smith, T.M.F.: Post Stratification. J. Roy. Statist. Soc./A, **142**, 33–46 (1979)
Jensen, F.V.: Bayesian Networks and Decision Graphs. Springer, New York (2001)
Lauritzen, S.L.: Graphical Models. Oxford University Press, Oxford (1996)
Lauritzen, S.L., Nilsson, D.: Representing and Solving Decision Problems with Limited Information. Management Science, **47**, 1235–1251 (2001)
Neapolitan, R.E.: Learning Bayesian Networks. Prentice Hall, Upper Saddle River (2004)
Pfeffermann, D.: The role of sampling weights when modelling survey data. International Statistical Review **61**, 317–337 (1993)
Rao, J.N.K., Wu.: Empirical Likelihood Methods. In: Rao, J.N.K. (eds.) Handbook of Statisics, Volume 29, Sample Surveys: Theory, Methods and Inference, pp. 189-207. Elsevier, Amsterdam (2010)
Thibaudeau, Y., Winkler, W.E. : Bayesian networks representations, generalized imputation, and synthetic micro-data satisfying analytic constraints. In: Research Report RRS2002/9 - 2002. U.S. Bureau of the Census (2002) Available via DIALOG www.census.gov/srd/papers/pdf/rrs2002-09.pdf

Part IV
Outliers and Missing Data

A Correlated Random Effects Model for Longitudinal Data with Non-ignorable Drop-Out: An Application to University Student Performance

Filippo Belloc, Antonello Maruotti, and Lea Petrella

Abstract Empirical study of university student performance is often complicated by missing data, due to student drop-out of the university. If drop-out is non-ignorable, i.e. it depends on either unobserved values or an underlying response process, it may be a pervasive problem. In this paper, we tackle the relation between the primary response (student performance) and the missing data mechanism (drop-out) with a suitable random effects model, jointly modeling the two processes. We then use data from the individual records of the faculty of Statistics at Sapienza University of Rome in order to perform the empirical analysis.

1 Introduction

The understanding of student performance is, at present, an issue of increasing concern among academics and policy makers. In this paper, we try to address empirically this issue. We use data from the individual students' records of the faculty of Statistics at Sapienza University of Rome and consider individual student's proficiency as the response variable.

To analyze the performance of university students, overlooking that some of them will conclude their degree course while some others drop-out before graduation,

F. Belloc
European University Institute, Fiesole (FI), Italy
e-mail: filippo.belloc@eui.eu

A. Maruotti (✉)
Dip. di Istituzioni Pubbliche, Economia e Società, Università di Roma Tre, Italy
e-mail: antonello.maruotti@uniroma3.it

L. Petrella
Dip. di Metodi e Modelli per l'Economia il Territorio e la Finanza, Sapienza Università di Roma, Italy
e-mail: lea.petrella@uniroma1.it

A. Di Ciaccio et al. (eds.), *Advanced Statistical Methods for the Analysis of Large Data-Sets*, Studies in Theoretical and Applied Statistics,
DOI 10.1007/978-3-642-21037-2_12,

may imply inconsistency of the estimated parameters. The factors that affect the performance of those students who retain at the university may differ, indeed, from the factors affecting the performance of those who drop-out. Moreover, both individual student's drop-out and performance may depend on the same unobservable characteristics, so that these characteristics simultaneously shape student performance and sample selection.

In order to tackle this problem, in this paper, we discuss, in a generalized linear mixed models (GLMMs) framework, a regression model for the analysis of longitudinal data in which the response outcome is observed over time and some units are lost to follow-up because of drop-out, causing missing data.

A missing data mechanism may be ignorable or non-ignorable. In this latter case, inference based on only the observed data may be not valid. In particular, non-ignorable missing data results from a drop-out mechanism dependent on either unobserved values or an underlying response process (Little and Rubin 2002). Therefore, if the drop-out mechanism is not incorporated in the analysis, the interpretation of the estimated parameters is misleading.

We construct a model by means of two specifications for, respectively, the primary outcome (the observed student performance) and the drop-out generation process. The primary response model has an autoregressive structure, so as to account for the serial correlation, while the drop-out model adopts a selection model approach. To model simultaneously the heterogeneous primary outcome and the drop-out process enables us to take into account the interdependence between the two processes. The model is extended to include random effects in the primary outcome equation in order to account for the population heterogeneity which complicates the likelihood function considerably. Among alternative maximum likelihood methods available to handle complicated models, we use the semiparametric approach proposed by Alfò and Maruotti (2009).

That student performance and retention cannot be studied independently from each other has been already pointed out in the literature. For example, Devadoss and Foltz (1996) develop a recursive system of two equations explaining student attendance and performance, in order to consider the correlation between the residuals of the equations. However they do not specifically model the drop-out and apply a seemingly unrelated regression (SUR) technique which does not cope for non-ignorable missing data. See also Zimmer and Fuller (1996) for a comprehensive survey of empirical studies on undergraduate performance in statistics.

At the best of our knowledge, this paper provides the first attempt at studying university student performance through a joint modeling of performance outcome and drop-out process. It is worth emphasizing, finally, that what we do is not proposing a tool to state the non-ignorability of the drop-out; differently, we discuss a method to cope with it.

The remainder of the paper is organized as follows. In Sect. 2, we discuss the statistical modeling. In Sect. 3, we describe data and variables used in the empirical analysis; the estimation results are presented in Sect. 4 while Sect. 5 concludes.

2 Statistical Modeling

In longitudinal studies the problem of drop-out of some of the observed individuals is an important one. The key question is whether those who drop out differ (in any way relevant to the analysis) from those who retain. Little and Rubin (2002) discusses two kind of models to handle non-ignorable missing data: namely, selection model and pattern mixture.

Selection model is intuitively more appealing when the object of interest is the marginal outcome distribution: a complete data model is defined for the primary response and augmented by a model describing the missing data mechanism conditional on the complete data. Selection model, indeed, makes it possible to study the marginal treatment effects and facilitates treatment comparisons. On the other hand, the pattern mixture model measures treatment effects conditional on different drop-out patterns, where, in order to estimate the marginal treatment effects, one needs to calculate the average of conditional likelihoods for the observed data across drop-out patterns.

Little (2008) defines a new class of likelihood-based models, namely mixed-effect hybrid models (MEHMs), based on a new factorization of the likelihood of the outcome process and the drop-out process. Unlike selection models and pattern-mixture models, MEHMs factorize the likelihood of both the outcome and the drop-out processes into the marginal distribution of random effects, the conditional distribution of the drop-out pattern given random effects, and the conditional distribution of the outcome given both random effects and the drop-out pattern. Differently from selection models, where the drop-out process is modeled directly, and from pattern mixture, where the sample is stratified by the missing data patterns and the outcome process is modeled over these patterns, the resulting MEHMs shares features of both. In fact, the MEHM directly models the missing data mechanism, as in the selection model, and shows computational simplicity, as in the pattern mixture model (Yuan and Little 2009).

Suppose to collect K repeated measurements of a count response variable Y and covariates X for each of the n individuals such that $\mathbf{Y}_i = (Y_{i1}, Y_{i2}, \ldots, Y_{iK})$ and $\mathbf{X}_i = (\mathbf{X}_{i1}, \mathbf{X}_{i2}, \ldots, \mathbf{X}_{iK})$ with $\mathbf{X}_{ik} = (X_{ik1}, X_{ik2}, \ldots, X_{ikp})$ denote the associated $K \times p$ covariates matrix. Since we consider only monotone missing data patterns, i.e. irretrievable drop-out, let D_i index drop-out patterns such that $D_i = K$ for complete cases and $D_i = k$ if the subject i drops-out between the kth and $(k+1)$th measurement time, for $k = 1, \ldots, K$; in formulas:

$$D_i = K - \sum_{k=1}^{K} R_{ik} = \sum_{k=1}^{K} (1 - R_{ik})$$

where $R_{ik} = 1$ if the i-th unit drops-out at any point within $(k-1, k)$, $k = 1, \ldots, K$, $R_{ik} = 0$ otherwise.

Let $\mathbf{b}_i$ be the random effects which model the correlation of repeated measurements on the same subject accounting for unobserved heterogeneity due to e.g. omitted covariates or overdispersion. In this way we deal with possible misspecification of the model, summarizing the joint effect of not considered covariates by including a set of unobserved variables. In the MEHM, the factorization of the joint distribution of $\mathbf{Y}_i$, b_i and D_i is:

$$f(D_i, \mathbf{Y}_i, \mathbf{b}_i|\mathbf{X}_i) = f_B(\mathbf{b}_i|\mathbf{X}_i) f_{D|B}(D_i|\mathbf{b}_i, \mathbf{X}_i) f_{Y|D,B}(\mathbf{Y}_i|\mathbf{b}_i, \mathbf{X}_i).$$

In particular, the first two factors model the drop-out process, a feature of mixed-effects selection models, and the third factor models the longitudinal outcome process conditional on the pattern of missing data, a feature of pattern-mixture models. In other words, the attrition is addressed in a straightforward way by the use of potential outcomes with a joint distribution (see e.g. Rubin 2000).

Let us assume that for some link function ζ the following model holds:

$$\zeta\,[E\,(D_i\,|\mathbf{b}_i\,)] = \mathbf{v}_i^{\mathrm{T}}\boldsymbol{\phi} + \mathbf{w}_i^T\mathbf{b}_i$$

where $\mathbf{v}_i$ is a (drop-out-specific) covariate vector, and $\boldsymbol{\phi}$ represents the corresponding vector of model parameters, while $\mathbf{w}_i$ is a (drop-out-specific) covariate whose effect is variable across subjects.

Without loss of generality, we will focus on random effect models, including some form of autoregression; this may help us to distinguish between sources of true and spurious contagion, i.e. between dependence on past outcomes and the effects of individual, unobserved, characteristics.

Let us assume that variables whose effects are fixed and variable across subjects are collected in $\mathbf{x}_{ik}$ and $\mathbf{z}_{ik}$ (respectively); responses Y_{ik}, $i = 1, \ldots, n$, $k = 1, \ldots, K$ are modelled using a linear mixed-effects model defined by the following linear function:

$$\theta_{ik} = \gamma d_i + \mathbf{x}_{ik}^{\mathrm{T}}\boldsymbol{\beta} + \alpha y_{i,k-1} + \mathbf{z}_{ik}^T\mathbf{b}_i, \quad k = 2, \ldots, K_i$$

where K_i is the number of measurements for each unit and θ_{ik} represents the logarithm of the Poisson distribution parameter vector. A different model structure is defined for the first occasion:

$$\theta_{i1} = \mathbf{x}_{i1}^{\mathrm{T}}\boldsymbol{\beta}^* + \mathbf{z}_{i1}^T\mathbf{b}_i^*$$

$\mathbf{b}_i^* = \lambda\mathbf{b}_i$, to account for potential overdispersion in the random effect distribution when the lagged term is not available.

This model specification allows for correlated random effects, as in multivariate mixture models (see e.g. Alfò and Trovato 2004). These (unobservable) latent characteristics control the association between repeated measures in the univariate profiles and the association between the primary response and the drop-out process.

We assume that the random effects are drawn from a bivariate distribution and that, conditional on the random effects, the response variable and the drop-out indicator are independent. These models are sometimes referred to as multivariate multi-factor models (Winkelmann 2000).

Monte Carlo EM algorithm for a linear mixed model with Gaussian random effects (Verzilli and Carpenter 2002) and Laplace approximation (Gao 2004) have been proposed to overcome the high-dimensional integration over the distribution of the random effects; further, numerical integration techniques, such as standard or adaptive Gaussian quadrature, can be used. Various alternative parametric specifications have been proposed for the random terms; however parametric specifications of the mixing distribution can be restrictive and a flexible specification is therefore to be preferred.

Thus, throughout the paper, we leave the distribution of the random effects unspecified and a non-parametric maximum likelihood (NPML) estimation of the mixing distribution can be achieved in a general finite mixture framework (for more details on the computational aspects see e.g. Aitkin 1999). Doing so, the model reduces to a finite mixture model, where the number of components G is unknown and needs to be estimated along with other model parameters. The use of finite mixtures has several significant advantages over parametric models; for instance, the discrete nature of the estimate helps to classify subjects in clusters characterized by homogeneous values of random parameters. This is particularly appealing in social and health sciences, where components can be interpreted as groups with similar behavior or propensity towards the event of interest. Computational aspects for a strictly related model are given in Alfò and Maruotti (2009).

Our operative model, finally, results in a three-equation structure. We model, respectively, the student performance, the drop-out process and, for taking into account the initial condition issue, the student performance in the first period at the university.

3 Data and Variables

In order to perform the empirical analysis, we use student level data provided by the administrative offices of the faculty of Statistics at Sapienza University of Rome. Specifically, our dataset covers 159 undergraduate students enrolled in the academic year 2003/2004 in a three-year bachelor program. The administrative data that we use collect all the students' information recorded at the moment of enrolling along with follow-up information on individual exams. This ensures that the data do not contain missing information, since all the information recorded must be provided by the student (see e.g. Belloc et al. 2010). Moreover, although such data lack information on parents' educational background and other potentially relevant information, they provide records of the students' exam marks, by means of which the student performance can be measured.

In this paper, we consider the student performance as the primary response. Of course, student performance may be defined in a variety of ways, what we decide to consider is the average mark as it is obtained considering all the exams passed by the individual student up till a certain moment. In particular, we calculate this indicator for four-month periods within the first three years of the degree course, so that we observe the variations of the individual student performance three times per year. Consequently, we obtain 9 observations for those students that conclude the third year of their degree course, while students that leave the degree program before the third year show a lower number of observations. Notice that the drop-out rate after the third year is ignorable. As the response variable we use an average mark index expressed as an average weighted by the number of credits that each exam assigns, where one credit is equivalent to 25 h of student's workload according to the European Credit Transfer System (ECTS).

In our model, we explicitly include the faculty drop-out and the related missing data generation process. Since our dataset refers to one single faculty, we do not make any distinction between students that withdraw from the university and those who change faculty within the athenaeum, though there can be some differences between the two groups (see Belloc et al. 2011), what however is not really matter of concern here. Using administrative data, furthermore, we can consider the effective drop-out, rather than the formal drop-out, since we can detect also those students that did not record their withdrawal to the administrative office but that did not renewed their registration in the following academic year, *de facto* dropping-out of the faculty.

In our analysis, we relate the student performance to personal characteristics of students. Thus, we include the following set of variables in the analysis: sex, place of residence, high school final mark, type of high school diploma, household income and the latency period. In particular, the household income is measured by means of a synthetic indicator of the household economic situation (ISEE), that is calculated as the sum of the household income and the 20% of the household wealth, weighted by the number of household members; while the latency period is defined as the number of years between the date of high school graduation and that of university enrollment. Finally, we include the one-period-lagged performance index as one of the covariates in the primary response equation.

Given this variables' definition, our dataset is composed as follows: 44% of the students are male, while 65% reside in Rome, where "Sapienza" is located. The average age of students is 22 years. With respect to the educational background, 56% of the students have attended a general high school (*liceo*) and the average high school mark is 85.45 (where minimum and maximum values are, respectively, 60 and 100). Moreover, students enroll to the university, on average, less than one year after their high school graduation, being the average latency period 0.73 years. Finally, in our dataset, only 63% of students conclude the third year of their degree course. Descriptive statistics are collected in Table 1, where also the distributions of students among different classes of ISEE and high school mark are considered.

Table 1 Data description

Variable	Percentage frequencies
University Performance	20.30 (mean)
Retention up to the 3rd Year	62.89
Sex: Male	44.02
Sex: Female	55.97
Place of Residence: Rome	65.40
Place of Residence: Other	34.59
Household Economic Situation: ISEE < 10000 euros	31.37
Household Economic Situation: 10000 euros < ISEE < 20000 euros	37.73
Household Economic Situation: 20000 euros < ISEE < 30000 euros	14.46
Household Economic Situation: ISEE > 30000 euros	16.35
Latency Period (years)	0.73 (mean)
Type of High School Diploma: General	56.60
Type of High School Diploma: Other Diploma	43.39
High School Mark: Very Low	15.72
High School Mark: Low	22.01
High School Mark: Middle	21.38
High School Mark: High	40.88

4 Results

In Table 2 we show the empirical results of the analysis. Results from the primary response equation, reported in the second and third columns, show interesting relations. Personal characteristics of students such as sex and place of residence do not have any statistically significant effect on the student performance, what contrasts previous evidence provided, for example, by Elmore and Vasu (1980) and Schram (1996) among others. Differently, the one-period-lagged response variable and the length of the retention at the university affect performance of students in a statistically significant way. So, two things may be argued. On the one hand, students that show high exam marks in a certain four-month period tend to continue to perform well. This means that students that systematically organize their workload have also a continued high level performance; conversely, students which show bad results in a certain moment of their degree course are probably trapped in a low performance steady state. On the other hand, those who perform better are also those who retain longer in the university; phrased differently, who are likely to conclude the third year of the degree course are those students that show the highest exam results. Income classes, moreover, are associated to negative sign, being the low income class the benchmark. The educational background, also, is relevant, given that having general high school diploma and showing good high school final marks positively affect the student performance at the university.

As one can notice from the results reported in the fourth and fifth columns of Table 2, our analysis unveils that the drop-out process is not an exogenous mechanism, but it is in fact dependent on individuals' characteristics. Specifically,

Table 2 Estimation results

Variable	Performance		Drop-out		First period performance	
	Coef.	Std.Err.	Coef.	Std.Err.	Coef.	Std.Err.
Lagged University Performance	0.757	0.020[c]				
Length of Retention	0.391	0.080[c]				
Sex (being male)	0.059	0.248	−0.129	0.213	0.127	0.589
Rome as Place of Residence	0.397	0.248	0.408	0.233[a]	−0.074	0.620
ISEE < 10000 euros	Benchmark		Benchmark		Benchmark	
10000 euros < ISEE < 20000 euros	−0.611	0.359[a]	−0.471	0.275[a]	0.002	0.814
20000 euros < ISEE < 30000 euros	−0.469	0.397	−0.559	0.317[a]	0.154	0.918
ISEE > 30000 euros	−1.063	0.372[c]	−0.334	0.273	−0.897	0.830
Latency Period	−0.016	0.052	0.133	0.043[c]	0.079	0.117
General High School Diploma	0.471	0.259[a]	−0.612	0.215[c]	1.053	0.607[a]
High School Mark: Very Low	Benchmark		Benchmark		Benchmark	
High School Mark: Low	2.724	0.531[c]	−0.352	0.307	0.931	1.059
High School Mark: Middle	2.753	0.550[c]	−0.477	0.332	0.916	1.085
High School Mark: High	2.636	0.533[c]	−0.899	0.326[c]	2.560	1.051[b]
Constant	0.360	0.763	−1.123	0.411[c]	15.232	1.524[c]

Note: statistical significance level: "*a*" 10%, "*b*" 5%, "*c*" 1%

the probability of dropping-out of the faculty increases when the student resides in Rome, where Sapienza is located, and when the latency period is long. Moreover the withdrawal probability is lower when students have attended general high schools rather than having other types of diploma, when they show good high school final marks and when they belong to the upper income classes. Income has negative effects in both equations; this is not surprising, since, as we have seen from preliminary analysis, a large fraction of students can be defined as *parking students* retaining university without giving exams. With respect to the student performance in the first period, finally, we find that it is positively affected by the educational background. So, the educational background of students seems to have a persisting effect on their proficiency throughout the higher education career, this keeping effectual the differences deriving from heterogeneous educational provenances.

As a by-product, we are able to distinguish three different types of student behavior with respect to the academic performance and the drop-out mechanism. As expected, those who are more likely to show a higher performance than the mean seem to be less likely to drop-out. In the same way, students with a lower propensity towards a good performance show a higher propensity to withdraw from the program. Interestingly, the third group identifies students who have a high propensity to retain even if this is associated with a low propensity to obtain good results in their career.

5 Conclusions

Our results allow us to argue that the student drop-out of the university may be generated by a latent process connected with the primary response mechanism. Indeed, those students who are lost to follow-up do not randomly drop-out of the sample, but seem to do so depending on observable and unobservable individuals' characteristics. Our empirical strategy, built on mixed-effect hybrid models (see Little 2008), tackles this problem and leads to parameter estimates that are more robust than those obtained otherwise, to the extent that we model the student performance equation jointly with an underlying drop-out mechanism. Thus, a misleading interpretation of the estimated parameters due to a (possibly) non-ignorable missing data is circumvented, and we are confident that our findings can be interpreted in a causal sense. Accordingly, we conclude suggesting that future research on university student performance should take into account also the student withdrawing. Evidence obtained through a joint modeling of the two processes, for other fields of study and at various stages of education, may be more informative than existing studies.

References

Aitkin, M. (1999). A General Maximum Likelihood Analysis of Variance Components in Generalized Linear Models. *Biometrics*, 55:117–128.

Alfò, M. and Maruotti, A. (2009). A selection model for longitudinal binary responses subject to non-ignorable attrition. *Statistics in Medicine*, **28**: 2435–2450.

Alfò, M. and Trovato, G. (2004). Semiparametric mixture models for multivariate count data, with application. *Econometrics Journal*, **2**:426–454.

Belloc, F., Maruotti, A. and Petrella, L. (2010). University drop-out: An Italian experience. *Higher Education*, 60: 127–138.

Belloc, F., Maruotti, A. and Petrella, L. (2011). How Individual Characteristics Affect University Students Drop-out: a Semiparametric Mixed-Effects Model for an Italian Case Study. *Journal of Applied Statistics*, 38(10): 2225–2239.

Devadoss, S. and Foltz, J. (1996). Evaluation of Factors Influencing Student Class Attendance and Performance, *American Journal of Agricultural Economics*, 78:499–507.

Elmore, P.B. and Vasu, E.S. (1980). Relationship between Selected Variables and Statistics Achievement: Building a Theoretical Model, *Journal of Educational Psychology*, 72:457–467.

Gao, S. (2004). A shared random effect parameter approach for longitudinal dementia data with non-ignorable missing data. *Statistics in Medicine*, **23**:211–219.

Little, R.J.A. (2008). Selection and Pattern-Mixture Models, in *Advances in Longitudinal Data Analysis*, G. Fitzmaurice, M. Davidian, G. Verbeke and G. Nolenberghs (eds.), London: CRC Press.

Little, R.J.A. and Rubin, D.B. (2002). *Statistical Analysis with Missing Data*, 2nd edition, New York: Wiley.

Rubin, D.B. (2000). The Utility of Counterfactuals for Causal Inference - Discussion of Causal Inference Without Counterfactuals by A. P. Dawid, *Journal of the American Statistical Association*, 95:435438.

Schram, C.M. (1996). A Meta-Analysis of Gender Differences in Applied Statistics Achievement, *Journal of Educational and Behavioral Statistics*, **21**:55–70.

Verzilli, C.J. and Carpenter, J.R. (2002). A Monte Carlo EM algorithm for random coefficiente-based dropout models. *Journal of Applied Statistics*, **29**:1011–1021

Yuan, Y. and Little, R.J.A. (2009). Mixed-Effect Hybrid Models for Longitudinal Data with Non-ignorable Dropout, *Biometrics*, **65**:478–486.

Winkelmann, R. (2000). Seemingly unrelated negative binomial regression. *Oxford Bullettin of Economics and Statistics*, **62**:553–560.

Zimmer, J. and Fuller, D. (1996). Factors Affecting Undergraduate Performance in Statistics: A Review of the Literature, paper presented at the Annual Meeting of the Mid-South Educational Research Association, Tuscalosa (AL), November.

Risk Analysis Approaches to Rank Outliers in Trade Data

Vytis Kopustinskas and Spyros Arsenis

Abstract The paper discusses ranking methods for outliers in trade data based on statistical information with the objective to prioritize anti-fraud investigation activities. The paper presents a ranking method based on risk analysis framework and discusses a comprehensive trade fraud indicator that aggregates a number of individual numerical criteria.

1 Introduction

The detection of outliers in trade data can be important for various practical applications, in particular for prevention of the customs fraud or data quality. From the point of view of a customs inspector, trade transactions detected as outliers may be of interest as due to possible on-going fraud activities. For example, low price outliers might indicate that the specific transaction is undervalued to evade import duties. As another example, low and high price outliers may be indicators of other frauds: VAT fraud or trade based money laundering.

The statistical algorithms used to detect outliers in large trade datasets typically produce high number of transactions classified as outliers (Perrotta et al. 2009). Large number of transactions flagged as suspicious are difficult to handle. Therefore the detected outliers must be ranked according to certain criteria in order to prioritize the investigation actions. Different criteria could be used for ranking purposes and they are derived from at least two very distinct information sources: statistical information of outlier diagnostics and customs in-house information systems.

V. Kopustinskas (✉) · S. Arsenis
European Commission, Joint Research Center, Institute for the Protection and Security of the Citizen, Via E. Fermi 2748, Ispra (VA), Italy
e-mail: vytis.kopustinskas@jrc.ec.europa.eu

A. Di Ciaccio et al. (eds.), *Advanced Statistical Methods for the Analysis of Large Data-Sets*, Studies in Theoretical and Applied Statistics,
DOI 10.1007/978-3-642-21037-2_13,

This paper discusses low price outliers ranking methods based on statistical information with the objective to prioritize anti-fraud investigation actions. The presented methodology to rank low price outliers is not generic, but can be extended to other type of fraud patterns by using the same principles.

2 Risk Analysis Framework

The risk analysis framework is applicable to the ranking problem of outliers in trade data. The fundamental questions in quantitative risk analysis are the following:

1. What can go wrong?
2. How likely it will happen?
3. If it happens, what consequences are expected?

To answer question 1, a list of initiating events should be defined. The likelihood of the events should be estimated and the consequences of each scenario should be assessed. Therefore, quantitatively risk can be defined as the following set of triplets (Kaplan and Garrick 1981):

$$R = < S_i, P_i, C_i >, \quad i = 1, \ldots, n. \tag{1}$$

where S_i–ith scenario of the initiating events; P_i – likelihood (probability or frequency) of the scenario i; C_i – consequence of the scenario; n – number of scenarios.

In case of outliers in trade data, the triplet can be interpreted in the following way: P – likelihood that an outlier is a real fraudulent transaction; C – consequence of the fraudulent trade transaction (e.g. unpaid taxes or duties). The interpretation of S can be important only if several methods are used to detect outliers or more than one fraud pattern is behind the observed outlier.

3 Approaches to Rank Low Price Outliers in Trade Data

There are several approaches to obtain a numerical ranking of low price outliers in trade data based on the risk analysis framework. The most suitable method is to multiply P and C, where C is an estimate of the loss to the budget (unpaid duties) and P is probability that the specific transaction is a fraud. The multiplication $R = P \times C$ provides an average fraud related damage estimate caused by specific trade activity and allows ranking them according to their severity.

In order to use the risk analysis framework (1) we have to estimate the probability (P) of fraud in the trade transaction. This is not an easy quantity to estimate, but we assume that p-value produced by statistical tests for outliers could be a suitable measure. It means that the lower the p-value, the more likely the transaction is fraudulent. In practice, this can also be a data typing error and not a fraud, but in general low price outliers are suspicious and should be investigated.

As the relationship between P and p-value is reverse, the transformation is used:

$$\begin{cases} P = \frac{-\log_{10}(pvalue)}{10}, & \text{if pvalue} \geq 10^{-10} \\ P = 1, & \text{if pvalue} < 10^{-10} \end{cases} \tag{2}$$

By transformation (2) the p-value is transformed into scale [0, 1]. The scale here is arbitrary and chosen mainly for the purpose of convenience, driven by the fact that extremely low p-values are no more informative for the ranking purposes.

The consequence part (C) of (1) can be estimated by multiplying the traded quantity (Q) and transaction unit price difference (ΔU) from the recorded to the estimated "fair" price: $C = Q \times \Delta U = Q \times (UF - U)$, where UF – the estimated "fair" transaction unit price determined by the regression after outliers have been removed; U – the unit price as recorded $(U = V/Q)$; V – value of the transaction as recorded. The interpretation of C is an average loss to the budget if the underlying transaction is fraudulent. In fact, (C) value already provides a ranking of outliers and such a ranking has been applied.

The fraud risk (RI) can be computed as follows: $\text{RI} = P \times Q \times \Delta U$. The indicator can also be transformed into the [0, 10] scale, as to make its use more standard for investigators. The investigator should start the investigation process from the outlying trade transactions with the highest values of RI.

The RI is a simple and easy to understand indicator, however the dataset of detected outliers contains additional statistical information. Table 1 provides a number of criteria which could be used for the development of a comprehensive ranking structure for low price outliers.

The criteria listed in Table 1 are all numerical and their higher value is associated with the higher impact to trade fraud risk indicator (FI). Most of the criteria $(I_1 - I_7)$ are easy to understand and compute as they reflect basic statistical information about the dataset. The criterion I_8 reflects inferential statistics from the method that was

Table 1 Numerical criteria for the development of ranking structure, indicating their original scale and rescaling method. V_{PO} – Trade value by aggregating all destinations; Q_{PO} – Trade quantity by aggregating all destinations; MaxN – maximum number of non-zero trade transactions

No	Criteria	Original scale	Rescaling
I_1	Quantity traded, Q	$[0, \infty]$	log and in-max translformation to [0, 1]
I_2	Value traded, V	$[0, \infty]$	log and min-max transformation to [0, 1]
I_3	Average loss, $Q \times \Delta U$	$[0, \infty]$	log and min-max transformation to [0, 1]
I_4	Ratio UF/U	$[0, \infty]$	log and min-max transformation to [0, 1]
I_5	Ratio V/V_{PO}	[0, 1]	No
I_6	Ratio Q/Q_{PO}	[0, 1]	No
I_7	Number of obs./MaxN	[0, 1]	No
I_8	P-value	[0, 0.1]	log transformation to [0, 1] as in (2)
I_9	Final goodness of fit R^2	[0, 1]	No
I_{10}	Final R^2/initial R^2	$[0, \infty]$	log and min-max transformation to [0, 1]

used to detect outliers. The ranking structure can be adapted to other price outlier detection methods by using the same principles.

The criteria I_9 and I_{10} take into account the model goodness of fit by using the coefficient of determination of the linear regression for Q versus V variables. The initial R^2 is computed on all the trade data in a particular trade flow assuming linear regression as the base model, while the final R^2 is computed on the remaining data points after removal of the outliers. The initial R^2 and the final R^2 are used as it might be important to reflect for the change in the model goodness of fit after removal of the outliers.

After rescaling as shown in Table 1, the individual criteria (I_i) are comparable among themselves. The log-transformation was used for a number of criteria to make the ranking smoother, because for many real data cases it is rather stair-like. The criteria when transformation is actually needed could be more elaborated in the future.The specific weights (w_i) must be assigned to each criterion to determine its relative impact to the final indicator score. The most popular method to combine different criteria into a single numerical indicator is to compute a weighted sum: $FI = \sum_{i=1}^{m} w_i \times I_i$, where m – number of individual criteria.

A complication arises from the fact that some criteria could be highly correlated and therefore their correlation matrix must be examined before assignment of weights to each criterion. Without prior knowledge, equal weights could be assigned to non-correlated criteria. However, the correlation matrix analysis is not enough and weights cannot be derived from statistical considerations only, but must by defined by subject matter experts and be closely related to the specific type of fraud in mind. One possible method is analytic hierarchy process (AHP) which provides a rational framework for integrating opinions of many subject matter experts into a single quantitative estimate (Zio 1996).

The list of possible criteria presented in Table 1 is not complete and could be modified in future applications. One possible type of analysis that could improve the ranking structure is correspondence analysis. As various associations between trade flow variables could be important and quantitative information about the existing links in the data could be integrated in the ranking: for example, quantitative relationship between products and exporting countries (for import datasets) among all the outliers could be important to determine whether fraudulent activities might be linked to specific origin countries or products.

The presented ranking methodology was developed for the situation when trade data represent a single population and low price outliers are detected within it assuming linear regression to be a model of the data. However, the problem of ranking becomes more interesting when outliers are detected in the mixtures of populations (Atkinson et al. 2004): when several populations of different price levels exist and it is not obvious from which populations the outliers are detected. It is an important problem in fraud detection, where fraudulent transactions are hidden within a mixture of several populations. For example, continuous systematic under-pricing of selected imports into one country could not be detected by doing single country analysis. In the case of mixed populations, the ranking structure needs to be further developed.

4 Application of the Ranking Criteria

The ranking was applied for the low price outliers detected in the monthly aggregated trade data of agricultural product imports into the EU 27 member states during 2006–2008 (dataset containing: product, reporting countries, volume and value). In total, 1,109 low price outliers were detected by using backward search based outlier detection method. The numerical criteria as shown in Table 1 were computed and their mutual correlation is presented in Table 2.

As evident from Table 2, several pairs are highly correlated (higher than 0.6). It is not surprising that quantity and value based numerical criteria (I_1, I_2, I_5 and I_6) are highly correlated because larger quantity trade transactions are associated with larger value transactions. Inclusion of all these criteria in the ranking at equal weights would have double counting effect on the total score. In customs fraud, misdeclaration of value happens to be much more frequent than misdeclaration of quantity. Considering this, the numerical criteria I_2 (value traded) and I_5 (ratio of value) should be eliminated from the ranking structure. In fact, the decision to eliminate them could have been done before the computations (following the reasoning as above).

The high correlation of quantity (I_1) and average loss (I_3) is also expected as average loss is a function of quantity. In this case the weight can be equally divided between the two criteria. The same approach can be used for the remaining two highly correlated numerical criteria – I_4 and I_{10}. This correlation is very interesting: ratio of fair price versus recorded price gives similar information as ratio of final model (without outliers) R^2 versus initial (all the data) model goodness of fit R^2. In this case, equal weights of 0.5 were applied.

Table 2 Correlation matrix of the numerical criteria $I_1 - I_{10}$ for the selected application

	I1	I2	I3	I4	I5	I6	I7	I8	I9	I10
*I*1	1.00	0.79	0.70	−0.07	0.20	0.13	0.12	0.13	0.05	−0.02
*I*2	**0.79**	1.00	0.74	−0.34	0.22	0.01	0.22	0.05	0.06	−0.20
*I*3	**0.70**	**0.74**	1.00	0.33	0.05	0.23	0.20	0.19	−0.15	0.25
*I*4	−0.07	−0.34	0.33	1.00	−0.20	0.37	−0.06	0.19	−0.27	0.69
*I*5	0.20	0.22	0.05	−0.20	1.00	0.76	−0.38	−0.06	0.12	−0.23
*I*6	0.13	0.01	0.23	0.37	**0.76**	1.00	−0.38	0.05	−0.07	0.22
*I*7	0.12	0.22	0.20	−0.06	−0.38	−0.38	1.00	0.22	−0.26	−0.06
*I*8	0.13	0.05	0.19	0.19	−0.06	0.05	0.22	1.00	0.07	0.16
*I*9	0.05	0.06	−0.15	−0.27	0.12	−0.07	−0.26	0.07	1.00	−0.24
*I*10	−0.02	−0.20	0.25	**0.69**	−0.23	0.22	−0.06	0.16	−0.24	1.00

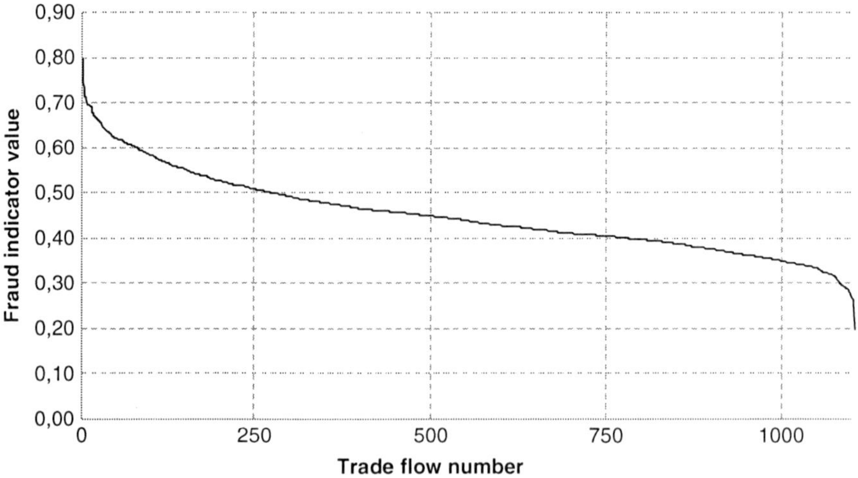

Fig. 1 Ranking of the detected outliers in the EU import trade data (sorted decreasingly)

Table 3 The weighting of the ranking structure.

No	Criteria	Weight w_i
I_1	Quantity traded, Q	0.5
I_2	Value traded, V	0
I_3	Average loss, $Q \times \Delta U$	0.5
I_4	Ratio UF/U	0.5
I_5	Ratio V/V_{PO}	0
I_6	Ratio Q/Q_{PO}	1
I_7	Number of obs/(maxN = 36)	1
I_8	P-value	1
I_9	Coefficient of determination R^2	1
I_{10}	Final R^2/initialR^2	0.5

The weights applied for the ranking procedure are shown in Table 3. The ranking indicator value can be further normalized to scale [0, 1] by dividing by the sum of weights.

The computed FI values are shown in Fig. 1. It reflects typical in risk rankings Pareto distribution, where the highest risk is associated with a small number of outliers, while the risk of the rest is distributed more smoothly. The highest and the lowest ranked trade outliers are shown in Figs. 2 and 3. The results of the ranking are as expected: the highest ranked outliers are severe outliers in terms of low price being far away from the regression line and the lowest ranked outliers – being close to it.

The next step to improve the ranking procedure would be to validate the ranking based on real fraud cases and involve fraud experts in the process of weight

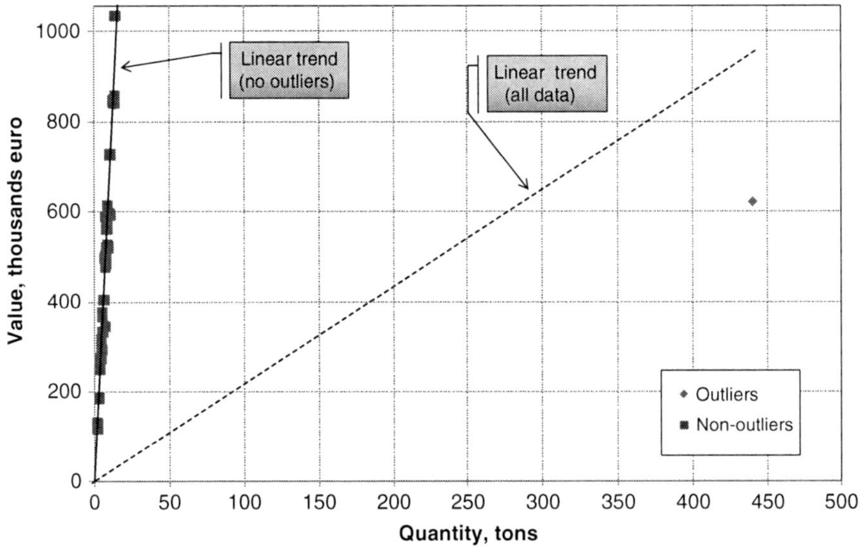

Fig. 2 The highest ranked low price outlier (EU import trade dataset)

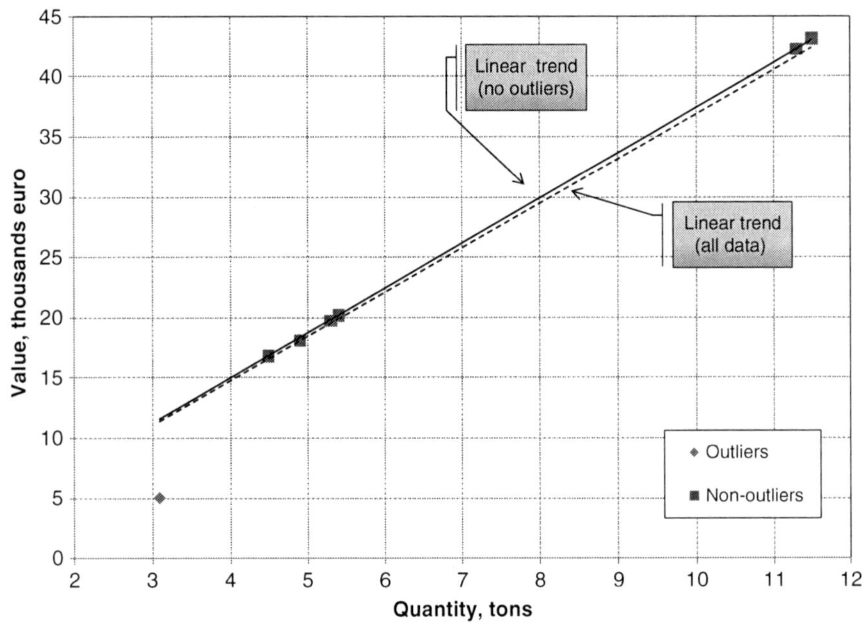

Fig. 3 The lowest ranked low price outlier (EU import trade dataset)

estimation. Both options require a lot of resources for implementation and especially feedback for the ranking validation.

Preliminary validation information suggests that severe price outliers could be more linked to data errors than fraudulent activities. Further development of the ranking structure by adding other indicators could address the data quality issues.

5 Final Remarks

The paper discusses the trade data outliers ranking methods with the objective to prioritize anti-fraud investigation actions. The risk analysis framework was used to develop a ranking structure based only on available statistical information in trade dataset. A comprehensive trade fraud risk indicator is discussed that combines a number of individual numerical criteria. An application study is presented that produced a ranking of the detected outliers in the aggregated European import trade data during 2006–2008. The ranking produced cannot be considered as final due to arbitrary weights that were used for the computations. Derivation of weights is an important part of the ranking methodology, however it cannot be produced only by statistical considerations. Subject matter expert opinions would be valuable to define the weights based on the type of fraud under investigation. The results of the test study show that even arbitrary weights can produce reasonable results. Further fine-tuning of the methodology is depended on feedback and judgments from fraud experts.

References

Atkinson A.C., Riani M., Cerioli A. (2004) *Exploring Multivariate Data With the Forward Search*, Springer, New York.

Perrotta D., Riani M. and Torti F. (2009) New robust dynamic plots for regression mixture detection. Advances in Data Analysis and Classification, 3(3), 263–279.

Kaplan, S., Garrick, B. J. (1981) On the quantitative definition of risk, *Risk Analysis*, 1(1), 11–27.

Zio E. (1996) On the use of the analytic hierarchy process in the aggregation of expert judgments, *Reliability Engineering and System Safety*, 53(2), 127–138.

Problems and Challenges in the Analysis of Complex Data: Static and Dynamic Approaches

Marco Riani, Anthony Atkinson and Andrea Cerioli

Abstract This paper summarizes results in the use of the Forward Search in the analysis of corrupted datasets, and those with mixtures of populations. We discuss new challenges that arise in the analysis of large, complex datasets. Methods developed for regression and clustering are described.

1 Introduction

Data are an overwhelming feature of modern life. As the amount of data increases so do the challenges facing the statistician in trying to extract information from ever larger data sets. We argue that larger data sets are also more complex and require flexible multiple analyses in order to reveal their structure. Only then can all information be efficiently extracted.

The analysis of large data sets may be complicated by the high dimensionality of responses, large numbers of observations and complexity of the choices to be made among explanatory variables. Although appreciable, these challenges to the statistician are not different in kind from those faced in the analysis of smaller data sets. We, however, focus on problems that become severe in large complex data sets, such as our inability to find a single model for all the data.

The simplest situation is that of a single model with possibly many outliers. In the presence of a core population and some isolated or clustered outliers, traditional robust methods (Maronna et al. 2006) can be successfully used to find proper models. However, when there are several populations and different sources of

M. Riani (✉) · A. Cerioli
Department of Economics, University of Parma, Italy
e-mail: mriani@unipr.it; andrea.cerioli@unipr.it

A. Atkinson
Department of Statistics, London School of Economics, London WC2A 2AE, UK
e-mail: a.c.atkinson@lse.ac.uk

A. Di Ciaccio et al. (eds.), *Advanced Statistical Methods for the Analysis of Large Data-Sets*, Studies in Theoretical and Applied Statistics,
DOI 10.1007/978-3-642-21037-2_14,

heterogeneity, traditional robust methods fail to recover the real structure of the data and more sophisticated procedures, such as those derived from the Forward Search (FS) (Atkinson and Riani 2000; Atkinson et al. 2004) are required. Some examples are presented in the next section. Section 3 introduces an example of complex data in a situation where automatic procedures need to be developed. We conclude with a longer example of robust model building.

2 Some Difficulties in Data Analysis

2.1 The Presence of Outliers

The presence of atypical observations may strongly and wrongly influence the output of statistical analyses. When the number of observations is large it is likely that there will be several atypical observations which mask one another. They will not be revealed by a single static analysis, although the dynamic analysis of many subsets of data through the FS will reveal such structure. However, the outliers should not be seen only as bad observations that estimation procedures must avoid; they may themselves contain valuable information. The discovery of the hole in the ozone layer is one example. In drug development, the existence of a subset of individuals with an adverse reaction to the drug might be one target of the analysis.

2.2 Calibration of Test Procedures

Particulary as the sample size grows, it is necessary to calibrate tests of the outlyingness of individual observations. The repeated application of statistical tests makes it necessary to calibrate for simultaneity. Use of procedures that are correctly calibrated to provide tests of the desired size will keep false alarms under control (Riani et al. 2009).

2.3 Subpopulations

Large datasets often contain hidden groups, which are not revealed by application of single population methods, even in their robustified forms. For multivariate data there are well established techniques of cluster analysis, which may work well for normal populations. However, automatic methods such as MCLUST (Fraley and Raftery 1999) for establishing cluster membership often indicate too many clusters. Standard clustering procedures are also badly affected by the presence of outliers. Procedures based on the forward search have been shown to work well in identifying clusters and establishing cluster membership, even in the presence of outliers, but are far from automatic, requiring appreciable input from the statistician.

3 An Example of Large Complex Corrupted Data

As an illustration of the problems involved with the analysis of complex data, consider the example given in Fig. 1 referred to the quantity (x) and the value (y) of 4719 import declarations of a specific technological product. This is an example of one of the thousands of datasets provided by the "Office Européen de Lutte Anti-Fraude" (OLAF) or by its partners in the Member States. The purpose is to find atypical transactions, which might correspond to cases of potential fraud (e.g. the evasion of import duties) or to potential money laundering activities.

The observations appear roughly distributed along three main lines departing from the origin of the coordinate axes. However, there seem also to be horizontal strips of concentrated data. It is certainly not clear how many groups are present in the data. Traditional methods which assume one single regression population will fail in revealing the real structure as will their robust counterparts. The general structure is of a mixture of linear models heavily contaminated by observations that do not follow the general pattern (Riani et al. 2008). Outliers may be isolated, originating from recording errors during the data collection process, or they may be clustered, when they represent some systematic behaviour. In the context of anti-fraud the outliers themselves are important. However, the size of any outlier tests needs to be calibrated: prosecutions which fail encourage fraudsters while law enforcement agencies will become discouraged.

Use of the "multiple random starts forward search" (Atkinson and Riani 2007) enables us to dissect these data into components and outliers. However, the clustering of regression lines is again a procedure that involves considerable statistical intervention. The context of anti-fraud indicates several important directions for statistical development.

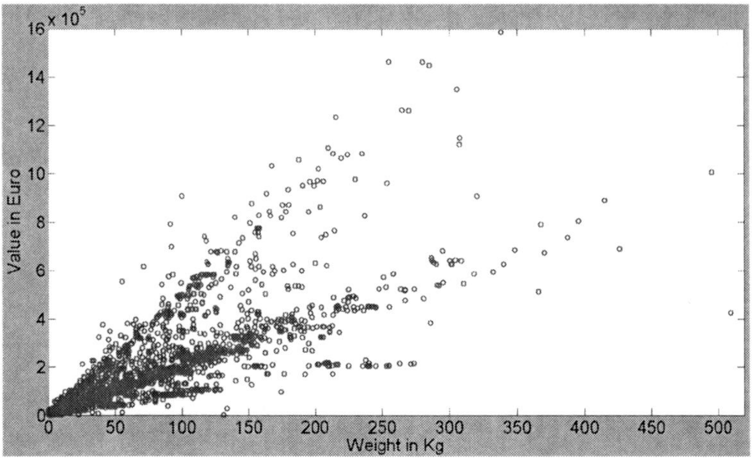

Fig. 1 An example of international trade data

4 Forward Directions for the Forward Search

4.1 Automatic Classification Procedures

While for a small number of datasets it is possible to envisage human intervention for each dataset, including the use of exploratory data analysis tools, in the presence of a huge amount of data only automatic procedures are feasible. These developments are required both for the clustering of multivariate data mentioned in Sect. 2 and for the mixtures of regression lines of Sect. 3.

4.2 Timeliness and On-Line Systems

The context of anti-fraud data analysis motivates the need for timeliness, which may only be achievable through on-line analysis. If a fraud is being committed it needs to be detected and prevented as quickly as possible. An important challenge in on-line analysis is to disseminate the results in a form that is again understandable by the final users. The importance of timeliness and on-line systems accentuates the need for research into the theoretical and practical aspects of dynamic updating methods.

4.3 Automatic Model Selection Procedures

For simple regression models with several explanatory variables the FS provides a robust form of the Cp statistic for selection of regression models. However, for high-dimensional data, the phrase "model selection" refers in addition to the choices of distribution of the responses and the functional form between the explanatory variables and the response.

5 Choosing Regression Models with Mallow's C_p

The analysis of the issues raised in the previous section requires book-length treatment. In this section we concentrate on the issue of model selection to illustrate how a robust flexible trimming approach (specifically that provided by the forward search), makes it possible to get inside the data in a manner impossible using standard statistical methods, be they robust or non-robust.

5.1 Background and Aggregate C_p

Mallows' C_p is widely used for the selection of a model from among many non-nested regression models. However, the statistic is a function of two residual sums of squares; it is an aggregate statistic, a function of all the observations. Thus C_p suffers from the well-known lack of robustness of least squares and provides no evidence of whether or how individual observations or unidentified structure are affecting the choice of model. In the remainder of this paper we describe a robust version of C_p that relies on the forward search to choose regression models in the presence of outliers. Theoretical details are given by Riani and Atkinson (2010). Here we provide a brief survey of the main results, before concentrating on a complex example.

There are n univariate observations y. For the linear multiple regression model $y = X\beta + \epsilon$, X is an $n \times p$ full-rank matrix of known constants, with ith row x_i^T. The normal theory assumptions are that the errors ϵ_i are i.i.d. $N(0, \sigma^2)$. The residual sum of squares from fitting this model to the data is $R_p(n)$.

In the calculation of C_p, σ^2 is estimated from a large regression model with $n \times p^+$ matrix X^+, $p^+ > p$, of which X is submatrix. The unbiased estimator of σ^2 comes from regression on all p^+ columns of X^+ and can be written $s^2 = R_{p^+}(n)/(n - p^+)$. Then

$$C_p = R_p(n)/s^2 - n + 2p = (n - p^+)R_p(n)/R_{p^+}(n) - n + 2p. \quad (1)$$

Provided the full model with p^+ parameters and the reduced model with p parameters yield unbiased estimates of σ^2, it follows that $\mathrm{E}(C_p)$ is approximately p.

Models with small values of C_p are preferred. Statements are often made that those models with values of C_p near p are acceptable. However, we find it helpful to use the distribution of the statistic which Mallows (1973) shows is a scaled and shifted F.

5.2 The Forward Search and Forward C_p

The forward search for a single regression model fits subsets of observations of size m to the data, with $m_0 \leq m \leq n$. Least squares on the subset of m observations yields a residual sum of squares $R_p(m)$. The C_p criterion (1) for all observations is a function of the residual sums of squares $R_p(n)$ and $R_{p^+}(n)$. For a subset of m observations we can define the forward value of C_p as

$$C_p(m) = (m - p^+)R_p(m)/R_{p^+}(m) - m + 2p. \quad (2)$$

For each m we calculate $C_p(m)$ for all models of interest.

Some care is needed in interpreting this definition. For each of the models with p parameters, the search may be different, and outliers, if any, may not enter in the same order for all models.

The distributional results of Mallows apply when C_p is calculated from the full sample. But, in the forward search with $m < n$ we order the observations during the search and take the central m residuals to calculate the sums of squares $R_{p^+}(m)$ and $R_p(m)$. These sums of squares are accordingly based on truncated samples and will have smaller expectations than those based on a full sample of m observations. However, Riani and Atkinson (2010) show that the full sample distribution holds to a good approximation with n replaced by m. That is

$$C_p(m) \sim (p^+ - p)F + 2p - p^+, \qquad \text{where} \qquad F \sim F_{p^+-p,m-p^+}, \tag{3}$$

which is Mallows' result with a change in the degrees of freedom of the F distribution.

6 Credit Card Data

6.1 Background and Aggregate Model Selection

As an example we analyse data on factors influencing the use of credit and other cards. The data are appreciably larger and more complex than those customarily used to illustrate aggregate C_p. There are 1,000 observations on the most active customers of a bank operating in the north of Italy. There is one response and nine explanatory variables which are listed in Appendix A.

Figure 2 gives the traditional C_p plot for these data, showing only those models with the smallest values of C_p for values of p from 3 to 9. The minimum value of C_p is for $p = 6$, for a model containing a constant and variables 1 2 4 5 and 6. There is a pleasing structure to this plot in that there is a well defined series of submodels that have the smallest C_p values. For $p = 5$ we have similar values for 1 2 4 and 5 and for 1 2 4 and 6. For $p = 3$ the best model seems to be 1 2 and 4, with 1 and 2 best for $p = 3$, although the value of C_p for this model lies above the 97.5% bound.

The models suggested by the plot are summarised in Table 1 in which the hierarchy of suggested models is clearly indicated. The table indicates that statistical procedures for checking the model should start with $p \leq 6$.

6.2 The Generalized Candlestick Plot

When we apply the forward search to model selection we obtain a forward plot of $C_p(m)$ for each model. Thus the points in Fig. 2 are replaced by the curves

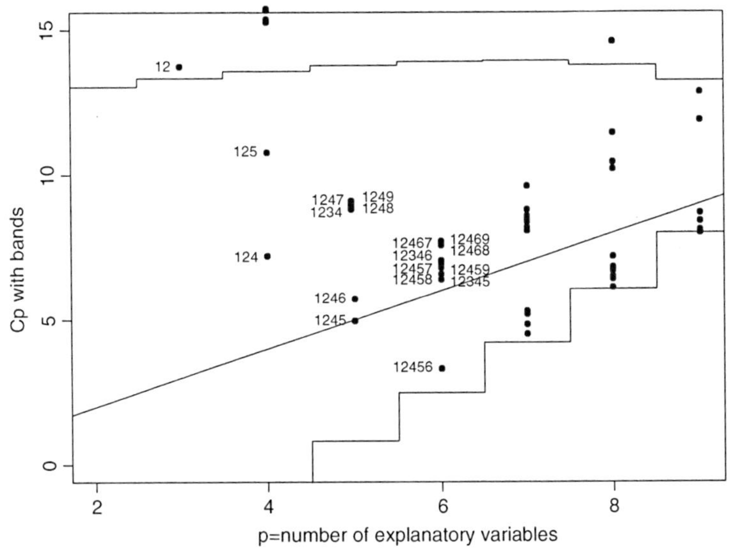

Fig. 2 Credit card data: C_p plot. There is a simple hierarchy of good models. Bands are the 2.5% and 97.5% points of the scaled and shifted F distribution of C_p

Table 1 Credit card data: some models selected by C_p and by $C_p(m)$ in the candlestick plot of Fig. 3

p Non-robust C_p	Variables	p $C_p(m)$	Variables
		7	1 2 3 4 5 8
6	1 2 4 5 6	6	1 2 3 4 5
5	1 2 4 5	6	1 2 3 5 8
5	1 2 4 6	5	1 2 3 5
4	1 2 4	4	1 2 5
3	1 2		

of forward plots for all values of m that are of interest. For example, Atkinson and Riani (2008) give separate plots of forward C_p for values of p from 4 to 7. The resulting quantity of graphical output can be overwhelming. We accordingly illustrate the plot introduced by Riani and Atkinson (2010) that cogently summarises this information.

The plot summarises, for each model, the information in the trajectory of the forward plots of $C_p(m)$. The starting point is the "candlestick" plot used to summarise such quantities as the high, low and closing values of stocks. Google provides many references. However, we need a generalization of this plot. Since we expect any outliers to enter the search towards the end, the last few values of $C_p(m)$ are of particular interest, as is the comparison of these values with earlier average behaviour.

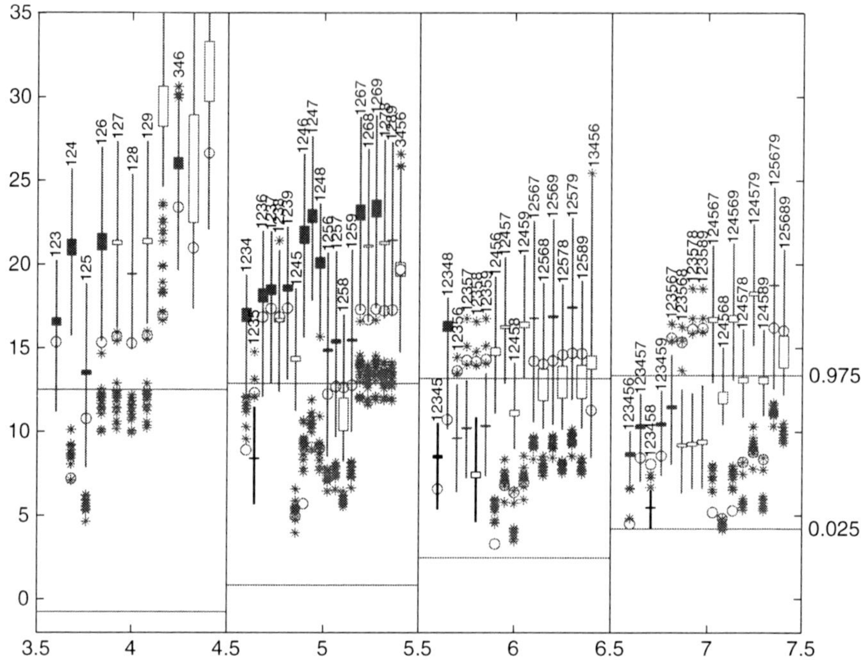

Fig. 3 Credit card data: generalized candlestick plot of $C_p(m)$ for the best models in the range $m = 900$–$1{,}000$. The last 20 observations to enter the individual searches are marked if they lie outside the candlestick. Models 1 2 3 5, 1 2 3 4 5, 1 2 3 5 8 and 1 2 3 4 5 8 are highlighted by a thick vertical line

Figure 3 gives the generalized candlestick plot for the values of $C_p(m)$ for the credit card data. The figure includes all models that were among the five best for $m \geq 900$, with symbols for the last 20 values if they are extreme. In order to check the indication of $p = 6$ as providing the largest model, we plot values in the range $p = 4$ to $p = 7$.

The plot depends on the range of values of m which define a "central part" of the plot. With 1,000 observations we take as the central part of the search values of m in the range 900–980. The figure includes all models that were among the five best for $m \geq 900$, with symbols for the last 20 values if they lie outside the "candle." The vertical lines in the plot summarise the values of $C_p(m)$ for each model in the central part of the search. The definition of the candlesticks is:

Lowest Value; minimum in the central part of the search;
Highest Value; maximum in the central part of the search;
Central Box; mean and median of the values in the central part of the search; filled if mean < median;
Stars; the values in steps "central part" + 1 to $n - 1$ if these lie outside the box;
Unfilled Circle; the final value.

Thus each point in the standard non-robust C_p plot such as Fig. 2 is replaced by a single vertical line and a series of extra symbols.

We start by looking at models for $p = 6$. The value of $C_p(m)$ for model 1 2 3 4 5 seems unaffected by exclusion of the last 20 observations. However, that for 1 2 4 5 6, which was the indicated model at the end of the search, increases to lie mostly above the bound when the observations are excluded. On the contrary, under the same conditions the values for 1 2 4 5 8 decrease, for it to become one of the two best models. If we now turn to $p = 7$, we see that the union of these models, that is 1 2 3 4 5 8, has a stable small value of $C_p(m)$.

The conclusions for $p = 5$ are straightforward: 1 2 3 5 is the only model which lies within the bounds for the central part of the search. This is a special case of the two models for $p = 6$ suggested above. Figure 3 indicates clearly that there is no satisfactory model with $p = 4$, although 1 2 5 is the best of a bad bunch. These models are also listed in Table 1.

The general shape of the plot in Fig. 3 is similar to that of the non-robust C_p plot in Fig. 2. However, for small values of p, many models have relatively small values of $C_p(m)$ only over the last values of m whereas, for larger p, there are many models with small values of $C_p(m)$ over most of the range. There is also a decrease in variability in the values of $C_p(m)$ as p increases. When p is too small, the values of $C_p(m)$ respond with extreme sensitivity to the addition of extra observations.

6.3 Outlier Detection

The ordering of observations by the forward search enables us to pinpoint the influential effect of individual observations. Table 1 shows appreciable change in the models selected as the last twenty observations are deleted. We accordingly now see whether there is evidence that some of these observations are outlying.

To detect outliers we use forward plots of minimum deletion residuals, with envelopes (Riani and Atkinson 2007). The left-hand panel of Fig. 4 is a forward plot of all such residuals for all 1,000 observations when the model fitted is 1 2 3 5. It is clear, from the exceedances of the upper threshold in the range m from 980 to 995, that there are some outliers, although the exact number is not obvious. With a large sample, the presence of several outliers has led to masking, so that departures are less extreme when $m = n$ than they are earlier in the search. Similar phenomena occur for multivariate data when forward plots of the minimum Mahalanobis distance are used for the detection of outliers. Riani et al. (2009) propose a rule that allows for masking and simultaneous inferences to provide an outlier detection rule with a size close to 1%. Torti and Perrotta (2010) amend the rule for regression. In the credit card data we detect between eight and ten outliers, the exact number depending on the model being fitted. Fortunately, the set of ten outliers contains that of nine or eight for all models of interest.

The forward plot of minimum deletion residuals for 990 observations is shown in the right-hand panel of Fig. 4. It shows that, for this model, there are no more

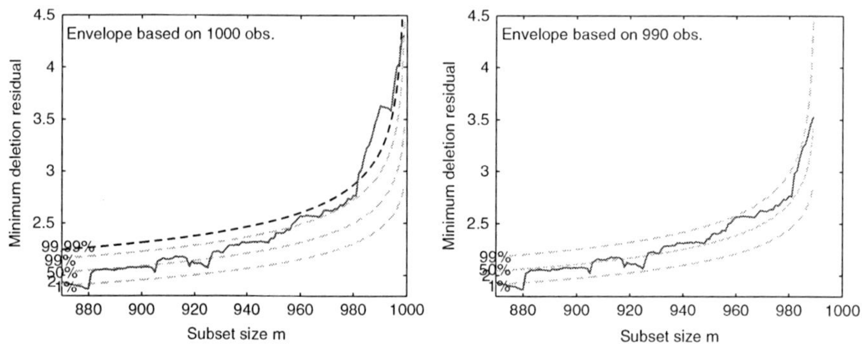

Fig. 4 Credit card data, outlier detection, model with variables 1 2 3 and 5. Upper panel, forward plot of minimum deletion residual $r_i(m)$, with 1%, 50%, 99% and 99.99% envelopes for $n = 1,000$. Lower panel, forward plot after deletion of ten observations; envelopes for $n = 990$

than ten outliers. Compared with the left-hand panel, the most extreme values occur at the end of the search, indicating that the observations are correctly ordered by the search and that there are no remaining outliers. The presence of these outliers explains the structure of the "starred" observations in Fig. 3. The outliers are causing the majority of the models to have small values of $C_p(m)$ towards the end of the search.

6.4 Model Building and Checking

The independent identification of outliers in the credit card data justifies the selected models listed in Table 1. It is interesting to investigate some of these models a little further.

Table 2 gives t values, when $n = 990$, for the terms of the best models in Table 1 for $p = 5$, 6 and 7. The models were selected by our interpretation of the generalized candlestick plot of Fig. 3. Model 1 2 3 4 5 8 is highlighted in the figure as the best model for $p = 7$. If we remove the least significant term, that for x_8, we obtain the stable model 1 2 3 4 5 with little change of the t values for the remaining terms. Now x_4 is the least significant term. Its omission, however, causes an increase in the t statistic for x_2 from 5.84 to 9.47. In this model all terms are significant at the 1% level.

Figure 5 give forward plots of $C_p(m)$ for a few selected models. This extends the information available from the generalized candlestick plot. For model 1 2 3 4 5 the values of $C_p(m)$ remain within the 97.5% bound throughout and are stable at the end of the search. But the figure shows how the values of $C_p(m)$ for the simpler model 1 2 3 5 are affected by the last few observations to enter. The plot also shows how model 1 2 4 5 6 was chosen by its small value of C_p at the very end of the search. However, values of $C_p(m)$ earlier in the search lie well above the 97.5% bound.

Table 2 Credit card data: t statistics of the three best models of Table 1 after removing outliers ($n = 990$)

Term	Model		
	1 2 3 5	1 2 3 4 5	1 2 3 4 5 8
Intercept	49.11	49.18	49.34
x_1	6.37	6.05	6.22
x_2	9.47	5.84	5.78
x_3	2.64	2.70	2.73
x_4	–	2.28	2.71
x_5	2.78	2.52	3.00
x_8	–	–	−2.29

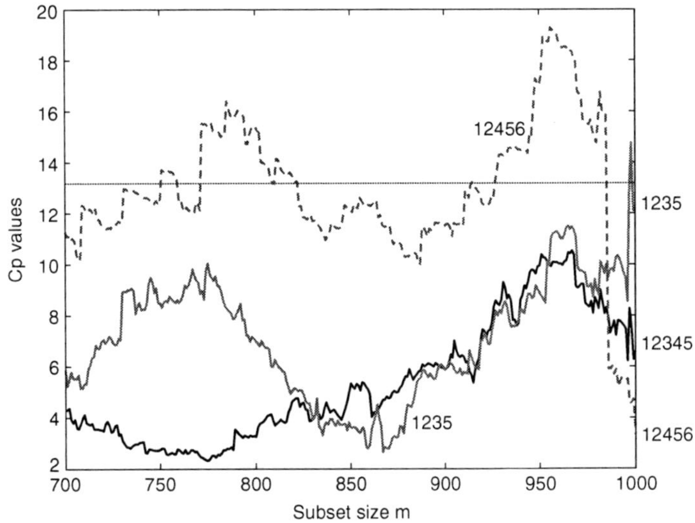

Fig. 5 Credit card data: forward plots of $C_p(m)$ for selected models from Table 1

A final comment on model choice is to compare 1 2 3 5 and 1 2 3 4 5 over the values of m from 700. Although the two models give very similar values of $C_p(m)$ for $m = 850$–980, the larger model is superior in the first part of the plot. Since the value of $C_p(m)$ is also unaffected by the outlying observations, we would recommend this as the chosen model.

7 Computation

The Matlab software used in this paper is part of the FSDA (Forward Search Data Analysis) toolbox which can be downloaded, free of charge, from the webpage `www.riani.it` in the section "Matlab code". Full documentation is included.

Appendix A: The Credit Card Data

Variables that are given as amount are in euros and are either annual totals or averages, depending on the nature of the variable.

y	Amount of use of credit, debit and pre-paid card services of the bank
x_1	Direct debts to the bank
x_2	Assigned debts from third parties
x_3	Amount of shares (in thousands of Euros)
x_4	Amount invested in investment funds (in thousands of Euros)
x_5	Amount of money invested in insurance products from the bank (in thousands of Euros)
x_6	Amount invested in bonds (in thousands of Euros)
x_7	Number of telepasses (Italian electronic toll collection system) of the current account holder
x_8	Number of persons from the bank dealing with the management of the portfolio of the customer (min=0, max=4). This variable has many zeroes
x_9	Index of use of point of sale services

In x_7 the telepass is a debit card issued for each car. Since other forms of payment are possible, this variable also contains many zeroes.

References

Atkinson, A. C. and M. Riani (2000). *Robust Diagnostic Regression Analysis*. New York: Springer–Verlag.

Atkinson, A. C. and M. Riani (2007). Exploratory tools for clustering multivariate data. *Computational Statistics and Data Analysis 52*, 272–285. doi:10.1016/j.csda.2006.12.034.

Atkinson, A. C. and M. Riani (2008). A robust and diagnostic information criterion for selecting regression models. *Journal of the Japanese Statistical Society 38*, 3–14.

Atkinson, A. C., M. Riani, and A. Cerioli (2004). *Exploring Multivariate Data with the Forward Search*. New York: Springer–Verlag.

Fraley, C. and A. E. Raftery (1999). Mclust: software for model-based cluster analysis. *Journal of Classification 16*, 297–306.

Mallows, C. L. (1973). Some comments on C_p. *Technometrics 15*, 661–675.

Maronna, R. A., D. R. Martin, and V. J. Yohai (2006). *Robust Statistics: Theory and Methods*. New York: Wiley.

Riani, M. and A. C. Atkinson (2007). Fast calibrations of the forward search for testing multiple outliers in regression. *Advances in Data Analysis and Classification 1*, 123–141. doi:10.1007/s11634-007-0007-y.

Riani, M. and A. C. Atkinson (2010). Robust model selection with flexible trimming. *Computational Statistics and Data Analysis 54*, 3300–3312. doi:10.1016/j.csda.2010.03.007.

Riani, M., A. C. Atkinson, and A. Cerioli (2009). Finding an unknown number of multivariate outliers. *Journal of the Royal Statistical Society, Series B 71*, 201–221.

Riani, M., A. Cerioli, A. Atkinson, D. Perrotta, and F. Torti (2008). Fitting mixtures of regression lines with the forward search. In F. Fogelman-Soulié, D. Perrotta, J. Piskorski, and R. Steinberger (Eds.), *Mining Massive Data Sets for Security*, pp. 271–286. Amsterdam: IOS Press.

Torti, F. and D. Perrotta (2010). Size and power of tests for regression outliers in the forward search. In S. Ingrassia, R. Rocci, and M. Vichi (Eds.), *New Perspectives in Statistical Modeling and Data Analysis*. Heidelberg: Springer-Verlag, pp. 377–384.

Tufte, E. (2001). *The Visual Display of Quantitative Information*. Cheshire, CT: Graphics Press.

Ensemble Support Vector Regression: A New Non-parametric Approach for Multiple Imputation

Daria Scacciatelli

Abstract The complex case in which several variables contain missing values needs to be analyzed by means of an iterative procedure. The imputation methods most commonly employed, however, rely on parametric assumptions.
In this paper we propose a new non-parametric method for multiple imputation based on Ensemble Support Vector Regression. This procedure works under quite general assumptions and has been tested with different simulation schemes. We show that the results obtained in this way are better than the ones obtained with other methods usually employed to get a complete data set.

1 Introduction

The impact of the missing data on the results of statistical analysis depends on the missing data mechanism. The sources of missingness of some variables can be numerous and their nature can vary from total randomness to a particular dependence from the variables.

The missing data mechanism standard definition, as presented by Little and Rubin (1987), can be classified into three distinct categories. The first missing category, called "missing at random" or MAR, occurs when the probability of response cannot depend on missing data values or on unobserved variables. The second missing category, called "missing completely at random" or MCAR, occurs when the response variables are unrelated to the values of any variable.

Little and Rubin referred to these two types of missing categories as "ignorable" in the sense that, the response variables distribution does not depend on the missing data mechanism and therefore it can be ignored. The last category, "missing not at

D. Scacciatelli (✉)
Dipartimento di Statistica, Probabilità e Statistiche Applicate, Sapienza Università di Roma, Italia
e-mail: daria.scacciatelli@gmail.com

A. Di Ciaccio et al. (eds.), *Advanced Statistical Methods for the Analysis of Large Data-Sets*, Studies in Theoretical and Applied Statistics,
DOI 10.1007/978-3-642-21037-2_15,

random" or MNAR, occurs when the mechanism is related to the unobserved values. The assumption for MCAR is very restrictive and often unrealistic.

In the late 1970s, Dempster, Laird and Rubin formalized the Expectation-Maximization algorithm, a computational method for efficient estimation from incomplete data. Until that time, the primary method used for treating missing value was case delation, by which any observations with missing values is discarded from the data set (Shafer and Olsen 1998). Recent advances in theoretical and computational statistics have produced a new generation of procedures, both parametric and non-parametric.

These new procedures employ Multiple Imputation (MI) (Rubin 1987), a technique in which every missing value is replaced by a set of $m > 1$ plausible values, which are bayesian draws from the conditional distribution of the missing observations given the observed data. In multiple imputation, some authors, in order to obtain several values for each missing value, use different approaches based on non-parametric methods with bootstrap, instead of using the Bayesian predictive distribution.

Rubin considered the so called Approximate Bayesian Bootstrap, an alternative non-parametric approach, based on Nearest Neighbors (Rubin and Schenker 1986).

The bootstrap was also employed in the missing data problem by Efron (1994), and recently Di Ciaccio proposed a multiple imputation approach based on bootstrap and on a non-parametric model, the Regression Tree with a Bagging algorithm (DiCiaccio 2008).

Here, we propose a new non-parametric approach for multiple imputation and variance estimation, based on Ensemble Support Vector Regression.

The aim of this work is to create a "complete" (imputed) data set, often requested by statistical agencies, that can be analyzed by any researcher on different occasions and using different techniques.

The article is structured as follows. In Sect. 2, definitions and basic assumptions about the Multiple Imputation approach are introduced. The proposed method and algorithm are described in Sect. 3. To obtain a complete data set, a simulation study in Sect. 4 evaluates the performance of our algorithm with the most used imputation techniques. Some concluding remarks are made in Sect. 5.

2 Multiple Imputation Approaches

By Rubin's Multiple Imputation m possible alternative versions of the complete data are produced and the variation among the m imputations reflects the uncertainty with which missing values can be predicted from observed ones.

Then each data set is analyzed by a complete data method. Rubin proposed a method to combine the results of m analyses. The Multiple Imputation need to fulfil certain conditions which is referred to as "proper" multiple imputations as defined by Rubin (1987), Shafer (1997), and Buuren et al. (2003).

To estimate a generic scalar parameter Q, we compute m different versions of the estimator $\widehat{Q}$ and its variance $\overline{U}$: $[\widehat{Q}^{(j)}, \widehat{U}^{(j)}]$, $j = 1, \ldots, m$.

Rubin's estimate is simply the average of m estimates:

$$\overline{Q} = \frac{1}{m} \sum_{j=1}^{m} \widehat{Q}^{(j)} \tag{1}$$

where the uncertainty in $\overline{Q}$ has two components:

- The within-imputation variance $\overline{U}$, which is the average of the m complete-data estimates:

$$\overline{U} = \frac{1}{m} \sum_{j=1}^{m} \widehat{U}^{(j)} \tag{2}$$

- The between-imputation variance B, the variance of the estimates themselves:

$$B = \frac{1}{m-1} \sum_{j=1}^{m} (\widehat{Q}^{(j)} - \overline{Q})^2 \tag{3}$$

Then the variance estimate associated to $\overline{Q}$ is the total variance T i.e. the sum of the above-mentioned two components, with B multiplied by an additional correction factor:

$$T = \overline{U} + \left(1 + \frac{1}{m}\right) B \tag{4}$$

If there were no missing data, then $\{\overline{Q}^{(1)}, \overline{Q}^{(2)}, \ldots, \overline{Q}^{(m)}\}$ would be identical, B would be 0 and T would simply be equal to $\overline{U}$.

The size of B relative to $\overline{U}$ is a reflection of how much information is contained in the missing part of the data relative to the observed part.

The statistic $T^{-\frac{1}{2}}(\overline{Q} - Q)$, by which confidence intervals and hypothesis tests can be calculated, is approximately distributed as a $t - Student$ with degrees of freedom (Rubin 1987) given by:

$$\nu_m = (m-1)\left[1 + \frac{\overline{U}}{(1+m^{-1})B}\right]^2 \tag{5}$$

The degree of freedom may vary from $m - 1$ to ∞ depending on the rate of missing information. When the number of degrees of freedom is large, the distribution is essentially normal, the total variance is well estimated, and there is little to be gained by increasing m. The estimated rate of missing information for Q is approximately $\frac{r}{r+1}$, where $r = (1 + m^{-1}B/\overline{U})$ is the relative increase in variance due to nonresponse.

The imputations of the incomplete data set can be generated by employing either the Expectation-Maximization (Dempster et al. 1977) or the Data Augmentation (Shafer 1997) algorithms or both.

In order to apply Rubin's Multiple Imputation, some conditions have to be satisfied: we must have missing at random mechanism, multivariate Normal distributions, and proper imputation model. The validity of this approach depends on these assumptions: if the imputations are proper, then $\overline{Q}$ is a consistent, asymptotically normal estimator, and the variance T is an unbiased estimator of its asymptotic variance.

Nevertheless, the evaluation of the properness of MI is analytically intractable, and therefore it is best done via simulations (Buuren et al. 2003). However a number of simulation studies have demonstrated that, if the amount of missing information is not large, MI is robust to violations of normality of the variables used in the analysis (Graham and Schafer 1999).

Besides, the standard multiple imputation approach is based on distributional assumptions and it is parametric. However, in general, there are many applications where fully parametric approaches may not be suitable.

For this reason it is important to focus-on semi-parametric or non-parametric imputation methods, without taking into account restrictive hypotheses (Durrant 2005; Di Zio and Guarnera 2008).

3 A New Approach Based on Ensemble Support Vector Regression

Support vector machines (SVM) are derived from Statistical Learning Theory and were first introduced by Vapnik (1999). They are tools for non-linear regression and classification and the terms Support Vector Classification (SVC) and Support Vector Regression (SVR) will be used for specification.

SVM may be likened to feed-forward Neural Networks, both are known as non-parametric techniques: they offer the efficient training characteristics of parametric techniques but have the capability to learn non-linear dependencies (Safaa 2009; Mallison and Gammerman 2003).

Given the training data set $TR = \{(\mathbf{x}_1, y_1), \ldots, (\mathbf{x}_\ell, y_\ell)\}$, in SVR the input vector $\mathbf{x}$ is first mapped to a higher dimensional feature space using some fixed (non-linear) mapping and then a linear model is constructed in this feature space.
The linear model (in the feature space) is given by:

$$f(\mathbf{x}) = (\mathbf{w} \cdot \phi(\mathbf{x})) + b \tag{6}$$

The functional form of the mapping $\phi(\cdot)$ does not need to be known because it is implicitly determined by the choice of a "kernel function" K (Boser et al. 1992):

$$K(\mathbf{x}_i, \mathbf{x}_j) = (\phi(\mathbf{x}_i), \phi(\mathbf{x}_j)) \tag{7}$$

the computation threshold b is done by exploiting Karush-Kuhn-Tucker (KKT) conditions (Smola and Schölkopf 1998), $\mathbf{w}$ is a constant vector that can be estimated by reducing the Empirical Risk and the complexity term.
In:

$$R(f) = \frac{1}{2}\|\mathbf{w}\|^2 + C \cdot \sum_{i=1}^{\ell} L_{\epsilon}(y_i, f(\mathbf{x})) \tag{8}$$

the first term $\|\mathbf{w}\|^2$ characterizes the complexity model, the second term represents the Empirical Risk, ℓ is the number of the examples of training set; C and ϵ are selected by users based on a priori knowledge and/or user expertise (Cherkassky et al. 2004); the parameter C depending on how much they want to penalize errors, meaning, for example, that a larger C corresponds to assign a higher penalty to errors.

The ϵ-insensitive loss function $L_{\epsilon}(y)$, proposed by Vapnik, is given by:

$$L_{\epsilon}(y, f(\mathbf{x})) = \begin{cases} 0 \;\; if \;\; |y - f(\mathbf{x})| \leq \epsilon \\ \mid y - f(\mathbf{x}) \mid - \epsilon \;\; otherwise \end{cases} \tag{9}$$

Formally, the SVR models the conditional expectation of the imputation variable, $E(y|\mathbf{x})$, where the conditional distribution is, usually, completely unknown.

The prediction algorithm SVR does not generate estimates for $P(y|\mathbf{x})$, but only estimates the conditional expectation.

One of the main reasons for the success of the algorithm is that it avoids density-estimation, as since the SVR does not model the conditional probability, $P(y|\mathbf{x})$, we cannot draw multiple values from this distribution as required for multiple imputation. A way to overcome this obstacle is to consider an ensemble of SVRs.

Some authors considered the use of ensemble of Support Vector Machines for Regression (wang and Yangh 2008; Kim et al. 2003), in order to improve the prediction accuracy. Ensemble methods are based on the idea of generating many so-called "weak" decision rules and aggregating their results. The main aspect in constructing the SVR ensemble is that each single SVR has to be as much as possible different from one another. This requirement can be obtained by using different training sets.

Some methods to generate the training samples are bagging (Breiman 1996) and boosting (Freud et al. 1997), that creates diversity in an ensemble by sampling and re-weighing the available training data. In this paper we considered the SVR ensemble by using the bagging method, essentially for its simplicity, where several SVRs are trained independently on bootstrap samples and then their outputs are aggregated with an appropriate combination technique.

Our non-parametric approach, based on Ensemble Support Vector Regression (SVRE) allows us to obtain m complete data sets: bootstrapping builds m replicate training sets $\{TR_{bootstrap_i} | i = 1, \ldots, m\}$, by which we can evaluate the additional variability due to missing values. Usually m is a number between 10 and 30.

Table 1 The non-parametric algorithm

STEP 1. For each variable: initialize missing data with the mean
STEP 2. Iterate (max num.of iteration = T)
STEP 3. Iterate (j = 1 to num. variables J)
STEP 4. Set the variable j as the target variable
STEP 5. Select cases which do not have missing value for the variable j
STEP 6. Estimate the predictor model based on Support Vector Regression Ensemble
STEP 7. Impute the missing values of variable j by the predictor model estimated in STEP 6
STEP 8. Update missing values of variable j. Go to STEP 3
STEP 9. If $\delta < 0.001$ then STOP else go to STEP 2.

Following the main objective of the work, after training each SVR in SVRE, we combine the outputs of these models by averaging to create the ensembles final output: the completed matrix (Raghunathan et al. 2001).

In the multivariate case, in which several variables contain missing values, the SVRE method initializes, for example, by substituting the missing values in the data set with the mean, and then, iteratively until convergence, considers each variable with missing data as the target variable. Some details of our method are described in Table 1.

To evaluate the convergence of the procedure at the iteration t, we use the index defined below:

$$\delta_t = \frac{1}{n_{mis}} \sum_{j=1}^{J} \frac{1}{\sigma_j^2} \sum_{i \in M_j} (\widehat{x}_{ij}^t - \widehat{x}_{ij}^{t-1})^2 \tag{10}$$

With $\widehat{x}_{ij}^t$ we indicate the estimated value, obtained applying to the missing value x_{ij} the model at the iteration t, M_j and σ_j^2 are respectively the set with missing values and the variance of variable j.

Then in STEP 8, to update the value $\widehat{x}_{ij}^t$, we use the formula:

$$\widehat{x}_{ij}^t \leftarrow \gamma \widehat{x}_{ij}^{t-1} + (1-\gamma)\widehat{x}_{ij}^t, \qquad \gamma \in (0,1) \tag{11}$$

4 Results

To evaluate the performance of the proposed SVRE algorithm we considered four different simulation schemes with both multinormal and non-multinormal data and different correlation levels between variables.

1. Multinormal independent data were simulated independently from five Normal distributions $X_i \sim N(\mu_i, 1)(i = 1, \ldots, 5)$, where μ_i was drawn from a Uniform distribution in the interval [0, 50].

2. Multinormal dependent data were simulated from a 5-dimensional distribution with $X_i \sim N(\mu_i, 1)(i = 1, \ldots, 5)$ marginals with Gaussian copula having a high correlation level among variables, where $\mu_i \sim U(0, 50)$.
3. Non-normal independent data were simulated from a 5-dimensional distribution with $Beta(3,2)$, $Beta(5,2)$, $Beta(2,5)$, $Beta(2,1)$, $Beta(2,2)$ marginals using a t-Student copula with low correlation level among variables.
4. Non-normal dependent data were simulated from a non-linear combination of 5 $Beta$ random variables with random parameters.

For each simulation scheme we generated 300 random populations, from each population we extracted one sample of 120 units, where we fixed randomly, two different (low and medium) proportion of missingness, 15% and 30%, for MAR type on the first two variables.

To simulate MAR type, we set X^1 and X^2 to have a probability of missingness five times higher for the units with $X^1 \cdot X^2$ above the median.

The predictor model used in STEP 6, was the SVRE algorithm, with 30 bootstrap replications. Gaussian kernels were chosen. Support Vector Regression model parameters σ, C and ϵ, were not estimated during training but were found through the grid search strategy (Cherkassky et al. 2004): $\sigma = 0.3, C = 500$ and $\epsilon = 0.8$. Moreover, we fixed $T = 50$ and $\gamma = 0.9$.

We compared the behavior of SVRE algorithm with Rubin's Multiple Imputation (MI) ($m = 30$), and, for a lower bound reference, also with the mean imputation (MEAN) naive algorithm.

We remind that Rubin's MI is here used in a different sense, with the only aim of obtaining a complete matrix (DiCiaccio 2008; Raghunathan et al. 2001).
We evaluated:

- The difference between the imputed data values and the known ones by the weighted sum of squares:

$$SSEW = \frac{1}{n_{miss}} \sum_{j=1}^{J} \frac{1}{\sigma_j^2} \sum_{i \in M_j} (\widehat{x}_{ij} - x_{ij})^2 \tag{12}$$

  Where M_j and σ_j^2 are respectively the set with missing values and the variance of variable j.

- the difference between Pearson's correlation coefficients on the population and on the completed sample (in particular we compared r_{12}, r_{13}, r_{14}).

Table 2 reports, for each simulation scheme and for each percentage of missing, the percentage of samples for which the considered methods have shown a smaller error with respect to the others.

With multinormal and non-normal independent data, both SVRE and MI were non effective: in these situations also an unsophisticated imputation method, such as the mean, performs better than the considered ones.

Table 2 Results of imputation: percentage of samples favorable to each method

	15% of missing			30% of missing		
	Mean	Svre	MI	Mean	Svre	MI
Multinormal independent	47	33	20	59	30	11
Multinormal dependent	0	10	90	0	9	91
Non-normal independent	71	23	6	80	17	3
Non-normal dependent	2	67	31	5	63	32

Table 3 Estimation of correlation coefficient: percentage of samples favorable to each method

	15% of missing			30% of missing		
	Mean	Svre	MI	Mean	Svre	MI
Multinormal independent	38	35	27	43	42	15
Multinormal dependent	1	3	96	1	6	93
Non-normal independent	59	22	20	71	21	8
Non-normal dependent	11	57	32	6	61	33

With multinormal dependent data, Rubin's MI produces best results as the distributional assumptions are satisfied and the interdependence among variables is verified. In the case of non-normal dependent data, our proposed imputation method SVRE appears preferable and produces good performances.

Table 3, for each simulation scheme and for 15% and 30% of missing, shows the percentage of samples for which each considered method was able to reproduce the correlation level among variables, in particular (r_{12}, r_{13}, r_{14}). The relationships among variables were preserved both in Rubin's MI with multinormal dependent data and in SVRE with non-normal dependent variables.

Mean imputation algorithm provided good results with both multinormal and non-normal independent data.

5 Conclusion

We proposed an iterative non-parametric method for multiple imputation, based on Ensemble Support Vector Regression. We considered the complex case in which several variables contain missing values.

If our objective is to obtain a complete data matrix, then Rubin's Multiple Imputation works well with multinormal dependent data. As the correlation level decreases, Rubin's method could be substituted by simpler models as, for instance, the mean imputation.

In order to obtain a complete data set in case of non-normal dependent data, our non-parametric method SVRE represents an efficient alternative to Rubin's Multiple Imputation. Moreover, our method gives good imputations also when the correlation level decreases slightly.

The results appear to be quite independent from the different fractions of missing data.

The obtained results indicate that, to select a suitable imputation method, it is convenient to check distribution assumptions carefully and to measure the relationship among variables.

In the future, we will consider other combination methods for the ensemble. Although, in our experiments we concentrated on the Gaussian kernel, that is widely used in many application areas, others kernels are available, and can be used for further study.

For further investigation it will be worth to compare confidence intervals estimated by Rubin's Multiple Imputation with confidence intervals estimated by Ensemble Support Vector Regression.

References

Breiman, L.: Bagging predictors. Machine Learning **26**, 123–140 (1996)

Boser, B., Guyon, I., Vapnik V.: A training algorithm for optimal margin classifiers. Proceedings of the Fifth Annual Workshop on Computational Learning Theory, 144–152 (1992)

Brand, J., Buuren, S., Groothuis-Oudshoorn, K., Gelsema, E. S.: A toolkit in SAS for the evaluation of multiple imputation methods. Statistical Neerlandica **57**, 36–45 (2003)

Cherkassky, V., Yunqian, M.: Practical selection of SVM parameters and noise estimation for SVM regression. Neural Networks **17(1)**, 113–126 (2004)

Dempster, A. P., Laird, N., Rubin, D. B.: Maximum likelihood from incomplete data via the EM algorithm. Journal of the Royal Statistical Society **39**, 1–38 (1977)

Di Ciaccio, A.: Bootstrap and Nonparametric Predictors to Impute Missing Data, Cladag,(2008)

Durrant, G. B.: Imputation methods for handling item-nonresponse in the social sciences: A methodological review. NCRM Working Paper Series, (2005)

Di Zio, M., Guarnera, U.: A multiple imputation method for non-Gaussian data. Metron - International Journal of Statistics **LXVI(1)**, 75–90 (2008)

Efron, B.: Missing data, imputation, and the bootstrap. Journal of the American Statistical Association **89**, 463–475 (1994)

Freud, Y., Shapire, R.: A decision theoretic generalization of online learning and an application to boosting. J. Comput.System Sci. **55(1)**, 119–139 (1997)

Graham, J.W., Schafer, J.L.: On the performance of multiple imputation for multivariate data with small sample size. Hoyle, R. (Ed.), Statistical Strategies for Small Sample Research. Sage, Thousand Oaks, CA, 1–29 (1999)

Kim, H.-C., Pang, S., Je, H.-M., Kim, D., Yang Bang, S.: Constructing support vector machine ensemble. Pattern Recongnition **36**, 2757–2767 (2003)

Little, R., Rubin, D.: Statistical Analysis with Missing Data. New York, Wiley (1987)

Mallinson, H., Gammerman, A.: Imputation Using Support Vector Machines. http://www.cs.york.ac.uk/euredit/ (2003)

Raghunathan, T. E., Lepkowski, J. M., Van Hoewyk, J., and Solenberger, P.: A Multivariate Technique for Multiply Imputing Missing Values Using a Sequence of Regression Models. Survey Methodology **27(1)**, 85–96 (2001)

Rubin, D.B.: Multiple Imputation for Nonresponse in Surveys. Jhon Wiley &Sons (1987)

Rubin, D.B., Schenker, N.: Multiple Imputation for interval estimatation from simple random samples with ignorable nonresponse. Journal of the American Statistical Association **81**, 366–374 (1986)

Safaa R. Amer: Neural Network Imputation in Complex Survey Design. International Journal of Electrical, Computer, and Systems Engineering **3(1)**, 52–57 (2009)
Shafer, J.L.: Analysis of Incomplete Multivariate Data. Chapman and Hall (1997)
Shafer, J.L., Olsen, M.K.: Multiple imputation for multivariate missing-data problems: a data analyst's perspective. Multivariate Behavioral Research **33**, 545571 (1998)
Smola, A.J., Schölkopf B.: A Tutorial on Support Vector Regression. NeuroCOLT, Technical Report NC-TR-98–030, Royal Holloway College, University of London, UK (1998)
Vapnik, V.N.: The Nature of Statistical Learning Theory. Springer Verlag (1999)
Wang, F., Yangh, H-Z.: epsilon-insensitive support vector regression ensemble algorithm based on improved adaboost. Computer Engineering and Applications **44**, 42–44 (2008)

Part V
Time Series Analysis

On the Use of PLS Regression for Forecasting Large Sets of Cointegrated Time Series

Gianluca Cubadda and Barbara Guardabascio

Abstract This paper proposes a methodology to forecast cointegrated time series using many predictors. In particular, we show that Partial Least Squares can be used to estimate single-equation models that take into account of possible long-run relations among the predicted variable and the predictors. Based on Helland (Scand. J. Stat. 17:97–114, 1990), and Helland and Almoy (J. Am. Stat. Assoc. 89:583–591, 1994), we discuss the conditions under which Partial Least Squares regression provides a consistent estimate of the conditional expected value of the predicted variable. Finally, we apply the proposed methodology to a well-known dataset of US macroeconomic time series (Stock and Watson, Am. Stat. Assoc. 97:1167–1179, 2005). The empirical findings suggest that the new method improves over existing approaches to data-rich forecasting, particularly when the forecasting horizon becomes larger.

1 Introduction

Recent advances in official statistics and informatics endow economic time series analysts with more and more variables and observations. While the availability of large data sets provides the opportunity to better understand macroeconomic dynamics, researchers need to resort to *ad hoc* statistical procedures to deal with high dimensional time series. Indeed, traditional tools in time series analysis require to invert the variance-covariance matrix of the predictors, which is numerically

G. Cubadda (✉)
University of Rome, "Tor Vergata", Italy
e-mail: gianluca.cubadda@uniroma2.it

B. Guardabascio
ISTAT, Rome, Italy
e-mail: guardabascio@istat.it

A. Di Ciaccio et al. (eds.), *Advanced Statistical Methods for the Analysis of Large Data-Sets*, Studies in Theoretical and Applied Statistics,
DOI 10.1007/978-3-642-21037-2_16, © Springer-Verlag Berlin Heidelberg 2012

cumbersome or even impossible when the number of variables, N, is large compared to the number of observations, T. Hence, the analysis of large dimensional dynamic systems has recently received large attention, both at the theoretical and empirical level.

The early focus of the large-N time series literature was on factors models, where factors are typically extracted by Principal Components [PCs], both in the time and frequency domain, see *inter alia* Stock and Watson (2002a, 2002b), Forni et al. (2004, 2005). More recently, the application of Bayesian regression [BR] has been advocated. Particularly Banbura et al. (2010) and De Mol et al. (2008) suggested the use of ridge-regression techniques to overcome the curse of dimensionality problem in traditional time series models.

However, both the above approaches ignore the possible presence of long-run equilibria among the variables. This information could be helpful in forecasting, particularly for large forecasting horizons. A similar point has recently been raised by Banerjee and Marcellino (2007), who put common trends and factor models together by means of the factor-augmented error correction model.

Based on Engle and Granger (1987), this paper proposes a forecasting model that includes both changes and levels of the predictors such that possible cointegration links are taken into account. Moreover, we resort to Partial Least Squares [PLS] in order to estimate the parameters of our model. PLS introduced by Wold (1985) is a statistical technique that aims to maximize the covariance between the target variable and a reduced number of orthogonal components that are obtained as linear combinations of the original predictors.

We notice that the use of PCs as a mean of extracting the relevant predictors from a large data set is inherently based on the following assumptions: (a) the explanatory variables have a factor structure; (b) N diverges to infinity. However, as noted by Bovin and Ng (2006), it is not guaranteed that increasing N is beneficial to the forecasting performances of PC regression. Hence, it is theoretically sensible to resort to PLS that, differently from PCs, explicitly takes into account the interaction between the target variables and the explanatory ones when projecting the latter into a small-dimension orthogonal subspace.

Indeed, Gröen and Kapetanios (2008) documented that PLS provides better forecasting performances when it is used in place of PCs for extracting common factors in the Stock and Watson (2002a, 2002b) framework. Moreover, Cubadda and Hecq (2011) have recently shown that a multivariate version of PLS is quite effective in identifying common autocorrelation in high-dimensional dynamic systems.

The remainder of this paper is organized as follows: the main features of our model are presented in Sect. 2. Section 3 discusses the conditions under which PLS provides a consistent estimator of the model parameters. Importantly, our approach does not require that N diverges to infinity in order to achieve consistency. A description of the data that we use in the empirical application is provided in Sect. 4. The various forecasting procedures, along with the empirical results, are detailed in Sect. 5. Finally Sect. 6 concludes.

2 The Model

Let us consider the vector error-correction model of an n-vector of time series z_t

$$\Gamma(L)\Delta z_t = \alpha\gamma' z_{t-1} + \varepsilon_t, \qquad (1)$$

where $\Delta = (1 - L)$, $\Gamma(L) = I_n - \sum_{i=1}^{p} \Gamma_i L^i$, α and γ are full-column rank $n \times r$ $(r < n)$ matrices such that $\alpha'_\perp \Gamma(1)\gamma_\perp$ has full rank, ε_t are i.i.d. $N_n(0, \Sigma_{\varepsilon\varepsilon})$. The process z_t is then cointegrated of order (1,1) and the matrix γ span the cointegrating space, see e.g. Johansen (1996). Moreover, we assume, without loss of generality, that deterministic elements have preliminarily been removed from time series z_t and that each element of both Δz_t and $\gamma' z_{t-1}$ has been standardized to unit variance.

It follows that each of series z_t is generated by a model with the following form:

$$y_t = \beta' x_{t-1} + \eta_t, \quad t = 1, \ldots, T \qquad (2)$$

where y_t is a generic element of Δz_t, η_t is the corresponding element of ε_t, and $x'_t = [z'_t\gamma, \Delta z'_t, \ldots, \Delta z'_{t-p+1}]$.

Suppose that $N = pn + r$, the number of elements of β, is too large to estimate model (2) by ordinary least squares. In order to reduce the number of parameters in (2), we follow Helland (1990) and Helland and Almoy (1994) by assuming that

$$\sigma_{xy} = \Upsilon_q \xi, \qquad (3)$$

where $\sigma_{xy} = \mathrm{E}(x_{t-1} y_t)$, Υ_q is a matrix formed by q eigenvectors of $\Sigma_{xx} = \mathrm{E}(x_t x'_t)$, and ξ is a vector with all the q elements different from zero. Notice that (3) implies

$$\beta = \Upsilon_q \Lambda_q^{-1} \xi$$

where Λ_q is the diagonal eigenvalue matrix associated with Υ_q. Hence, model (2) has the following factor structure:

$$y_t = \xi' f_{t-1} + \eta_t,$$

where $f_t = \Lambda_q^{-1} \Upsilon'_q x_t$. Notice that $y_{t+h|t} = \mathrm{E}(y_{t+h} | f_t)$ is a linear combination of the factors f_t for $h > 0$.

3 Estimation of the Factors

In order to estimate model (2), we resort to the PLS algorithm proposed by Wold (1985). Particularly, we define $y \equiv (y_{p+1}, \ldots, y_T)'$, $V_0 = X \equiv (x_{p+1}, \ldots, x_T)'$, and

$$V_i = V_{i-1} - \widehat{f}_i \widehat{\phi}'_i = X - \sum_{j=1}^{i} f_j \widehat{\phi}'_j, \qquad (4)$$

for $i = 1, .., N$, where

$$\widehat{f}_i = V_{i-1}\widehat{\omega}_i, \tag{5}$$

$$\widehat{\omega}_i = V'_{i-1}y, \tag{6}$$

$$\widehat{\phi}_i = (\widehat{f}'_i\widehat{f}_i)^{-1}V'_{i-1}\widehat{f}_i, \tag{7}$$

Based on Helland (1990) and Helland and Almoy (1994), we argue that, under condition (3), the PLS estimator for β,

$$\widehat{\beta}_{PLS} = \widehat{\Omega}_q(\widehat{\Omega}'_q X'X\widehat{\Omega}_q)^{-1}\widehat{\Omega}'_q X'y,$$

where $\widehat{\Omega}_q \equiv (\widehat{\omega}_1, \ldots, \widehat{\omega}_q)$, is consistent as $T \to \infty$. Indeed, Helland (1990) proved that, at the population level, the PLS weights are obtained by the recursive relation.

$$\omega_{i+1} = \sigma_{xy} - \Sigma_{xx}\Omega_i(\Omega'_i\Sigma_{xx}\Omega_i)^{-1}\Omega'_i\sigma_{xy} \tag{8}$$

where $\omega_1 = \sigma_{xy}$ and $\Omega_i = (\omega_1, \ldots, \omega_i)$ for $i = 1, \ldots, N-1$. It follows by induction that Ω_q lies in the space spanned by

$$(\sigma_{xy}, \Sigma_{xx}\sigma_{xy}, \ldots, \Sigma_{xx}^{q-1}\sigma_{xy}),$$

from which it is easy to see that condition (3) implies $\omega_{q+1} = 0$ and

$$\beta_{PLS} \equiv \Omega_q(\Omega'_q\Sigma_{xx}\Omega_q)^{-1}\Omega'_q\sigma_{xy} = \Sigma_{xx}\sigma_{xy}. \tag{9}$$

We notice that β_{PLS} is a continuous function of the elements of the variance-covariance matrix of $(y_t, x'_{t-1})'$. Since the PLS algorithm by Wold (1985) builds the weight matrix $\widehat{\Omega}_q$ such that it lies in the space spanned by

$$(\widehat{\sigma}_{xy}, \widehat{\Sigma}_{xx}\widehat{\sigma}_{xy}, \ldots, \widehat{\Sigma}_{xx}^{q}\widehat{\sigma}_{xy}),$$

where $\widehat{\sigma}_{xy}$ and $\widehat{\Sigma}_{xx}$ are, respectively, the natural estimators of σ_{xy} and Σ_{xx}, we conclude that $\widehat{\beta}_{PLS}$ is a consistent estimator of β by the Slutsky's theorem.

Remarkably, the weight matrix $\widehat{\Omega}_q$ is computed without requiring any covariance matrix inversion. Hence, this method is very attractive when N is large compared to the sample size T.

A problem arises about the estimation of the cointegrating matrix γ. Since r is unknown and we can not resort to a maximum likelihood approach (Johansen 1996) due to the large predictor number N, we use the first lag of the principal components of z_t in place of $\gamma' z_{t-1}$ in model (2).

The rationale of resorting to principal components is that Snell (1999) proved that the eigenvectors associated with the r smallest eigenvalues of $z'z$ are a super-consistent estimator (up to an identification matrix) of the cointegrating vectors γ whereas the other $(n-r)$ principal components converge to I(1) processes. Hence, the principal components associated with the $(n-r)$ largest eigenvalues of $z'z$ will be asymptotically uncorrelated with the I(0) target variable y_t.

4 The Data-Set

We employ the same data set as Stock and Watson (2005), which contains 131 US monthly time series covering a broad range of economic categories such as industrial production, income, unemployment, employment, wages, stock prices, and consumer and producer prices. The time span is from 1959.01 through to 2003.12.

We apply logarithms to most of the series with the exception of those already expressed in rates. All variables are transformed to achieve stationarity.

Moreover, we compute the principal components of the 103 time series that appear to be I(1), possibly after differencing. The first lag of these principal components are then included in the predictor set in place of the unknown error correction terms. All the predictors are then standardized to have unit variance.

Finally, the predicted variables are Industrial Production (IP), Employment (EMP), Federal Funds Rate (FYFF), and Consumer Price Index (CPI).

5 Estimation and Forecasting

Following Stock and Watson (2002a, 2002b) and De Mol et al. (2008), the PLS h-step ahead forecasts, for $h = 1, 3, 6, 12$, of variable $i = \text{IP}, \text{EMP}, \text{FYFF}, \text{CPI}$, are obtained as

$$(1 - L^h)\widehat{z}_{i,t+h} = \widehat{\xi}'_{i,h}\widehat{f}_t,$$

where the PLS factors are computed as detailed in the previous section and the loadings are estimated by GLS, allowing for both heteroskedasticity and autocorrelation of order $(h-1)$.

Both the PLS factors and their loadings are estimated using a rolling window of 299 observations.

The number of PLS factors, q, and the number, p, of lags of Δz_t to be included in x_{t-1}, p, are fixed by minimizing the 6-step head mean square forecast error (MSFE) that is computed using the training sample 1959.01–1979.12 and the validation sample 1980.01–1984.12.

In order to check out the best number of long-run relations r to be considered in the model, we apply a second training procedure. First of all, we reorder the principal components according to magnitude of their PLS regressions coefficients, then we run n PLS regressions such that the $i-th$ model includes i reordered principal components for $i = 1, \ldots, n$.

Finally, we choose the number of cointegrating vectors r that is associated with the model that achieves the lowest 6-step head MSFE, given the values of p, q previously selected.

We choose BR as competitor of PLS. In particular we compare directly our results with the ones obtained by De Mol et al. (2008), as they provided convincing evidence that BR forecasts outperforms those of dynamic factor models using the same data-set that we are considering here.

The h-steps ahead BR forecasts are obtained by running the same code as in De Mol et al. (2008).[1] Both the shrinking parameter and the rolling window are the same chosen by the authors. More in detail, we consider a rolling window of 120 observations while the shrinking parameter is selected such that the fit in the training sample explains a given fraction, from 0.1 up to 0.9, of the variance of the variable to be forecasted.

In the following tables we report the MSFE relative to the naive random walk forecast of both PLS and BR for the evaluation period 1985.01–2002.12.

In order to evaluate the relevance of the long-run term in the PLS forecasting exercise we consider also the cases in which $r = 0$ and $r = n$ (Tables 1, 2, 3, 4).

The empirical findings suggest that, starting from a data-set of 131 variables, only a few number of PLS factors are sufficient for obtaining a good forecasting performance.

The comparison reveals that the PLS forecasts are often more accurate than those obtained by BR, particularly when the forecasting horizon is large. Anyway PLS

Table 1 IPI, Relative MSFE[a]

Models	$h = 1$	$h = 3$	$h = 6$	$h = 12$
PLS($r = 0$)	0.81	0.78	0.89	1.04
PLS($r = 54$)	0.80	0.74	0.84	1.07
PLS($r = n$)	0.80	0.74	0.85	1.09
BR($k = 0.1$)	2.52	1.40	1.10	1.31
BR($k = 0.2$)	1.45	0.95	0.94	1.10
BR($k = 0.3$)	1.05	0.80	0.89	1.02
BR($k = 0.4$)	0.91	0.76	0.89	0.99
BR($k = 0.5$)	0.85	0.76	0.89	0.98
BR($k = 0.6$)	0.83	0.78	0.90	0.98
BR($k = 0.7$)	0.83	0.82	0.92	0.98
BR($k = 0.8$)	0.85	0.86	0.94	0.98
BR($k = 0.9$)	0.90	0.92	0.97	0.97

[a]MSFE are relative to a Random Walk forecast. PLS forecasts are obtained using $p = 1, q = 5$; the cross validation procedure choose $r = 54$. BR forecasts are obtained using different values of k where $1-k$ is the fraction of explained variance in the training sample

[1]The replication files are available on the web page http://homepages.ulb.ac.be/dgiannon/.

Table 2 EMP, relative MSFE[a]

Models	$h = 1$	$h = 3$	$h = 6$	$h = 12$
PLS($r = 0$)	0.54	0.45	0.53	0.66
PLS($r = 103$)	0.48	0.39	0.46	0.58
PLS($r = n$)	0.48	0.39	0.46	0.58
BR($k = 0.1$)	1.84	0.46	0.58	1.11
BR($k = 0.2$)	0.87	0.38	0.53	0.91
BR($k = 0.3$)	0.59	0.38	0.54	0.82
BR($k = 0.4$)	0.51	0.42	0.57	0.78
BR($k = 0.5$)	0.48	0.47	0.62	0.77
BR($k = 0.6$)	0.49	0.54	0.67	0.77
BR($k = 0.7$)	0.53	0.63	0.74	0.79
BR($k = 0.8$)	0.61	0.73	0.81	0.83
BR($k = 0.9$)	0.75	0.85	0.90	0.89

[a]MSFE are relative to a Random Walk forecast. PLS forecasts are obtained using $p = 8$, $q = 5$; the cross validation procedure choose $r = 103$. BR forecasts are obtained using different values of k where $1 - k$ is the fraction of explained variance in the training sample

Table 3 FYFF, relative MSFE[a]

Models	$h = 1$	$h = 3$	$h = 6$	$h = 12$
PLS($r = 0$)	0.70	0.55	0.54	0.61
PLS($r = 97$)	0.69	0.53	0.52	0.57
PLS($r = n$)	0.69	0.53	0.53	0.58
BR($k = 0.1$)	5.26	2.73	1.36	1.52
BR($k = 0.2$)	3.00	1.56	0.99	1.30
BR($k = 0.3$)	2.09	1.01	0.88	1.17
BR($k = 0.4$)	1.55	0.78	0.83	1.08
BR($k = 0.5$)	1.18	0.70	0.81	1.01
BR($k = 0.6$)	0.92	0.68	0.81	0.96
BR($k = 0.7$)	0.76	0.71	0.84	0.94
BR($k = 0.8$)	0.70	0.77	0.87	0.93
BR($k = 0.9$)	0.77	0.87	0.93	0.95

[a]MSFE are relative to a Random Walk forecast. PLS forecasts are obtained using $p = 3$, $q = 5$; the cross validation procedure choose $r = 97$. BR forecasts are obtained using different values of k where $1 - k$ is the fraction of explained variance in the training sample

sometime improves over BR even for short forecasting horizon particularly when the target variable is more volatile (i.e. Industrial Production, Federal Funds Rate).

In the few cases when PLS is outperformed by BR, the former methods is still a close competitor.

Turning to the choice of the number of cointegrating vectors, r, we notice that the lowest relative mean square forecast error is obtained, with few exceptions, by fixing r according to the cross validation results. In general, the inclusion of the error corrections terms in the predictor set seems beneficial for PLS forecasting but for the case of CPI, for which the cross-validation procedure suggests to fix r equal to 1.

Table 4 CPI, relative MSFE[a]

Models	$h = 1$	$h = 3$	$h = 6$	$h = 12$
PLS($r = 0$)	0.89	0.91	0.83	0.71
PLS($r = 1$)	0.90	0.91	0.83	0.71
PLS($r = n$)	0.89	0.94	0.86	0.72
BR($k = 0.1$)	1.94	2.69	1.81	1.62
BR($k = 0.2$)	1.29	1.65	1.37	1.43
BR($k = 0.3$)	1.03	1.29	1.17	1.31
BR($k = 0.4$)	0.90	1.11	1.06	1.23
BR($k = 0.5$)	0.83	1.01	0.99	1.14
BR($k = 0.6$)	0.80	0.95	0.95	1.08
BR($k = 0.7$)	0.81	0.94	0.94	1.02
BR($k = 0.8$)	0.84	0.93	0.94	0.97
BR($k = 0.9$)	0.90	0.95	0.95	0.95

[a]MSFE are relative to a Random Walk forecast. PLS forecasts are obtained using $p = 1, q = 5$; the cross validation procedure choose $r = 1$. BR forecasts are obtained using different values of k where $1 - k$ is the fraction of explained variance in the training sample

6 Conclusions

In this paper we have assessed the forecasting performances of a single equation error correction model in a data rich environment. In order to overcome the curse of dimensionality problem, we have suggested the use of PLS for estimating the parameters of the model. Having discussed the theoretical aspects of our approach, we have applied our method to the well-known data set of 131 US macro time series. The empirical comparison with the BR approach proposed by De Mol et al. (2008) has revealed that the suggested method often outperforms the competitor, especially when the forecasting horizons become larger.

Another empirical finding that we emphasize is that, taking into account the number of long-run relations, generally, helps to achieve better performances for large forecast horizons.

Acknowledgements We thank Alain Hecq as well as an anonymous referee for useful comments and corrections. The usual disclaimers apply.

References

Banbura M, Giannone D, Reichlin L (2010) Large bayesian vector auto regressions. Journal of Applied Econometrics 25:71–72.

Banerjee A, Marcellino M (2007) Factor augmented error correction models. In: Castle J, Shepard N (eds) The methodology and practice of econometrics: a festschrift for David Hendry, pp. 227–254, Oxford University Press, Oxford

Boivin J, Ng S (2006) Are more data always better for factor analysis? Journal of Econometrics 132:169–194

Cubadda G, Hecq A (2011) Testing for common autocorrelation in data rich environments, Journal of Forecasting 30:325–335. DOI: 10.1002/for.1186
De Mol C, Giannone D, Reichlin L (2008) Forecasting using a large number of predictors: is bayesian regression a valid alternative to principal components? Journal of Econometrics 146:318–328
Engle R, Granger C (1987) Co-integration and error correction: representation, estimation and testing. Econometrica 55:2:251–276
Forni M, Lippi M, Reichlin L (2004) The Generalized factor model: consistency and rates. Journal of Econometrics 119:231–255
Forni M, Lippi M, Reichlin L (2005) The generalized dynamic factor model: one-sided estimation and forecasting. Journal of the American Statistical Association 100:830–840
Gröen JJ, Kapetanios G (2008) Revisiting useful approaches to data-rich macroecnomic forecasting. Federal Reserve Bank of New York Staff Reports 327
Helland S (1990) Partial least squares regression and statistical models. Scandinavian Journal of Statistics 17:97–114
Helland S, Almoy T (1994) Comparison of prediction methods when only a few components are relevant. Journal of the American Statistical Association 89:583–591
Johansen S (1996) Likelihood-based inference in cointegrated vector autoregressive models. Oxford University Press, Oxford
Snell A, (1999) Testing for r versus $r-1$ cointegrating vectors. Journal of Econometrics 88: 151–191
Stock JH, Watson MW (2002a) Forecasting using principal components from a large number of predictors. Journal of the American Statistical Association 97:1167–1179
Stock JH, Watson MW (2002b) Mocroeconomic forecasting using diffusion index. Journal of Business and Economics Statistics 20:147–162
Stock JH, Watson MW (2005) An empirical comparison of methods for forecasting using many predictors, Manuscript, Princeton University, Princeton
Wold H (1985) Partial least squares. In Kotz S, Johnson NL (eds) Encyclopedia of the Statistical Sciences 6:581–591. John Wiley & Sons, New York

Large-Scale Portfolio Optimisation with Heuristics

Manfred Gilli and Enrico Schumann

Abstract Heuristic optimisation techniques allow to optimise financial portfolios with respect to different objective functions and constraints, essentially without any restrictions on their functional form. Still, these methods are not widely applied in practice. One reason for this slow acceptance is the fact that heuristics do not provide the "optimal" solution, but only a stochastic approximation of the optimum. For a given problem, the quality of this approximation depends on the chosen method, but also on the amount of computational resources spent (e.g., the number of iterations): more iterations lead (on average) to a better solution. In this paper, we investigate this convergence behaviour for three different heuristics: Differential Evolution, Particle Swarm Optimisation, and Threshold Accepting. Particular emphasis is put on the dependence of the solutions' quality on the problem size, thus we test these heuristics in large-scale settings with hundreds or thousands of assets, and thousands of scenarios.

1 Overview

The popularity of mean–variance optimisation (Markowitz 1952) in finance stems to a considerable extent from the ease with which models can be solved numerically. The model requires that investors decide on their portfolios solely on the basis of the first two moments of return distribution. Given the non-Gaussian nature of financial time series (Cont 2001), alternative selection criteria have been proposed that take into account empirical regularities like asymmetric return distributions and "fat tails". But unfortunately, these criteria are rarely used in realistic portfolio selection problems, mainly because the optimisation becomes much more difficult. Heuristics

M. Gilli (✉) · E. Schumann
Department of Econometrics, University of Geneva, Switzerland
e-mail: Manfred.Gilli@unige.ch; EnricoSchumann@yahoo.de

A. Di Ciaccio et al. (eds.), *Advanced Statistical Methods for the Analysis of Large Data-Sets*, Studies in Theoretical and Applied Statistics,
DOI 10.1007/978-3-642-21037-2_17,

have been shown to work well for such problems that are completely infeasible for standard techniques like gradient-based methods (for heuristics in general, see Michalewicz and Fogel (2004); for financial applications, see Maringer (2005)).

In this short paper, we will discuss the application of heuristics to such portfolio selection problems. The emphasis will be on large-scale problems: while mathematically the properties of a given optimisation problem may not change with its size, the same does not hold true numerically. The paper analyses the performance of three particular methods, Differential Evolution, Particle Swarm Optimisation, and Thresholds Accepting, and how their performance changes with the size of the problem. The paper is structured as follows: Sect. 2 discusses the model and different optimisation techniques; Sect. 3 gives results for two problems; Sect. 4 concludes.

2 Methodology and Techniques

2.1 A One-Period Model

There are $n_{\mathscr{A}}$ risky assets available. We are endowed with an initial wealth v_0, and wish to select a portfolio $x = [x_1\ x_2\ \ldots\ x_{n_{\mathscr{A}}}]'$ of the given assets (the x represent weights). The chosen portfolio is held for one period, from time 0 to time T. End-of-period wealth is given by

$$v_T = v_0 \sum_{n_{\mathscr{A}}} (1 + y^j) x_j$$

where the vector y holds the assets' returns between time 0 and T. Since these returns are not known at the time when the portfolio is formed, v_T will be a random variable, following an unknown distribution that depends on x. We will often rescale v_T into returns.

The problem is then to choose the x according to some selection criterion Φ which may be a function of v_T (final wealth), or the path that wealth takes $\{v_t\}_0^T$. In the latter case, we still have a one period problem since we do not trade between 0 and T. The basic problem can be written as

$$\begin{aligned} &\min_x\ \Phi\,, \\ &\textstyle\sum_{n_{\mathscr{A}}} x = 1\,. \end{aligned} \tag{1}$$

Typical building blocks for Φ may be central moments (e.g., variance), also partial or conditional moments, or quantiles. Criteria that depend on the path of wealth are often based on the drawdown, see Gilli and Schumann (2009) for a discussion of different possibilities. In this paper we will only discuss Φ that depend on v_T.

We will conduct scenario optimisation, i.e., we assume we can obtain a sample of v_T. This is equivalent to directly working with empirical distribution function of returns. It is not restrictive: if we preferred a parametric approach, we could always estimate parameters from our scenarios. Given a sample of portfolio returns, we can easily evaluate any objective function, for instance compute moments.

The only restriction included in the problem above is the budget constraint. The solution to this problem would thus not be very relevant in practice, we will need more constraints: we may introduce minimum and maximum holding sizes x^{inf} and x^{sup} for the assets included in the portfolios, and also cardinality constraints that set a minimum and maximum number of assets. We can also include other kinds of constraints like exposure limits to specific risk factors, or sector constraints.

2.2 Techniques

Model (1) combines combinatorial (select a number of assets from the set of all $n_{\mathscr{A}}$ assets) and continuous elements (the objective function, maximum holding size constraints); it can generally not be solved with standard methods since it is not convex and exhibits many local minima. As an example, Fig. 1 shows the search space (i.e., the mapping from portfolio weights into the objective function values) for a three asset problem with the objective to minimise Value-at-Risk, i.e., a quantile of the distribution of final wealth. The third asset's weight is implicitly defined by the budget constraint. A deterministic method would stop at the first local

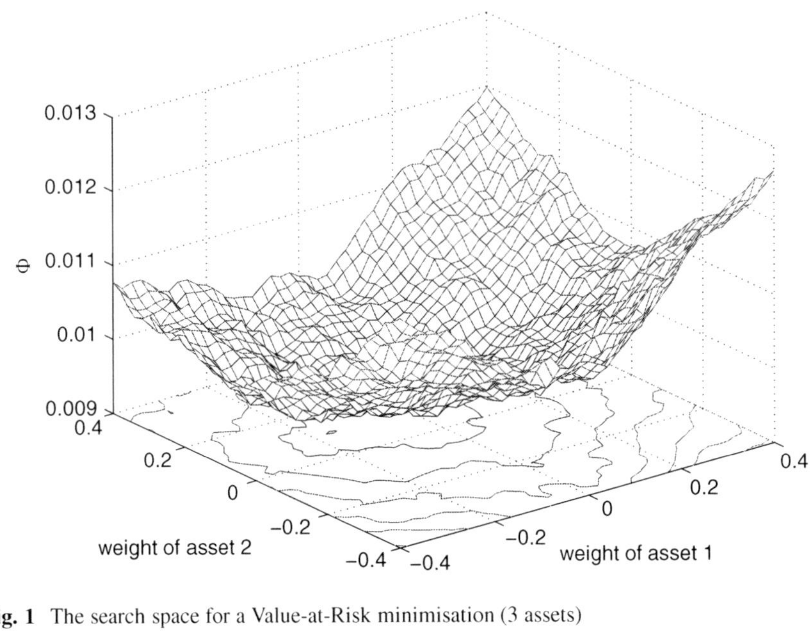

Fig. 1 The search space for a Value-at-Risk minimisation (3 assets)

minimum encountered. Heuristics on the other hand are designed to escape local minima and thus are appropriate methods for such problems. We compare three different techniques: Differential Evolution (DE; Storn and Price (1997)), Particle Swarm Optimisation (PS; Eberhart and Kennedy (1995)), and Threshold Accepting (TA; Dueck and Scheuer (1990); Moscato and Fontanari (1990)). The main emphasis of the paper is on the performance of these methods for large-scale instances of model (1) under constraints.

2.2.1 Differential Evolution

DE evolves a population of n_P solutions, stored in real-valued vectors of length p (i.e., a vector of portfolio weights). The population P may be visualised as a matrix of size $p \times n_P$, where each column holds one candidate solution. In every iteration (or 'generation'), the algorithm goes through the columns of this matrix and creates a new candidate solution for each existing solution $P^{(0)}_{\cdot,i}$. This candidate solution is constructed by taking the difference between two other solutions, weighting this difference by a parameter F, and adding it to a third solution. Then an element-wise crossover takes place with probability CR between this auxiliary solution $P^{(v)}_{\cdot,i}$ and the existing solution $P^{(0)}_{\cdot,i}$ (the symbol ζ represents a random variable that is uniformly distributed between zero and one). If this final candidate solution $P^{(u)}_{\cdot,i}$ is better than $P^{(0)}_{\cdot,i}$, it replaces it; if not, the old solution $P^{(0)}_{\cdot,i}$ is kept. Algorithm 1 describes an unconstrained optimisation procedure. We include

Algorithm 1 Differential Evolution

initialise parameters n_P, n_G, F and CR
initialise population $P^{(1)}_{j,i}$, $j = 1, \ldots, p$, $i = 1, \ldots, n_P$
for $k = 1$ to n_G **do**
  $P^{(0)} = P^{(1)}$
  for $i = 1$ to n_P **do**
    generate $\ell_1, \ell_2, \ell_3 \in \{1, \ldots, n_P\}$, $\ell_1 \neq \ell_2 \neq \ell_3 \neq i$
    compute $P^{(v)}_{\cdot,i} = P^{(0)}_{\cdot,\ell_1} + \mathrm{F} \times (P^{(0)}_{\cdot,\ell_2} - P^{(0)}_{\cdot,\ell_3})$
    for $j = 1$ to p **do**
      if $\zeta < \mathrm{CR}$ **then** $P^{(u)}_{j,i} = P^{(v)}_{j,i}$ **else** $P^{(u)}_{j,i} = P^{(0)}_{j,i}$
    end for
    if $\Phi(P^{(u)}_{\cdot,i}) < \Phi(P^{(0)}_{\cdot,i})$ **then** $P^{(1)}_{\cdot,i} = P^{(u)}_{\cdot,i}$ **else** $P^{(1)}_{\cdot,i} = P^{(0)}_{\cdot,i}$
  end for
end for

constraints as follows: any solution $P_{\cdot,j}$ is first checked for elements smaller than $x^{\inf}$, these are set to zero. With this approach it is necessary to have a sufficient population size, lest too many solutions converge prematurely on zero elements. Then, the budget constraint is enforced by dividing every element by the sum of the x. All other constraints are included through penalty terms.

We ran a number of preliminary experiments to find appropriate values for F and CR. F should be chosen rather small (in the range 0.1 to 0.3). CR had less influence on results, but here again low values worked best, implying that only few x_j should be changed at a time. Both parameters suggest that the step size – the change from one solution to a new candidate solution – should be small.

2.2.2 Particle Swarm Optimisation

In PS, we have again a population that comprises n_P solutions (portfolio vectors). In every generation, a solution is updated by adding another vector called velocity v_i. Velocity changes over the course of the optimisation, the magnitude of change is the sum of two components: the direction towards the best solution found so far by the particular solution, $Pbest_i$, and the direction towards the best solution of the whole population, $Pbest_{\text{gbest}}$. These two directions are perturbed via multiplication with a uniform random variable ζ and constants $c_{(\cdot)}$, and summed, see Statement 7. The vector so obtained is added to the previous v_i, the resulting updated velocity is added to the respective solution. Algorithm 2 describes the procedure. The efficiency

Algorithm 2 Particle Swarm

1: initialise parameters n_P, n_G, δ, c_1 and c_2
2: initialise particles $P_i^{(0)}$ and velocity $v_i^{(0)}$, $i = 1, \ldots, n_P$
3: evaluate objective function $F_i = \Phi(P_i^{(0)})$, $i = 1, \ldots, n_P$
4: $Pbest = P^{(0)}$, $Fbest = F$, $Gbest = \min_i(F_i)$, $gbest = \text{argmin}_i(F_i)$
5: **for** $k = 1$ to n_G **do**
6:     **for** $i = 1$ to n_P **do**
7:         $\Delta v_i = c_1 \times \zeta_1 \times (Pbest_i - P_i^{(k-1)}) + c_2 \times \zeta_2 \times (Pbest_{\text{gbest}} - P_i^{(k-1)})$
8:         $v_i^{(k)} = \delta v^{(k-1)} + \Delta v_i$
9:         $P_i^{(k)} = P_i^{(k-1)} + v_i^{(k)}$
10:     **end for**
11:     evaluate objective function $F_i = \Phi(P_i^{(k)})$, $i = 1, \ldots, n_P$
12:     **for** $i = 1$ to n_P **do**
13:         **if** $F_i < Fbest_i$ **then** $Pbest_i = P_i^{(k)}$ and $Fbest_i = F_i$
14:         **if** $F_i < Gbest$ **then** $Gbest = F_i$ and $gbest = i$
15:     **end for**
16: **end for**

of PS can often be improved by systematically reducing velocities over time; this is achieved by setting the parameter δ to a value smaller than unity. We implement constraints in the same way as for DE which insures better comparability. Here again, we ran experiments to find appropriate values for c_1, c_2 and δ. PS appears relatively insensitive to parameter settings for this kind of problem: c_1 should have a higher weight than c_2, but performance differences were small. For both DE and PS the number of objective function evaluations is $n_P \times n_G$.

2.2.3 Threshold Accepting

TA is a local search method; other than DE and PS, it evolves only one solution at a time. In every iteration, a new solution x^n is proposed that results from a slight perturbation of the old solution x^c. A classical local search accepts such a new solution only if it improves on the old solution, or is at least not worse. But TA may also move uphill in the search space. More specifically, it accepts new solutions that are inferior when compared with the current solution, as long as the deterioration does not exceed a specified threshold. (Simulated Annealing (Kirkpatrick et al. 1983), to which TA is closely related, employs a stochastic acceptance criterion.) Over time, this threshold decreases to zero, and so TA turns into a classical local search. Algorithm 3 describes the procedure. For an in-depth description see Winker (2001); for portfolio applications, see Gilli and Schumann (2010b). For each of the

Algorithm 3 Threshold Accepting

```
initialise n_Rounds and n_Steps
compute threshold sequence τ
randomly generate current solution x^c
for r = 1 : n_Rounds do
    for i = 1 : n_Steps do
        generate x^n ∈ N(x^c)  and compute  Δ = Φ(x^n) − Φ(x^c)
        if Δ < τ_r then x^c = x^n
    end for
end for
x^sol = x^c
```

n_{Rounds} thresholds, stored in the vector τ, the algorithm performs n_{Steps} iterations, so the number of objective function evaluations is $n_{\text{Rounds}} \times n_{\text{Steps}}$. We set n_{Rounds} to 10, even though TA is robust for other choices (Gilli and Schumann 2010a). The thresholds are then computed with the approach described in Gilli et al. (2006).

The neighbourhood $\mathcal{N}$ can be intuitively implemented for TA: randomly select one asset from the portfolio, sell a small quantity z; then select another asset from all available assets, and invest z into this asset. This neighbourhood automatically ensures that the budget constraint is not violated. Other constraints are implemented by penalty terms.

2.3 *Data and Computational Complexity*

In this paper we focus on the efficacy of specific optimisation techniques – not on the empirical, out-of-sample performance of the selected portfolios. We thus create our data through simulation. The following characteristics are desirable for a data-generating model:

- The data model should be easily scalable in terms of observations $n_{\mathcal{O}}$ and numbers of assets $n_{\mathcal{A}}$.
- The generated data should exhibit certain characteristics that are empirically relevant, in particular should the data series be correlated.
- The data series should exhibit sufficient variability in the cross-section. A qualitative aim of portfolio optimisation is to select "good" assets from a potentially large universe. Numerically, many similar assets make the optimisation harder; but empirically, there is little interest in searching flat landscapes: any solution is about as good as any other.

We are only interested in the cross-section of returns (our Φ depends only on v_T), not in temporal dependencies. So we model the jth asset's return by a linear factor model

$$y^j = \beta_0^j + \beta_M^j f_M + \epsilon^j . \tag{2}$$

The β_M-values are randomly drawn for each asset; they are distributed normally around 1 with a standard deviation of 0.2; β_0 is zero for all assets. By drawing market returns f_M and errors ϵ from specific distributions (here we use standard Gaussian variates) and inserting them into (2), we create a scenario set Y which is a matrix of size $n_{\mathcal{O}} \times n_{\mathcal{A}}$. Each row of Y holds asset returns for one scenario, each column holds the returns of a specific asset. (The so-created returns roughly resemble actual daily equity returns.)

For mean–variance optimisation, we estimate the means, variances and covariances of all assets, independently of specific portfolio weights. All information is then condensed into a mean-return vector of size $n_{\mathcal{A}} \times 1$, and a variance–covariance-matrix of size $n_{\mathcal{A}} \times n_{\mathcal{A}}$. Hence, computing time will increase with the number of selectable assets, but not with the number of observations. The heuristics described above evolve whole portfolios. In every iteration then, we evaluate new candidate solutions by first computing a set of portfolio returns for the respective solution, i.e., $y^{\mathrm{P}} = Yx$. Thus, the computing time will rise with the number of assets, but also with the number of scenarios. The matrix Y may be large, hence the multiplication is expensive. For our TA implementation, we can update the multiplication, since we have

$$\begin{aligned} x^{\mathrm{n}} &= x^{\mathrm{c}} + x^{\Delta} \\ Yx^{\mathrm{n}} &= Y(x^{\mathrm{c}} + x^{\Delta}) = \underbrace{Yx^{\mathrm{c}}}_{\text{known}} + Yx^{\Delta} . \end{aligned}$$

Hence, let Y_* denote the submatrix of changed columns (size $n_{\mathcal{O}} \times 2$) and x_*^{Δ} the vector of weight changes (size 2×1), then we can replace Yx^{Δ} by $Y_* x_*^{\Delta}$. This makes the matrix multiplication practically independent of the number of assets. In principle, this approach could also be applied to DE and PS; but we would need to make sure that only few asset weights are changed in every iteration (e.g., we could no longer repair the budget constraint by rescaling).

3 Convergence Results

Heuristic methods are, with only few exceptions, stochastic algorithms. Every restart of the algorithm can thus be regarded as the realisation of a random variable with some unknown distribution $\mathscr{D}$. For a given problem, this distribution will depend on the chosen method: some techniques may be more appropriate than others, and thus give less variable and better results. But the stochastics stem not only from the method, but also from the problem. Think of the search space in Fig. 1: even if we used a non-stochastic method, the many local minima would make sure that repeated runs from different starting points would result in different solutions. In other words, repeated runs of a technique can be perceived as draws from $\mathscr{D}$. In the case of heuristics, the shape of $\mathscr{D}$ is influenced by the amount of computational resources (i.e., the number of iterations) spent. Since heuristics can move away from local minima, allowing more iterations leads on average to better solutions. (There are convergence proofs for several heuristic methods; unfortunately, these proofs are useless for practical applications.) In this section, we compare this convergence behaviour for all our methods. We will look into two specific cases.

3.1 Minimising Squared Variation

An advisable test is to run newly-implemented algorithms on problems that can be solved by other, reliable techniques. This helps to find mistakes in the implementation, and builds intuition of the stochastics of the solutions. The prime candidate for such a benchmark is a mean–variance problem with few restrictions (budget constraint, and minimum and maximum holding sizes) such that an exact solution is computable by quadratic programming (we use Matlab's `quadprog` function). The problem can be summarised as

$$\begin{aligned} &\min_x \ \Phi \\ &\textstyle\sum_{n_{\mathscr{A}}} x = 1\,, \\ &0 \leq x_j \leq x^{\text{sup}} \qquad \text{for } j = 1, 2, \ldots, n_{\mathscr{A}}\,. \end{aligned} \tag{3}$$

We set x^{sup} to 3%. Φ is the squared return which is very similar to variance. The data-set consists of 250 assets and 2 500 observations.

We conduct 500 optimisation runs with 20 000 and 200 000 function evaluations (FE). For each solution, we record the solution quality, that is, the obtained objective function value. The distributions $\mathscr{D}$ of these values are pictured in Fig. 2 (the upper

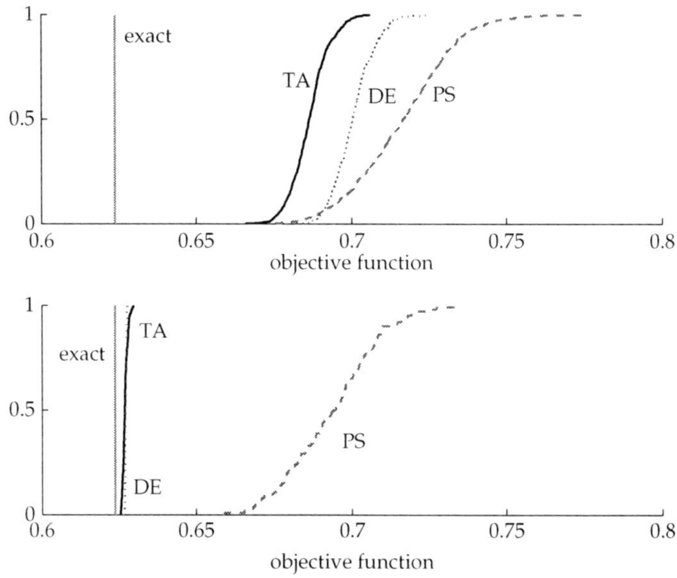

Fig. 2 Distributions $\mathscr{S}$ of solutions for minimising squared returns

panel for 20 000 function evaluations, the lower panel 200 000 function evaluations). The objective function is given in percentage points (e.g., 0.6 is 0.6%). The distributions of DE and TA come rapidly close to the exact solution, but it takes long until they converge. PS in particular converges very slowly. However, studies on real data-sets suggest that the remaining suboptimality of heuristics is negligible when compared with estimation error for portfolio selection problems. For a discussion of this point, see Gilli and Schumann (2011).

We could actually speed up convergence: the best algorithm will be the one that resembles a gradient search as closely as possible. For instance, if we set all thresholds for TA to zero, the solution would converge faster. Heuristics deliberately employ strategies like randomness, or acceptance of inferior solutions. These strategies are necessary to overcome local minima, but make heuristics inefficient for well-behaved (i.e., convex) problems when compared with classical methods. But changing problem (3) only slightly already renders classic methods infeasible.

3.2 *Minimising Losses*

We replace the objective function by a partial moment and set minimum and maximum holding sizes only for those assets included in the portfolio. Then our optimisation problem becomes

$$\begin{aligned} &\min_x \ \Phi \\ &\textstyle\sum_{n_{\mathscr{A}}} x = 1\,, \\ &x^{\inf} \leq x_j \leq x^{\sup} \qquad \text{for all } j \in \mathscr{J}, \end{aligned} \tag{4}$$

where $\mathscr{J}$ is the set of those assets in the portfolio. We set $x^{\inf}$ to 0.5%, and $x^{\sup}$ to 3%. The size of the set $\mathscr{J}$ is thus implicitly defined: we need at least 34 (smallest integer greater than $1/3\%$) assets, and at most 200 assets ($1/0.5\%$). Data is generated as before; we create 2 500 scenarios, and now vary the number of assets. Selection criterion is the lower partial moment of order one, which we compute as

$$\Phi = -\frac{1}{n_{\mathscr{O}}} \sum_{n_{\mathscr{O}}} \min(y^{\mathrm{P}}, 0).$$

As benchmarks, we use randomly-chosen equal-weight portfolios with the correct cardinality.

Results are given in Fig. 3 for 250 assets, and in Fig. 4 for 2 000 assets. The benchmark distributions are given in grey (unlabelled). DE converges very rapidly to good solutions, TA follows. Though we have no proof to have found the global minimum, it is encouraging that both methods converge on similar solutions. PS however, converges only very slowly. In particular for 2 000 assets, it requires a far greater number of FE to provide solutions comparable with those of DE and TA.

A final comment, on computing time. All algorithms were implemented and tested in Matlab 2007b. Run times for the last problem but with 20 000 scenarios and 100 000 FE are given in the following table (in seconds; processor was an Intel P8700 (single core) at 2.53GHz with 2GB RAM).

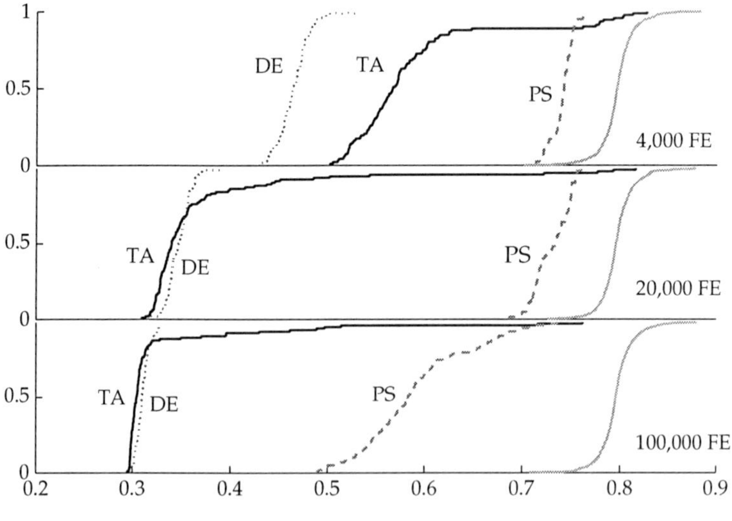

Fig. 3 Distributions $\mathscr{D}$ of solutions for minimising a lower partial moment (250 assets)

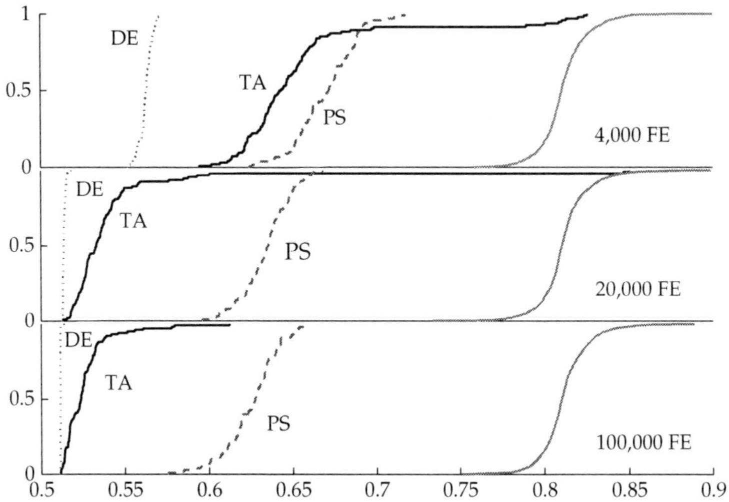

Fig. 4 Distributions $\mathscr{D}$ of solutions for minimising a lower partial moment (2 000 assets)

	500 assets	1500 assets
DE	228	590
PS	221	569
TA	49	58

Indeed, the computing time of TA does hardly change when we increase the number of assets.

4 Conclusion

In this paper, we have briefly detailed several optimisation heuristics and their application to portfolio selection problems. Our computational experiments suggest that DE and TA are well-capable of solving the problems, whereas PS needed more computing time to provide comparably good solutions.

References

Rama Cont. Empirical Properties of Asset Returns: Stylized Facts and Statistical Issues. *Quantitative Finance*, 1: 223–236, 2001.

Gunter Dueck and Tobias Scheuer. Threshold Accepting. A General Purpose Optimization Algorithm Superior to Simulated Annealing. *Journal of Computational Physics*, 90(1): 161–175, September 1990.

Russell C. Eberhart and James Kennedy. A new optimizer using particle swarm theory. In *Proceedings of the Sixth International Symposium on Micromachine and Human Science*, pages 39–43, Nagoya, Japan, 1995.

Manfred Gilli and Enrico Schumann. An Empirical Analysis of Alternative Portfolio Selection Criteria. *Swiss Finance Institute Research Paper No. 09-06*, 2009.

Manfred Gilli and Enrico Schumann. Distributed Optimisation of a Portfolio's Omega. *Parallel Computing*, 36(7): 381–389, 2010a.

Manfred Gilli and Enrico Schumann. Portfolio Optimization with "Threshold Accepting": a Practical Guide. In Stephen E. Satchell, editor, *Optimizing Optimization: The Next Generation of Optimization Applications and Theory*. Elsevier, 2010b.

Manfred Gilli and Enrico Schumann. Optimal Enough? *Journal of Heuristics*, 17(4): 373–387, 2011.

Manfred Gilli, Evis Këllezi, and Hilda Hysi. A Data-Driven Optimization Heuristic for Downside Risk Minimization. *Journal of Risk*, 8(3): 1–18, 2006.

S. Kirkpatrick, C. D. Gelatt, and M. P. Vecchi. Optimization by Simulated Annealing. *Science*, 220(4598): 671–680, May 1983.

Dietmar Maringer. *Portfolio Management with Heuristic Optimization*. Springer, 2005.

Harry M. Markowitz. Portfolio selection. *Journal of Finance*, 7(1): 77–91, March 1952.

Zbigniew Michalewicz and David B. Fogel. *How to Solve it: Modern Heuristics*. Springer, 2004.

Pablo Moscato and J.F. Fontanari. Stochastic Versus Deterministic Update in Simulated Annealing. *Physics Letters A*, 146(4): 204–208, 1990.

Rainer M. Storn and Kenneth V. Price. Differential Evolution – a Simple and Efficient Heuristic for Global Optimization over Continuous Spaces. *Journal of Global Optimization*, 11(4): 341–359, 1997.

Peter Winker. *Optimization Heuristics in Econometrics: Applications of Threshold Accepting*. Wiley, 2001.

Detecting Short-Term Cycles in Complex Time Series Databases

F. Giordano, M.L. Parrella and M. Restaino

Abstract Time series characterize a large part of the data stored in financial, medical and scientific databases. The automatic statistical modelling of such data may be a very hard problem when the time series show "complex" features, such as nonlinearity, local nonstationarity, high frequency, long memory and periodic components. In such a context, the aim of this paper is to analyze the problem of detecting automatically the different periodic components in the data, with particular attention to the short term components (weakly, daily and intra-daily cycles). We focus on the analysis of real time series from a large database provided by an Italian electric company. This database shows complex features, either for the high dimension or the structure of the underlying process. A new classification procedure we proposed recently, based on a spectral analysis of the time series, was applied on the data. Here we perform a sensitivity analysis for the main tuning parameters of the procedure. A method for the selection of the optimal partition is then proposed.

1 Introduction

Time series data mining has attracted great attention in the statistical community in recent years. The automatic statistical modelling of large databases of time series may be a very hard problem when the time series show complex features, such as nonlinearity, nonstationarity, high frequency, long memory and periodic components. Classification and clustering of such "complex objects" may be particularly beneficial for the areas of model selection, intrusion detection and

F. Giordano (✉) · M.L. Parrella · M. Restaino
Department of Economics and Statistics, University of Salerno, Via Ponte Don Melillo, 84084, Fisciano (SA), Italy
e-mail: giordano@unisa.it; mparrella@unisa.it; mlrestaino@unisa.it

A. Di Ciaccio et al. (eds.), *Advanced Statistical Methods for the Analysis of Large Data-Sets*, Studies in Theoretical and Applied Statistics,
DOI 10.1007/978-3-642-21037-2_18, © Springer-Verlag Berlin Heidelberg 2012

pattern recognition. As a consequence, time series clustering represents a first mandatory step in temporal data mining research.

In this paper we consider the case when the time series show periodic components of different frequency. Most of the data generated in the financial, medical, biological and technical fields present such features. A clustering procedure useful in such a context has been proposed in Giordano et al. (2008). It is based on the detection of the dominant frequencies explaining large portions of variation in the data through a spectral analysis of the time series. The approach considered here for the classification of the time series is completely different from those proposed so far (see, for example, Corduas and Piccolo 2008; Alonso et al. 2006). Like all clustering methods, also our procedure depends on some kind of *tuning parameters*, which directly influence the results of the classification. The purpose of this paper is to perform a sensitivity analysis on such parameters, and to control if there is a particular behaviour in the results which could help in the determination of the optimal partition and in the dynamical detection of cycles in the database.

The next section describes briefly the clustering procedure, with particular attention to the role played by the tuning parameters. In the third section we propose a method for the selection of the best partition for the clustering procedure. In the last section we present an application of the clustering procedure to a real database. It is aimed at showing some interesting results about the determination of the optimal partition.

2 The Clustering Algorithm and the Tuning Parameters

Consider a stationary process $\{X_t; t = 1, 2, \ldots\}$ and denote by $g_X(\omega)$ its spectrum. Based on the Wold decomposition, the spectrum of a stationary process may be viewed as the sum of two components

$$g_X(\omega) = g_V(\omega) + g_Z(\omega), \qquad -\pi \leq \omega \leq \pi, \tag{1}$$

where the first (the discrete component) is associated with a harmonic process V_t and the second (the continuous component) is associated with a linear process Z_t. In particular, the harmonic process V_t is given by a combination of (say m) sinusoidal functions,

$$V_t = \sum_{j=1}^{m} A_j \cos(\omega_j t + \phi_j), \qquad 0 \leq \omega_j \leq \pi, \tag{2}$$

and it identifies the cycles of the process. For a detailed and technical description of (1) and (2) we remand to the book of Priestly (1981). Here we give only the fundamentals to explain the rationale of our proposal.

The function $g_V(\omega)$ measures the variability explained by each cycle ω_j appearing in (2). Given the uncorrelation of V_t and Z_t, the variance of the process is

$$Var(X_t) = Var(V_t) + Var(Z_t).$$

We argue that, for the considered process, a large portion of the variability in the data is due to the component V_t. Therefore, a clustering procedure for this kind of time series may be naturally based on the explanation of $Var(V_t)$. As said before, this is connected with the identification of g_V, which therefore becomes the main goal of the clustering procedure. Anyway, this is not a trivial point. We want to stress here that we are not interested in the estimation of the exact value of $g_V(\omega_j)$. We only want to identify the relevant discrete frequencies, i.e. those ω_j in (2) for which $g_V(\omega_j) \neq 0$. This is equivalent to test if $A_j \neq 0$ in (2), $\forall j$. To this aim, we use the *Whittle test*, which is specifically devoted to deal with the problem of identifying the discrete component of a mixed spectrum.

The proposed algorithm is based on the following main steps (for the details and some remarks, see Giordano et al. 2008, and references therein):

1. For each time series, denoted with $u = 1, .., T$, estimate the spectrum by using any consistent estimator (see, for example, Priestley (1981))

$$\hat{g}^u_X(\omega_j), \qquad 0 \le \omega_j = \frac{2\pi j}{n} \le \pi;\ j = 0, 1, \ldots, m;\ m = \left[\frac{n}{2}\right],$$

   where $[x]$ denotes the integer part of x.
2. Use the Whittle test (with a Bartlett window) to test the hypothesis H_0: $A_j = 0$, $\forall j$. Derive the *relevant discrete frequencies* for the user u, as those frequencies ω_j for which H_0 is rejected. Denote these frequencies with $\hat{\omega}^u_l, l = 1, 2, \ldots$.
3. For each user u, extract the most important frequency $\hat{\omega}^u_{(1)}$, called the *dominant frequency*. For an easier interpretation, this frequency is converted into the correspondent period $\hat{P}$ (hours, days, etc...).
4. Derive the distribution of the *dominant periods* $\hat{P}_i$. Denote with δ_i the percentage of users which have $\hat{P}_i$ as estimated dominant period. Let $\delta_1 \ge \delta_2 \ge \ldots \ge \delta_r$.
5. For a fixed integer h_1, consider the first *most observed* dominant periods, and denote with $\Delta_{h_1} = \sum_{i=1}^{h_1} \delta_i$ the percentage of dominant periods globally covered by the first h_1 positions. The parameter h_1 is the *first tuning parameter*.

Dominating periods (hours, days, etc.)	% of users δ_i	Global % Δ_i
P_1	δ_1	Δ_1
P_2	δ_2	Δ_2
...	...	...
P_{h_1}	δ_{h_1}	Δ_{h_1}

6. Consider a fixed integer value h_2. Define the binary matrix

$$\mathbf{D} = \{d_{u,s}\} \qquad u = 1, \ldots, T; s = 1, \ldots, h_1,$$

whose generic element $d_{u,s}$ is equal to one if the period P_s, derived in the previous step, represents one of the relevant discrete periods for the user u; otherwise $d_{u,s}$ is equal to zero. Remember that the relevant discrete periods for each user u are derived in step 2. Anyway, now we consider only the first h_2 "most important ones" (i.e. those cycles with a stronger evidence against H_0 in the test). The matrix **D** acts as a dissimilarity matrix, and h_2 represents the *second tuning parameter*.

7. By considering the different combinations of the relevant periods P_i, $i = 1, \ldots, h_1$, derive the $k = 2^{h_1}$ clusters of users by associating the rows ($=$ users) of the matrix **D** with the same sequence of zeroes/ones. Note that some clusters could be empty, due to possible exclusive disjunction relationships between some of the periodic components. Denote with $k^* \leq k$ the number of *positive clusters*.
8. Derive the "best partition" of clusters by exploring different values for the tuning parameters h_1 and h_2.

Note that the first tuning parameter h_1 has the role of setting the partition, since it determines the number of dominant periods and so the number of clusters $k = 2^{h_1}$. Once fixed h_1, the second tuning parameter h_2 has the role of determining the distribution of the users among the clusters, and therefore it only reflects on the number of positive clusters k^*.

It is clear that the identification of the relevant cycles trough the Whittle test represents the crucial step of the classification algorithm (see the remarks in Giordano et al. 2008). For this reason, in order to enforce the faith in correctly identifying the discrete part of the spectrum, only the periods with a "strong impact" in the database are then considered (i.e. we take only the first most observed periods in the database ($\delta_1 \geq \delta_2 \geq \ldots, \delta_{h_1}$). This consideration must be taken into account when determining a threshold value for δ_{h_1} (or equivalently for Δ_{h_1}).

3 A Proposal for the Selection of the Optimal Partition

In order to analyze the effects on the clusters of a change in the tuning parameters h_1 and h_2, we apply the clustering procedure to a real database. The analyzed database was supplied from a large electric company. It collects the time series of energy consumption (*load curves*) observed hourly in the year 2006 for a particular category of customers. Since we look at the short term periodic components, we concentrate on the data from February to May, so the dimension of the database is 64,522 users $\times$ 2,880 observations. For each observed time series, we perform two different estimations of the spectrum. The first is based on the hourly observations, and is aimed to capture the daily and intra-daily periodic components, whereas the second is based on the aggregated daily observations, and is aimed to better identify the weekly cycles.

The selection of the "optimal" values for h_1 and h_2 will be made in two steps, by following different criteria: first of all we will select the optimal h_2 by considering mainly the movements among the clusters for different values of h_2; than we will select the optimal h_1 by considering also the time series aggregated in the clusters. The rationale behind this is that the parameter h_1 directly influences the number of potential clusters $k = 2^{h_1}$, and so the optimal partition must be selected by taking into account the efficiency in classifying the time series among the k clusters. On the other hand, the parameter h_2 only reflects on the number of positive clusters k^* and on the distribution of the T users among such clusters. So, for the selection of h_2, we consider mainly the stability of the clusters.

When moving h_1 and h_2, some useful indicators may show the features of the particular configuration. We consider four types of values, all based on well known indicators.

In order to evaluate *one single* configuration, we use:

1. The number of clusters k and the number of positive clusters k^*.
2. The (normalized) Shannon index, given by

$$S = -\sum_{i=1}^{k^*} \frac{n_i}{T} \log\left(\frac{n_i}{T}\right) / \log\left(2^{h_1}\right),$$

where n_i, $i = 1, \ldots, k^*$, indicates the number of users in the i-th (non empty) cluster. The Shannon index tends to be equal to 0 when all the users are in the same cluster, and to 1 when the users are equally distributed among the k clusters of the partition.

In order to compare *two alternative* configurations, we use:

3. The number of changes, i.e. the number of "migrations" when passing from one configuration to another.
4. The (normalized) Rand index, given by

$$R = 1 - \frac{\sum_{r=1}^{k} n_{r.}^2 + \sum_{c=1}^{k} n_{.c}^2 - 2\sum_{r=1}^{k}\sum_{c=1}^{k} n_{rc}^2}{T(T-1)},$$

where $n_{r.}$ is the number of users classified in the r-th cluster of the first partition, $n_{.c}$ is the number of users classified in the c-th cluster of the second partition, and n_{rc} is the number of users classified in the r-th cluster of the first partition which fall in the c-th cluster of the second partition. The Rand index tends to be equal to 0 when the two compared alternatives are completely different (i.e. every two users that are classified in the same cluster in the first configuration are separated in the second one), and to 1 when the two compared alternatives are equivalent.

Remark 1. There is a connection between the indicators under 1. and 2., since the Shannon entropy is directly influenced by the percentage of positive clusters. Note that we point to maximize the Shannon index and to minimize the number of clusters k and k^*.

Remark 2. The Rand index is directly connected with the number of movements among the clusters. Note that in this case we point to minimize the number of changes and to maximize the Rand index.

4 The Results on the Observed Database of Time Series Energy Consumption

Tables 1 and 2 report the results for the observed time series database for daily data and hourly data, respectively. In particular, the indicators 1. and 2. (the Shannon index and the number of clusters k and k^*) refer to the configuration of

Table 1 Performance indicators for different values of the parameters h_1 and h_2 (daily data)

	h_2	*1*	2	*3*	*4*	5	*6*	*7*	*8*
$h_1 = 1$ Δ_1: **0.869**)									
Rand index (R)		–	0.87	0.94	1	1	1	1	1
Shannon index (S)		0.85	0.93	0.96	0.96	0.96	0.96	0.96	0.96
Number of clusters k^*	(max $k = 2$)	2	2	2	2	2	2	2	2"
Number of changes		–	4402	1806	38	0	0	0	0
Mean(R,S)		–	0.9	0.95	**0.98**	0.98	0.98	0.98	0.98
$h_1 = 2$ Δ_2: **0.952**)									
Rand index (R)		–	0.87	0.94	1	1	1	1	1
Shannon index (S)		0.51	0.56	0.57	0.58	0.58	0.58	0.58	0.58
Number of clusters k^*	(max $k = 4$)	3	3	3	3	3	3	3	3
Number of changes		–	4778	1942	41	0	0	0	0
Mean(R,S)		–	0.71	0.76	**0.79**	0.79	0.79	0.79	0.79
$h_1 = 3$ Δ_3: **0.976**)									
Rand index (R)		–	0.86	0.94	1	1	1	1	1
Shannon index (S)		0.36	0.4	0.41	0.41	0.41	0.41	0.41	0.41
Number of clusters k^*	(max $k = 8$)	4	4	4	4	4	4	4	4
Number of changes		–	4847	1951	41	0	0	0	0
Mean(R,S)		–	0.63	0.67	**0.70**	0.70	0.70	0.70	0.70
$h_1 = 4$ Δ_4: **0.988**)									
Rand index (R)		–	0.86	0.94	1	1	1	1	1
Shannon index (S)		0.28	0.31	0.31	0.31	0.31	0.31	0.31	0.31
Number of clusters k^*	(max $k = 16$)	5	5	5	5	5	5	5	5
Number of changes		–	4880	1955	41	0	0	0	0
Mean(R,S)		–	0.59	0.63	0.66	0.66	0.66	0.66	0.66
$h_1 = 5$ Δ_5: **0.992**)									
Rand index (R)		–	0.86	0.94	1	1	1	1	1
Shannon index (S)		0.23	0.25	0.25	0.25	0.25	0.25	0.25	0.25
Number of clusters k^*	(max $k = 32$)	6	6	6	6	6	6	6	6
Number of changes		–	4886	1957	41	0	0	0	0
Mean(R,S)		–	0.56	0.6	0.63	0.63	0.63	0.63	0.63
$h_1 = 6$ (Δ_6: **0.995**)									
Rand index (R)		–	0.86	0.94	1	1	1	1	1
Shannon index (S)		0.19	0.21	0.21	0.21	0.21	0.21	0.21	0.21
Number of clusters k^*	(max $k = 64$)	7	8	8	8	8	8	8	8
Number of changes		–	4909	1959	41	0	0	0	0
Mean(R,S)		–	0.54	0.58	0.61	0.61	0.61	0.61	0.61

Table 2 Performance indicators for different values of the parameters h_1 and h_2 (hourly data)

h_2	*1*	*2*	*3*	*4*	*5*	*6*	*7*	*8*
$h_1 = 1$ Δ_1: **0.780**)								
Rand index (R)	–	0.75	0.89	0.94	0.97	0.98	0.99	0.99
Shannon index (S)	0.94	0.75	0.62	0.55	0.51	0.49	0.47	0.46
Number of clusters k^* (max $k = 2$)	2	2	2	2	2	2	2	2
Number of changes	–	8955	3702	1789	818	498	329	160
Mean(R,S)	–	0.75	**0.76**	0.75	0.74	0.73	0.73	0.73
$h_1 = 2$ Δ_2: **0.950**)								
Rand index (R)	–	0.66	0.77	0.86	0.93	0.96	0.97	0.98
Shannon index (S)	0.64	0.87	0.79	0.72	0.68	0.65	0.63	0.62
Number of clusters k^* (max $k = 4$)	3	4	4	4	4	4	4	4
Number of changes	–	29896	11905	6400	2792	1779	1155	682
Mean(R,S)	–	0.77	0.78	0.79	**0.81**	0.80	0.80	0.80
$h_1 = 3$ Δ_3: **0.969**)								
Rand index (R)	–	0.64	0.78	0.86	0.90	0.92	0.93	0.94
Shannon index (S)	0.46	0.72	0.76	0.75	0.75	0.74	0.73	0.73
Number of clusters k^* (max $k = 8$)	4	7	8	8	8	8	8	8
Number of changes	–	35792	18704	10998	6755	5003	3915	3252
Mean(R,S)	–	0.68	0.77	0.80	**0.83**	0.83	0.83	0.83
$h_1 = 4$ Δ_4: **0.975**)								
Rand index (R)	–	0.63	0.8	0.87	0.91	0.93	0.94	0.95
Shannon index (S)	0.35	0.62	0.74	0.77	0.78	0.77	0.77	0.76
Number of clusters k^* (max $k = 16$)	5	11	15	16	16	16	16	16
Number of changes	–	42838	26327	17068	11228	8513	6849	5550
Mean(R,S)	–	0.63	0.77	0.82	0.84	**0.85**	0.85	0.85
$h_1 = 5$ Δ_5: **0.978**)								
Rand index (R)	–	0.63	0.82	0.9	0.93	0.94	0.95	0.95
Shannon index (S)	0.29	0.54	0.7	0.77	0.79	0.79	0.78	0.76
Number of clusters k^* (max $k = 32$)	7	18	26	31	32	32	32	32
Number of changes	–	46114	33496	25507	17583	14216	11305	9111
Mean(R,S)	–	0.58	0.76	0.83	0.86	**0.87**	0.86	0.86
$h_1 = 6$ Δ_6: **0.980**)								
Rand index (R)	–	0.63	0.82	0.9	0.93	0.95	0.95	0.96
Shannon index (S)	0.24	0.47	0.61	0.7	0.74	0.77	0.78	0.77
Number of clusters k^* (max $k = 64$)	8	24	42	56	62	64	64	64
Number of changes	–	46901	34894	27676	21077	18679	16119	13551
Mean(R,S)	–	0.55	0.72	0.8	0.84	0.86	0.86	0.87

clusters identified by a given pair (h_1, h_2). The indicators 3. and 4. (the number of changes and the Rand index) refer to the comparison between the configuration under (h_1, h_2) and the "adjacent one", i.e. the partition identified by the pair $(h_1, h_2 - 1)$ (this explains why these values are not available for $h_2 = 1$). Note that the comparisons can be made only by row, when considering the same value of h_1, because different values of h_1 determine different partitions which may not be directly compared, although the indexes are normalized). The mean value between the Rand index and the Shannon index is a normalized index which takes into account both the criteria "movements among the clusters" and "percentage of positive clusters". The max value for each row (written in bold in the tables) identifies the optimal value of h_2 for a given value of h_1.

For the selection of the parameter h_1, a first evaluation may be made by considering the value of $\Delta_{h_1} = \sum_{i=1}^{h_1} \delta_i$, which is reported in bold in Tables 1 and 2. We could fix an absolute threshold or we could consider a minimum relative increment for increasing values of h_1. For example, if we fix the threshold of 95%, we should consider for the classification the first two daily periodic components and the two first hourly periodic components. If we consider a minimum relative increment of 2%, we should consider the first three daily cycles and the first two hourly cycles, and so on. Basing on the previous results, we decided to set the following smoothing parameters:

- $h_1 = 4$ ($h_1 = 2$) for the daily (hourly) data.
- $h_2 = 4$ ($h_2 = 5$) for the daily (hourly) data.

So we considered the first four daily cycles, denoted with D_1, D_2, D_3, D_4, and the first two hourly cycles, denoted with H_1 and H_2, as summarized in Table 3.

Note that the full partition includes max $k = 2^6 = 64$ clusters. Anyway, it may be that some of the selected cycles could be neglected. To confirm the selection, we derived all the possible combinations of one or more daily cycles (D_1, D_2, D_3, D_4) with one or more hourly cycles (H_1, H_2). These amounts to 45 different partitions, which are summarized in Table 4, by crossing the rows and the columns. We introduce a normalized index (a variant of the Jaccard index) which measure the "goodness" of the classification by evaluating the agreement between the characteristics of the cluster and the structure of the aggregated time series. More

Table 3 Cycles identified in the energy consumption database, for the optimal configuration of h_1 and h_2

Dominant periods	Approximate time	% of users on the total δ_i (%)
H_1	4 h	78
H_2	1 day	17
D_1	3 weeks	86.9
D_2	10 days	8.3
D_3	1 week	2.4
D_4	1 (short) week	1.2

Table 4 "Goodness" index for the 45 compared partitions (normalized). The value in bold show the best partition of clusters to consider

Cycles	D_1	D_2	D_3	D_4	D_1 D_2	D_1 D_3	D_1 D_4	D_2 D_3	D_2 D_4	D_3 D_4	D_2 D_3 D_4	D_1 D_3 D_4	D_1 D_2 D_4	D_1 D_2 D_3	D_1 D_2 D_3 D_4
H_1	0.5	0.75	1	0.75	0.57	0.71	0.57	0.86	0.71	0.86	0.8	0.7	0.6	0.7	0.69
H_2	0.75	1	1	0.75	0.86	0.86	0.71	1	0.86	0.86	0.9	0.8	0.8	0.9	0.85
H_1, H_2	0.83	1	1	0.83	0.9	0.9	0.8	**1**	0.9	0.9	0.93	0.86	0.86	0.93	0.89

Table 5 Distribution of the users among the $k = 2^4$ clusters of the optimal partition. The number of positive clusters is equal to $k^* = 12$

Cluster	Cycles	Users	%
C0	None	3148	5.0
C1	$D2$	271	0.4
C2	$D3$	99	0.2
C3	$D2, D3$	0	0.0
C4	$H2$	13199	21.1
C5	$H2, D2$	1010	1.6
C6	$H2, D3$	284	0.5
C7	$H2, D2, D3$	0	0.0
C8	$H1$	3439	5.5
C9	$H1, D2$	88	0.1
C10	$H1, D3$	29	0.0
C11	$H1, D2, D3$	0	0.0
C12	$H1, H2$	39984	64.0
C13	$H1, H2, D2$	794	1.3
C14	$H1, H2, D3$	156	0.2
C15	$H1, H2, D2, D3$	0	0.0

precisely, for each partition we derive the aggregated time series of the clusters by summing the data observed for all the users classified in the same cluster. We expect that the aggregated time series will be characterized by the presence of the same periodic components as those which characterize the cluster. If not, it would mean that there are masking effects in the aggregation which denote that the partition is not efficient. The goodness index is given by

$$J = \frac{\alpha}{\nu},$$

where α is the number of cycles expected for each cluster of a particular partition which are effectively detected in the aggregated time series of the clusters, and ν is the total number of comparisons made for that partition. This index tends to be equal to the maximum value 1 in the case of maximum agreement, i.e. when (all) the expected cycles are present in the aggregated time series. It tends to the minimum

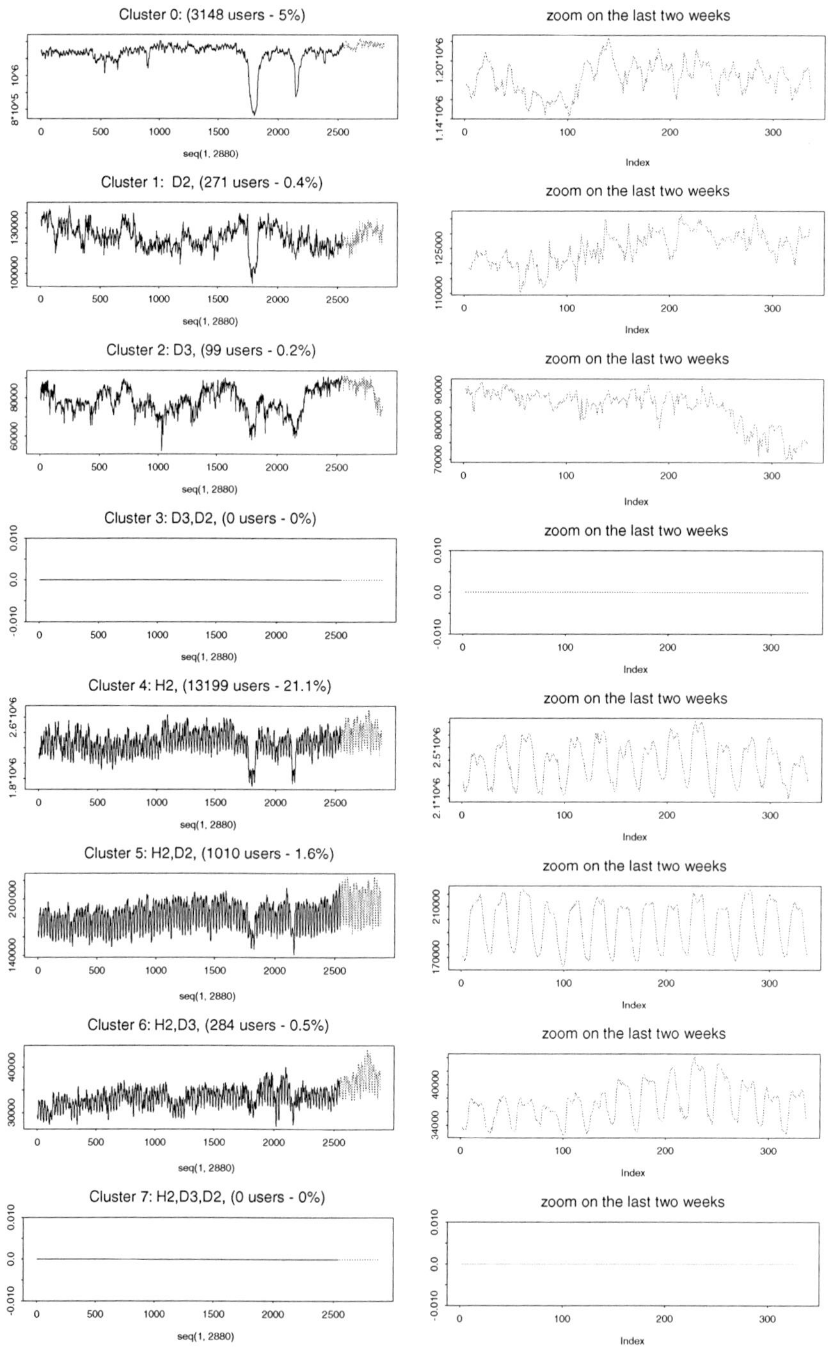

Fig. 1 Plots of the first 7 clusters of the optimal partition, based on the combinations of the cycles D_2, D_3, H_1 and H_2. On the left side, the plots of the aggregated time series for each cluster. On the right side, the zoom on the last two weeks

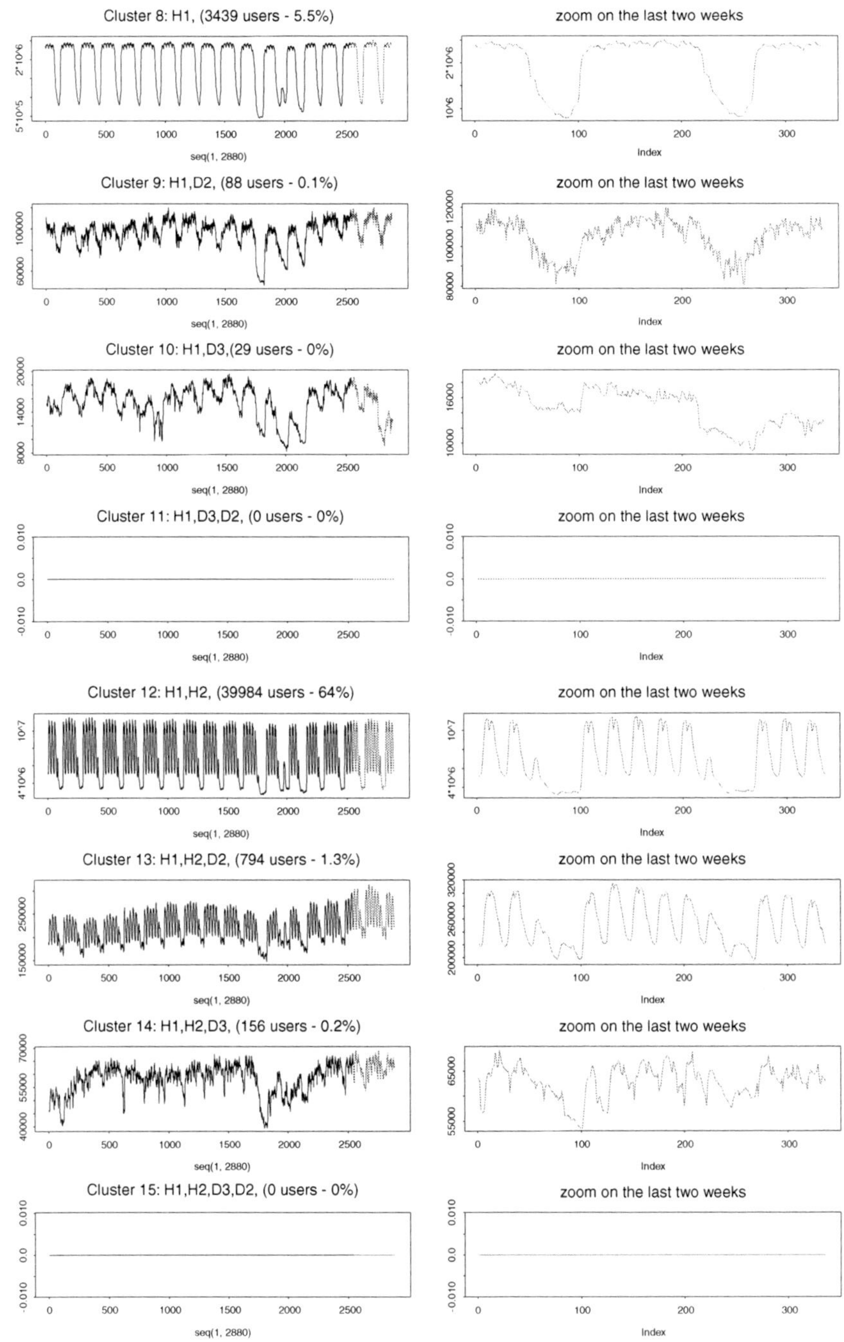

Fig. 2 Plots of the last 7 clusters of the optimal partition, based on the combinations of the cycles D_2, D_3, H_1 and H_2. On the left side, the plots of the aggregated time series for each cluster. On the right side, the zoom on the last two weeks

value 0 in the case of minimum agreement, that is when (some of) the expected cycles are not detected in the aggregated time series.

In Table 4 we report the results for such goodness index for the 45 partitions. Again we head for the maximum value, which identifies the "best partition". It is interesting to note that there are 7 candidates, all of which involves the same periodic components (D_2, D_3, H_1, H_2). Among these, it is natural to select as the best candidate the partition which include the maximum number of cycles (reported in bold in the Tables 1 and 2). Therefore, the best configuration is the one which includes the periodic components D_2, D_3, H_1 and H_2, for a total of $k = 2^4$ potential clusters. Table 5 summarizes the features of the best partition, while Figs. 1 and 2 show the plots of the clusters.

References

Alonso, A.M., Berrendero, J.R., Hernández, A., Justel, A.: Time series clustering based on forecast densities. Computational Statistics & Data Analysis **51**, 762–776 (2006)

Corduas, M., Piccolo, D.: Time series clustering and classification by the autoregressive metric. Computat. Stat. & Data Analysis **52**, 1860–1872 (2008)

Giordano, F., La Rocca, M., Parrella, M.L.: Clustering Complex Time Series Databases. Submitted to *SFC-CLADAG Proceedings* (2008)

Priestley, M.B.: Spectral Analysis and Time Series. Academic Press, London (1981)

Assessing the Beneficial Effects of Economic Growth: The Harmonic Growth Index*

Daria Mendola and Raffaele Scuderi

Abstract In this paper we introduce the multidimensional notion of *harmonic growth* as a situation of diffused well-being associated to an increase of per capita GDP. We say that a country experienced a *harmonic growth* if during the observed period all the key indicators, proxies of the endogenous and exogenous forces driving population well-being, show a significantly common pattern with the income dynamics. The notion is operationalized via an index of time series harmony which follows the functional data analysis approach. This *Harmonic Growth Index* (HGI) is based on comparisons between the coefficients from cubic B-splines interpolation. Such indices are then synthesized in order to provide the global degree of harmony in growth inside a country. With an accurate selection of the key indicators, the index can be used also to rank countries thus offering a useful complementary information to the Human Development Indexes from UNDP. An exemplification is given for the Indian economy.

*The present study is the result of a common idea of both authors: nevertheless, paragraphs 1, 3 and 4.1 are attributable to Daria Mendola, while paragraphs 2, 4.2 and 5 to Raffaele Scuderi. Authors wish to thank dr. Valerio Lacagnina who provided precious suggestions for the improvement of the index here proposed. The authors are solely responsible for any error and inconsistency in this paper.

D. Mendola (✉)
Department of Economics, Business, and Finance, University of Palermo, Viale delle Scienze ed. 13, 90128, Palermo, Italy
e-mail: daria.mendola@unipa.it

R. Scuderi
School of Economics and Management, Free University of Bozen/Bolzano, Italy

A. Di Ciaccio et al. (eds.), *Advanced Statistical Methods for the Analysis of Large Data-Sets*, Studies in Theoretical and Applied Statistics,
DOI 10.1007/978-3-642-21037-2_19, © Springer-Verlag Berlin Heidelberg 2012

1 Motivation

Social and economic growth is what we should look at when we intend to monitor the well-being of a country. This kind of growth taking place across both social and economic aspects would require to be measured by more than one indicator, each one proxying several aspects of living standards. In many studies such aspects are claimed to be measured by per capita GDP which is assumed to be a proxy for the aggregated level of the individuals' living standard. Nonetheless it is well known and deeply debated in the literature that GDP does not take into account some relevant aspects of country well-being, such as pollution, improvement of people's health, education levels, and so on.

The pressing need to find an index alternative to GDP in order to put into account further aspects of the living standards is likely the main reason of the success of measures like Human Development Indexes since 1990 (UNDP various years), which nowadays compete with GDP as headline indicators of development. Literature reports a wide debate about pros and cons of HDIs and their role as a country's well-being indicators (see among others Anand and Sen 1994). We will not investigate on such issue, but we share the conviction of many scholars about the poor informative power of these measures, and at the same time we recognize that they are an important improvement of the information provided by the sole GDP.

A further typical issue concerning well-being is connected to the level of diffusion of growth within the population. The most widely used measure of income spread is Gini income concentration index, which again measures the diffusion of per capita income only.

Indeed, the explanation of the controversial relationship between growth and its inequality has a long tradition in literature, and in our opinion the study of well-being issues, particularly in developing countries, has to concern about it. At this regard, Persson and Tabellini (1994) suggest that "inequality is harmful for growth" and assess that in developed countries with democratic political settings there is a "significant and large negative relation between inequality and growth". But the opposite conclusion is argued by Forbes (2000), who proves that "in short and medium term an increase in a country's level income inequality has a significant positive relation with subsequent economic growth" even if this result does not apply to very poor countries.

Our paper introduces a measure of growth based on the concept of *harmony* between the temporal patterns of social and economic indicators. Our basic idea is that the joint use of social and economic variables with per capita GDP can better characterize growth, and can also give useful information about the changes of the living conditions associated with the variation of income over time. Here the informative power of GDP is not substituted by or implemented with further indicators, as it instead happens with HDIs where it accounts for one third of its value. In fact according to our concept of *harmonic growth* GDP remains the main indicator but plays the role of benchmark to check whether the time series of other social and economic indicators evolve in the same way. Obviously the indicators expected to be *harmonic* with GDP should carry desirable effects to the population. Thus life expectancy at birth is *harmonic* with GDP when an increase of income leads to the increase of the average age an individual would expect at birth, in so far as it represents the enhancement of living conditions in the country. Same reasoning refers to *harmony* with education levels, quality of air, and so on. Note that we can observe *harmony* in growth even when a decrease of GDP is associated with negative changes in social indicators. On the whole, if an indicator has the same growth pattern of GDP, it has grown *harmonically* with the latter.

Let us now point out some aspects of the concept introduced above in order to better characterize it. First, the choice of indicators appears to be crucial and has to be made in order to associate their raise with the spread of "growth". Secondly, it is not a matter of "rough" comparison between global trends but rather a search of similar local patterns in the dynamics of chosen indicators: this is why in the next section we provide a brief review of the most commonly used measures of similarity among time series which could be useful to measure *harmony*, and then present a formalization of the concept of *harmonic growth*. Section 3 proposes the rationale, the methodology, and the properties of our index of *harmonic growth* at country level: here we propose also a composite indicator of *global harmonic growth*. Section 4 shows a case study on a developing country, while Sect. 5 concludes.

2 A Formal Definition of Harmonic Growth

Our search for a measure of growth is driven by the need to verify whether a series grows in the same way as the benchmark and also to quantify the degree of similarity in those patterns, that is their *harmony*. Literature on time-series clustering (see the review of Liao 2005) offers helpful approaches. As a matter of fact it is important to point out that the mere identification of clusters within a set of time series is not in our interest, but rather we are concerned with that part of the clustering methodology which is aimed at assessing "how close" the time patterns of two series are. Thus we specifically make use of the notion of *similarity*.

Linear correlation and Euclidean distance are two of the most widely used similarity criteria for splitting a set of time series into homogeneous groups. Their main limitation is that they are invariant to transformations that alter the order of observations over time (i.e. the temporal indexing of data), and therefore they do not take into account information deriving from the typical autocorrelation structure. We agree with Douzal Chouakria et al. (2007) that the similarity criterion has to be a combination between the "strength of the monotonicity" and the "closeness of the growth rates"; i.e., two series show similar behavior if, in a fixed time interval, they "increase or decrease simultaneously (monotonicity), with a same growth rate".

Thus the methodology that allows evaluating *harmony* in a more proper way has to embody the features strictly related to the sequencing of the observed values over time. In this sense the seminal paper by Piccolo (1990) presents a method based on stochastic processes where similarity is evaluated by the distance between autoregressive coefficients of an ARIMA model. However the requirement of long time series for the estimation of an ARIMA model is a feature which is not consistent with the short length of most available time series of social indicators, especially for the developing countries.

A less data-demanding approach is offered by some clustering procedures within the functional data analysis framework (Ramsay and Silverman 2005), which treats the data stream over the time as deriving from a continuous function. The paper by Abraham et al. (2003) introduces a curve clustering approach for time series using cubic B-splines coefficients. Such approach takes advantage of the functional structure of data and keeps the information deriving from their time sequence. Splines fitting (see De Boor 1978 and Schumaker 1981) is also a more flexible and robust procedure than linear models (in particular than polynomials estimating the overall trend) for it is less sensitive to outliers. Moreover, unlike AR metrics, the smoothing of splines reduces the measurement error in data, if any. The index we propose in Sect. 3 employs such functional analysis approach via B-splines.

Now suppose that we deal with P time series of T years each, referring to P social and economic indicators, and that in addition we consider a series numbered with "0" playing the role of benchmark for all the remaining ones (the GDP in our case). Fit a proper statistical model to each series $X_p = \{x_{p,t}\}$, where $p = 0, 1, \ldots, P$, and $t = 1, 2, \ldots, T$, producing a set of N coefficients so that $\boldsymbol{c}_p$ is the coefficients' vector of dimension N describing the pth series (i.e. if we fit a polynomial then N corresponds to its degree, and $N \leq T-1$). Each $\boldsymbol{c}_p$ assesses the shape of X_p in such a way that even local trends are detected. Suppose that vectors $\boldsymbol{c}_0, \ldots, \boldsymbol{c}_P$ are not affected by scale and unit of measurement effects. A discussion about the adequate model to fit is reported in the following section.

Definition 1. Time series X_0 and X_p, $p \neq 0$, are *perfectly harmonic* when the equality of coefficients implies that their values are equal up to the following linear transformation:

$$\mathbf{c}_0 - \mathbf{c}_p = \mathbf{0} \Rightarrow x_{0,t} = a + x_{p,t},$$
$$\text{where} \quad a \in \mathbb{R}, \quad \forall t = 1, 2, \ldots, T \quad \text{and} \, \forall p = 1, 2, \ldots, P \tag{1}$$

From Definition 1 the perfect *harmony* between two series happens if $x_{0,t} = x_{p,t}$ (the series are equal), or $x_{0,t} = a + x_{p,t}$ (the series have the same pattern but different intercepts).

Definition 2. Time series X_0 and $X_p, p \neq 0$, are *perfectly disharmonic* when coefficients are equal but have opposite sign, for each t, that is:

$$\mathbf{c}_0 + \mathbf{c}_p = \mathbf{0} \Rightarrow x_{0,t} = a - x_{p,t},$$
$$\text{where} \quad a \in \mathbb{R}, \quad \forall t = 1, 2, \ldots, T \quad \text{and} \, \forall p = 1, 2, \ldots, P \tag{2}$$

The perfect *disharmony* happens when $x_{0,t} = -x_{p,t}$ (values of the series are equal but have opposite signs) or $x_{0,t} = a - x_{p,t}$ (series differ by the additive constant a).

Note that, there could even be situations of *lagged harmony* where definitions are true if series X_p is lagged with respect to the benchmark series; or that two series can be harmonic after a proper transformation which makes them comparable (for instance accommodating for different variability, or whatever).

3 The Harmonic Growth Index

Let X_0 be the GDP series (benchmark) and X_p, $p \neq 0$, for instance the enrollment in education rate series. Suppose that X_0 and X_p can be described by spline functions of degree M indexed by m. Since every knot interval originates M coefficients[1], for each series we have $\boldsymbol{C}_0 = \{c_{0mt}\}$ and $\boldsymbol{C}_p = \{c_{pmt}\}$, two matrices of coefficients of dimension $M \times (T-1)$. If two series present a perfectly *harmonic* pattern, then $\boldsymbol{C}_0 - \boldsymbol{C}_p = \boldsymbol{0}$. Changes in the dynamics of the series are mirrored by differences in the coefficients, which are as greater as the patterns are dissimilar. Here we propose the *Harmonic Growth Index (HGI)*:

$$HGI = 1 - \sum_{m=1}^{M} \sum_{t=1}^{T-1} \frac{\left|c_{0mt} - c_{pmt}\right|}{2\max\left(\left|c_{0mt}\right|, \left|c_{pmt}\right|\right)} w_{mt},$$

$$\text{where} \quad p \neq 0; \quad c_{0mt} \neq 0 \quad \text{or} \quad c_{pmt} \neq 0 \tag{3}$$

where each absolute difference between coefficients, $c_{0mt} - c_{pmt}$, is compared to its theoretical maximum which corresponds to the perfect *disharmony*, and at least one of the two compared coefficients differs from zero given m and t. The term w_{mt} is assumed to be constant and equals $1/(M(T-1)-r)$, where r is the number of times where both coefficients equal zero given m and t. An alternative formulation of *HGI* allows for different kind of weights w_{mt}. For instance w_{mt} may increase as the degree of the monomial to which coefficients c_{0mt} and c_{pmt} belong raises, and be in the form $w_{mt} = 2m(T-1)/M(M+1)$. That is, the higher the order of the monomial inside the polynomial of the spline function, the higher the importance given to a possible divergence in the compared coefficients, and thus higher its contribution to *disharmony* in the common pattern of two series.

Note that raw OLS estimates are influenced by (i) the average magnitude of the series, (ii) their observed range, and (iii) their units of measurement. So before comparing homologous coefficients through the polynomials it is necessary to find a transformation allowing for proper comparisons. According to classical regression theory we standardize series according to *z-score*, for their estimated coefficients account for the three aspects mentioned above.

HGI spans over [0, 1], where 0 corresponds to the absence of *harmony* (that is the perfect *disharmony* in time series patterns, as stated in Definition 2) and 1 results

[1] We ignore the constant term of the splines.

from perfectly *harmonic* patterns (see Definition 1). The intermediate situation of $HGI = 0.5$ is the one between a time series with constant values and a non-constant one. Since *harmonic* growth is a symmetric concept, the *harmony* between X_0 and X_p is analogous to the *harmony* between X_p and X_0. This is why we used the absolute value operator at the numerator of the index and, consistently, also at the denominator.

In the following we will refer to cubic spline functions (that is $M = 3$) because they allow for a good smoothing through a small number of parameters, and avoid the Runge's phenomenon. Of course there can be alternative functional forms to describe the series such as the unique polynomial over the whole time range of each series. B-Splines are nevertheless more robust than polynomials to outliers or small changes in data. Thus the latter can be considered if we assume that data are not affected by measurement errors or if small changes are to be highlighted by the index.

Indices of *harmonic growth* upon all the couples of time series (i.e. between GDP and each of the P indicators) can be synthesized into a *Global Harmonic Growth Index – GHGI* – according to a selected aggregation function. The one we propose here is the "weakest link criterion" or the "minimum criterion":

$$HGI = \min(HGI_1, \ldots, HGI_i, \ldots, HGI_P) \tag{4}$$

where $HGI_i (i = 1, 2, \ldots, P)$ is the index of *harmony* between GDP and each well-being indicator. Also *GHGI* spans in [0, 1], where 0 is the absence of global *harmony* and 1 is the perfect *harmonic growth* situation in a multidimensional set. The idea is that the lowest level of *harmony* expresses the frailty of the economic growth in terms of diffusion of well-being. Alternative aggregation functions could be the arithmetic mean between the P indices if one can suppose that a *disharmony* between a couple of series could be compensated by the higher *harmony* of another couple. Otherwise one could also consider to take into account the median of the P indexes if there are outliers or anomalous values within the set of *HGI*s.

GHGI provides a measure of the total degree of multidimensional *harmony* in growth inside a country. Index (4) can be used also to rank countries according to their level of development, and it offers complementary information to the Human Development Index.

4 Testing the Harmonic Growth Hypothesis: The Case of India

4.1 Data

We propose an application of our measure of growth using data for India which is a country where the problem of the diffusion of the beneficial effects of the growth is actual and topical (Datt and Ravallion 1998, Dev 2002). Our main scope is to show how *HGI* describes our extended concept of growth. Nevertheless many and different difficulties raise in constructing holistic measures of economic well-being, both in finding appropriate data and in single out the relationship among the components of the socio-economic growth of a country. This is dramatically true when we look at developing countries. Following Hobijn and Franses (2001) we say that the main issue is "the actual choice of the indicators that are used to evaluate the standard of living. Sen (1987) argues that one should consider indicators of both functionings, that is the actual outcome of peoples' decisions, like life expectancy, and capabilities, that is the opportunities that people have, like their available per capita income".

In this paper we present our analyses using free available statistical databases, resorting to proxies and partial indicators where necessary, in order to carry out our investigation on *harmonic growth* in India. We selected a period of thirteen consecutive years for which acceptable quality data are available without any missing value. The eight selected indicators are related to beneficial effects of the economic growth[2] and are compared in dynamics with GDP per capita in constant local currency unit (The World Bank 2005). Each one is transformed according to the *z-score* (see above), and when necessary the original index has been transformed once more so that its increase corresponds to beneficial effects for the population. This last is the case of CO_2 emissions and Urban population, as it will be explained in the following. All the selected indicators are well known in literature and thus we spend only few words to describe them.

- LEX: Life expectancy at birth (UNDP various years) – as highlighted in Anand et al. (1994), "life expectancy can be thought to be both valuable in itself and also helpful for pursuing other objectives. It encompasses measurements of the state of the sanitary/health system in a country, the living conditions of individuals, the hygienic conditions of the house, the availability of appropriate food, drinkable water, and so on".
- EDU: Combined gross enrolment ratio in education, % (UNDP various years) – education plays a fundamental role as indicator of development of a nation as it can be seen both as a determinant and an effect of individual poverty.
- CO2: Carbon dioxide emissions, metric tons of CO_2 per capita (United Nations 2009) – they are one of the most monitored detrimental effects of the current industrialization model. They enter with the inverse sign in the computation of the index since their decrease is beneficial to living standards.
- EMP: Employment-to-population ratio, both sexes, %; and EMW: Employment-to-population ratio, women, % (United Nations 2009) – we are aware that, as highlighted by Dev (2002), "for economic growth to generate the kind of

[2]In this paper we do not address the issue about the relation between GDP and such variables. Furthermore we do not discuss about causes and effects of the socioeconomic development of a country which play simultaneously as endogenous and exogenous forces, as it is well known by well-being studies: from both an interpretative and computational point of view this is not a problem for us due to the symmetry of *HGI*.

employment that contributes directly to poverty alleviation, it must be in sectors that have relatively high elasticities of employment [...]; workers must share in the benefits of increased labour productivity [...]; and the jobs created must be relatively unskilled in order to be accessible to the poor". Unfortunately at this stage of analysis we have availability only of the above mentioned EMP and EMW, where the latter is also used as a proxy of gender issue and fair opportunities.

- TEL: Number of telephone lines (United Nations 2009) – it refers to both mainlines and cellulars. It is a classical indicator of development and as part of the Technology Achievement Index in the Human Development Report it proxies the spread of technology. It is expected to grow with GDP.
- COM: Combustible renewables (The World Bank 2005) – comprises solid biomass, liquid biomass, biogas, industrial waste, and municipal waste utilized as source of primary energy (expressed as percentage of total energy).
- NUP: Non Urban population, % of total population (The World Bank 2005) – it is related to the issue of urban overcrowding in developing countries and urban poverty issues. As stated in Datt and Ravallion (1998), comparing the effects of urban and rural growth on poverty in India shows that growth in urban incomes has no effect on rural poverty, but also only a modest effect on urban poverty. On the other hand, rural growth reduces rural poverty and reduces urban poverty. This is why, among available indicators, we observe whether raise of GDP is associated with the increase of non-urban population. The latter is obtained as ratio between nonurban population (obtained by subtracting the amount of urban population from total population) and the total population.

4.2 Results

Figure 1 shows pairwise comparisons between non-lagged[3] time series indicators of India over 1991–2003 and real per capita GDP. Values of the *Harmonic Growth Indices* in (3) are computed using $w_{mt} = 1/(M(T-1) - r)$. The global index *GHGI* equaling 0.356 indicates that the growth of India has been quite *disharmonic* during the observed period. Graphs report that GDP reveals a similar dynamics as TEL and LEX, whose values of *HGI*s are near to 0.6. Indices show that *harmony* has not been equally strong in these two cases. The highest value reported by the proxy of technology spread (TEL) suggests that the growth of GDP is characterized by a *harmonic* increase of the spread of technology. Also life expectancy at birth (LEX) revealing a concurdant overall time pattern with the benchmark series but

[3] Although we are conscious that the raise of GDP could have delayed effects on indicators such as life expectancy at birth, education and so on, we decided to use non lagged time series in order to create a measure which can be comparable to the Human Development Index, which adopts the same rationale.

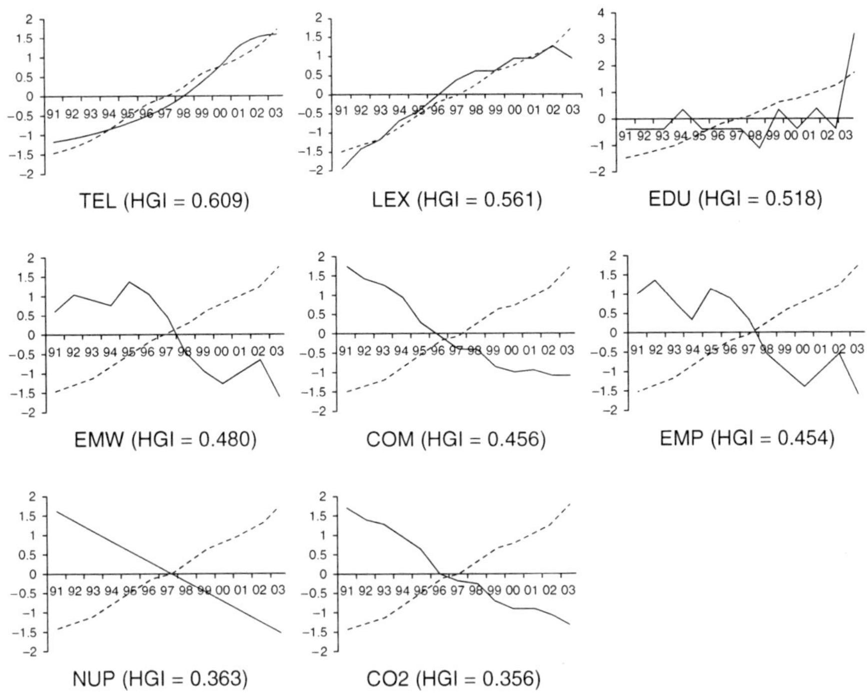

Fig. 1 Plot of selected indicators (continuous line) for India with respect to GDP (dotted line), and the correspondent HGI values (standardized series)

less *harmonic* local trends. A substantially medium *harmony* is expressed by the gross enrolment rate (EDU). Specular remarks can be made for EMW, COM, EMP, NUP and CO_2, where even global trends have opposite sign than GDP. As remarked above the index reveals that the *harmonic* growth is not a mere matter of comparison between the global trends, but rather a way to see how similar time paths of series have been through the detection of common local patterns. While from the plot we would expect, for instance, that the *harmony* between GDP and COM would be very near to zero, local trends affect the measure in such a way that time paths are more similar to the condition of medium *harmony* rather than to the *disharmony*. The relevance of local trends also emerges if we compare the final years of CO_2 with COM: while the values of both indices related to the environmental sustainibility show *disharmony* with GDP, the more diverging path of the former than the latter is detected by a lower value of the index. *Harmony* of Indian growth is also low for the overall employment to population ratio (EMP), which appears to be more *disharmonic* than the one pertaining only the women (EMW). Finally, the *HGI* of NUP suggests that GDP growth is characterized by the non *harmonic* pattern of the share of population living in nonurban areas.

5 Conclusions

This paper addresses the issue of the relationship between economic growth and social development of a country. Our basic idea is that there is an inclusive (*harmonic*) growth if the raising of the GDP is associated with a higher diffusion of the wealth and some beneficial effects on the living standards in the country. We assess that a country experienced *harmonic growth* if all the indicators of well-being show a significantly common pattern with the GDP. The macro level index here proposed, called *Harmonic Growth Index*, derives its formulation from the functional data analysis in so far as it exclusively focuses on the values of the coefficients of the B-splines approximating time series.

Nevertheless many and different difficulties raise in constructing holistic measures of economic well-being, both in finding appropriate indicators and data, and in choosing proper functions, and this is dramatically true when we look at developing countries. Just for the sake of exemplification, we proposed an application of the HGI in Indian economy. As expected the country shows that *harmony* in growth does not appear in all the considered social and economic dimensions of the development, even if there appears to be a remarkable improvement in the spread of technology and health conditions.

We believe that future studies on pro-poor growth and of well-being could benefit by the use of *HGI* and *GHGI*, providing alternative or complementary information to more famous indices such as the Human Development Index.

References

Abraham, C., Cornillon, P.A., Matzner-Lober, E., Molinari, N.: Unsupervised curve clustering using B-splines. Scand. J. Stat. **30**, 581–595 (2003)

Anand, S., Sen, A.K.: Human development index: methodology and measurement. Human Development Report Office, occasional Papers, **12** (1994)

Datt, G., Ravallion, M.: Why have some Indian states done better than others at reducing rural poverty? Economics **65**, 17–38 (1998)

De Boor, C.: A practical guide to splines. Springer, New York (1978)

Dev, S.M.: Pro-poor growth in India: what do we know about the employment effects of growth 1980–2000? Overseas Dev. Inst. Workp. 161 (2002)

Douzal Chouakria, A., Diallo, A., Giroud, F.: Adaptive clustering of time series. Proc. IASC Conf. “Stat. Data Min. Learn. and Knowl. Extr.”. Aveiro, Portugal (2007)

Forbes, K.J.: A reassessment of the relationship between inequality and growth. Am. Econ. Rev. **90**, 869–887 (2000)

Hobijn, B., Franses, P.: Are living standards converging? Struct. Chang. Econ. Dyn. **12**, 171–200 (2001)

Liao T.W.: Clustering of time series data - a survey. Pattern Recognit. **38,** 1857–1874 (2005)

Persson, T., Tabellini G.: Is inequality harmful for growth? Am. Econ. Rev. **84**, 600–621 (1994)

Piccolo D.: A distance measure for classifying ARIMA models. J. of Time Ser. Anal. **11**, 153–163 (1990)

Ramsay, J.O., Silverman, B.W.: Functional data analysis, 2nd edition, Springer, Berlin (2005)

Schumaker, L. L.: Spline functions: basic theory. Wiley, New York (1981)

Sen, A.: Standard of living. Cambridge University Press, New York (1987)

The World Bank: World Development Indicators. http://www.worldbank.org/ (2005). Accessed 18 July 2005

UNDP: Human development report, Palgrave Macmillan, New York. http://hdr.undp.org (various years). Accessed 24 November 2009

United Nations: Millennium Development Goals Indicators. http://mdgs.un.org/unsd/mdg/ (2009). Accessed 24 November 2009

Time Series Convergence within I(2) Models: the Case of Weekly Long Term Bond Yields in the Four Largest Euro Area Countries

Giuliana Passamani

Abstract The purpose of the paper is to suggest a modelling strategy that can be used to study the process of pairwise convergence within time series analysis. Moving from the works of Bernard (1992) and Bernard and Durlauf (1995), we specify an I(1) cointegrated model characterized by broken linear trends, and we identify the driving force leading to convergence as a common stochastic trend, but the results are unsatisfactory. Then we deal the same question of time series convergence within I(2) cointegration analysis, allowing for broken linear trends and an I(2) common stochastic trend as the driving force. The results obtained with this second specification are encouraging and satisfactory. The suggested modelling strategy is applied to the convergence of long-term bond markets in the Economic and Monetary Union (EMU), that we observe during the years covering the second stage, that is the period from 1993 to the end of 1998, before the introduction of euro. During the third stage, started in 1999 and continuing, the markets show a tendency to move together and to behave similarly.

1 Introduction

The scenario of interest is the one in which the observed time series are characterized by a non-stationary behaviour driven by common stochastic trends, but with different linear deterministic trends, over a first sub-period, and a non-stationary behaviour, common stochastic trends and no deterministic linear trends over the remaining period. A typical example of such scenario is represented in Fig. 1, where the ten-years zero-coupon bond yields of the four largest euro area countries show a clear convergence behaviour, though following different trend paths, as if a common

G. Passamani (✉)
Department of Economics, University of Trento, Via Inama, 5 - 38122 Trento, Italy
e-mail: giuliana.passamani@economia.unitn.it

A. Di Ciaccio et al. (eds.), *Advanced Statistical Methods for the Analysis of Large Data-Sets*, Studies in Theoretical and Applied Statistics,
DOI 10.1007/978-3-642-21037-2_20,

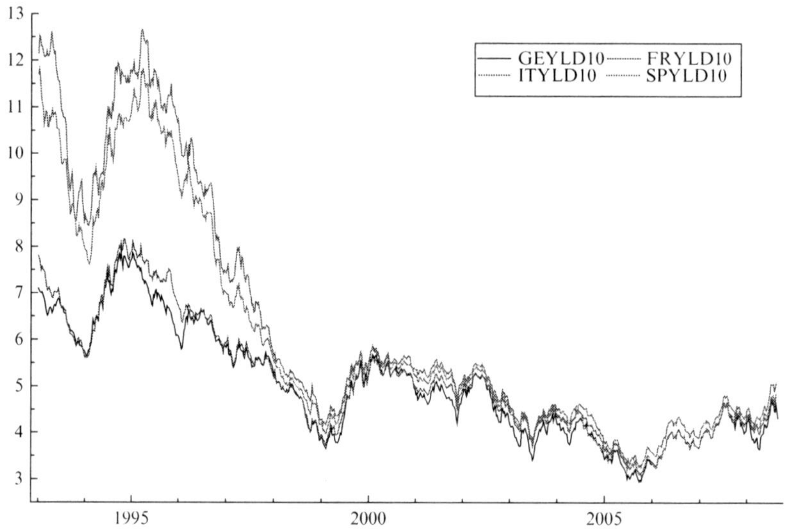

Fig. 1 Ten-year zero-coupon bond yields for Germany, France, Italy and Spain
Source: Ehrmann et al. (2007)

driving force had led them to a point from which they have moved together, still sharing the same driving force. The point corresponds to the time of the introduction of euro. Therefore, the idea of convergence we have in mind is the one according to which series, following different dynamic behaviours and time paths, get to converge through narrowing their differences.

It's to note that from the figure it's possible to see some evidence, at least over the first sub-period, of a certain similarity with the behaviour of zero-coupon bonds yields of different maturities, where the German and French series variability resembles the behaviour of yields of long-term maturities, whereas the Italian and Spanish series resemble the behaviour of short-term yields, although the long-term yields should typically be higher in level than the short-term yields. If we applied the expectations theory of the term structure to investigate the relations among the series represented in the figure, we would expect to find evidence of some broken trend stationary relations representing the co-movements between pairs of series, where the German series can be considered as the benchmark series to which referring the others.

The data chosen for the empirical analysis are weekly long-term zero-coupon bond yields for France, Germany, Italy and Spain. They cover the period from 1993 to 2008, that is through the years leading to monetary union in 1999, and to monetary unification in 2002, and before the financial crisis of 2008.[1]

[1]The persistence properties of each observed variable have been analyzed in terms of the characteristic roots of its autoregressive polynomial. Allowing for a certain number of significant

2 The Analysis Within the I(1) Model

In order to analyze the process of convergence, we assumed the observed vector time series $\mathbf{x}_t$ to be generated by a cointegrated vector auto-regression (CVAR) model with broken linear trends, both in the data and in the cointegrating relations (Juselius 2006, p.297):

$$\Delta \mathbf{x}_t = \sum_{k=1}^{k-1} \mathbf{\Gamma}_k \Delta \mathbf{x}_{t-k} + \alpha \tilde{\beta}' \tilde{\mathbf{x}}_{t-1} + \mathbf{\Phi} \mathbf{D}_t + \mu_0 + \varepsilon_t, \;\; \varepsilon_t \sim N_p(\mathbf{0}, \mathbf{\Omega}), \qquad (1)$$

where $\tilde{\beta}' = [\beta', \beta_{11}, \beta_{12}]$, $\tilde{\mathbf{x}}_t' = [\mathbf{x}_t', t, tD_{t^*}]$ and $t = 1993:01:08, \ldots, 2008:07:25$. The observation times correspond to the last day of the working weeks.

When applying[2] this cointegration model to our data set, made up of $p = 4$ variables, according to our expectations we should find one common stochastic trend driving the system of variables towards convergence, and possibly three stationary cointegrating relations. Considering t^* as the first week of 1999, which corresponds to the beginning date of the third stage of EMU, and analyzing the data[3] within model (1), the simulated critical values of the rank test statistic (Johansen 1997) gave some evidence, at 10%, of the existence of three cointegrating relations. When trying to identify them as spreads between bond yields, the relative hypotheses were rejected. But, when considering them as simple stationary relations between pairs of bond yields and identifying them through the restriction that the deterministic trends have the same coefficients with opposite sign over the second sub-period, that is just a broken linear trend in the first sub-period and no trend component since the beginning of 1999, the relative hypotheses were accepted with a p-value $= 0.191$. The identified relations are represented in Fig. 2, where they

lags (lags 1 and 15 for both Italy and Spain; lags 1, 2, 3, 4, 6, 7, 8, 10 and 12 for France; lags 1, 2, 13 and 14 for Germany), the modulus of the largest root, ρ_1, satisfies $\rho_1 = 1.0$, and the next root, ρ_2, is less than 1.0, but not too far from it, i.e.: 0.89 for Italy, 0.90 for Spain, 0.85 for both France and Germany. As Juselius (2010, p.9) observes: "... whether a characteristic root can be interpreted as evidence of persistent behaviour or not depends both on the sample period and the observational frequency." Therefore our univariate series can be considered as generated either by an I(1) process, or by a near I(2) process.

[2]The empirical analysis was performed using the subroutine CATS, which needs the software RATS to be run (Dennis 2006).

[3]The number $K = 2$ of lags chosen is the one suggested by the information criteria and the LR lag reduction tests, when starting from 5 lags. $\mathbf{D}_t$ is a vector of three impulse dummies, which take the value one the weeks ending on 1995:03:10, 1996:03:29 and 2003:03:21. A shift dummy is introduced by the program when specifying the broken trend in the cointegrating relations. The misspecification tests for the unrestricted VAR(2) model with dummies, take the following values: the *LM*(1) test for first order autocorrelation is equal to 20.175 with a p-value of 0.212, while the *LM*(2) test for second order autocorrelation is equal to 21.778 with a p-value of 0.151. The tests for normality and ARCH effects show some problems, but adding more dummies is not a solution. As VAR results are reasonably robust anyway, we continue with this specification.

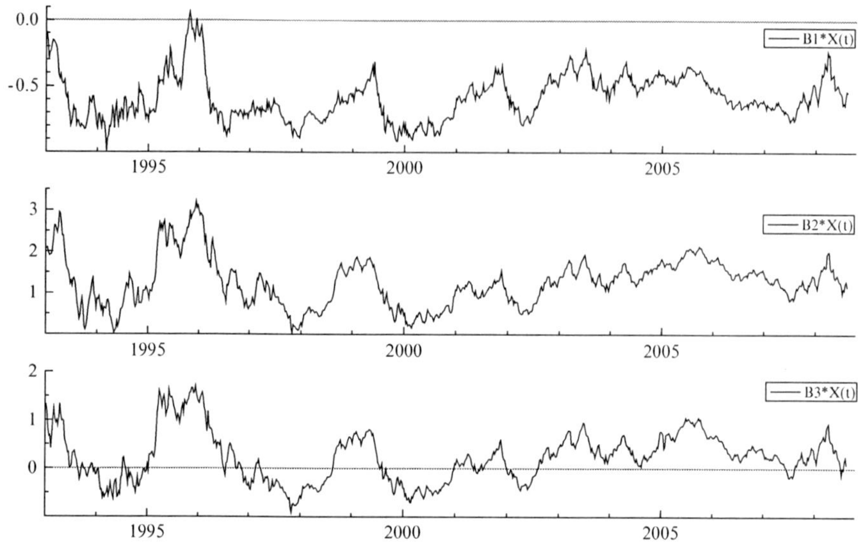

Fig. 2 The three identified cointegrating relations within the I(1) model

show some evidence of non-stationary behaviour, at least over the first part of the observation period. Such evidence of non stationarity can be attributed to the fact that the choice of a cointegration rank $r = 3$, leaves in the model a root very close to unit[4], which means that one very persistent component in the data has been considered as stationary.

Therefore, in a further analysis, we chose a cointegration rank $r = 2$, that is two cointegrating relations and $(p-r) = 2$ common stochastic trends, corresponding to the two near unit roots. Trying to identify the long-run structure underlying the two relations, a reasonable supposition we made is that some linear combinations of the spreads could emerge as stationary, instead of the spreads themselves, as we made for rank $r = 3$. The final structure is given by the following normalized relations, accepted with a p-value $= 0.141$:

$$\hat{\tilde{\beta}}'_1\tilde{\mathbf{x}}_t = 0.837(FRYLD10_t - GEYLD10_t) - 0.163(ITYLD10_t - RYLD10_t) + 0.002t_{08:01:1999} - 0.002t$$

$$\hat{\tilde{\beta}}'_2\tilde{\mathbf{x}}_t = 0.823(FRYLD10_t - GEYLD10_t) - 0.177(ITYLD10_t - FRYLD10_t) + 0.002t_{08:01:1999} - 0.002t$$

These relations show that, when corrected for the slope coefficients of the broken deterministic trends, weighted differences between pairs of spreads seem to be

[4]The largest characteristic roots of the unrestricted VAR are: 0.985, 0.966, 0.938, 0.840, 0.292.

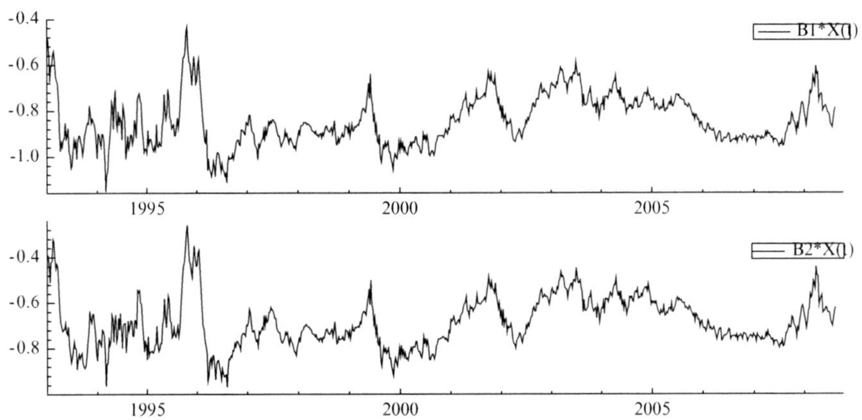

Fig. 3 The two identified cointegrating relations within the I(1) model

stationary, and it's quite interesting to note that the spread between France and Germany and either the one between Italy and France, or the one between Spain and France, result to cointegrate, as if French yields were the linking factor. Similar results have been found by Giese (2008), where she analyzes monthly US treasury zero-coupon bond yields of different maturities. Giese found that weighted differences between pairs of spreads of short, medium and long-term maturities become stationary with the medium-term yields as the linking factor.

The two final relations are represented in Fig. 3. As we can see, they provide evidence of some stationarity in the first part, though some long swings are still present. Analyzing the estimated adjustment dynamics of the system, we found that both Italian and Spanish yields show significant adjusting behaviours to the two identified relations, as we would expect, whereas German yields satisfy the restriction of being weakly exogenous with respect to the system of variables, with a p-value $= 0.312$, that is the German series results to be non-equilibrium correcting within the system and its cumulated residuals form a common trend.

The results obtained within the I(1) model are, anyway, unsatisfactory from different points of view: the number of unit or near unit roots larger than expected and, therefore, a smaller number of cointegrating relations; the non-stationarity of the yields spreads by themselves, whereas linear combinations of the spreads are much more stationary; the results of recursive tests, suggested by Hansen and Johansen (1999) showing that coefficients in the restricted model are not really stable; finally, the clear indication, coming from the last relations, that we should consider the possibility that spreads, which are non stationary over the observation period, could become stationary when combined with other spreads or with non stationary variables, which could be the first differences of the same yields making up the spreads, or any other combination of the same yields.

Table 1 The trace test for the determination of the I(2) rank indices

$(p-r)$	r	$s_2=4$	$s_2=3$	$s_2=2$	$s_2=1$	$s_2=0$
2	2			111.510	49.118	27.878
				(0.000)	(0.001)	(0.026)
1	3				**17.076**	**5.995**
					(0.129)	(0.471)

3 The Analysis Within the I(2) Model

The methodological approach of analysis just described, that is trend-adjusting for the change in regime and treating the series as I(1), has proven unsatisfactory in terms of both identification and stationarity of the cointegrating relations. In particular, their graphs, together with the graphs of the data in first and second differences, and the values of the model characteristic roots, are clear signals that we should consider also the possibility of a double unit roots in the time series, that is analysing them as I(2) variables[5]. As Juselius (2006, p. 293) explains: "... the typical smooth behaviour of a stochastic I(2) trend can often be approximated with an I(1) stochastic trend around a broken linear deterministic trend ..." Moreover, Juselius (2010, p.7) argues: "... in a p-dimensional VAR model of $\mathbf{x}'_t = [x_{1,t}, \ldots, x_{p,t}]$, the number of large roots in the characteristic polynomial depends on the number of common stochastic trends pushing the system, $(p-r)$, and whether they are of first order, s_1, or second order, s_2. To determine s_1 and s_2, we can use the I(2) test procedure ..." Table 1 reports the I(2) trace test results for both choices, $r=2$ and $r=3$.

As we can see from the table, the sequential testing of the joint hypothesis (r, s_1, s_2) for all values of r, s_1 and s_2, has given as the first non rejection[6] $r=3$, $s_1=0$ and $s_2=1$ with a $p-$ value $=0.129$. Therefore, the two unit roots of the model should be captured by an I(2) common stochastic trend.

These results suggest to model the regime change stochastically within a cointegrated I(2) model, as follows (Juselius (2006, p.319):

$$\Delta^2\mathbf{x}_t = \boldsymbol{\alpha}(\tilde{\boldsymbol{\beta}}'\tilde{\mathbf{x}}_{t-1} + \tilde{\boldsymbol{\delta}}'\Delta\tilde{\mathbf{x}}_{t-1}) + \boldsymbol{\zeta}\tilde{\boldsymbol{\tau}}'\Delta\tilde{\mathbf{X}}_{t-1} + \boldsymbol{\Phi}\mathbf{D}_t + \boldsymbol{\varepsilon}_t,\ \boldsymbol{\varepsilon}_t \sim N_p(\mathbf{0}, \boldsymbol{\Omega}), \quad (2)$$

where the number r of stationary polynomially cointegrating relations - the relations within round brackets in (2) -, the number s_1 of I(1) common stochastic trends and the number s_2 of the I(2) ones among the $(p-r)$ non stationary components, have been determined by the trace test. We started with the estimation of the unrestricted model and, quite surprisingly, the iterative procedure converged to the

[5]A formal treatment of I(2) models and relative tests can be found in Johansen (1997).

[6]The second non rejection is just the case $r=3$, $(p-r)=1$.

final estimates in few iterations, while for any other choice for r, s_1 and s_2 the number of iterations taken was really very high.

Before proceeding with the identification of the polynomially cointegrating relations, we followed the approach adopted by Johansen et al. (2008), of testing a set of non-identifying hypotheses.

First we tested the restriction whether a linear trend is needed in the sub-period going from the beginning of 1993 to the end of 1998, but not in the following sub-period, that is the deterministic trend characterizing the relations is just a broken trend ending in 1998. This is a test of the hypothesis that the variables t and tD_{t^*} have got the same coefficient with opposite sign since the beginning of 1999. The results show that for each polynomially cointegrating relation the hypothesis is largely accepted with an overall p-value $= 0.296$. This can be interpreted as a clear signal that the convergence in the long-term bond yields has been achieved by January 1999. After that date, the data show no significant linear trends and the eventual deterministic components imply only that the equilibrium means are different from zero. Then we tested separately four hypotheses, each of which, if not rejected, implies that the variable in question is at most I(1). The hypothesis was borderline rejected for German and for French yields, but strongly rejected for Italian and for Spanish yields, implying that the last two variables can be considered I(2), while the other two are only borderline I(2).

Another interesting hypothesis is the one of no long-run levels feed-back from the variable considered, that is the variable is rather pushing the system than adjusting. The testing results were such that the null hypothesis was accepted with a p-value $= 0.312$ for the German long term bond yields and with a p-value $= 0.134$ for the French ones. As regards purely adjusting variables, Spanish yields seem such a variable with a p-value $= 0.696$, while Italian yields are such only borderline, with a p-value $= 0.073$.

Moving to the identification of the polynomially cointegrating relations, we were interested to see whether relations between pairs of yields, corrected for short-run dynamics and deterministic trends, could be considered stationary within the I(2) model. Therefore, we identified the long-run structure by imposing the restrictions that each vector making up the $\tilde{\beta}$ matrix represents a relation between pairs of yields, with a broken linear trend whose effect ends at the beginning of 1999. The LR test statistic on the over-identifying restrictions gave the value $\chi^2_3 = 3.698$, with a p-value $= 0.296$, making the restrictions largely accepted. The estimated identified dynamic long-run equilibrium relations are the following:

$$\begin{aligned}\hat{\tilde{\beta}}_1'\tilde{\mathbf{x}}_t + \hat{\tilde{\delta}}_1'\Delta\tilde{\mathbf{x}}_t &= FRYLD10_t - 1.119GEYLD10_t - 0.0005t_{02:01:1999} + 0.0005t \\ &\quad + 3.111\Delta GEYLD10 + 3.481\Delta FRYLD10 + 5.307\Delta ITYLD10 \\ &\quad + 5.147\Delta SPYLD10 - 0.646\Delta t_{04:01:1999} + 0.793\Delta t\end{aligned}$$

$$\begin{aligned}\hat{\tilde{\beta}}_2'\hat{\mathbf{x}}_t + \hat{\tilde{\delta}}_2'\Delta\tilde{\mathbf{x}}_t &= ITYLD10_t - 1.711GEYLD10_t - 0.012t_{02:01:1999} + 0.012t \\ &\quad + 11.656\Delta GEYLD10 + 13.044\Delta FRYLD10 + 19.927\Delta ITYLD10\end{aligned}$$

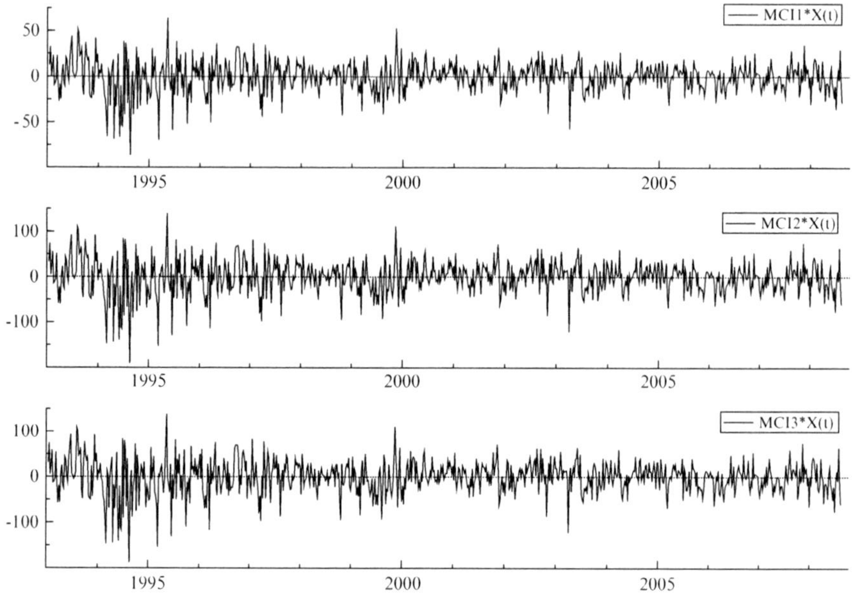

Fig. 4 The three polynomially cointegrating relations within I(2) model

$$+\,19.319\Delta SPYLD10 - 2.184\Delta t_{04:01:1999} - 0.820\Delta t$$

$$\begin{aligned}\hat{\tilde{\beta}}'_3\hat{x}_t + \hat{\tilde{\delta}}'_3\Delta\tilde{\mathbf{x}}_t &= SPYLD10_t - 1.659GEYLD10_t - 0.009t_{02:01:1999} + 0.009t \\ &+ 10.329\Delta GEYLD10 + 11.558\Delta FRYLD10 + 17.647\Delta ITYLD10 \\ &+ 17.111\Delta SPYLD10 - 1.970\Delta t_{04:01:1999} - 0.179\Delta t\end{aligned}$$

These equations show the existence of significant stationary relations between Germany yields and any other country yields if in the same relations we allow for the presence of differenced variables, that is the resulting polynomially cointegrated relations need the differenced variables to become stationary. The estimated relations are plotted in Fig. 4.

If we compare this figure with Fig. 2, we can see the importance of the differenced variables in making the cointegrating relations stationary over the sample period.

We then identified the long-run structure by imposing the restrictions that each vector making up the $\tilde{\beta}$ matrix represents a spread between pairs of yields, in particular, a spread between German yields and any other country's yields. The LR test statistic on the over-identifying restrictions gave the value $\chi^2_6 = 10.571$, with a p-value $= 0.103$, making the restrictions still accepted. As the results in terms of estimated identified dynamic long-run equilibrium relations, were very similar to the relations already plotted in Fig. 4, we don't report them.

Table 2 The estimated matrix of adjustment coefficients $\hat{\alpha}$ and the estimated vector $\hat{\alpha}_{\perp 2}$. Bold numbers denote coefficients significant at 5%

	$\hat{\alpha}_1$	$\hat{\alpha}_2$	$\hat{\alpha}_3$	$\hat{\alpha}_{\perp 2}$
DDGEYLD	−0.030	0.018	−0.020	**1.000**
DDFRYLD	**−0.076**	0.018	−0.009	**−0.494**
DDITYLD	0.076	**−0.070**	0.040	−0.046
DDSPYLD	0.020	**0.069**	**−0.100**	−0.175

When trying to analyze the adjustment dynamics of the system, we have to keep in mind that the adjustment structure embodied by the I(2) model is much more complex than the one embodied by the I(1) model. Within the polynomially cointegrating relations, if $\delta_{ij}\beta_{ij} > 0$, then the differences $\Delta x_{i,t}$ are equilibrium correcting to the levels $x_{i,t}-1$. It is, therefore, interesting to note that first differences in German, French, Italian and Spanish bond yields, are equilibrium correcting to the levels of the normalized variable in each relation. Within the model, the estimated $\hat{\alpha}$ matrix contains information on the adjustment dynamics between $\Delta x^2_{i,t}$ and the variables in levels and in differences, in particular if $\alpha_{ij}\delta_{ij} < 0$, then $\Delta x^2_{i,t}$ is equilibrium correcting in $\Delta x_{i,t}$. As we can see from the values in Table 2, French yields are significantly equilibrium correcting in the first relation, Italian yields are significantly equilibrium correcting in the second and Spanish are equilibrium correcting in the third, as expected.

From the table we can have interesting information also on the I(2) stochastic trend component which can be considered as the driving force for the system. The estimated vector $\hat{\alpha}_{\perp 2}$ shows, in fact, that it is primarily the twice cumulated shocks to the German long-term bond yields which have generated the I(2) stochastic trend. A significant contribution has been given also by the twice cumulated shocks to the French yields. This result confirms that the convergence process has been mainly led by German long-term bond yields and in part by French yields, to which the other yields have been adjusting. As regards the weights with which the I(2) trend have influenced the variables, the results have shown that Italian and Spanish yields were the most influenced, followed, in order, by French and German ones.

4 Conclusions

According to theory, the bond market unification process that we have analysed would imply a single latent factor, that is a single common stochastic trend as a driving force underlying the co-movements of yields of the same maturities across different countries' market. Analyzing weekly observations of long term bond yields relative to France, Germany, Italy and Spain within the I(1) model we have detected some clear signals that the convergence process is more complex than expected and that, in order to investigate the empirical regularities behind the swings in the series, we have to use an I(2) model. In fact, analyzing the observations within an

I(2) model, we have found evidence of the existence of a common stochastic trend given mainly by the twice cumulated shocks to the German long-term bond yields, but also by the twice cumulated shocks to the French ones. Such results indicate the importance of modelling the smoothing behaviour shown by the series in the process of convergence using an approach which takes into accounts the curvature of the co-movements in the series, as well as the level and the slope. As a conclusion, it's reasonable to state that the chosen I(2) approach has allowed a better modelling of the convergence process than within the I(1) model, giving evidence to expected characteristics that the I(1) model wouldn't show.

Paper financed by PRIN/MIUR 2007: "Proprietà macroeconomiche emergenti in modelli multi-agente: il ruolo della dinamica industriale e della evoluzione della struttura finanziaria"

The author thanks Refet S. Gürkaynak, Department of Economics, Bilkent University, and former Economist, Federal Reserve Board, Division of Monetary Affairs, for kindly providing data.

References

Bernard, A. B.: Empirical Implications of the Convergence Hypothesis. Working Paper MIT, Cambridge, MA (1992)

Bernard, A. B., Durlauf, S. N.: Convergence in international output. J. App. Econom. **10**, 97–108 (1995)

Dennis J. G.: Cats in rats cointegration analysis of time series, version 2. Estima, Evanston (2006)

Ehrmann M., Fratzscher M., Gürkaynak R., Swanson E.: Convergence and anchoring of yield curves in the Euro area. FRBSF Working Paper 2007–24 (2007)

Giese J. V.: Level, slope, curvature: characterising the yield curve in a cointegrated VAR Model. E-Economics **2**, 2008–28 (2008)

Johansen S.: likelihood based inference in cointegrated vector auto-regressive models. Oxford University Press, Oxford (1997)

Juselius K.: The cointegrated var model. methodology and applications. Oxford University Press, Oxford (2006)

Juselius K.: Testing exchange rate models based on rational expectations versus imperfect knowledge economics: a scenario analysis. Working Paper, Department of Economics, University of Copenhagen (2010)

Johansen S., Juselius K., Frydman R., Goldberg M.: Testing hypotheses in an I(2) model with piecewise linear trends. An analysis of the persistent long swings in the Dmk/$ rate. J. Economet., forthcoming (2008)

Hansen H., Johansen S.: (1999) Some tests for parameter constancy in the cointegrated VAR. Economet. J. **2**, 306–333 (1999)

Part VI
Environmental Statistics

Anthropogenic CO_2 Emissions and Global Warming: Evidence from Granger Causality Analysis

Massimo Bilancia and Domenico Vitale

Abstract This note reports an updated analysis of global climate change and its relationship with Carbon Dioxide (CO_2) emissions: advanced methods rooted in econometrics are applied to bivariate climatic time series. We found a strong evidence for the absence of Granger causality from CO_2 emissions to global surface temperature: we can conclude that our findings point out that the hypothesis of anthropogenically-induced climate change still need a conclusive confirmation using the most appropriate methods for data analysis.

1 Introduction

There is an increasing evidence that global climate change is occurring during the Anthropocene (Crutzen, 2002): this conclusion is strongly supported in the Fourth Assessment Report of the Intergovernmental Panel on Climate Change (IPCC 2007), in which it is stated that the warming of the climate system is unquestionable, considering the increase in global average land and sea surface temperatures and the massive decrease of snow and polar ice extent, with the consequent rising of the global average sea level. In the same report, eleven of the twelve years since 1995–2006 were ranked among the warmest years since 1850; for global mean surface temperatures an increase of 0.74°C ± 0.18°C (when estimated by a linear trend over the last 100 years, 1906–2005) was reported. According to the IPCC panel, the rate of warming over the last 50 years is almost double that over the last 100 years (0.13°C ± 0.03°C vs. 0.07°C ± 0.02°C per decade).

Both authors conceived the study: Massimo Bilancia wrote Sects. 1 and 3 and Domenico Vitale wrote Sects. 2 and 4.

M. Bilancia (✉) · D. Vitale
Department of Statistical Sciences "Carlo Cecchi" – University of Bari, Via C. Rosalba n.53, 70124 Bari, Italy
e-mail: mabil@dss.uniba.it; domenicovitale1981@libero.it

A. Di Ciaccio et al. (eds.), *Advanced Statistical Methods for the Analysis of Large Data-Sets*, Studies in Theoretical and Applied Statistics,
DOI 10.1007/978-3-642-21037-2_21,

Even if controversies exist about the causes, many authors agree about the fact that Carbon Dioxide (CO_2) emissions play a positive and predominant role in global climate warming. The perturbation to Radiative climate Forcing (RF) that has the largest magnitude and the least scientific uncertainty is the one related to changes in long-lived and well mixed GreenHouse gases (GHGs), in particular CO_2, Methane (CH_4), Nitrous Oxide (N_2O) and halogenated compounds (mainly Chlorofluorocarbons, CFCs). These conclusion are based on the data compiled by IPCC, which recommended expressions to convert GHGs variations with respect to 1,750 baseline to instantaneous RFs; these empirical expressions were derived from atmospheric radiative transfer models and generally have an uncertainty of about 10%, leading to a value of 1.66 Wm^{-2} (Watt per Square Meter) for CO_2 versus 0.48 Wm^{-2} for CH_4 (or 0.16 Wm^{-2} for N_2O and even smaller values for CFCs, IPCC (2007)).

The hypothesized causal relationship between anthropogenic CO_2 emissions and global warming cannot be easily verified in practice, for the reason that it is difficult to establish a feasible definition of causality in a non-experimental setting. Tests based on the stochastic view of time series behavior are based on the assumption that temporal ordering of events can be used to make an empirical distinction between leading and lagging variables, a distinction which is the basis of the well-know concept of causality that was introduced in Granger (1969). Assume that $\{U_t\}$ and $\{Z_t\}$ represent covariance-stationary time series, and let the information set I_t have the form $I_t = (U_t, Z_t, U_{t-1}, Z_{t-1}, \ldots, U_1, Z_1)$: we say that $\{Z_t\}$ is Granger causal for $\{U_t\}$ with respect to I_t if the optimal linear predictor of U_{t+k} based on I_t has smaller variance for any forecasting horizon k than the optimal linear predictor of U_{t+k} based on $(U_t, U_{t-1}, \ldots, U_1)$. In other words $\{Z_t\}$ is Granger causal for $\{U_t\}$ if $\{Z_t\}$ helps to linearly predict $\{U_t\}$ at some stage in the future. It is worth noting, however, that Granger causality is not causality in a deep sense of the word: it just suggests whether one thing happens before another or not. A breakthrough in causal statistical analysis of global climatic time series occurred after Kaufmann and Stern (1997), where Granger causality testing was applied by the first time; however, the assumption that one or more unit roots are present in observed temperature and CO_2 series requires careful testing and it is based on strong hypotheses (which means that non-stationary volatility and structural breaks must be properly addressed): the lack of consensus regarding the adequate representation of non-stationarities obscures the possibilities of finding universally accepted statistical relationship between global temperature and GHGs (Gay-Carcia et al. 2009). For these reasons, in this paper we re-analyze the whole issue in the light of recent developments of time series econometrics.

2 The Datasets

In order to test whether information about CO_2 emissions may be useful to predict global warming, we will use two distinct temperature dataset: this approach is a sort of sensitivity analysis, for the reason that it is a safeguard against inaccuracies that might have been introduced in the data and that could bias the results. The first one is the gridded HadCRUT3v temperature anomaly (relative to the 1961–1990 reference period means) time series based on land and marine data, discussed in Brohan et al. (2006): global annual time-series are produced by averaging the gridded data.

The second dataset is the global annual temperature index (anomalies relative to the 1951–1980 base period) derived from the NASA Goddard Institute for Space Studies (GISS) temperature gridded dataset (Hansen et al., 1999): the raw input data for the land component are the unadjusted data from the Global Historical Climatological Network (Peterson and Vose, 1997), United States Historical Climatological Network (USHCN) data and Scientific Committee on Antarctic Research (SCAR) data from Antarctic Stations.

Finally, the Carbon Dioxide Information Analysis Center (CDIAC) annual global CO_2 emission data were estimated on the ground of summary compilation of coal, brown, peat and crude oil by nation and year, and data about fossil fuel trade, solid and liquid fuel imports and exports (Boden et al., 2009). The 1950 to present CO_2 emission estimates are derived primarily from energy statistics published by the United Nations using the methods introduced in Marland and Rotty (1984). Data from the U.S. Department of Interior's Geological Survey were used to estimate CO_2 emitted during cement production. Further details on the contents and processing of the historical energy statistics are provided in Andres et al. (1999). The three time series used throughout our statistical analyses (covering the 1880–2007 period) are shown in Fig. 1.

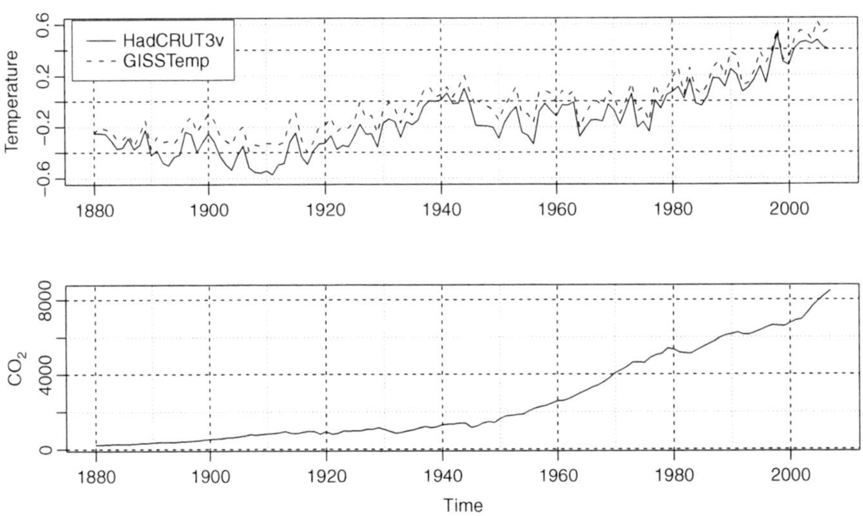

Fig. 1 Global temperature anomalies and CO_2 emission series (see the text for further details)

3 Unit Root Testing

Our approach considers testing for Granger causality in a stationary VAR model (see Sect. 4): as we said in Sect. 1, the obvious primary precondition is that data must be a realization from a covariance stationary process, which means that the first and second statistical moments of each variable do not vary with time.

For dealing with non-stationarities in mean, assuming that the series $\{Y_t\}$ contains at most one unit root, we test for the order of integration by exploiting the Augmented Dickey-Fuller machinery (ADF, Wolters and Hassler (2005)). The test equation in the presence of a deterministic time trend is given by the following trend-stationary AR(p) process, reparametrized according to the Sims–Stock–Watson canonical form

$$\Delta Y_t = \alpha + \delta t + \gamma Y_{t-1} + \sum_{j=1}^{p-1} \zeta_j \Delta Y_{t-j} + \epsilon_j \tag{1}$$

with Gaussian i.i.d. innovations for small samples. Test for the pair of hypothesis $H_0 : \gamma = 0$ versus $H_1 : \gamma < 0$ is based on the t-statistic from an ordinary least squares estimation of (1): it is well known that the limiting distribution of the t-statistics is nonstandard and that it can be expressed as a suitable functional of standard Brownian motion on $[0, 1]$ (last, but not least, it depends on deterministic terms included in the test regression under H_0). Critical values for small samples have been simulated, among the others, in Davidson and MacKinnon (1993). Order selection of lagged differences was based on three different criteria: (a) the deterministic (Schwert) rule introduced in Schwert (1989), where $p - 1 = int\{c(T/100)^{1/d}\}$ with $c = 4$ and $d = 4$; (b) Akaike and Bayesian Information Criteria (AIC and BIC respectively), which are data-dependent rules that allow to choose model order by minimizing an objective function that trades off parsimony against goodness of fitting (see Lütkepohl (2006) for a detailed exposition of such criteria); (c) the General-to-Specific (GtS) scheme discussed in Ng and Perron (1995). Since the hypothesis that the stochastic process generating each series is driven by at most $k = 2$ unit roots cannot be excluded by visual inspection (see Fig. 1), we followed the principle stated in Dickey and Pantula (1987) (see also Haldrup and Lildholdt (2002)), in which the null hypothesis of k unit roots against the alternative of $k - 1$ unit roots is shown to be a winner (in terms of simplicity and power) versus a sequence of unit root tests in the traditional order. For these reasons we tested I(2) versus I(1) first by applying the test regression (1) to $\Delta^2 Y_t$, and then I(1) versus I(0) if I(2) versus I(1) test rejects.

Another key problem in unit root testing is that tests are conditional on the presence of deterministic regressors and test for the presence of deterministic regressors are conditional to the presence of unit roots: as too few or too many deterministic regressors may dramatically reduce the power, each test was carried out according to the sequential procedure introduced in Dolado et al. (1990) and popularized in (Enders, 2003; Pfaff, 2008).

Table 1 Results of the ADF unit root test for the HadCRUT3v temperature series: only the final model, chosen according to the Dolado's procedure, is reported

Deterministic terms	Lag length selection[a]	Lag length $(p-1)$	Test statistics	5% Crit. Val.
I(2) versus I(1) – HadCRUT3v				
$\alpha + \delta t$	Schwert	4	$\tau_t = -7.1026$ ($H_0 : \gamma = 0$ given $\delta = 0$)	−3.43
$\alpha + \delta t$	BIC	1	$\tau_t = -11.395$ ($H_0 : \gamma = 0$ given $\delta = 0$)	−3.43
$\alpha + \delta t$	AIC	1	$\tau_t = -11.3950$ ($H_0 : \gamma = 0$ given $\delta = 0$)	−3.43
$\alpha + \delta t$	GtS	2	$\tau_t = -10.0483$ ($H_0 : \gamma = 0$ given $\delta = 0$)	−3.43
I(1) versus I(0) – HadCRUT3v				
None	Schwert	4	$\tau = -0.8355$ ($H_0 : \gamma = 0$)	−1.95
$\alpha + \delta t$	BIC	1	$\tau_t = -3.9842$ ($H_0 : \gamma = 0$ given $\delta = 0$)	−3.43
$\alpha + \delta t$	AIC	1	$\tau_t = -3.9842$ ($H_0 : \gamma = 0$ given $\delta = 0$)	−3.43
None	GtS	3	$\tau = -0.6929$ ($H_0 : \gamma = 0$)	−1.95

[a] Legend: GtS = General-to-Specific

Results for the HadCRUT3v temperature datasets are given in Table 1: by repeatedly applying the ADF test inside the steps prescribed by the Dolado's algorithm, we can conclude that first differences ΔY_t may be described by ARMA(p,q) process that can be approximated by an AR(p) autoregressive structure of suitable order. Once a tentative lag was determined, diagnostic checking of test regression residuals was conducted (numerical results are available by the authors upon request): the AIC and BIC lag length selection criteria did not permit to appropriately capture the actual error process (so that the standard error of γ could not be consistently estimated), unlike the Schwert method which suggests that $p - 1 = 4$ lags have to be used in the test regression, or the GtS procedure which prescribes $p - 1 = 3$ ensuring as much uncorrelated residuals. Similar inferences apply to the GISS temperature series: the Schwert method gives appropriate results by choosing $p - 1 = 4$ to take account of serial correlation in the disturbances (Table 2).

Results concerning the CO_2 series are slightly less stable: we conclude that (Table 3) first differences are stationary with non zero drift. The Schwert method reasonably accounts for serial correlation with $p - 1 = 4$, but a mild-to-moderate ARCH effect up to $\ell = 5$ lags is undoubtedly present in the I(1) versus I(0) test regression residuals (see Table 4). It is worth noting that both AIC and BIC criteria lead to the conclusion that the driving process is difference-stationary around a linear trend: anyway, the chosen number of lags ($p - 1 = 1$) resulted to be inadequate to control for serial correlation.

Testing for unit roots in the presence of ARCH/GARCH residuals is notoriously difficult: some powerful results are given in Phillips (1987), where mild forms

Table 2 Results of the ADF unit root test for the GISS temperature series: only the final model, chosen according to the Dolado's procedure, is reported

Deterministic terms	Lag length selection[a]	Lag length $(p-1)$	Test statistics	5% Crit. Val.
I(2) vs. I(1) – GISS Temp				
$\alpha + \delta t$	Schwert	4	$\tau_t = -7.8691$ ($H_0 : \gamma = 0$ given $\delta = 0$)	−3.43
$\alpha + \delta t$	BIC	1	$\tau_t = -11.9301$ ($H_0 : \gamma = 0$ given $\delta = 0$)	−3.43
$\alpha + \delta t$	AIC	1	$\tau_t = -11.9301$ ($H_0 : \gamma = 0$ given $\delta = 0$)	−3.43
$\alpha + \delta t$	GtS	4	$\tau_t = -7.8691$ ($H_0 : \gamma = 0$ given $\delta = 0$)	−3.43
I(1) vs. I(0) – GISS Temp				
None	Schwert	4	$\tau = -0.1162$ ($H_0 : \gamma = 0$)	−1.95
$\alpha + \delta t$	BIC	1	$\tau_t = -4.2303$ ($H_0 : \gamma = 0$ given $\delta = 0$)	−3.43
$\alpha + \delta t$	AIC	1	$\tau_t = -4.2303$ ($H_0 : \gamma = 0$ given $\delta = 0$)	−3.43
None	GtS	3	$\tau = -0.0045$ ($H_0 : \gamma = 0$)	−1.95

[a]Legend: GtS = General-to-Specific

Table 3 Results of the ADF unit root test for the CO_2 emission series: only the final model, chosen according to the Dolado's procedure, is reported

Deterministic terms	Lag length selection[a]	Lag length $(p-1)$	Test statistics	5% Crit. Val.
I(2) versus I(1) – CO_2				
$\alpha + \delta t$	Schwert	4	$\tau_t = -4.6698$ ($H_0 : \gamma = 0$ given $\delta = 0$)	−3.43
$\alpha + \delta t$	BIC	1	$\tau_t = -6.7462$ ($H_0 : \gamma = 0$ given $\delta = 0$)	−3.43
$\alpha + \delta t$	AIC	1	$\tau_t = -6.7462$ ($H_0 : \gamma = 0$ given $\delta = 0$)	−3.43
$\alpha + \delta t$	GtS	2	$\tau_t = -4.6204$ ($H_0 : \gamma = 0$ given $\delta = 0$)	−3.43
I(1) versus I(0) – CO_2				
α	Schwert	4	$z = 3.0323$ ($H_0 : \gamma = 0$ given $\alpha \neq 0$)	$p > 0.99$
$\alpha + \delta t$	BIC	1	$z = 0.3079$ ($H_0 : \gamma = 0$ given $\delta \neq 0$)	$p > 0.62$
$\alpha + \delta t$	AIC	1	$z = 0.3079$ ($H_0 : \gamma = 0$ given $\delta \neq 0$)	$p > 0.62$
$\alpha + \delta t$	GtS	1	$z = 0.3079$ ($H_0 : \gamma = 0$ given $\delta \neq 0$)	$p > 0.62$

[a]Legend: GtS = General-to-Specific

Table 4 Testing for the presence of ARCH effects in the residuals of the test regression of each final model (chosen according to the Dolado's procedure)

Test	Lag length selection[a]	$\ell = 5$	$\ell = 10$	$\ell = 15$	$\ell = 20$
CO_2					
I(1) vs I(0)	Schwert	0.0469	0.2399	0.4224	0.6807
I(1) vs I(0)	BIC	0.0130	0.1689	0.2178	0.5574
I(1) vs I(0)	AIC	0.0130	0.1689	0.2178	0.5574
I(1) vs I(0)	GtS	0.0130	0.1689	0.2178	0.5574

[a] Legend: GtS = General-to-Specific

of unconditional heteroscedasticity are allowed as long as these residuals vanish asymptotically, satisfy a weak dependence condition (strong mixing) and the finiteness of the fourth moment is warranted. Unfortunately, if we assume a GARCH(1,1) disturbance $\{u_t\}$ in the test regression (1) i.e.

$$\begin{aligned} u_t &= \epsilon_t h_t \\ h_t &= \sqrt{\omega + \theta_1 \epsilon_{t-1}^2 + \beta_1 h_{t-1}} \end{aligned} \tag{2}$$

with $\omega, \theta_1, \beta_1 > 0$ to warrant the existence of conditional variance (in addition Boussama (1998) proves that under mild regularity conditions from the standard assumption $\theta_1 + \beta_1 < 1$ follows that $\{u_t\}$ is strictly stationary and strongly mixing), the finiteness of the fourth moment is clearly a more restrictive requirement. The adjusted Phillips and Perron $Z(\tau_\mu)$ test conducted with a test regression where the only drift term is present confirmed the existence of a unit root ($Z(\tau_\mu) = 0.8954$, 5% critical value -2.88), but the finiteness of the fourth moment is necessary in this case as well (Phillips and Perron, 1988). Anyway, assuming as reasonable the existence of at most one unit root (as the I(2) versus I(1) tests are likely to be valid), we produced a covariance-stationary $\{Z_t\}$ series by assuming the following model obtained from (1) conditionally to $\gamma = \delta = 0$

$$\begin{aligned} \Delta Y_t &= \alpha + \sum_{j=1}^{p-1} \zeta_j \Delta Y_{t-j} + u_t \\ u_t &\sim GARCH(1,1) \end{aligned} \tag{3}$$

with $p - 1 = 4$, and filtering the time-varying conditional volatility in the following way

$$Z_t = \Delta Y_t / h_t \tag{4}$$

Standardized residuals from the AR(4)/GARCH(1,1) model (3) showed good properties (no significant autocorrelations or ARCH effects were found): filtered series Z_t is shown in Fig. 2. It is worth noting that the physical meaning of our

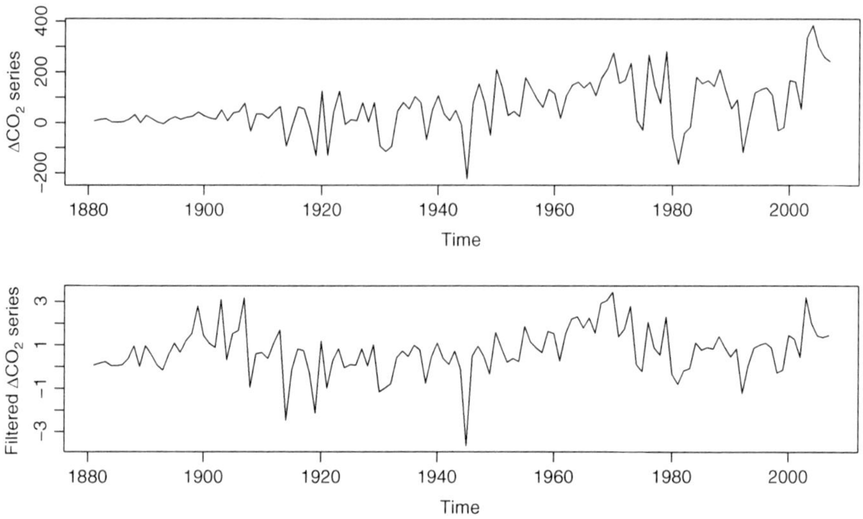

Fig. 2 Raw ΔCO_2 (*top*) and AR(4)/GARCH(1,1) filtered ΔCO_2 series (*bottom*)

method is, at present, unknown: we don't know whether short term cycles in the CO_2 levels with varying amplitudes exist and may be justified on the ground of human activities (e.g. economic cycles), or estimation based on indirect proxies introduces some artifacts in the data.

4 Testing for Granger Causality

The implications of Granger causality can be expressed in a form that is feasible for direct statistical testing; for example, $\{Z_t\}$ fails to Granger-cause $\{U_t\}$ if in a bivariate stationary VAR(m) (Vector AutoRegression) describing $\{U_t\}$ and $\{Z_t\}$ the coefficient matrices are lower triangular (Lütkepohl, 2006)

$$\begin{bmatrix} U_t \\ Z_t \end{bmatrix} = \begin{bmatrix} c_1 \\ c_2 \end{bmatrix} + \begin{bmatrix} \phi_{11,1} & 0 \\ \phi_{21,1} & \phi_{22,1} \end{bmatrix} \begin{bmatrix} U_{t-1} \\ Z_{t-1} \end{bmatrix} + \cdots + \begin{bmatrix} \phi_{11,m} & 0 \\ \phi_{21,m} & \phi_{22,m} \end{bmatrix} \begin{bmatrix} U_{t-m} \\ Z_{t-m} \end{bmatrix} + \begin{bmatrix} \epsilon_{1t} \\ \epsilon_{2t} \end{bmatrix} \quad (5)$$

for the reason that under this condition it follows that

$$U_t(k|\{(U_s, Z_s)|s \leq t\}) = U_t(k|\{U_s|s \leq t\}), \quad k = 1, 2, 3, \dots \quad (6)$$

where $U_t(k|I_t)$ is the best linear predictor of U_t at horizon k given the information set available at time t.

Even if a variety of Granger causality tests have been proposed, a simple approach uses directly the autoregressive specification (5): assuming a particular lag-length m and given the information set $I_t = (U_t, Z_t, U_{t-1}, Z_{t-1}, \dots, U_1, Z_1)$, we estimate by OLS the following (unrestricted) linear model

$$\begin{aligned} U_t &= c_1 + \alpha_1 U_{t-1} + \alpha_2 U_{t-2} + \cdots + \alpha_m U_{t-m} \\ &\quad + \beta_1 Z_{t-1} + \beta_2 Z_{t-2} + \cdots + \beta_m Z_{t-m} + \epsilon_t \end{aligned} \tag{7}$$

where ϵ_t is an i.i.d. disturbance. The zero constraints for coefficients translate into the null hypothesis $H_0 : \beta_1 = \beta_2 = \ldots = \beta_m = 0$, which may be tested by means of a standard F statistics having an exact F distribution for fixed regressor and Gaussian disturbances. For lagged dependent regressors an asymptotically equivalent test is given by Hamilton (1994)

$$F^\star = \frac{T(RSS_0 - RSS_1)}{RSS_1} \tag{8}$$

for a time series of length T, where RSS_0 and RSS_1 are the residual sum of squares respectively under the restricted (under H_0) and the unrestricted model (7): under the null hypothesis we have that $F^\star \rightarrow \chi^2_m$. Finally, it must be carefully taken into account the fact that any empirical test of Granger causality is very sensitive to the choice of the lag length m or the method used to deal with the potential non stationarity of the series (Christiano and Ljungqvist, 1988).

As in the previous section, the lag length m was estimated according to several data-dependent criteria such as the AIC and the BIC, the HQ (Hannan-Quinn) and the FPE (Final Prediction Error) (see Pfaff (2008) for more details). To verify whether model assumptions are satisfied or not, for each VAR(m) model we tested for the absence of serial correlation (Portmanteau test) and time-varying conditional volatility (multivariate ARCH test) (Pfaff, 2008). Given the model selection results (available by the authors upon request) we tested for Granger causality at lags $m = 3, 5$ and $m = 3, 6$ for the two pair of series respectively. For feedback assessment, we assumed the global temperatures as the leading variable. The test results are shown in Table 5: high p-values suggest a strong evidence for the absence of causality from CO_2 emissions to global surface temperatures.

We do not intend to question in a radical way the idea, now widely acquired, that CO_2 emissions are causing a rise in global temperature. Rather, we emphasize that the analysis of deterministic and stochastic properties of global climatic time series is complex; an alternative to our point of view is contained in Liu and Rodriguez (2005), where a structural analysis of cointegration between the temperatures and anthropogenic forcings is presented. However, even in this case, the correct identification of the order of integration is essential: it is a well known fact that estimation and hypothesis testing in a non-stationary VAR model in which variables are at most I(1) becomes quite different when these variables can be considered as at most I(2). The presence of non stationary conditional volatility

Table 5 Results of the Granger causality test. Here $Z_t \mapsto U_t$ means that Z_t is a lagging and U_t is a leading variable: the null hypothesis is that the past values of Z_t are not useful to predict the future values of U_t

Null Hypothesis	Lag length m	p-value
HadCRUT3v and filtered CO_2 first differences		
$CO_2 \mapsto$ HadCRUT3v	3	0.8206
HadCRUT3v $\mapsto CO_2$	3	0.3852
$CO_2 \mapsto$ HadCRUT3v	5	0.4164
HadCRUT3v $\mapsto CO_2$	5	0.2188
GISS Temp and filtered CO_2 first differences		
$CO_2 \mapsto$ GISS Temp	3	0.8192
GISS Temp $\mapsto CO_2$	3	0.2696
$CO_2 \mapsto$ GISS Temp	6	0.1382
GISS Temp $\mapsto CO_2$	6	0.3193

makes the task even more difficult: to ensure that the VAR model used in Sect. 4 is not inconsistent for data description, future research will address the issue of the presence of cointegration in the framework of the error correction model with non stationary volatility recently proposed in Cavaliere et al. (2010). Therefore, for the foreseeable future, the interchange between developments of econometric theory and climatology will provide significant advantages and exciting developments.

References

Andres, R.J., Fielding, D.J., Marland, G., Boden, T.A., Kumar, N.: Carbon dioxide emissions from fossil-fuel use, 1751-1950. Tellus, **51B**, 759–65 (1999). Doi: 10.1034/j.1600-0889.1999.t01-3-00002.x

Boden, T.A., Marland, G., Andres, R.J.: Global, Regional, and National Fossil-Fuel CO2 Emissions. Carbon Dioxide Information Analysis Center, Oak Ridge National Laboratory, U.S. Department of Energy, Oak Ridge, Tenn., USA (2009). Doi 10.3334/CDIAC/00001

Boussama, F.: *Ergodicité, mélange et estimation dans le modelés GARCH*. Ph.D. Thesis, Université 7 Paris (1998)

Brohan, P., Kennedy, J.J., Harris, I., Tett, S.F.B., Jones, P.D.: Uncertainty estimates in regional and global temperature changes: a new dataset from 1850. J. Geophys. Res., **111**, D12106 (2006). Doi:10.1029/2005JD006548

Cavaliere, G., Rahbek, A., Taylor, A.M.R.: Testing for co-integration in vector autoregressions with non-stationary volatility. J. Econometr., **158**, 7–24 (2010) doi:10.1016/jeconom.2010.03.003

Christiano, L.J., Ljungqvist, L.: Money does Granger-cause output in the bivariate money-output relation. Journal of Monetary Economics, **22**, 217–235 (1988) doi:10.1016/0304-3932(88)90020-7

Crutzen , P.J.: Geology of mankind. Nat., **415**, 23 (2002) doi:10.1038/415023a

Davidson, R., MacKinnon, J.G.: *Estimation and Inference in Econometrics*. Oxford University Press, New York, (1993) ISBN: 0-19-506011-3

Dickey, D.A., Pantula, S.G.: Determining the order of differencing in autoregressive processes. J. Bus. Econ. Stat., **5**, 455–461 (1987) http://www.jstor.org/stable/1391997

Dolado, J.J, Jenkinson, T., Sosvilla-Rivero, S.: Cointegration and unit-roots. J. Econ. Surv., **4(3)**, 249–273 (1990) 10.1111/j.1467-6419.1990.tb00088.x

Enders, W.: *Applied Econometric Time Series*, 2nd Ed. Wiley, New York (2003) ISBN: 978-0-471-23065-6

Gay-Carcia, C., Estrada, F., Snchez, A.: Global and hemispheric temperatures revisited. Climatic Change, **94**, 333–349 (2009). doi: 10.1007/s10584-008-9524-8

Granger, C.W.J.: Investigating causal relations by econometric models and cross-spectral methods, Econometrica, **37**, 424–438 (1969) http://www.jstor.org/stable/1912791

Haldrup, N., Lildholdt, P.: On the robustness of unit root tests in the presence of double unit roots. J. Time Ser. An., **23(2)**, 155–171 (2002) doi: 10.1111/1467-9892.00260

Hamilton, J.D.: *Time Series Analysis*, Princeton University Press, Princeton, New Jersey (1994)

Hansen, J., Ruedy R., Glascoe J., Sato. M.: GISS analysis of surface temperature change. J. Geophys. Res., **104(D24)**, 30997–31022 (1999) doi:10.1029/1999JD900835

IPCC (Intergovernmental Panel on Climate Change): Synthesis Report, Fourth Assessment Report (2007) http://www.ipcc.ch/pdf/assessment-report/ar4/syr/ar4_syr.pdf

Kaufmann R.K, Stern D.I.: Evidence for human influence on climate from hemispheric temperature relations. Nat., **388**, 38–44 (1997) doi:10.1038/40332

Liu H., Rodriguez G.: Human activities and global warming: a cointegration analysis. Environ. Model. Soft., **20**, 761–773 (2005) doi: 10.1016/j.envsoft.2004.03.017

Lütkepohl, H.: *New Introduction to Multiple Time Series*. Springer, Berlin (2006) ISBN: 978-3-540-26239-8

Marland, G., Rotty, M.R.: Carbon dioxide emissions from fossil fuels: a procedure for estimation and results for 1950-82. Tellus, **36(B)**, 232–61 (1984)

Ng, S., Perron, P.: Unit root tests in ARMA models with data-dependent methods for the selection of the truncation lag. J. Am. Stat. Ass., **90**, 268–281 (1995) http://www.jstor.org/stable/2291151

Peterson, T.C., Vose, R.S.: An overview of the Global Historical Climatology Network temperature database. Bull. Amer. Meteorol. Soc., **78**, 2837–2849 (1997)

Phillips, P.C.B.: Time series regression wuth a unit root. Econometrica, **55(2)**, 277–301 (1987) http://www.jstor.org/stable/1913237

Phillips, P.C.B., Perron, P.: Testing for a unit root in time series regression. Biometrika, **75**, 335–46 (1988) http://www.jstor.org/stable/2336182

Pfaff, B.: *Analysis of Integrated and Cointegrated Time Series with R*, 2nd Ed. Springer (2008) ISBN: 978-0-387-75966-1

Schwert, J.: Tests for unit roots: a Monte Carlo investigation. J. Bus. Econ. Stat., **7**, 147–160 (1989) http://www.jstor.org/stable/1391432

Wolters, J., Hassler U.: Unit root testing. Allg. Stat. Archiv, **90(1)**, 43–58 (2005) doi: 10.1007/s10182-006-0220-6

Temporal and Spatial Statistical Methods to Remove External Effects on Groundwater Levels

Daniele Imparato, Andrea Carena, and Mauro Gasparini

Abstract This paper illustrates a project on monitoring groundwater levels elaborated jointly with officers from Regione Piemonte. Groundwater levels are strongly affected by external predictors, such as rain precipitation, neighboring waterways or local irrigation ditches. We discuss a kriging and transfer function approach applied to monthly and daily series of piezometric levels to model these neighboring effects. The aims of the study are to reconstruct a groundwater virgin level as an indicator of the state of health of the groundwater itself and to provide important regulatory tools to the local government.

1 Introduction

This work is a description of an ongoing project on groundwater monitoring, a joint effort with the environmental department of Regione Piemonte, the local government of a large part of Northwestern Italy.

Water located beneath the ground surface in the fractures of lithologic formations and in soil pore spaces is called groundwater. It is generally recharged from, and eventually flows to, the surface naturally.

Groundwater is drawn and used for agricultural and industrial purposes by means of extraction wells. Thus, groundwater monitoring is an important issue in upcoming environmental legislation and governance. The regulatory framework is given, within the European Union (EU), by European Union (2006), which states:

D. Imparato (✉)
Department of Economics, Università dell'Insubria, via Monte Generoso, 71 21100 Varese, Italy
e-mail: daniele.imparato@polito.it

A. Carena · M. Gasparini
Department of Mathematics, Politecnico di Torino, C.so Duca degli Abruzzi 24, 10129 Torino, Italy
e-mail: andrea.carena@studenti.polito.it; mauro.gasparini@polito.it

A. Di Ciaccio et al. (eds.), *Advanced Statistical Methods for the Analysis of Large Data-Sets*, Studies in Theoretical and Applied Statistics,
DOI 10.1007/978-3-642-21037-2_22,

> Groundwater is a valuable natural resource and as such should be protected from deterioration and chemical pollution. This is particularly important for groundwater-dependent ecosystems and for the use of groundwater in water supply for human consumption.

and it is used to set guidelines for local governments.

Groundwatering is a complex phenomenon in which different external variables are involved, primarily the presence of other water sources, rain precipitation and evapotranspiration. In order to make correct decisions, the contribution of such natural predictors should be identified at some point. Due to the geography of the area and the locations of the monitoring stations considered, in this project the level of neighboring rivers and rain precipitations are taken as the most important predictors of groundwater levels. The aim of this exercise is to reconstruct a groundwater "virgin" level as an indicator of the state of health of the groundwater itself, by removing the effects of the external predictors which, from a theoretical standpoint, are viewed as confounders of the underlying level.

To this end, a transfer function approach, based on the observation of neighboring rivers and rain precipitation, is chosen in order to isolate and quantify the influence of predictors on groundwater levels. Different models, with and without the effect of the predictors, are then estimated.

It should be noted that the way predictors affect the groundwater basin strictly depends on the area considered, that is, on its geography, on the soil characteristics, on the basin depth, and so on. Depending on the basin conformation, the effects of groundwater levels may develop over different time lags and time scales. Moreover, different time scales may capture different modes of action of the predictors on the groundwater. Therefore, two classes of statistical models are estimated in this project: the first on a monthly and the second on a daily time scale. Selected examples of the two classes of models are presented in this paper. The examples bring us to conclude that the daily time scale captures immediate local natural phenomena, whereas the monthly scale is more appropriate for modelling effects on a larger scale, such as the influence of the whole hydrographic basin.

In Sect. 2, results are shown for some monitoring stations where the effects of predictors are estimated on a monthly scale. Then, an example of daily scale is discussed in Sect. 3, with application to a portion of the Po river plain.

As for the predictors, the values of the neighboring waterways are observed directly, whereas cumulative rain amounts are calculated based on point precipitation data at the rain monitoring stations. The reconstruction is done through a kriging prediction approach described in Sect. 3.

2 Data Analysis on a Monthly Scale

A *piezometer* is a special artificial well designed to measure the underground height of groundwater and it represents the core of a monitoring station. It consists of a small-diameter observation well, which is able to measure the water-table level, that is, the depth at which soil pore spaces become completely saturated with water. Piezometric levels are conventionally measured on a negative scale, where zero denotes the ground-level.

Since 2004, Regione Piemonte has been rapidly increasing the number of piezometers throughout its area, in order to capture information about groundwater resources and comply with the above mentioned EU regulations.

2.1 Data Pre-processing

On a monthly scale, rain effects are generally found not to be significant and the analysis in this section concerns mainly piezometers which can be found near waterways and are primarily affected by them. The waterways turn out to be rivers in all cases we consider.

Piezometric measurements are recorded irregularly, that is, the number of measurements available per day has been changing through the years. Moreover, piezometers are rather fragile tools, so that measurements may not be available for several days. We consider therefore monthly time series of groundwater levels derived from the original series by averaging the data observed in a given month. Neighboring river level time series are processed in the same way.

Several statistical issues arise, since many observed series still present missing values and appear to be highly nonstationary. For example, the Montecastello groundwater time series still shows a long period of missing values even after the monthly averaging operation. Imputation of missing data is required. We use here single imputation based on the correlated time series of the river Tanaro levels, which is used as a predictor with different time-lags in a linear regression model. The reconstruction obtained can be found in Fig. 1.

The levels of the neighboring river will be used again in the transfer function approach described in the next section; this double use of the data is an inevitable weakness of the procedure. However, this kind of imputation is preferred to other proposed methods based on ARIMA and state space models, that give less realistic reconstructions (not shown here).

Following a standard approach in time series analysis, we finally take first and/or second differences, according to need, to achieve stationarity of all time series involved.

2.2 Modelling the Effects of Neighboring Waterways via Transfer Function Models

Let (Y_t) be the monthly time series of groundwater piezometric levels and let (X_t) be the time series of the corresponding levels of the next waterway, once the

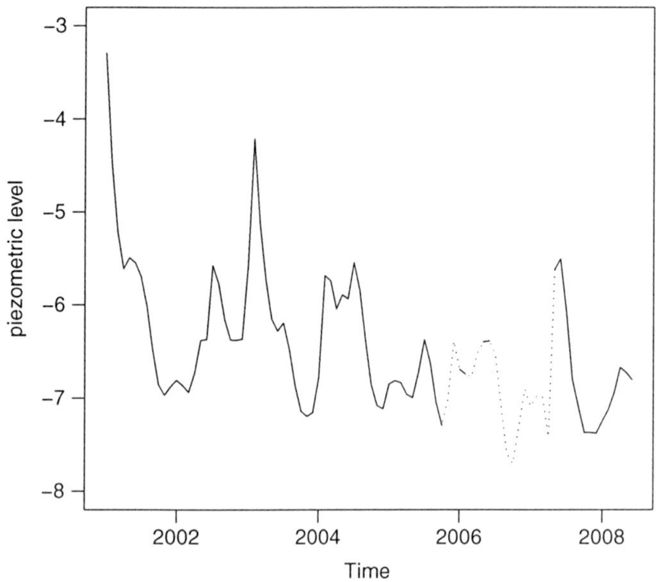

Fig. 1 Pre-processing of Montecastello piezometric levels: monthly missing value imputation (*pointed line*) via regression on the Tanaro river levels

pre-processing steps described in the previous section have been taken. We assume (X_t) and (Y_t) are regularly spaced stationary time series and we follow a transfer function approach. Define the lag operator $B(X_t) = X_{t-1}$. A linear relationship is postulated between the two processes through a linear filter, that is

$$Y_t = \sum_{s=0}^{\infty} \nu_s X_{t-s} + N_t := H(B)X_t + N_t \tag{1}$$

where $H(B) = \omega(B)/\delta(B)$ is the ratio of two finite degree polynomials in B and (N_t) is the error process (not necessarily a white noise), representing the part of Y_t which cannot be explained by a linear relationship with (X_t). The two processes (N_t) and (X_t) are assumed uncorrelated to each other. Notice that the presence of $\delta(B)$ is equivalent to using (Y_t) lags. Identification and estimation for model (1) is described, for example, in Battaglia (2007). The number of significant time lags in the model is determined by pre-whitening both (X_t) and (Y_t) through ARMA modelization of the input series and through the evaluation of the corresponding residuals for both series. Next, the appropriate transfer function is identified using the sample cross-correlation between such residuals.

As an example, let us consider here the results concerning the Carignano station. Written in an equivalent way, the final model (1) estimated for this case is

$$Y_t = 0.47(1 + B)X_t + (1 + 0.35B)e_t, \tag{2}$$

where (X_t) denotes the times series of the Po river levels next to Carignano and (e_t) is a white noise. Notice that the error component of our fitting is

$$(N_t) = (1 + 0.35B)e_t,$$

a moving average of an independent white noise process.

In order to remove the river effect on groundwater levels, the following proxy process $\tilde{Y}_t$ is then considered:

$$\tilde{Y}_t = (1 + 0.35B)\hat{e}_t, \tag{3}$$

where $\hat{e}_t$ are the fitted residuals from model (2). Finally, the proxy $\tilde{Y}_t$ is integrated in order to get groundwater virgin levels, in which the Po immediate effects have been removed. A new trend component is finally estimated from such a series through a LOWESS regression. LOWESS is a more empirical but less rigid approach than other methods based on low degree polynomial interpolation. It is chosen to show trends easily interpretable by practitioners. A comparison between the original trend component and this new one can be seen in Fig. 2.

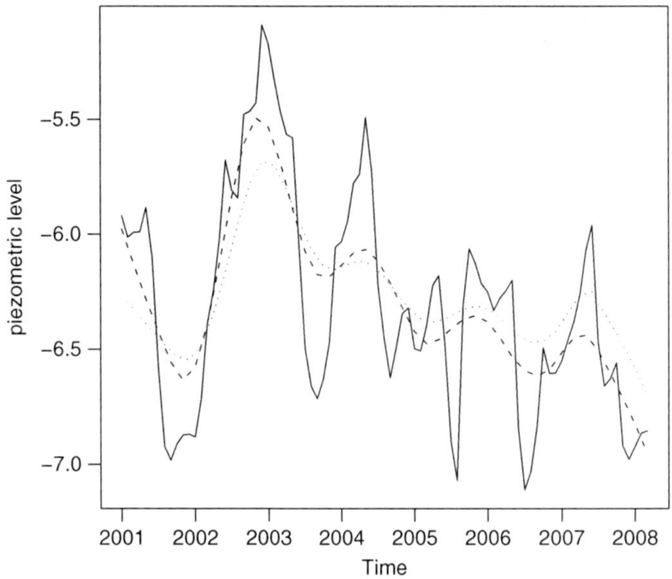

Fig. 2 Carignano piezometer: comparison between original trend (*dashed line*) and virgin trend without the Po effect (*pointed line*): fluctuation effects due to the river have been removed

3 Data Analysis on a Daily Scale

The $AL04$ area, sketched in Fig. 3, includes a plain between the two rivers Orba and Scrivia where the city of Alessandria is located.

The aim of this part of the study is to discuss both rain and river effects on piezometric measurements in the $AL04$ area and to reconstruct the underlying hydrological virgin level for groundwater upon removal of these effects.

The groundwater here is measured by 8 different piezometers, while rain precipitations are measured by 5 rain gauges scattered in the region. The area also contains several rivers, the levels of which are measured by 7 river gauges. The $AL04$ area is chosen because it exhibits a fairly regular time series of piezometric levels, without many missing data.

3.1 The *Al*04 Study: Rain Predictions Based on Kriging

As it appears reasonable to geological experts, the effect of a single rain gauge station, even when located next to the groundwater station, is negligible, due to the

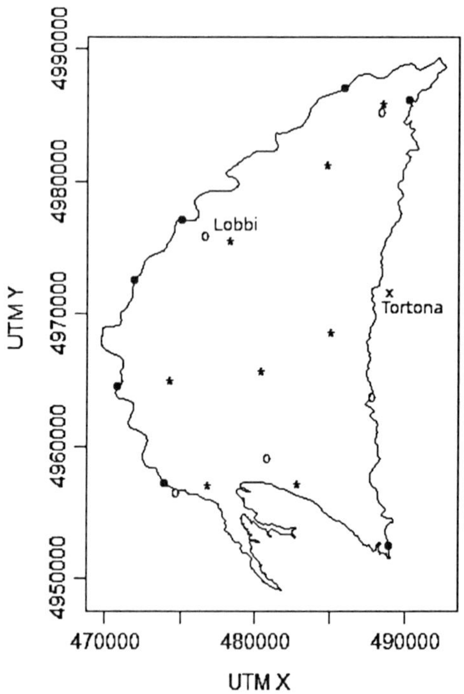

Fig. 3 The *Al*04 area: the groundwater basin of the daily scale study. 8 piezometers (*star*), 5 rain gauges (*circle*) and 7 river gauges (*bullet*)

particularly large hydrographic basin considered. To obtain a more relevant analysis, cumulative rain measurements over the whole basin must be taken into account.

In order to catch the rain effects on the piezometric levels, the study is conducted at a daily scale. Rain data are collected daily from the 5 rain gauges – as they have been collected regularly for many years. However, to find a cumulative measure of rain precipitation, a *kriging* reconstruction over the whole area is necessary.

Kriging is a statistical tool used to make predictions on unobserved space points and, more generally, to estimate the surface of the values taken by a variable of interest over a region. The methodology, first proposed in the 1950s by the mining engineer Krige, is still widely used because of its ability to fit suitably the prediction surface on the area considered, no matter what the geometric locations of the point observations are.

The spatial approach described in Diggle and Ribeiro (2007) is used to reconstruct the rain surface of the $AL04$ region. Let $R(x, y)$ denote a stationary Gaussian process with constant mean μ_0 and variance σ^2 describing the true underlying rain precipitation at point (x, y) in $AL04$. The region is assumed to be a two-dimensional portion of the plane, due to its small size (relative to earth diameter) and to its flatness. For the same reasons, no spatial trend specification is required. The spatial correlation structure of the process is modeled through the Matérn correlation function

$$\rho(d; k, \phi) = (d/\phi)^k K_k(d/\phi)\{2^{k-1}\Gamma(k)\}^{-1},$$

where $K_k(\cdot)$ denotes the second order modified Bessel function and k and ϕ represent the shape and scale parameters, respectively.

The rain gauges in the region provide observable variables Z_i, $i = 1, \ldots, 5$, the point rain levels. By applying a Box-Cox transformation, the following model is considered for them:

$$\sqrt{Z}_i := y_i = R(x_i, y_i) + N_i,$$

where N_i are i.i.d. normal random variables $\mathcal{N}(0, \tau^2)$ and τ^2 represents the so-called *nugget* effect. This implies that the random vector $[y_1, \ldots, y_5]$ is multivariate Gaussian:

$$\mathbf{y} \sim \mathcal{MN}(\mu, \sigma^2 H(\phi) + \tau^2 I),$$

where $\mu := [\mu_0, \ldots, \mu_0]$ is a constant vector, H is a function of the scale parameter ϕ and I is the identity matrix.

Unlike other work regarding rain prediction, such as Ravines et al. (2008), in our case the small number of rain gauges available in the region does not allow for a graphical analysis of empirical variograms in order to estimate the model parameters. We estimate instead these parameters by maximizing the likelihood

$$L(\mu, \tau^2, \sigma^2, \phi|\mathbf{y}) = -1/2\{n\log(2\pi) + \log(|\sigma^2 H(\phi) + \tau^2 I|) +$$
$$+ (\mathbf{y} - \mu)^T(\sigma^2 H(\phi) + \tau^2 I)^{-1}(\mathbf{y} - \mu)\}.$$

Once parameters have been estimated, the ordinary kriging method is used to predict the observed rain precipitation levels on a regular grid of $AL04$. The ordinary kriging predicts a rain value $\hat{R}(x, y)$ at the point (x, y) as a solution to the following constrained minimum problem:

$$\begin{cases} \min & E[\{\hat{R}(x,y) - R(x,y)\}^2] \\ \text{sub} & \hat{R}(x,y) = \sum_i w_i(x,y) z_i \\ \text{sub} & \sum_i w_i(x,y) = 1. \end{cases}$$

As a result, with the ordinary kriging the prediction is expressed as a weighted linear combination of the point observations, in such a way to minimize the mean squared prediction error. Moreover, the mathematical constraint that the weights must sum to one allows us to make spatial predictions without estimating the mean process μ_0. In our case, such an estimation would not be reliable due to the small number of rain gauges. Finally, cumulative rain values for the whole area are obtained through two-dimensional integration of the predicted surface along the whole domain. The sample mean of all the predicted values of rain levels on the regular grid gives a good approximation of this integral.

3.2 *Modelling the Joint Effects of Rain and Neighboring Rivers*

A pre-processing plus transfer function approach, similar to the one described in Sect. 2, is now used in order to remove the effects of external predictors for the piezometers in $AL04$.

As a first example, the results concerning the piezometer near Tortona is shown in Fig. 4, where the observed piezometric levels are compared with the reconstructed piezometric levels, after removal of the rain effects. No near waterway is found to be a relevant predictor in this case. Let (Y_t) and (W_t) represent, respectively, the pre-processed series of the piezometric levels near the city of Tortona and the series of cumulative rain precipitations, reconstructed with the methods described in Sect. 3.1. The resulting estimated model is

$$Y_t = 0.013 W_t + 0.008 W_{t-1} + \eta_t,$$

where η_t is the residual time series of the model. This time series represents the reconstructed virgin model, as described in Sect. 2.

A second model is discussed for the piezometer near the village of Lobbi. In this case, the river Tanaro is found to be a significant predictor, to be added to the cumulative rain amounts, providing the final estimated model

$$Y_t = 0.00082 W_t + 0.04201 X_t - 0.04186 X_{t-1} + \varepsilon_t,$$

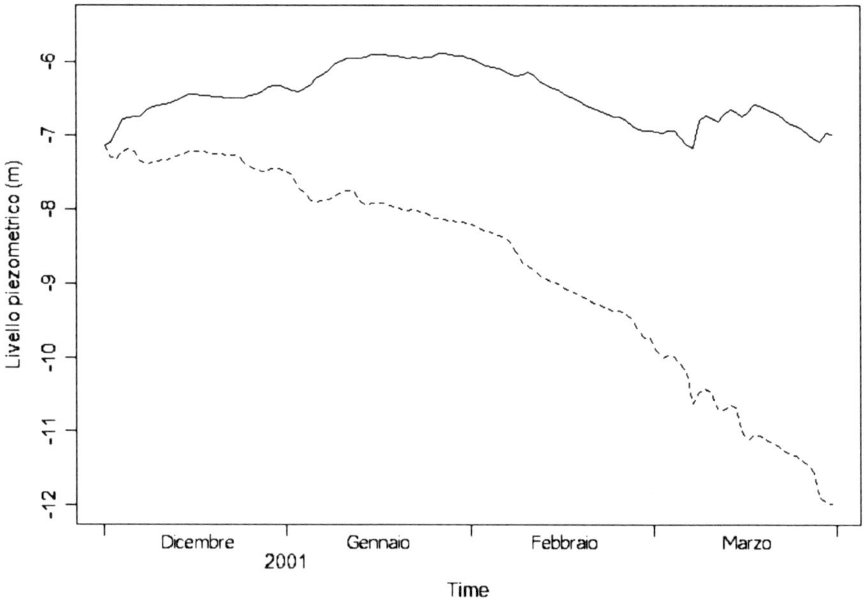

Fig. 4 Tortona piezometer: comparison between daily trend of the piezometric levels (*solid line*) and virgin trend without rain effects (*dashed line*)

where, X_t is the time series of Tanaro, preprocessed in a similar way, and ε_t is the residual term, which is interpreted as the final virgin level.

The results are shown in Fig. 5, in which the observed piezometric levels are plotted together with different reconstructed trends. Trends are estimated based on the reconstructed piezometric levels using a LOWESS interpolation. The dashed trend refers to the model in which only the rain effect was removed; the dashed-pointed one was obtained by removing the Tanaro effect; finally, the pointed one is the reconstructed "virgin" trend, where both the effects are removed. From a geological point of view, an interesting similarity seems to hold between the estimated groundwater virgin levels obtained in this way and the so-called *exhaustion curves* of water springs.

4 Discussion

In this paper we describe the statistical modelling of hydrological external contributions to groundwater levels through a transfer function approach. To this end, the neighboring rivers and rain precipitations are considered as the main predictors. Removing these external contributions in order to restore a virgin groundwater level makes our work different from other literature. Groundwater time series are widely

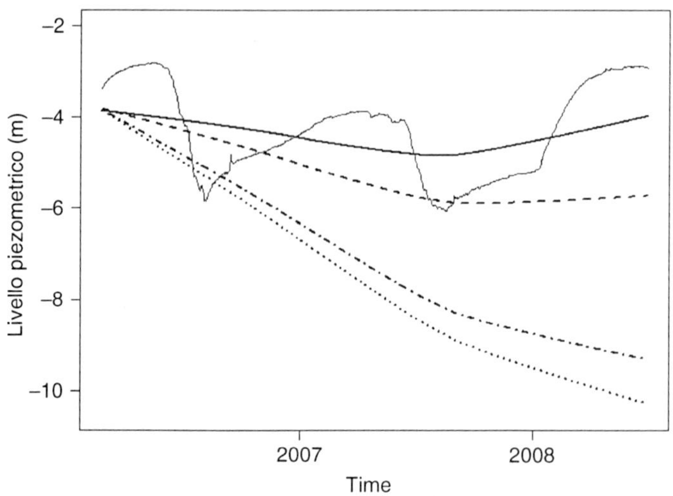

Fig. 5 Lobbi piezometer: comparison between different trend reconstructions of the piezometric levels: original trend (*solid line*), without rain (*dashed line*), without Tanaro river (*dashed-pointed line*), without both rain and Tanaro (*pointed line*)

discussed in Ravines et al. (2008), where a Bayesian approach is taken, and in Yi and Lee (2004). In the latter, a Kalman filtering approach is used on irregularly spaced data to regularize the original time series. In both references, very different geographical and hydrological situations are considered.

In Sect. 2, the monthly scale study is considered and only the river effect is found to be of some interest. As can be argued from Fig. 2, the new estimated trend appears smoother than the original one: fluctuation effects due to the biological rhythm of the Po river have been successfully removed.

In order to deal instead with local rain effects, the daily scale is considered in Sect. 3. In this case, when rain and river contributions are removed, the estimated trend shows a significantly slow and continuous exhaustion of the groundwater levels, similar to exhaustion curves of water springs. From the study of these curves, the health of the groundwater can be evaluated more accurately. This analysis depends on many factors, such as the geological conformation and the hydrological network near the piezometer. In fact, the Tortona example shows a faster decay of the virgin level than the Lobbi case, where a similar effect is evident only after a two-year monitoring. Moreover, the Lobbi piezometer shows a stronger seasonality than in Tortona. This is due to the presence of the nearby river, and, probably, to a more complex hydrological underground equilibrium.

The statistical exercise presented here is the starting point for several possible actions the local government could take, according to EU guidelines:

- Construction of reliable *nowcasting* predictions: according to the geological officers involved, alarm thresholds for such predictions may be discussed, aimed at building semi-automatic procedures for controlling locally the health of the groundwater, in a way similar to on line process control in industrial quality control.

- Careful modelling of the local water cycle: stochastic models could be built to replace more rigid existing deterministic models based on partial differential equations.
- Improved control over private irrigation: the incumbency of water depletion may suggest that actions be taken to discourage private tapping.

Acknowledgements Work partially supported by Regione Piemonte, RSA Project CIPE 2004: "Rilevamento di inquinanti in diverse matrici ambientali". We thank an anonymous referee for useful suggestions and comments.

References

Battaglia, F: Metodi di previsione statistica. Springer Verlag Italia, Milano (2007)

Diggle, P.J, Ribeiro, P.J. Jr.: Model-based Geostatistics. Springer Verlag, New York (2007)

European Union: Directive 2006/118/ec on the protection of groundwater against pollution and deterioration. Official Journal of the European Union, L372/19–L372/31 (2006)

Ravines, R.R.,. Schmidt, A.M., Migon, H.S., Rennó, C.D.: A joint model for rainfall-runoff: the case of Rio Grande Basin. Journal of Hydrology **353**, 189–200 (2008)

Yi, M., Lee, K.: Transfer function-noise modelling of irregularly observed groundwater heads using precipitation data. Journal of Hydrology **288**, 272–287 (2004)

Reduced Rank Covariances for the Analysis of Environmental Data

Orietta Nicolis and Doug Nychka

Abstract In this work we propose a Monte Carlo estimator for non stationary covariances of large incomplete lattice or irregularly distributed data. In particular, we propose a method called "reduced rank covariance" (RRC), based on the multiresolution approach for reducing the dimensionality of the spatial covariances. The basic idea is to estimate the covariance on a lower resolution grid starting from a stationary model (such as the Mathérn covariance) and use the multiresolution property of wavelet basis for evaluating the covariance on the full grid. Since this method doesn't need to compute the wavelet coefficients, it is very fast in estimating covariances in large data sets. The spatial forecasting performances of the method has been described through a simulation study. Finally, the method has been applied to two environmental data sets: the aerosol optical thickness (AOT) satellite data observed in Northern Italy and the ozone concentrations in the eastern United States.

1 Introduction

The analysis of many geophysical and environmental problems requires the application of interpolation techniques based on the estimation of covariance matrices. Due to the non stationary nature of the data and to the large size of the data set it the usual covariance models can not be applied. When the spatial dimension of the sample is very large, the operations of reducing the size of covariance matrices need to be

O. Nicolis (✉)
Department of Information Technology and Mathematical Methods, University of Bergamo, Italy
e-mail: orietta.nicolis@unibg.it

D. Nychka
Institute for Mathematics Applied to Geosciences, National Center for Atmospheric Research, Boulder, CO (USA)
e-mail: nychka@ucar.edu

A. Di Ciaccio et al. (eds.), *Advanced Statistical Methods for the Analysis of Large Data-Sets*, Studies in Theoretical and Applied Statistics,
DOI 10.1007/978-3-642-21037-2_23,

applied to make their calculation feasible. Many approaches have been proposed in literature, mainly based on multiresolution analysis, on tapering methods or on the approximating the likelihood function (Cressie and Johannesson 2008; Matsuo et al. 2008; Zhang and Du 2008; Banerjee et al. 2008; Fuentes 2007; Stein 2008).

In this work we proposed a non parametric method for computing the covariance matrices of massive data sets based on the multiresolution approach introduced by Nychka et al. (2003). In particular, this method is based on the wavelet decomposition of covariance matrix as follows.

Let **y** be the m data points of the field on a fine grid and Σ the $(m \times m)$ covariance matrix among grid points. By the multiresolution approach (Nychka et al. 2003), a spatial covariance matrix Σ can be decomposed as

$$\Sigma = WDW^T = WHH^TW^T \tag{1}$$

where W is a matrix of basis functions evaluated on the grid, D is the matrix of coefficients, H is a square root of D, and the apex T denotes transposition. Unlike the eigenvector/eigenvalue decomposition of a matrix, W need not be orthogonal and D need not be diagonal. Since for massive data sets Σ may be very large, some authors (Nychka et al. 2003) suggested an alternative way of building the covariance by specifying the basis functions and a matrix H. The basic idea of this work is to estimate in a iterative way the matrix H on a lower resolution grid starting from a stationary model for Σ. The evaluation of the wavelet basis on a fine grid in (1) provides a reduced rank covariance matrix.

The method can be used for the estimation of covariance structures of irregularly distributed data points and lattice data with many missing values.

In this paper, the multiresolution method based on the reduced rank covariance is applied to two environmental data sets: the AOT satellite data (Nicolis et al. 2008) and to daily ozone concentrations (Nychka 2005).

Next section discusses the multiresolution approach for the analysis of observational data. Section 3 describes the Reduced Rank Covariance (RCC) algorithm for the estimation of conditional variance in large data sets. Section 4 shows some simulation results. Applications to satellite and ozone data are described in Sect. 5. Section 6 presents conclusions and further developments.

2 Modelling Observational Data

In many geophysical applications, the spatial fields are observed over time and one can exploit temporal replication to estimate sample covariances. In this section we focus on this case and also for gridded data with the goal of deriving a estimator that scale to large problems. Suppose that the point observations **y** are samples of a centered Gaussian random field on a fine grid and are composed of the observations at irregularly distributed locations $\mathbf{y}_o$, and the missing observations $\mathbf{y}_m$. In other words, we assume that the grid is fine enough in resolution so that any observation can registered on the grid points (as in Fig. 1a). Hence,

$$\mathbf{y} = \begin{pmatrix} \mathbf{y}_o \\ \mathbf{y}_m \end{pmatrix} \sim MN(0, \Sigma) \tag{2}$$

where $\Sigma = WDW^T$ is the covariance fine gridded data described in (1), D is a non diagonal matrix and W is a matrix of non-orthogonal scaling and wavelet functions. Although the non-orthogonality property of wavelet basis provide off diagonal coefficients in the matrix D, the localized support of these functions ensures that many covariance terms in D will be close to zero, reducing the computational complexity in the interpolation problems of surfaces. An important class of compactly supported basis function, used in the applications of this work, is the Wendland family proposed by Wendland (1995). The Wendland functions are also implemented in the R statistical language (http://www.r-project.org) in the fields package (Nychka 2005). Figure 2 provides an example for the matrices D and H, obtained by (1),

$$D = W^{-1} \Sigma W^{-T}, \tag{3}$$

where Σ is the covariance resulting from the fitting of a Matérn model to a regular grid, W is a matrix whose columns contain Wendland functions, and W^{-T} is the transpose of the inverse of W.

The observational model (2) can be written as $\mathbf{z}_o = K\mathbf{y} + \varepsilon$ where ε is a multivariate normal $MN(0, \sigma^2 I)$, $\mathbf{z}_o$ is a vector of m observations, and $\mathbf{y}$ is the underlying spatial field on the grid. The matrix K denotes an incidence matrix of ones and zeroes with a single one in each row indicating the position of each observation with respect to the grid. The conditional distribution of $\mathbf{y}$ given $\mathbf{z}_o$ is Gaussian with mean

$$\Sigma_{o,m}(\Sigma_{o,o})^{-1}\mathbf{z}_o \tag{4}$$

and variance

$$\Sigma_{m,m} - \Sigma_{m,o}(\Sigma_{o,o})^{-1}\Sigma_{o,m} \tag{5}$$

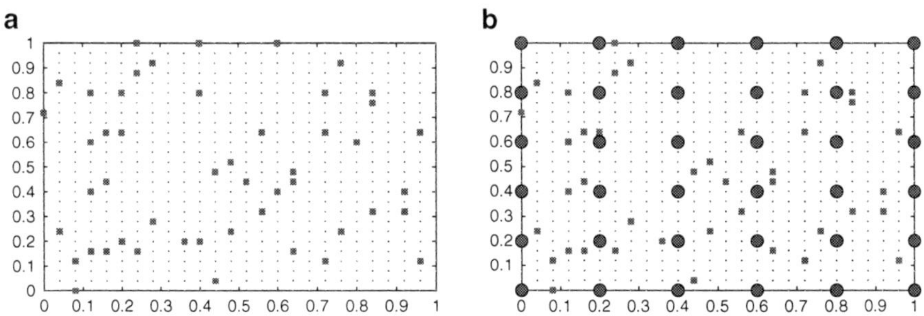

Fig. 1 Gridded data: irregularly distributed data (*squares*); missing data (*small points*) and knots (*circles*)

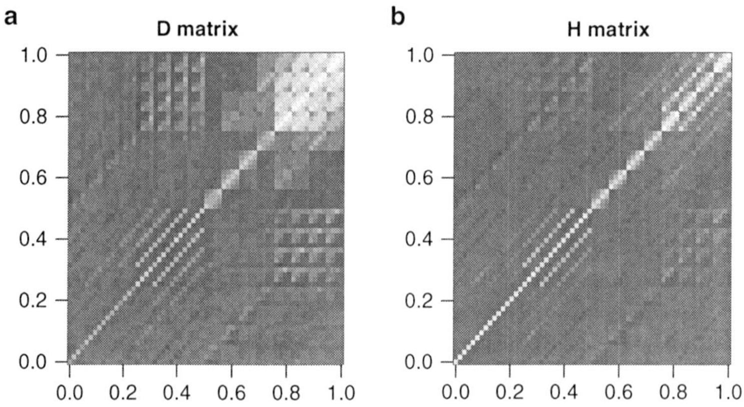

Fig. 2 Example of D (**a**) and H matrix (**b**) on a 8×8 grid data (**b**) using wavelet-based non stationary covariance

where $\Sigma_{o,m} = W_o H H^T W_m^T$ is the cross-covariance between observed and missing data, $\Sigma_{o,o} = W_o H H^T W_o^T + \sigma^2 I$ is covariance of observed data and $\Sigma_{m,m} = W_m H H^T W_m^T$ is the covariance of missing data. The matrices W_o and W_m are wavelet basis evaluated at the observed and missing data, respectively. For a chosen multiresolution basis and a sparse matrix H there are fast recursive algorithms for computing the covariance Σ. Matsuo et al. (2008) proposed a method that allows for sparse covariance matrices for the basis coefficients. However the evaluation of wavelet coefficients can be slow for large data sets.

3 The Reduced Rank Covariance (RRC) Method

In this section we propose an estimation method for Σ based on the evaluation of a reduced rank matrices. We denote by "*knots*" the spatial points on a lower resolution grid $\mathcal{G}$ of size $(g \times g)$, where $g \leq m$. The idea is to estimate the matrix H on the grid of knots starting from a stationary model for Σ and using the Monte Carlo simulation for providing an estimator for the conditional covariance. A flexible model of stationary covariance is the Matérn covariance given by

$$C(h) = \sigma^2 \frac{1}{\Gamma(\nu) 2^{\nu-1}} \left(2\sqrt{\nu}\frac{h}{\theta} \right) \mathcal{K}_\nu \left(2\sqrt{\nu}\frac{h}{\theta} \right), \quad \theta > 0, \nu > 0$$

where h is the distance, θ is the spatial range and $\mathcal{K}_\nu(\cdot)$ is the Bessel function of the second kind whose order of differentiability is ν (smoothing parameter). Since W is fixed for a chosen basis, the estimation procedure for the conditional covariance is given by the estimation of the matrix H after a sequence of approximations.

Following this approach the covariance in (1) can be approximated as

$$\Sigma \approx W\tilde{H}_g\tilde{H}_g^T W^T, \tag{6}$$

where $\tilde{H}_g$ is an estimate of the matrix H on the grid $\mathcal{G}$. The RRC estimation algorithm can be described by the Monte Carlo EM algorithm in the following steps.

1. Find Kriging prediction on the grid $\mathcal{G}$:

$$\hat{\mathbf{y}}_g = \Sigma_{o,g}(\Sigma_{o,o})^{-1}\mathbf{z}_o,$$

   where $\tilde{H}_g = (W_g^{-1}\Sigma_{g,g}W_g^T)^{1/2}$ and $\Sigma_{g,g}$ is stationary covariance model (es. Matern).
2. Generate synthetic data: $\mathbf{z}_o^s = K\mathbf{y}_g^s + \varepsilon$ where $\mathbf{y}_g^s = W_g\tilde{H}_g a$ with $a \sim N(0,1)$.
3. Compute Kriging errors:

$$\mathbf{u}^* = \mathbf{y}_g^s - \hat{\mathbf{y}}_g^s,$$

   where $\hat{\mathbf{y}}_g^s = \Sigma_{o,g}(\Sigma_{o,o})^{-1}\mathbf{z}_o^s$.
4. Find conditional field $\mathbf{y}_m|\mathbf{z}_o$:

$$\hat{\mathbf{y}}_u = \hat{\mathbf{y}}_g + \mathbf{u}^*.$$

5. Compute the conditional covariance on T replications, $\Sigma_u = COV(\hat{\mathbf{y}}_u)$ and use the new $\tilde{H}_g$ in the step 1.

Performing this several times will give an ensemble of fields and, of course, finding the sample covariance across the ensemble provides a Monte Carlo based estimate of the conditional covariance.

3.1 Simulation Study

The purpose of this study is to investigate the forecasting ability of the proposed RRC method in two different contexts: (a) approximation of stationary covariance models and (b) estimation of non-stationary covariances. In order to study the properties of approximation of the RRC method, we simulated $n = 20$ Gaussian random fields on a 20×20 grid using a Matèrn model with parameters $\theta = 0.1$ and $\nu = 0.5$. In order to generate the missing data we removed randomly 50% of the simulated data. An example of simulated random field with missing data is shown in Fig. 3.

For each simulated random field we estimated the missing data using the RRC method on a grid of 8×8 knots and then we computed the root mean square errors on the predictions. The parameters of the Matèrn model used in the step 1. of the algorithm has been chosen by cross validation. Figure 4a compares the RMSE for each simulated random field for the Matèrn model and the non-stationary

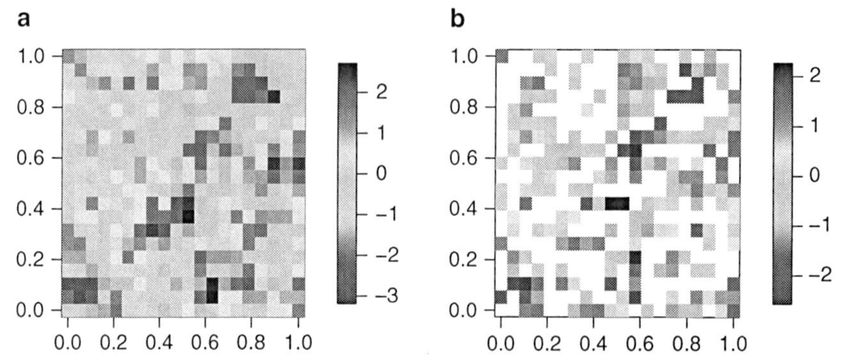

Fig. 3 Simulated Gaussian random field on a 20 × 20 grid with Matèrn covariance ($\theta = 0.1$ and $\nu = 0.5$) without (**a**) and with (**b**) 50% of missing values

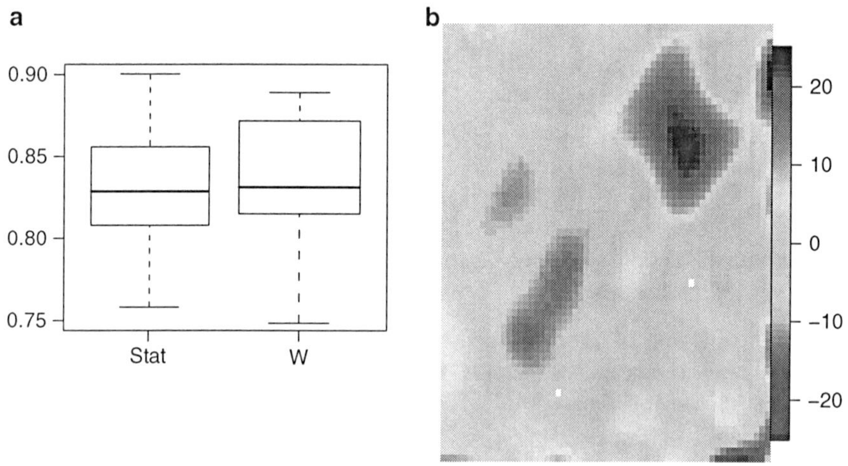

Fig. 4 (**a**) RMSE of the 50% of missing values. The estimates are obtained using a Matèrn model (Stat) and RRC method (W) with five iterations; (**b**) covariance between the point indicated by black circle and the rest of grid points

wavelet-based covariance model (RRC). The similarity of the two boxplots indicates a good approximation of the proposed method to the stationary model. The covariance between a specific point and the rest of grid points shown in Fig. 4b highlight the higher correlation between neighboring points.

In order to estimate non-stationary covariance by RRC method we generated $n = 20$ synthetic non-stationary data on a grid of 40 × 40 points with 50% of missing values. These spatial data has been obtained from a mixture of two dependent stationary random fields as in Matsuo et al. (2008) with Matèrn covariances ($\Sigma_1(\nu = 1, \theta = 0.125)$ and $\Sigma_2(\nu = 0.5, \theta = 0.1)$), and a weight function $w(\mathbf{u}) = \Phi((u_x - 0.5)/.15)$ where $\Phi(\cdot)$ is the normal cumulative distribution function and u_x is the horizontal coordinate. Figure 5a shows an example of simulated

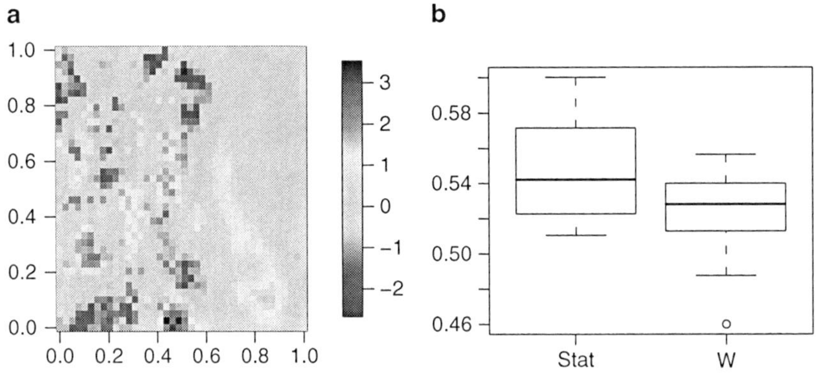

Fig. 5 Non-stationary random field simulated on a 40 × 40 grid with Matèrn covariance ($\theta = 0.1$ and $\nu = 0.5$) (**a**) and RMSE results for 50% of missing values using a Matérn model and a RRC method on a grid of 8 × 8 knots

non-stationary random field. The RRC method with a grid of 8 × 8 knots has been applied for forecasting 50% of the missing values. In this case the RMSE results (Fig. 5b) indicates that RRC method provides better estimates than a stationary model.

4 Applications

4.1 Satellite Data

Satellite data are very important in the analysis of environmental data as they cover wide monitoring areas and can sometimes be easily downloaded from specialized Internet web-sites. However, their statistical analysis often requires special techniques to cope with large data sets or to treat other exogenous variables that affect the satellite measurements in the atmosphere. The satellite aerosol optical thickness (AOT) data is an example. The study of these measurements is particularly important to assess the amount of fine particulate matters in the air and the consequent risk to human health. However, the meteorological conditions determine the AOT retrieval since the availability of the data is restricted to cloud-free conditions. Indeed, the cloudy coverage causes a large number of missing data, sometimes making problematic the application of traditional correlation models. In this work we consider the column aerosol optical thickness (AOT)[1] data derived from the Moderate Resolution Imaging SpectroRadiometer (MODIS) on

[1] AOT measurements can be downloaded from the NASA web page http://disc.sci.gsfc.nasa.gov/.

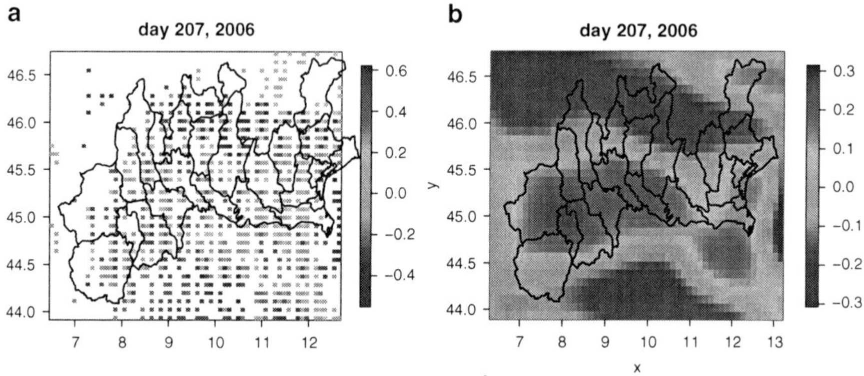

Fig. 6 Satellite AOT measurements (**a**) and kriging predictions (**b**) using RRC method on a grid of 8 × 8 knots

the Terra/Aqua satellites in Northern Italy for the period April 1st - June 30th, 2006 (Nicolis et al. 2008). The Terra satellite crosses Europe near 10:30 local solar time (morning orbit), while Aqua crosses Europe near 13:30 local solar time (afternoon orbit). Hence, at least two observations of any place in Europe are obtained per day during daylight hours. These data are based on analyzing 20 × 20 pixels at 500 m resolution and reported at 10 × 10 km^2 resolution. The reflectivity measured at 2.1 μm at the top-of-atmosphere is used to infer surface reflectivity at that wavelength. Figure 6a shows the daily average of AOT data on July 26, 2006 in Northern Italy measured on a grid of 54×32 locations. Figure 7 shows the non-stationary structure of the estimated conditional covariance, $W \tilde{H}_g \tilde{H}_g^T W^T$, for four points obtained after five iterations of MC simulations using a grid of 8 × 8 knots. It is important to note that the correlation structure is slightly different for the four graphs which indicates that the model is able to capture the non-stationarity of data. The Kriging prediction for the day July 26, is shown in Fig. 6b. These results indicate that higher estimates of AOT values are in the Western part of Northern Italy around the cities of Turin and Milan. Similar results were found by Fassò et al. (2007) and Sahu and Nicolis (2008) for the analysis of fine particulate matters (PM_{10}) in Northern Italy.

4.2 *Ozone Data*

In order to apply the RRC method to irregular data, we considered ozone concentrations, included in `Fields` package of `R` software The database consists of daily 8-hour average ozone concentration measured in parts per billion (PPB) for 153 sites in the Midwestern US over the period June 3, 1987 through August 31, 1987 (89 days). Many of these station have incomplete records both in time and space. Figure 8 shows ozone concentrations on July 19, 1987. The application of RRC

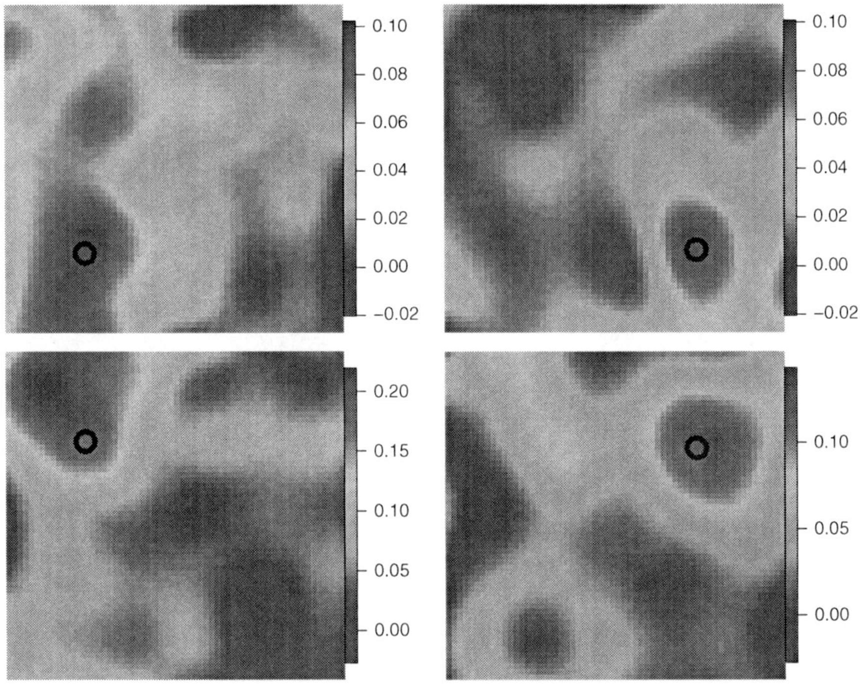

Fig. 7 Non-stationary covariance $W\widetilde{H}_g\widetilde{H}_g^T W^T$ obtained after five iterations of MC simulations. Each panel shows the covariance between the point indicated by black circle and the rest of grid points

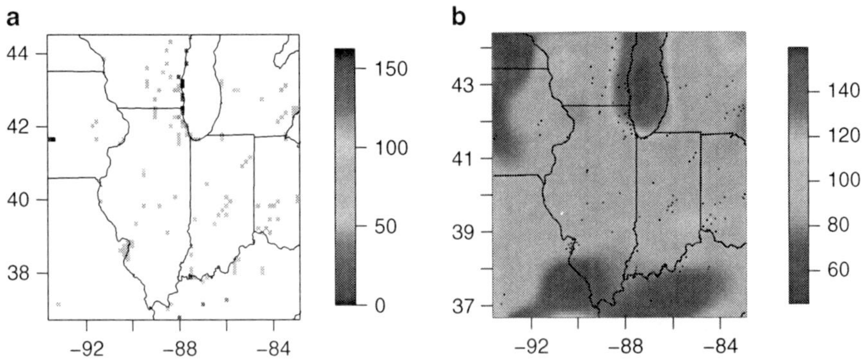

Fig. 8 (**a**) Daily ozone concentrations on June 18, 1987; (**b**) Kriging map using the RRC method on a grid of 8 × 8 knots

method allows to estimate the ozone concentrations on a fine grid with resolution (100 × 100) using a lower resolution grid of 8 × 8 knots. Figure 9 shows the non-stationary of the estimated reduced rank covariance.

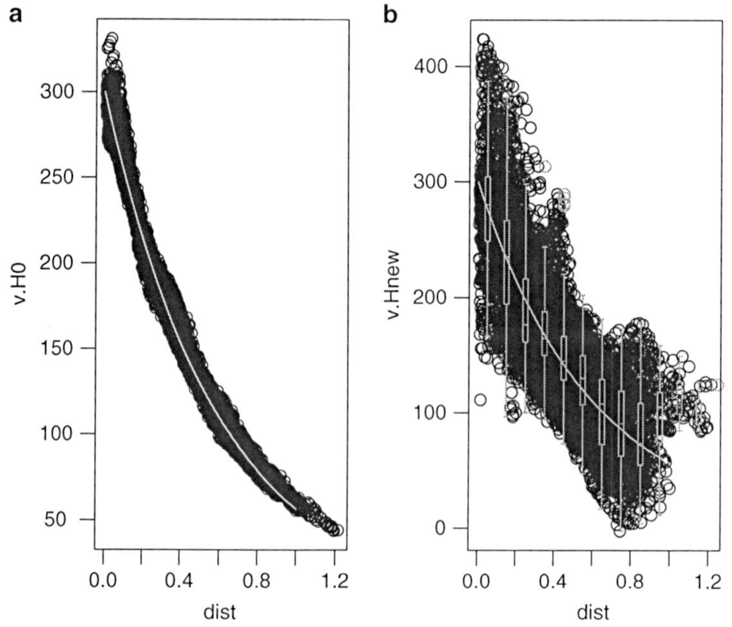

Fig. 9 Covariances vs distances: RRC after one iteration (*circles*) and Matérn covariance (*green line*) (on the left); RRC after five iteration (*circles*) and Matérn covariance (*green line*) (on the right). Boxplot plot on the right indicates the distribution of the estimated reduced rank covariance

5 Conclusions and Further Developments

We have proposed a practical method for approximating stationary covariance models and estimating non-stationary covariances. The method based on empirical estimation the reduced rank matrix provides an efficient tool for handling large data sets. Although this method is still in its preliminary stages, the results from the simulation study are very encouraging. We believe that RRC method can be used for a preliminary analysis of the spatial covariance structure and can be developed for the covariance estimation of space time models. We also intend to find a parametrization for the RRC matrix and using an EM algorithm for the estimation of the model with missing data.

References

Banerjee, S., Gelfand, A.E., Finley, A.O., Sang, H.: Gaussian predictive process models for large spatial datasets. Journal of the Royal Statistical Society Series B, **70**, 825–848 (2008).

Cressie, N., Johannesson, G.: Fixed rank kriging for very large spatial data sets. Journal of the Royal Statistical Society: Series B (Statistical Methodology), **70** (1), 209–226 (2008).

Fassò, A., Cameletti, M., Nicolis O.: Air quality monitoring using heterogeneous networks. Environmetrics, **18**, 245–264 (2007).

Fuentes, M.: Approximate likelihood for large irregularly spaced spatial data. J. Amer. Statist. Assoc. **102**, 321–331 (2007).

Matsuo, T., Nychka, D., Paul, D.: Nonstationary covariance modeling for incomplete data: Smoothed monte carlo em approach. Computational Statistics & Data Analysis, **55** (6), (2011).

Nicolis, O., Fassò, A., Mannarini, G.: Aot calibration by spatio-temporal model in northern italy. Proceedings of Spatio-Temporal Modelling (METMA 5) Workshop, 24-26 September, Alghero, Sardinia (2008).

Nychka, D.: Tools for spatial data. National Center for Atmospheric Research, Boulder, CO (2005).

Nychka, D., Wikle, C., Royle, J.A.: Multiresolution models for nonstationary spatial covariance functions. Statistical Modeling, **2**, 315–332 (2003).

Sahu, S., Nicolis, O.: An evaluation of european air pollution regulations for particulate matter monitored from a heterogeneous network. Environmetrics, **20**, 943–961 (2008).

Stein, M. L.: A modeling approach for large spatial datasets. J. Korean Statist. Soc. **37**, 3–10 (2008).

Wendland H.: Piecewise polynomial, positive definite and compactly supported radial functions of minimal degree. Adv. Comput. Math., **4**, 389–396 (1995).

Zhang, H. and Du, J. Covariance tapering in spatial statistics. In Positive definite functions: From Schoenberg to space-time challenges, eds. Mateu, J. and Porcu, E., Grficas Casta, s.l. (2008)

Radon Level in Dwellings and Uranium Content in Soil in the Abruzzo Region: A Preliminary Investigation by Geographically Weighted Regression

Eugenia Nissi, Annalina Sarra, and Sergio Palermi

Abstract Radon is a noble gas coming from the natural decay of uranium. It can migrate from the underlying soil into buildings, where sometimes very high concentration can be found, particularly in the basement or at ground floor. It contributes up to about the 50% of the ionizing radiation dose received by the population, constituting a real health hazard. In this study, we use the geographically weighted regression (GWR) technique to detect spatial non-stationarity of the relationship between indoor radon concentration and the radioactivity content of soil in the Provincia of L'Aquila, in the Abruzzo region (Central Italy). Radon measurements have been taken in a sample of 481 dwellings. Local estimates are obtained and discussed. The significance of the spatial variability in the local parameter estimates is examined by performing a Monte Carlo test.

1 Introduction

Radon is a naturally-occurring radioactive noble gas, produced by the decay of radium which, in turn, is a member of the radioactive chain starting from uranium. Both radium and uranium are ubiquitous trace components of the earth's crust, and their concentrations vary with the geological setting. Due to dilution in the air, outdoor radon levels are usually very low. On the other hand, radon may accumulate in buildings where it enters advectively across openings in the substructure, driven by pressure differences between the soil gas and the indoor air (Nazaroff 1992). Indoor radon is a serious health concern to humans as prolonged exposure to

E. Nissi (✉) · A. Sarra
Dipartimento di Metodi Quantitativi e Teoria Economica, Università "G.d'Annunzio"
e-mail: nissi@dmqte.unich.it

S. Palermi
Agenzia Regionale per la Tutela dell'Ambiente dell'Abruzzo (ARTA)-Pescara

A. Di Ciaccio et al. (eds.), *Advanced Statistical Methods for the Analysis of Large Data-Sets*, Studies in Theoretical and Applied Statistics,
DOI 10.1007/978-3-642-21037-2_24,

high concentrations of this gas and related progeny (namely, radioactive short lived isotopes of polonium, lead and bismuth) results in an increased risk of developing lung cancer (Darby et al. 2005; WHO 2009). Many developed countries have performed systematic studies aimed to assessing the residential exposure to radon and identifying geographical areas where there is an increased probability of exceeding reference values and for which remedial actions are necessary (*radon-prone areas*). A lognormal probability distribution of data is usually assumed, whichever the size of the geographic unit under investigation (Gunby et al. 1993). There have been several efforts to use explanatory variables of geological nature (lithological type as indicator variable, uranium/radium content in soil, permeability etc.), along with information on house construction and air exchange rate, to predict indoor radon concentrations or to classify areas as radon prone (Appleton et al. 2008). For example, Smith and Field 2007 suggest a geostatistical hierarchical Bayesian approach for the joint modelling of indoor radon and soil uranium data. Their model allows for spatial prediction considering geological data and housing characteristics. In these papers as well as in many others (e.g. Price et al. 1996; Apte et al. 1999; Bossew et al. 2008), global estimation techniques have been employed, assuming spatial stationarity of the relationships under study. We hypothesize that there could be a significant and interesting spatial variation in the relationship between indoor radon data and the radioactivity content of soil. Accordingly, in order to capture local variations of that phenomenon, we adopt a local regression known as geographically weighted regression (GWR). Despite it being a relatively new technique, GWR has become a standard tool for constructing spatial models in a variety of disciplines, recognising that the relationship between variables, measured at different geographic points, might not be constant over space. GWR analysis has been successfully used to quantify spatial relationships between different variables in the field of human and economical geography (McMillen 1996; Pavlov 2000; Fotheringham et al. 2002), in social sciences (Cahill and Mulligan 2007), in remote sensing (Foody 2003), in health sciences (Nakaya et al. 2005), etc. Actually, to address the issue of spatial non-stationarity, researchers have developed many other local techniques, such as the spatial expansion method (Jones and Casetti 1992), spatially adaptive filtering (Trigg and Leach 1968), spatial regression models (Ord 1975). GWR is used in this study because it is based on an intuitive traditional framework. In addition, the GWR produces useful information and outputs: local standard errors, tests to assess the significance of spatial variation in the local parameter estimates, tests to determine if the local model performs better than the global one. These local, specific results may provide a more detailed perspective on underlying relationships, allowing refinements in the global specification. Therefore this paper empirically investigates, by using GWR analysis, the spatial association between indoor radon concentrations, collected in a sample of 481 dwellings of Provincia dell'Aquila (Central Italy) and some naturally occurring radioactive elements (uranium, thorium, potassium) which characterize the underlying soil. Superficial uranium and/or radium concentrations are a useful indicator of radon potential in a certain area, being directly related to the rate of generation of radon gas in the soil (Nazaroff 1992). We believe that measuring the local relationships, via GWR, between indoor radon and surface uranium can provide useful insights in the radon mapping process, particularly in areas suspected to be radon prone. The remainder of this paper is organised as follows: Sect. 2 gives a short description of Geographically Weighted Regression technique. Section 3 presents data and the modelling setting. Results as well as their interpretation can be found in Sect. 4.

2 Methodology

GWR extends the traditional global regression by allowing local rather than global parameters to be estimated. A typical model of GWR can be written as:

$$y_i = \beta_0(u_i, v_i) + \sum_k \beta_k(u_i, v_i)x_{ik} + \varepsilon_i. \tag{1}$$

where y_i is the dependent variable at location i; $\beta_k(u_i, v_i)(k = 1, 2, \ldots, M)$ are the regression coefficients for each location i and each variable k; x_{ik} is the value of the kth explanatory variable at location i; β_{0i} is the intercept variable at location i; (u_i, v_i) denotes the coordinates of ith point in space and ε_i are error terms at location i, which follows an independent normal distribution with zero mean and homogeneous variance. Equation (1) creates a continuous surface of estimated parameters values. The local parameters $\beta_k(u_i, v_i)$ are estimated by a weighted least-squares estimator, given by:

$$\hat{\boldsymbol{\beta}}(i) = (\mathbf{X}^T\mathbf{W}(i)\mathbf{X})^{-1}\mathbf{X}^T\mathbf{W}(i)\mathbf{y}. \tag{2}$$

where $\mathbf{W}(i) = \mathrm{diag}[\mathrm{w}_1(i), \ldots, \mathrm{w}_\mathrm{n}(i)]$ is the diagonal weights matrix that varies for any calibration location i and applies weights to n observations; $\mathbf{X}$ is the matrix of explanatory variables with a first column of 1's for the intercept; $\mathbf{y}$ is the vector of dependent variables and $\hat{\boldsymbol{\beta}}(i)$ is the vector of local regression coefficients at location i. In (2) the weight matrix is no longer constant but varies according to the location of point i. It is worth noting that localised parameters can be derived for any point in space regardless of whether or not that point is one at which data are measured. In GWR an observation is weighted in accordance with its proximity to regression point i: data points near to location $i(u_i, v_i)$ will be assigned higher weights in the model then data points farther away. Hence, an important step in calculating GWR estimates is the choice of a weighting scheme: this requires the specification of a kernel shape and bandwidth. Several weighting functions ("kernels") can be considered and calibrated, although they tend to be Gaussian or "Gaussian-like", reflecting the type of dependency found in many spatial processes. Whichever weighting function is used, the results are very sensitive to the bandwidth. This is a measure of distance-decay in the weighting function and indicates the extent to which the resulting local calibration results are smoothed. One can use a constant bandwidth around every regression point. However, it is desirable to have larger bandwidths where data are sparse and smaller bandwidths where data are plentiful. This consideration justifies the employment of a variable (or *adaptive*) weighting function. Consequently, the calibration of the model involves also the choice of the number of data point to be included in the estimation of local parameters (N). Different methods are traditionally used in order to define the finest bandwidth value or the appropriate value of N. Among them, there are the Aikake Information Criterion (AIC) and the *cross-validation* procedure. The cross-validation criterion is a least square approach which can be formalised as follows:

$$CV = \sum_{i=1}^{n} [y_i - \hat{y}_{\neq i}(b)]^2 \tag{3}$$

where $\hat{y}_{\neq i}(b)$ is the fitted value of y_i with the observation for point i omitted from the calibration process. In order to test whether the GWR model describes the relationship significantly better than an ordinary global model an ANOVA test has been formulated. It is an approximate likelihood ratio test based on F-test defined as follows:

$$F = \frac{RSS_0/d_0}{RSS_1/d_1} \tag{4}$$

where RSS_1 and RSS_0 are the residual sum of squares of GWR model and global regression model (OLS) while d_1 and d_0 are the degrees of freedom for GWR and global models, respectively. This test relies on the result that the distribution of residual sum of squares of the GWR model divided the effective number of parameters may be reasonably approximated by a χ^2 distribution with effective degrees of freedom equal to the effective number of parameters. In the non-parametric framework of GWR, the concept of "number of parameters" and "degree of freedom" are fairly meaningless (see Fotheringham et al. 2002). Besides the non-stationarity of all regression coefficients one can check whether any of the local parameter estimates are significantly non-stationary. This concern is addressed via a Monte Carlo significance test as well as a test proposed by Leung et al. 2000. Since in the GWR the regression equation is calibrated independently for each observation, the main output is a set of local parameter estimates. Local statistics, generated by the calibration of a GWR, include: the local β value of model parameters with their associated t-values, indicating the significance of individual parameters, and the local R-square. Essentially, these statistics measure how well the model calibrated at regression point i can replicate the data in the vicinity of point i. However, the local R-square cannot be interpreted with the same confidence as the global measure. Finally, other diagnostic measures such as local standard errors and local standardised residuals arise from a GWR model. For a comprehensive overview about GWR methodology see Fotheringham et al. 2002.

3 Data and Modelling Setting

3.1 Geologic Setting of the Area

In this paper, we refer to Provincia dell'Aquila (AQ), identified in previous studies as the highest indoor radon risk area of Abruzzo (Palermi and Pasculli 2008). AQ is a territory of about 5000 km^2, moderately populated (60 inhab per km^2). A large fraction of this area is covered by mountains belonging to the Apennine chain, a Cenozoic thrust and fold belt which has been undergoing uplift and extension since the late Pliocene, intersected by karstic plateaus and intermontane basins of tectonic origin; the latter are filled with Quaternary fluvial–lacustrine sediments (Cavinato and De Celles 1999). The lithology is mainly carbonate, with limestone alternating with marls and sandstone. Structures of this kind usually show a low content of natural radioactivity; however anomalous occurrence of uranium/thorium bearing minerals have been observed within the Quaternary sediments (Bellotti et al. 2007). The source of these minerals, mostly present in the western sector of AQ, must be looked for in the volcanic ashes and pyroclastics coming from the Tuscan–Latial magmatic Plio-Pleistocenic province.

3.2 Indoor Radon Data

The Regional Agency for Environment Protection has carried out various indoor radon surveys in Abruzzo since early nineties, using passive etched-track detectors. This has resulted in a geo-referenced database of annual average radon concentrations (expressed in Bq/m^3) measured in more than 1900 buildings, mainly homes, mostly at ground level. In the area of interest (AQ) 481 data are available. As the soil is the main source of indoor radon, measurements made at ground level are more indicative of the strength of this source. Thus, data not measured at ground level have been normalized to a virtual ground level condition, by means of multiplicative factors estimated from the data itself. Further details on this normalization procedure, as well as more information about the surveys, e.g. the adopted sampling strategies, can be found in Palermi and Pasculli 2008.

3.3 Soil Radiometric and Climate Data

The radiometric data, namely potassium (K), equivalent uranium (eU), and equivalent thorium (eTh) concentrations (eU and eTh are expressed in units of parts per million, i.e. ppm, whereas K is expressed in percent, i.e. %) determined from gamma

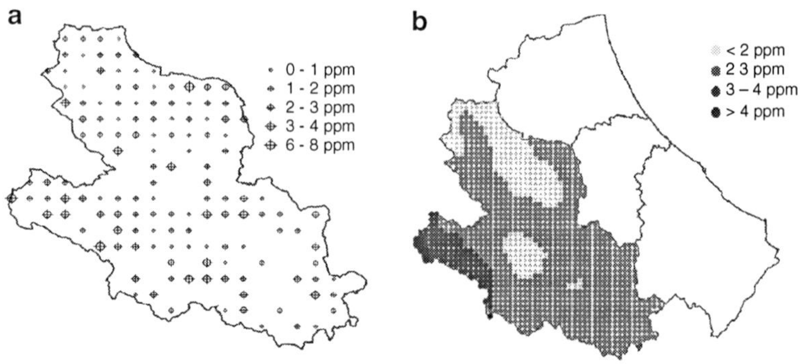

Fig. 1 (**a**) Map of eU data (**b**) Predicted eU data by block kriging

signal emitted by ^{40}K, ^{214}Bi and ^{208}Tl, respectively, present in the top 20 cm or so of the ground, are provided by Bellotti et al. 2007, who carried out systematic laboratory measurements of γ-ray activity on almost 200 samples of soils collected in the territory of AQ, following a grid-based spatial sampling design (Fig. 1a).[1] Following Scheib et al. 2006, we have included K in the regressors set, as this element is known to be a good indicator of the clay content and, consequently, of the permeability of bedrock in areas which are characterised by limestones, siltstones and sandstones. Permeability of underlying rock strata and/or sub-soil is an important controlling factor in the relationship between eU and indoor radon; the same level of uranium generally gives rise to higher indoor radon in presence of high permeability features such as fissures, fractures, faults and joints. These geological structures are known to enhance the ascending radon migration, as they provide pathways along which the advection of groundwater and soil gas is facilitated. In the territory under investigation, they might be very important as a possible cause of local radon anomalies, linked to the widespread occurrence of tectonic faults (Ciotoli et al. 2007). Climate exerts a strong control over the temperature and moisture content of soils, thus affecting radon emanation and diffusion (Nazaroff 1992). Indoor Radon concentrations are also directly influenced by outdoor temperature, as people reduce ventilation and enhance heating in cold weather conditions, thereby increasing radon entry from the soil and its accumulation indoors. For this reason, we also add the altitude in our models as it acts as a rough surrogate for mean annual temperature and other climate-related parameters.

[1] ^{214}Bi and ^{208}Tl belong to the short lived progeny of ^{222}Rn (radon itself, coming from ^{238}U) and ^{220}Rn (the isotope coming from ^{232}Th), respectively. It should be noted that the eU and eTh data do not imply that uranium and thorium are actually present in soil samples, since these elements can be leached away while radium remains in situ, causing a breakdown of secular equilibrium assumption.

Table 1 Model semivariograms for radiometric variables

Variable	Model type	Nugget	Partial sill	Range (m)
eU	Exponential	1.03	0.47	13,000
K	Exponential	0.19	0.084	14,000
eTh	Exponential	55	79	4,560

3.4 Data Processing and Model Setting

As mentioned before, several other studies have used superficial uranium data as a predictor of indoor radon. Often, this kind of data are available at aggregate levels as averages, in particular when deriving from airborne gamma spectrometry (Price et al. 1996; Scheib et al. 2006). In contrast, in this study, we can rely on point-referenced measurements (see Fig. 1a); furthermore, these data, except altitude, are not available at the same sites where radon measurements are taken. In order to overcome this limitation, predicted eU, K and eTh data must be used. Accordingly, the interpolation through the block kriging method (Olea 1999), which provides estimates of the linear average inside square blocks centered on the nodes of a 2 km spaced grid, is employed to simulate radiometric areal data (see Fig. 1b for predicted eU data). A radius of 15 km delimits the size of the local neighborhood in which to look for data points for kriging calculations. In Table 1 we summarise the model semivariograms for the variables of interest, selected by means of leave-one-out cross validation.[2]

It can be noted that eU and K have comparable model parameters (namely, the range and the ratio σ^2/τ^2), unlike eTh; this could be due to similar response to weathering and other alteration processes.

With regard to the Geographically Weighted Regression model, we use the adaptive weighted scheme (fixed number of nearest neighbours for each local model) and the common bisquare function of calculating the weights. The number of nearest neighbour data points to be included within the calibration of the local model is determined from the cross-validation criteria, as formalised in (3). The GWR model was fitted using a computer software program, GWR 3.0, detailed information can be downloaded from the website http://www.ncl.ac.uk/geography/GWR (Fotheringham et al. 2002).

[2]The semivariograms $\gamma(d)$ have been modelled by means of isotropic exponential functions as:$\gamma(d) = \tau^2 + \sigma^2(1 - \exp(-d/R))$ where d is the modulus of Euclidean distance between pairs of data points, calculated from the geographical coordinates of each dwellings, and τ^2, σ^2 and R are parameters known as, respectively, nugget, partial sill and range (Olea 1999).

4 Results and Discussion

In this section we present the outcomes arising from the application of both Ordinary Least Squares (OLS) and GWR models to the available data. After examining collinearity diagnostics and stepwise regression, only the uranium (eU) remained in the linear regression: all other predictor variables in both models (OLS and GWR) were not significant.

In Table 2 we summarise the global regression results. The global relationship between the natural logarithm of indoor radon level and uranium is significantly positive with a t-value of 3.20. The GWR model, in which the spatial information has been incorporated, leads to an improvement of the model's predictability. We observe a reduction for the AIC values and an increase of global R^2 (as displayed in Table 3).

According to the outputs of ANOVA test (in Table 4) we can reject the null hypothesis that GWR technique has no significant improvement over the global regression, even if at quite high significance level. Local findings suggest that there are meaningful spatial variations in the relationships between radon measurements and uranium soil content ($p-\text{value} = 0.0000$ for Monte Carlo significance test for spatial variability of parameters).

It is worth noting that the low values of R^2 (Table 3) in both models are found in many other studies (Apte et al. 1999) and may be justified invoking indoor radon multifactorial dependency, as discussed below. Casewise diagnostics, expressed in terms of local R^2 (Fig. 2a), show values ranging from 0.13 to 0.47 which can be deemed quite good in comparison with the global R^2 obtained by the OLS method.

In Fig. 2b we show the local parameter estimates for the eU variable. The largest circle highlights the major strength of the relationship between variables. It is

Table 2 The results of global regression analysis

Variable	Parameter estimate	Std Err	T test
Intercept	3.64	0.15	23.38
eU	0.21	0.06	3.20

Table 3 AIC and R^2 for radon regression models

Model	AIC		R^2	
	OLS	GWR	OLS	GWR
log (radon) vs. eU	1107	1094	0.01	0.11

Table 4 Goodness-of-fit test for improvement in model fit of GWR over global model (OLS)

Source	SS	DF	MS	F
Global residuals	316.8	2		
GWR improvement	29.4	16.58	1.77	
GWR residuals	287.4	462.42	0.62	2.85

SS sum of squares, *DF* degree of freedom, *MS* residual mean square, F-statistic

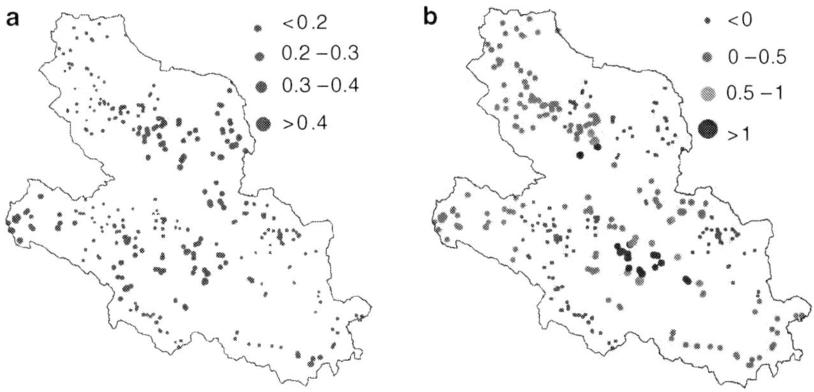

Fig. 2 (**a**) Local *R*-square (**b**) GWR estimates for eU coefficients

interesting to note that in some parts of the study region we have higher indoor radon levels than would be expected from uranium concentrations. In fact, analysing the variability of the local R^2 over the territory one could infer that the model explains well the observed values of indoor radon where factors apart from surface uranium (mainly soil permeability) do not show so much variability to significantly perturb the relationship between radon and its primary source. Comparatively high R^2 values occur near the western boundary with the Latium region, in an area with high values of eU (see Sect. 1): in this case, eU coefficients are positive, as expected. On the other hand, fairly elevated values of R^2 coexist with negative eU coefficients in the North-East area (including the eastern sector of the Gran Sasso massif), pointing out some anomalous and counterintuitive relationship between radon and eU. In other areas (such as in the intermontane sedimentary basins of Valle Peligna and Fucino/Valle Roveto) the local outputs of GWR (Fig. 2) suggest a lack of significance of the relationship. In all these cases, we believe that other factors do have a strong influence on this linkage. Indeed, apart from the building related factors, soil parameters such as moisture content and permeability (not considered in our study) have a notable impact on indoor radon levels. For example high permeability enhances radon flux from the soil, uranium concentration being equal; on the contrary, low permeability can reduce the indoor radon concentration even if uranium is high.

We observe that the GWR highlights interesting spatial patterns that would be missed completely if we relied solely on the global analysis. As known, one of the main advantages of this technique, in comparison with other spatial methods (geostatistical and Bayesian hierarchical models), is its ability in revealing the presence of spatial non-stationarity allowing parameters to vary over space. Additionally, the GWR has a greater prediction accuracy because the model being fitted locally is more tuned to local circumstances. In that way we are forced to investigate the nature of the spatial heterogeneity and understand local factors playing an important role in affecting radon distribution, distinguishing those related to its geochemical source (soil uranium content) from those related to its transport through rocks and soil and its entry into buildings (permeability, moisture, faults and fractures). For these reasons we believe that the use of GWR in radon studies, where this technique represents a novelty, offers a noticeable elucidation of the facets of linkage between indoor radon and soil uranium content. Although our experimental results are encouraging we are aware that they are preliminary findings. Some issues need to be addressed: the sensitivity of GWR results to the choices made in implementing the block kriging scheme for the radiometric variables (grid-size, semivariogram model, extension of the research area for the kriging), the increase of the number of experimental sites and a more accurate standardization of indoor radon measurement aimed at filtering out, as much as possible, the building related factors.

References

Appleton, J.D., Miles, J.C.H., Green, B.M.R., Larmour, R.: Pilot study of the application of Tellus airborne radiometric and soil geochemical data for radon mapping. J. Env. Rad. **99**, 1687–1697 (2008)

Apte, M.G., Price P.N., Nero, A.V., Revzan, R.L.: Predicting New Hampshire indoor radon concentrations from geological information and other covariates. Env. Geol. **37,** 181–194 (1999)

Bellotti, E., Di Carlo, G., Di Sabatino, D., Ferrari, N., Laubenstein, M., Pandola, L., Tomei, C.: γ-Ray spectrometry of soil samples from the Provincia dell'Aquila (Central Italy). Appl. Rad. Isot. **65**, 858–865 (2007)

Bossew, P., Dubois, G., Tollefsen, T.: Investigations on indoor Radon in Austria, part 2: geological classes as categorical external drift for spatial modelling of the radon potential. J. Env. Rad. **99**, 81–97 (2008)

Cahill, M., & Mulligan, G.: Using geographically weighted regression to explore local crime patterns. Soc. Sci. Comput. Rev. **25,** 174–193 (2007)

Cavinato, G.P., De Celles, P.G.: Extensional basins in the tectonically bimodal central Apennines fold-thrust belt, Italy: response to corner flow above a subducting slab in retrograde motion. Geology. **27**, 955–958 (1999)

Ciotoli, G., Lombardi, S., Annunziatellis, A.: Geostatistical analysis of soil gas data in a high seismic intermontane basin: Fucino Plain, central Italy. J. Geophys. Res. **112**, B05407 (2007). doi:10.1029/2005JB004044.

Darby, S., Hill, D., Auvinen, A., Barros-Dios, J.M., Baysson, H., Bochicchio, F. et al. Radon in homes and risk of lung cancer: collaborative analysis of individual data from 13 Europeancase-control studies. Brit. Med. J. **330** (7485): 223 (2005). doi:10.1136/bmj.38308.477650.63.

Foody, G.M.: Geographical weighting as a further refinement to regression modelling: an example focused on the NDVI-rainfall relationship. Rem. Sens. Environ. **88**, 283–293 (2003)

Fotheringham, A.S., Brunsdon, C., and Charlton, M.: Geographically weighted regression: the analysis of spatially varying relationships. Chichester: Wiley (2002)

Gunby, J.A., Darby, S.C., Miles, J.C.H., Green, B.M., Cox, D.R.: Factors affecting indoor radon concentrations in the United Kingdom. Health Phys. **64**, 2–11 (1993)

Jones, J., Casetti, E.: Applications of the expansion method. London: Routledge (1992)

Leung, Y., Mei, C.L., Zhang, W.X.: Statistical tests for spatial non-stationarity based on the geographically weighted regression model. Env. Plan. A **32**, 9–32 (2000)

McMillen, D.P.: One hundred fifty years of land values in Chicago: a non-parametric approach. J. Urban Econon. **40**, 100–124 (1996)
Nakaya, T., Fotheringham, A.S., Brundson, C., Chartlton, M.: Geographically weighted poisson regression for disease association mapping. Stat. Med. **24,** 2695–2717 (2005)
Nazaroff, W.W.: Radon transport from soil to air. Rev. Geophys. **30**, 137–160 (1992)
Olea, R.A.: Geostatistics for engineers and earth scientists, Kluwer (1999)
Ord, K.: Estimation methods for models of spatial interaction. J. Am. Stat. Assoc. **70**, 120–127 (1975)
Palermi, S., Pasculli, A.: Radon mapping in Abruzzo, Italy. Proceedings of 4th Canadian Conference on Geohazards Québec City Canada, May 20–24th (2008)
Pavlov, A. D.: Space-varying regression coefficients: a semi-parametric approach applied to real estate markets. Real Estate Econon. **28**, 249–283 (2000)
Price, P.N, Nero, A.V., Gelman A.: Bayesian prediction of mean indoor radon concentrations for Minnesota Counties. Health Phys. **71**, 922–936 (1996)
Scheib, C., Appleton, J.D., Jones, D., Hodgkinson, E.: Airborne gamma spectrometry, soil geochemistry and permeability index data in support of radon potential mapping in Central England. In: Barnet, I., Neznal, M., Pacherova, P. (Eds.), Proceedings of the 8th international workshop on the geological aspect of radon risk mapping, 26–30 September 2006, Prague, Czech Republic. Czech Geological Survey, RADON Corp., Prague, pp. 210–219.
Smith, B.J, Field, R.W.: Effect of housing factors and surficial uranium on the spatial prediction of residential radon in Iowa. Environmetrics **18**, 481–497 (2007)
Trigg, D., Leach, D.: Exponential smoothing with an adaptive response rate. Oper. Res. Quart. **18**, 53–59 (1968)
World Health Organization, WHO Handbook on indoor radon: a public health perspective, edited by H. Zeeb and F. Shannoun, Geneva (2009)

Part VII
Probability and Density Estimation

Applications of Large Deviations to Hidden Markov Chains Estimation

Fabiola Del Greco M.

Abstract Consider a Hidden Markov model where observations are generated by an underlying Markov chain plus a perturbation. The perturbation and the Markov process can be dependent from each other. We apply large deviations result to get an approximate confidence interval for the stationary distribution of the underlying Markov chain.

1 Introduction

Hidden Markov models (HMMs) describe the relationship between two stochastic processes: An observed one and an underlying hidden (unobserved) process. These models are used for two purposes. The first one is to make inferences or predictions about an unobserved process based on the observed one. A second reason is to explain variation in the observed process on variation in a postulated hidden one. For these reasons, HMMs became quite popular and many are its applications (i.e. biology, speech recognition, finance, etc.). For a general reference on these models, see Cappé et al. (2005).

More precisely, suppose a phenomena is driven by a discrete time finite state space Markov chain $\mathbf{X}$. Due to measurements errors we can observe only perturbed values. Suppose we have a set of observations $\mathbf{Y} := \{Y_i,\ i \in \mathbb{N}\}$, which take values in $\mathbb{R}^d$, and the following relationship holds

$$Y_i = X_i + \varepsilon_i, \quad \forall i \geq 1,$$

F. Del Greco M. (✉)
Institute of Genetic Medicine, EURAC research, Viale Druso 1, 39100 Bolzano, Italy
e-mail: fabiola.delgreco@eurac.edu

A. Di Ciaccio et al. (eds.), *Advanced Statistical Methods for the Analysis of Large Data-Sets*, Studies in Theoretical and Applied Statistics,
DOI 10.1007/978-3-642-21037-2_25, © Springer-Verlag Berlin Heidelberg 2012

where the processes $\mathbf{X} := \{X_j,\ j \in \mathbb{N}\}$ and $\varepsilon := \{\varepsilon_j, j \in \mathbb{N}\}$ satisfy the following assumptions.

The process $\mathbf{X}$ is a homogeneous Markov chain with a finite known state space $\Omega := \{x_1, \ldots, x_m\}$, with $x_i \in \mathbb{R}^d$. Let P^* be the true transition matrix of $\mathbf{X}$, whose entries are

$$P^*_{i,j} := \mathbb{P}(X_1 = x_j \mid X_0 = x_i).$$

Of course P^* is unknown and we assume it is irreducible and aperiodic. This assumption implies the existence and uniqueness of the stationary distribution, which we denote by $\pi^* = (\pi^*_1, \pi^*_2, \ldots, \pi^*_m)$. Moreover, we assume that given the value of the Markov process at time j, the error ε_j is independent from the past. The errors are allowed to be continuous random variables.

In the literature we found three different methods for computing confidence intervals (CIs) of Hidden Markov chains parameters, namely likelihood profiling, bootstrapping and CIs based on finite-differences approximation of the Hessian. The problem of interval estimate for parameters of a HMM has been studied in Visser et al. (2000). Their conclusion is that likelihood profiling and bootstrapping provide similar results, whereas the finite-differences intervals are mostly too small. There is no detailed analysis of the true coverage probability CIs in the context of HMMs. We propose a CI for the stationary distribution of the underlying Markov chain using a large deviations approach. Roughly, we estimate the rate of convergence of the frequencies of times that the observations fall in a certain interval to its limit, which is a linear transform of the unknown stationary distribution.

Nowadays, the theory of large deviations is rapidly expanding and is applied in many areas, such as statistics, engineering and physics; the reader can find few references in Varadhan (2008). It has been applied to a wide range of problems in which detailed information on rare events is required. Of course, one could be interested not only in the probability of rare events but also in the characteristic behavior of the system as the rare event occurs.

This paper is organized as follows. In Sect. 2 we define the framework of the study; in Sect. 3 we construct the confidence interval and explore its properties.

2 Study Framework

The Markov chain $\mathbf{X} := \{X_i\}_{i \geq 1}$ is observed through a perturbation sequence $\varepsilon := \{\varepsilon_j\}_{j \geq 1}$ of random variables. Assume that given $\{X_n = x_j\}$, ε_n has distribution $\mathbb{Q}_j$, with $j \in \{1, 2, \ldots, m\}$, and is independent from $\{\varepsilon_l,\ X_l,\ l \in \{1, 2, \ldots, n-1\}\}$.

For any subset $C \subset \mathbb{R}^d$ and $\mathbf{x} \in \mathbb{R}^d$ let $C - \mathbf{x} := \{\mathbf{y} \in \mathbb{R}^d : \mathbf{y} + \mathbf{x} \in C\}$. Let $\mathcal{U}$ the collection of partitions $U := (U_1, U_2, \ldots, U_m)$ with $U_i \subset \mathbb{R}^d$, satisfying the following properties. The sets $U_1, U_2, \ldots, U_m$ are disjoint, with non–empty interior set (i.e. the largest open set contained in U_i is not the empty set), $\bigcup_{j=1}^m U_j = \mathbb{R}^d$ and the matrix

$$Q_U = \begin{bmatrix} q_{1,1}^{(U)} & q_{1,2}^{(U)} & \cdots & q_{1,m}^{(U)} \\ \vdots & \vdots & \ddots & \vdots \\ q_{m,1}^{(U)} & q_{m,2}^{(U)} & \cdots & q_{m,m}^{(U)} \end{bmatrix}$$

has full rank, where $q_{i,j}^{(U)} := \mathbb{Q}_j(U_i - x_j)$. We also suppose that each entry of the matrix is strictly positive. Denote by Q_U^{-1} the inverse matrix of Q_U. We assume that the measures $\mathbb{Q}_j$, with $j \in \{1, 2, \ldots, m\}$, make $\mathscr{U}$ non-empty.

For a vector $\mathbf{x} \in \mathbb{R}^d$ we use the notation $\mathbf{x} \gg 0$ to indicate that each coordinate of $\mathbf{x}$ is positive. For any vector $\mathbf{u}$ set $\operatorname{diag}(\mathbf{u})$ to be the diagonal matrix whose (i, i) element is u_i, and let I_m be the $m \times m$ identity matrix. Define

$$\mathscr{H}_U := \Big\{\mathbf{x}\colon \det\big[P^* \operatorname{diag}(\mathbf{x}Q_U) - I_m\big] = 0\Big\}, \qquad J_U(\mu) := \sup_{\mathbf{x}\in\mathscr{H}_U:\mathbf{x}\gg 0} \sum_{k=1}^{m} \mu_k \log x_k,$$

and $B_r(\mathbf{x}) := \{\mathbf{y}\colon \|\mathbf{y} - \mathbf{x}\| \geq r\}$, where $\|\cdot\|$ stands for the Euclidean norm in $\mathbb{R}^d$. Set $\overline{B}_r = B_r(Q_U\pi^*)$. We assume that there exists a partition U such that $\inf_{\mu\in\overline{B}_r^c} J(\mu) - m + 1$ has a known lower bound H_r which is strictly positive for $r > 0$.

Denote by $\widehat{d}_i^{(n)} := \frac{1}{n}\sum_{j=1}^{n} 1\!\!1_{\{Y_j\in U_i\}}$ and let $\widehat{d}^{(n)}(U) := (\widehat{d}_1^{(n)}, \widehat{d}_2^{(n)}, \ldots, \widehat{d}_m^{(n)})$, where $1\!\!1_A$ is the indicator function of the event A.

Theorem 1 (Confidence Interval). *Fix $\alpha > 0$ and choose the smallest r such that $e^{-nH_r} \leq \alpha$. Then, the set $A_r = Q_U^{-1}\big(B_r(\widehat{d}^{(n)}(U))\big)$ is an approximate $(1-\alpha)$-confidence interval.*

Our approach relies on the fact that $\widehat{d}_n(U)$ converges to $Q_U\pi^*$ and does it quite fast. The large deviations principle makes this statement rigorous. Hence we use H_r to lower estimate the rate function.

3 Construction of the Confidence Interval

Definition 1 (Large deviations principle). A rate function is a function which is non-negative and lower semicontinuous. A sequence of random variables Z_n, $n \in \mathbb{N}$, satisfies a large deviations principle with rate function $I(\cdot)$ if we have

$$\limsup_{n\to\infty} \frac{1}{n} \log \mathbb{P}(Z_n \in F) \leq -\inf_{x\in F} I(x), \qquad \text{for any closed set } F \text{, and}$$

$$\liminf_{n\to\infty} \frac{1}{n} \log \mathbb{P}(Z_n \in A) \geq -\inf_{x\in A} I(x), \qquad \text{for any open set } A.$$

Proposition 1. *Define*

$$\Lambda_j^{(U)}(\lambda) := \mathbb{E}\Big[\exp\Big\{\sum_{k=1}^{m} \lambda_k \, 1\!\!1_{\{\varepsilon_1 \in U_k - x_j\}}\Big\}\Big].$$

For any Borel set $U \subset \mathbb{R}^d$, $\widehat{d}^{(n)}(U)$ satisfies a large deviations principle with rate function

$$I_U(z) := \sup_{\lambda \in \mathbb{R}^m} \big\{\langle \lambda, z\rangle - \log \rho(P_\lambda)\big\},$$

where $\langle \cdot \rangle$ is the usual inner product, P_λ is the matrix whose (i,j) entry is $P^*_{i,j}\Lambda_j^{(U)}(\lambda)$ and for any irreducible matrix A the scalar $\rho(A)$ is the so called Perron–Frobenius eigenvalue, with the properties $\rho(A) \in (0,\infty)$ and $|\theta| \leq \rho(A)$ for any eigenvalue θ of A.

Proof. Define

$$f(X_i,\varepsilon_i) = \Big(\sum_{k=1}^{m} 1\!\!1_{\{X_i = x_k\}} 1\!\!1_{\{\varepsilon_i \in U_1 - x_k\}}, \sum_{k=1}^{m} 1\!\!1_{\{X_i = x_k\}} 1\!\!1_{\{\varepsilon_i \in U_2 - x_k\}}, \ldots, \sum_{k=1}^{m} 1\!\!1_{\{X_i = x_k\}} 1\!\!1_{\{\varepsilon_i \in U_m - x_k\}}\Big).$$

The sequence $\{f(X_i,\varepsilon_i), 1 \leq i \leq n\}$, given a realization $\{X_i = \eta_i, 1 \leq i \leq n\}$, with $\eta_i \in \Omega$ for each i, is composed by independent random variables. This random function meets the hypothesis of Exercise 3.1.4 of Dembo and Zeitouni (1998) which yields this proposition. □

Proposition 2. $\widehat{d}^{(n)}(U)$ *converges a.s. to $Q_U\pi^*$.*

Proof. We can write $1\!\!1_{\{Y_j \in U_i\}} = \sum_{s=1}^{m} 1\!\!1_{\{X_j = x_s\}} 1\!\!1_{\{\varepsilon_j \in U_i - x_s\}}$. Hence,

$$\widehat{d}_i^{(n)} = \sum_{s=1}^{m} \Big(\frac{1}{n}\sum_{j=1}^{n} 1\!\!1_{\{X_j = x_s\}} 1\!\!1_{\{\varepsilon_j \in U_i - x_s\}}\Big).$$

Consider the markov chain $(X_n, \sum_{i=1}^{m} i \, 1\!\!1_{\{\varepsilon_n \in U_i - X_n\}})$, taking values in the set $\Omega \times \{1,2,\ldots,m\}$. This chain is irreducible and aperiodic, and our result follows by the ergodic theorem. □

Denote by $\mathcal{N}(Q_U)$ the set of linear combinations, with non-negative coefficients, of the rows of Q_U. More precisely

$$\mathcal{N}(Q_U) := \big\{\mathbf{x}Q_U : \mathbf{x} \gg 0\big\}.$$

For any $\mathbf{t} := (t_1, t_2, \ldots, t_m)$, with $\mathbf{t} \gg 0$, define

$$L_k(\mathbf{t}) := \frac{t_k}{\sum_{i=1}^m t_i P^*_{i,k}}, \qquad \text{for } k \in \{1,\ldots,m\},$$

and $L(\mathbf{t}) := \big(L_1(\mathbf{t}), L_2(\mathbf{t}), \ldots, L_m(\mathbf{t})\big)$. Notice that for any positive scalar a we have $L(a\mathbf{t}) = L(\mathbf{t})$, i.e. is homogeneous of degree 0. Denote by $\mathscr{M}_1$ the set of probability measures on Ω. We consider $\mathscr{M}_1$ as a subset of $\mathbb{R}^m$, that is the m-dimensional simplex. Let $q_U^{(-1)}(i,j)$ the (i,j)-th entry of the matrix Q_U^{-1}. Denote by $\mathscr{K}$ the image $L\big(\mathscr{M}_1\big) \subset \mathbb{R}^m$.

Define the map $G_U : \mathscr{M}_1 \to \mathbb{R}$

$$G_U(\mu) := \sup_{z \in \mathscr{K} \cap \mathscr{N}(Q_U)} \sum_{k=1}^m \mu_k \log \sum_{i=1}^m z_i q_U^{(-1)}(i,k).$$

Proposition 3. *For $\mu \notin \mathscr{M}_1$ we have $I(\mu) = \infty$, and if $\mu \in \mathscr{M}_1$ then*

$$I_U(\mu) \geq G_U(\mu) - (m-1).$$

Proof. Notice that

$$\begin{aligned}
\Lambda_j^{(U)}(\lambda) &= \mathbb{E}\Big[\exp\Big\{\sum_{k=1}^m \lambda_k \mathbb{1}_{\{\varepsilon_1 \in U_k - x_j\}}\Big\}\Big] \\
&= \sum_{k=1}^m e^{\lambda_k + m - 1} \mathbb{P}(\varepsilon_1 \in U_k - x_j) \\
&= e^{m-1} \sum_{k=1}^m e^{\lambda_k} q_{k,j}^{(U)}.
\end{aligned}$$

Fix $\mathbf{t}$ such that $L(\mathbf{t}) \in \mathscr{N}(Q_U)$. For a vector $\mathbf{x} \gg 0$ denote by $\log(\mathbf{x})$ the vector whose j-th entry is $\log x_j$. The vector $\lambda^* := \log\big[e^{1-m}\big(L(\mathbf{t})Q_U^{-1}\big)\big]$ satisfies

$$\Lambda_j(\lambda^*) = \frac{t_j}{\sum_{i=1}^m t_i P^*_{i,j}}.$$

Now the argument is standard. In fact $\mathbf{t}$ satisfies $\mathbf{t}P_{\lambda^*} = \mathbf{t}$, which together with the fact $\mathbf{t} \gg 0$, implies $\rho(P_{\lambda^*}) = 1$. Hence

$$\begin{aligned}
I(\mu) &= \sup_{\lambda \in \mathbb{R}^m} \big\{\langle \lambda, z\rangle - \log \rho(P_\lambda)\big\} \\
&\geq \langle \lambda^*, \mu\rangle - \log \rho(P_{\lambda^*}) \\
&= \sum_{k=1}^m \mu_k \log \sum_{i=1}^m e^{1-m} L_i(\mathbf{t}) q_U^{(-1)}(i,j).
\end{aligned}$$

It remains to take the suprema over the set of $\mathbf{t}$ satisfying $L(\mathbf{t}) \in \mathcal{N}(Q)$ to get our result. □

Proposition 4. $G_U = J_U$.

Proof. Notice that $\mathbf{x}Q_U \in \mathcal{K}$ if and only if

$$\sum_{i=1}^{m} x_i Q_{i,k}^{(U)} = \frac{t_k}{\sum_{j=1}^{m} t_j P_{j,k}^*},$$

which can be reformulated as $\mathbf{t}P^*\text{diag}(\mathbf{x}Q_U) = \mathbf{t}$. This is verified if and only if the matrix $P^*\text{diag}(\mathbf{x}Q_U)$ as eigenvalue 1, and is equivalent to

$$\det\left[P^*\text{diag}(\mathbf{x}Q_U) - I_m\right] = 0.$$ □

Hence

$$\mathbb{P}(\widehat{d}^{(n)}(U) \in \overline{B}_r^C) \leq e^{-n \inf_{\mu \in \overline{B}_r^C} \left(J_U(\mu) - (m-1) + o(1)\right)}.$$

We get

$$\begin{aligned}\mathbb{P}(\pi^* \notin A_r) &= \mathbb{P}(\widehat{d}^{(n)}(U) \in \overline{B}_r^C) \\ &\leq e^{-n \sup_{U \in \mathcal{U}} \inf_{\mu \in \overline{B}_r^C} \left(J_U(\mu) - (m-1) + o(1)\right)} \\ &\leq e^{-n\left(H_r + o(1)\right)},\end{aligned}$$

which proves Theorem 1.

4 Conclusions

We end this paper by considering few of the properties of the confidence interval. The coverage of this set is quite good, in fact, for any $p > 0$, there exists a constant $c(p)$ such that

$$\mathbb{P}(\pi^* \notin A_r) \leq \alpha + \frac{c(p)}{n^p}.$$

The advantage of using the large deviations approach stands in the fact that for fixed r, the probability that A_r does not contain π^* decreases exponentially fast as n increases. On the other hand, the measure of the confidence interval is less than $\|Q_U^{-1}\| r$, where $\|.\|$ here denotes the usual matrix norm, and r is the ray of B_r defined in the introduction.

Acknowledgements The author is grateful to anonymous referees for their valuable comments which led to improvements in this paper.

References

Cappé O, Moulines E, Rydén T (2005) Inference in Hidden Markov Models. Springer
Dembo A and Zeitouni O (1998) Large deviations techniques and applications. Springer
Varadhan S R S (2008) Special invited paper large deviations. Ann Probab 36 : 397 – 419
Visser I, Raijmakers M, and Molenaar P (2000) Confidence intervals for Hidden Markov models parameters. Brit J Math Stat Psychol 53 : 317 – 327

Multivariate Tail Dependence Coefficients for Archimedean Copulae

Giovanni De Luca and Giorgia Rivieccio

Abstract We analyze the multivariate upper and lower tail dependence coefficients, obtained extending the existing definitions in the bivariate case. We provide their expressions for a popular class of copula functions, the Archimedean one. Finally, we apply the formulae to some well known copula functions used in many financial analyses.

1 Introduction

The different relationship between positive and negative extreme events occurring in financial markets has been studied through several papers, see e.g. Engle (2002), Tse and Tsui (2002), Longin and Solnik (2001). In particular, in bear markets the strength of the correlation between returns is higher than in bull markets and it tends to increase when the markets are more volatile (Campbell et al. 2002). This suggests a significant dependence in the tails of the joint distribution of asset returns to analyze with an asymmetric model. A first approach involves the multivariate Extreme Value Theory (EVT), e.g. Coles et al. (1999), Pickands (1981), Einmahl et al. (2001), Hall and Tajvidi (2000), Peng (1999), a second one is based on the non-parametric estimation techniques of the concordance between rare events, e.g. Dobríc and Schmid (2005), Schmidt and Stadmüller (2005). Finally, a popular way of proceeding is to model the whole dependence structure between assets with a copula function and to measure the relationship in the tails of the joint distribution using the upper and lower tail dependence coefficients (see e.g. Embrechts et al. (2003)). However, the literature on this argument suggests to make use of the upper and lower tail dependence coefficients only to measure the association between

G. De Luca · G. Rivieccio (✉)
Department of Statistics and Mathematics for Economic Research, *Parthenope* University, via Medina 40, 80133 Napoli, Italy
e-mail: giovanni.deluca@uniparthenope.it; giorgia.rivieccio@uniparthenope.it

A. Di Ciaccio et al. (eds.), *Advanced Statistical Methods for the Analysis of Large Data-Sets*, Studies in Theoretical and Applied Statistics,
DOI 10.1007/978-3-642-21037-2_26,

extreme returns of a couple of assets modelled by a bivariate copula function (e.g. Schmidt (2005) and Cherubini et al. (2004)).

In this paper, we focus on the tail dependence in a multivariate context providing the general expressions of the coefficients for Archimedean copulae in terms of their generator function. Finally, we apply these formulae to some multivariate biparametric (MB) Archimedean copulae.

2 Archimedean Survival Copulae

A copula function C is a multivariate distribution function defined on the unit cube $[0,1]^n$, with uniformly distributed margins (see Joe 1997 and Nelsen 2006). The well known Sklar's Theorem shows the relationship between the copula function, C, and the n-dimensional function H of the random variables $X_1, \ldots, X_n$ with domain $\bar{R}^n$, such that H is grounded, n-increasing and $H(\infty, \ldots, \infty) = 1$, defined as

$$H(x_1, \ldots, x_n) = P(X_1 \leq x_1, \ldots, X_n \leq x_n).$$

According to the theorem, there exists an n-dimensional copula function C such that for all $\mathbf{x}$ in $\bar{R}^n$

$$H(x_1, \ldots, x_n) = C(u_1, \ldots, u_n) \tag{1}$$

where u_i is the uniform marginal distribution $F_i(X_i)$. Moreover, by Sklar's theorem, it is possible to derive the relationship between the survival joint function, $\bar{H}(x_1, \ldots, x_n)$, and the survival copula function, $\hat{C}(1-u_1, \ldots, 1-u_n)$.

The survival joint function is given by

$$\bar{H}(x_1, \ldots, x_n) = P(X_1 > x_1, \ldots, X_n > x_n)$$

and it follows from (1) that

$$\bar{H}(x_1, \ldots, x_n) = \hat{C}(1-u_1, \ldots, 1-u_n).$$

For $n = 2$, it has been shown that the survival copula function

$$\begin{aligned}\hat{C}(1-u_1, 1-u_2) = [1 - P(U_1 \leq u_1)] + [1 - P(U_2 \leq u_2)] \\ -[1 - P(U_1 \leq u_1, U_2 \leq u_2)]\end{aligned} \tag{2}$$

is strictly related to the copula function through the relationship

$$\hat{C}(1-u_1, 1-u_2) = 1 - u_1 - u_2 + C(u_1, u_2). \tag{3}$$

Following the same reasoning, it is easy to derive the expression of the survival copula function in a multivariate framework, showing, in particular, the relationships for Archimedean copulae. The Archimedean copulae family (see Cherubini et al. 2004) can be built starting from the definition of a generator function $\Phi : I = [0,1]$, continuous, decreasing and convex, such that $\Phi(1) = 0$ and where $\Phi^{[-1]}(t)$ is defined the "pseudo-inverse" of $\Phi(t) : \Phi^{[-1]}(t) = \Phi^{-1}(t)\ \forall t \in [0, \Phi(0)]$ and $\Phi^{[-1]}(t) = 0$ for $t \geq \Phi(0)$. If $\Phi(0) = \infty$, then $\Phi(t)$ and C are said to be strict; if $\Phi(0) < \infty$, $\Phi(t)$ and C are non-strict (for a definition, see Nelsen 2006). Let $\Phi^{-1}(t)$ be the inverse of $\Phi(t)$ a strict generator of an Archimedean copula; then an Archimedean copula can be expressed as

$$C(u_1, \ldots, u_n) = \Phi^{-1}(\Phi(u_1) + \ldots + \Phi(u_n)).$$

Archimedean copulae share the important features to be symmetric and associative.

For our purposes, we generalize the expression of the survival copula $\hat{C}$ for an Archimedean copula C with generator function Φ, that is

$$\hat{C}(1-u_1, \ldots, 1-u_n) = \sum_{i=0}^{n} \left\{ \binom{n}{n-i} (-1)^i \left[\Phi^{-1}\left(\sum_{j=0}^{i} \Phi(u_j) \right) \right] \right\}, \tag{4}$$

where $\Phi^{-1}(\Phi(u_0)) = 1$ and $\Phi^{-1}(\Phi(u_1)) = u_1$.

3 Tail Dependence

The tail dependence is a measure of concordance between less probable values of variables. This concordance tends to concentrate on the lower and upper tails of the joint distribution.

Definition 1. In a bivariate context, let $F_i(X_i)$, $i = 1, 2$, be the marginal distribution functions of two random variables X_1 and X_2 and let u be a threshold value; then the upper tail dependence coefficient, λ_U, is defined as

$$\lambda_U = \lim_{u \to 1^-} P(F_1(X_1) > u \,|\, F_2(X_2) > u) = \lim_{u \to 1^-} P(U_1 > u \,|\, U_2 > u).$$

When $\lambda_U \in (0, 1]$, X_1 and X_2 are asymptotically dependent on the upper tail; if λ_U is null, X_1 and X_2 are asymptotically independent.

Since

$$\begin{aligned} P(U_1 > u \,|\, U_2 > u) &= \frac{P(U_1 > u, U_2 > u)}{P(U_2 > u)} \\ &= \frac{1 - P(U_1 \leq u) - P(U_2 \leq u) + P(U_1 \leq u, U_2 \leq u)}{1 - P(U_2 \leq u)}, \end{aligned}$$

then, from (2) and (3) it follows that the upper tail dependence coefficient can be also expressed in terms of survival copula functions, that is

$$\lambda_U = \lim_{u \to 1^-} \frac{\hat{C}(1-u, 1-u)}{1-u} = \lim_{u \to 1^-} \frac{1 - 2u + C(u,u)}{1-u}.$$

In a similar way, the lower tail dependence coefficient, λ_L, is defined as

$$\begin{aligned}\lambda_L &= \lim_{u \to 0^+} P(F_1(X_1) \leq u | F_2(X_2) \leq u) = \lim_{u \to 0^+} P(U_1 \leq u | U_2 \leq u) \\ &= \lim_{u \to 0^+} \frac{P(U_1 \leq u, U_2 \leq u)}{P(U_2 \leq u)}\end{aligned}$$

and in terms of copula function as

$$\lambda_L = \lim_{u \to 0^+} \frac{C(u,u)}{u}.$$

It is easy to show that for an Archimedean copula each tail dependence coefficient can be derived using the generator function (e.g. Cherubini et al. 2004). In fact, if the first derivative of the inverse of the generator function $\Phi^{-1'}(0)$ is finite, then an Archimedean copula does not have upper tail dependence; conversely, λ_U is given by

$$\lambda_U = \lim_{u \to 1^-} \frac{1 - 2u + C(u,u)}{1-u} = \lim_{u \to 1^-} \frac{1 - 2\Phi^{-1}(\Phi(u)) + \Phi^{-1}(2\Phi(u))}{1 - \Phi^{-1}(\Phi(u))},$$

where

$$C(u,u) = \Phi^{-1}(\Phi(u) + \Phi(u)) = \Phi^{-1}(2\Phi(u))$$

and

$$1 - u = 1 - \Phi^{-1}(\Phi(u)).$$

To solve the limit, it is necessary to apply de L'Hôpital theorem,

$$\begin{aligned}\lambda_U &= \lim_{u \to 1^-} \frac{-2\Phi^{-1'}(\Phi(u))(\Phi'(u)) + \Phi^{-1'}(2\Phi(u))2(\Phi'(u))}{-\Phi^{-1'}(\Phi(u))\Phi'(u)} \\ &= \lim_{u \to 1^-} \frac{-2\Phi^{-1'}(\Phi(u))(\Phi'(u))}{-\Phi^{-1'}(\Phi(u))\Phi'(u)} + \frac{\Phi^{-1'}(2\Phi(u))2(\Phi'(u))}{-\Phi^{-1'}(\Phi(u))\Phi'(u)} \\ &= 2 - 2 \lim_{u \to 1^-} \frac{\Phi^{-1'}(2\Phi(u))}{\Phi^{-1'}(\Phi(u))} = 2 - 2 \lim_{t \to 0^+} \frac{\Phi^{-1'}(2t)}{\Phi^{-1'}(t)}.\end{aligned} \quad (5)$$

The lower tail dependence coefficient is given by

$$\lambda_L = \lim_{u \to 0+} \frac{C(u,u)}{u} = \lim_{u \to 0+} \frac{\Phi^{-1}(2\Phi(u))}{\Phi^{-1}(\Phi(u))}.$$

By applying de L'Hôpital theorem, the solution of the limit is

$$\lambda_L = \lim_{u \to 0+} \frac{\Phi^{-1'}(2\Phi(u))(2\Phi'(u))}{\Phi^{-1'}(\Phi(u))\Phi'(u)} = 2 \lim_{t \to \infty} \frac{\Phi^{-1'}(2t)}{\Phi^{-1'}(t)}. \tag{6}$$

Note that the associative property implies that

$$\lambda_U = \lim_{u \to 1^-} P(U_1 > u | U_2 > u) = \lim_{u \to 1^-} P(U_2 > u | U_1 > u)$$

and

$$\lambda_L = \lim_{u \to 0+} P(U_1 \leq u | U_2 \leq u) = \lim_{u \to 0+} P(U_2 \leq u | U_1 \leq u).$$

We propose to extend the definition introducing the multivariate upper and lower tail dependence coefficients for Archimedean copulae.

Definition 2. A multivariate generalization of the tail dependence coefficients consists in to consider h variables and the conditional probability associated to the remaining $n - h$ variables, given by

$$\begin{aligned}\lambda_U^{1\ldots h|h+1\ldots n} &= \lim_{u \to 1^-} P(F_1(X_1) > u, \ldots, F_h(X_h) > u | F_{h+1}(X_{h+1}) \\ &\quad > u, \ldots, F_n(X_n) > u) \\ &= \lim_{u \to 1^-} \frac{\hat{C}_n(1-u, \ldots, 1-u)}{\hat{C}_{n-h}(1-u, \ldots, 1-u)}.\end{aligned}$$

From our definition given in (4) the upper multivariate tail dependence coefficient is

$$\lambda_U^{1\ldots h|h+1\ldots n} = \lim_{u \to 1^-} \frac{\sum_{i=0}^{n} \left\{ \binom{n}{n-i} (-1)^i \left[\Phi^{-1}\left(i\Phi(u)\right)\right] \right\}}{\sum_{i=0}^{n-h} \left\{ \binom{n-h}{n-h-i} (-1)^i \left[\Phi^{-1}\left(i\Phi(u)\right)\right] \right\}}$$

and after applying de L'Hôpital theorem, we obtain

$$\lambda_U^{1\ldots h|h+1\ldots n} = \lim_{t \to 0+} \frac{\sum_{i=1}^{n} \left\{ \binom{n}{n-i} i \, (-1)^i \left[\Phi^{-1'}(it)\right] \right\}}{\sum_{i=1}^{n-h} \left\{ \binom{n-h}{n-h-i} i \, (-1)^i \left[\Phi^{-1'}(it)\right] \right\}}. \tag{7}$$

The corresponding multivariate lower tail dependence coefficient is defined as

$$\lambda_L^{1\dots h|h+1\dots n} = \lim_{u\to 0+} P(F_1(X_1) \le u, \dots, F_h(X_h) \le u | F_{h+1}(X_{h+1}) \le u, \dots, F_n(X_n) \le u)$$
$$= \lim_{u\to 0+} \frac{C_n(u,\dots,u)}{C_{n-h}(u,\dots,u)} = \lim_{u\to 0+} \frac{\Phi^{-1}(n\Phi(u))}{\Phi^{-1}((n-h)\Phi(u))}.$$

Exploiting de L'Hôpital theorem, the result is

$$\lambda_L^{1\dots h|h+1\dots n} = \frac{n}{n-h} \lim_{t\to\infty} \frac{\Phi^{-1'}(nt)}{\Phi^{-1'}((n-h)t)}. \tag{8}$$

For the associative property, (7) and (8) hold for each of the $n!$ coefficients in correspondence of the $n!$ permutations of the variables $X_1, \dots, X_n$.

An example. When $n = 4$ variables and $h = 1$, the upper tail dependence coefficient is

$$\lambda_U^{1|234} = \lim_{u\to 1^-} P(F_1(X_1) > u | F_2(X_2) > u, F_3(X_3) > u, F_4(X_4) > u)$$
$$= \lim_{u\to 1^-} \frac{\hat{C}_4(1-u, 1-u, 1-u, 1-u)}{\hat{C}_3(1-u, 1-u, 1-u)}$$
$$= \lim_{u\to 1^-} \frac{1 - 4u + 6C(u,u) - 4C(u,u,u) + C(u,u,u,u)}{1 - 3u + 3C(u,u) - C(u,u,u)}$$
$$= \lim_{u\to 1^-} \frac{1 - 4\Phi^{-1}(\Phi(u)) + 6\Phi^{-1}(2\Phi(u)) - 4\Phi^{-1}(3\Phi(u)) + \Phi^{-1}(4\Phi(u))}{1 - 3\Phi^{-1}(\Phi(u)) + 3\Phi^{-1}(2\Phi(u)) - \Phi^{-1}(3\Phi(u))}.$$

Applying de L'Hôpital theorem,

$$\lambda_U^{1|234} = \lim_{u\to 1^-} \frac{-4\Phi^{-1'}(\Phi(u))\Phi'(u) + 6\Phi^{-1'}(2\Phi(u))2\Phi'(u) - 4\Phi^{-1'}(3\Phi(u))3\Phi'(u)}{-3\Phi^{-1'}(\Phi(u))\Phi'(u) + 3\Phi^{-1'}(2\Phi(u))2\Phi'(u) - \Phi^{-1'}(3\Phi(u))3\Phi'(u)} +$$
$$+ \frac{\Phi^{-1'}(4\Phi(u))4\Phi'(u)}{-3\Phi^{-1'}(\Phi(u))\Phi'(u) + 3\Phi^{-1'}(2\Phi(u))2\Phi'(u) - \Phi^{-1'}(3\Phi(u))3\Phi'(u)}$$
$$= \lim_{t\to 0+} \frac{-4\Phi^{-1'}(t) + 12\Phi^{-1'}(2t) - 12\Phi^{-1'}(3t) + 4\Phi^{-1'}(4t)}{-3\Phi^{-1'}(t) + 6\Phi^{-1'}(2t) - 3\Phi^{-1'}(3t)}. \tag{9}$$

In a similar way, the lower tail dependence coefficient, with $n = 4$ and $h = 1$, is

$$\lambda_L^{1|234} = \lim_{u\to 1^-} P(F_1(X_1) \le u | F_2(X_2) \le u, F_3(X_3) \le u, F_4(X_4) \le u)$$
$$= \lim_{u\to 0+} \frac{C(u,u,u,u)}{C(u,u,u)} = \lim_{u\to 0+} \frac{\Phi^{-1}(4\Phi(u))}{\Phi^{-1}(3\Phi(u))}$$

and applying de L'Hôpital theorem,

$$\lambda_L^{1|234} = \lim_{u \to 0+} \frac{\Phi^{-1'}(4\Phi(u))(4\Phi'(u))}{\Phi^{-1'}(3\Phi(u))(3\Phi'(u))}$$
$$= \frac{4}{3} \lim_{t \to \infty} \frac{\Phi^{-1'}(4t)}{\Phi^{-1'}(3t)}. \qquad (10)$$

The associative property guarantees the invariance of the coefficients (9) and (10) when we permute the four variables.

4 MB Copula Functions

Two-parameter families can be used to capture different types of dependence structure, in particular lower or upper tail dependence or both. We consider two multivariate bi-parametric (MB) copula functions, extending BB1 and BB7 copulae, analyzed by Joe (1997) in the bivariate case. They are characterized by non-null upper and lower tail dependence. The formulation of a MB copula is

$$C(u_1, \ldots, u_n) = \psi(-\log K(e^{-\psi^{-1}(u_1)}, \ldots, e^{-\psi^{-1}(u_n)}))$$

where K is max-infinitely divisible and ψ belongs to the class of Laplace Transforms. Two-parameter families result if K and ψ are parametrized, respectively, by parameters κ and θ. If K has the Archimedean copula form, then also C has the same form.

4.1 MB1 Copula

The MB1 copula is obtained letting K be the Gumbel family and ψ the Laplace Transform B (Joe 1997, p. 375), then

$$C(u_1, \ldots, u_n) = \left\{ 1 + \left[\sum_{i=1}^{n} (u_i^{-\theta} - 1)^{\kappa} \right]^{1/\kappa} \right\}^{-1/\theta}$$

where $\theta > 0$, $\kappa \geq 1$. For $\kappa = 1$ we get the popular Clayton copula. The generator function is

$$\Phi(t) = (t^{-\theta} - 1)^{\kappa}$$

and its inverse is

$$\Phi^{-1}(t) = (1 + t^{1/\kappa})^{-1/\theta}.$$

The bivariate upper and lower tail dependence coefficients (5) and (6) are, respectively,

$$\lambda_U = 2 - 2^{1/\kappa}$$

and

$$\lambda_L = 2^{-1/(\kappa\theta)}.$$

The general expressions of the coefficients are given (see (7) and (8)) by

$$\lambda_U^{1\ldots h|h+1\ldots n} = \frac{\sum_{i=1}^{n}\left[\binom{n}{n-i}(-1)^i\,(i)^{1/\kappa}\right]}{\sum_{i=1}^{n-h}\left[\binom{n-h}{n-h-i}(-1)^i\,(i)^{1/\kappa}\right]}$$

and

$$\lambda_L^{1\ldots h|h+1\ldots n} = \left(\frac{n}{n-h}\right)^{-1/\kappa\theta}.$$

4.2 MB7 Copula

In the second case, let K be the Clayton family and ψ be Laplace Transform C (Joe 1997, p. 375), then

$$C(u_1,\ldots,u_n) = 1 - \left(1 - \left[\sum_{i=1}^{n}(1-(1-u_i)^{\theta})^{-\kappa} - (n-1)\right]^{-1/\kappa}\right)^{1/\theta}$$

is the MB7 copula, also known as Joe-Clayton copula, where $\kappa > 0$, $\theta \geq 1$. For $\theta = 1$ we get again the Clayton copula.

The generator function is

$$\Phi(t) = [1-(1-t)^{\theta}]^{-\kappa} - 1$$

while its inverse is

$$\Phi^{-1}(t) = 1 - [1-(1+t)^{-1/\kappa}]^{1/\theta}.$$

The bivariate upper tail dependence coefficient, see (5), is

$$\lambda_U = 2 - 2^{1/\theta},$$

while the bivariate lower tail dependence coefficient, see (6), is

$$\lambda_L = 2^{-1/\kappa}.$$

The generalization of the results leads to

$$\lambda_U^{1\ldots h|h+1\ldots n} = \frac{\sum_{i=1}^{n}\left[\binom{n}{n-i}(-1)^i (i)^{1/\theta}\right]}{\sum_{i=1}^{n-h}\left[\binom{n-h}{n-h-i}(-1)^i (i)^{1/\theta}\right]}$$

and

$$\lambda_L^{1\ldots h|h+1\ldots n} = \left(\frac{n}{n-h}\right)^{-1/\kappa}.$$

5 Concluding Remarks

We have analyzed the upper and lower tail dependence coefficients in a multivariate framework, providing their expressions for a widely used class of copula function, the Archimedean copulae. The tail dependence measures can be of interest in many fields. For instance, given n financial asset returns, the upper (lower) tail dependence coefficient can be interpreted as the probability of very high (low) returns for h assets provided that very high (low) returns have occurred for the remaining $n-h$ assets. In a risk management perspective, the implementation of a strategy of risk diversification can be helped by the knowledge of these coefficients, especially in a financial crisis scenario.

References

Campbell, R., Koedijk, K., Kofman, P.: Increased correlation in bear markets. Financ. Anal. J., January-February, 87–94 (2002)

Cherubini, U., Luciano, E., Vecchiato, W.: Copula Methods in Finance. John Wiley & Sons, (2004)

Coles, S., Heffernan, J., Tawn, J.: Dependence measures for extreme value analysis. Extremes **2**, 339–366 (1999)

Dobríc, J., Schmid, F.: Non parametric estimation of the lower tail dependence λ_L in bivariate copulas. J. Appl. Stat. **32**, 387–407 (2005)

Einmahl, J., de Haan, L., Piterbarg, V.: Multivariate extremes estimation. Ann. Stat. **29**, 1401–1423 (2001)

Embrechts, P., Lindskog, F., McNeil, A.: Modelling Dependence with Copulas and Applications to Risk Management. In Rachev S. (eds), *Handbook of Heavy Tailed Distributions in Finance*, pp. 329-384 (2003)

Engle, R.F.: Dynamic conditional correlation: a simple class of multivariate generalized autoregressive conditional heteroscedasticity models. J. Bus. Econ. Stat. **20**, 339–350 (2002)

Hall, P., Tajvidi, N.: Disribution and dependence-function estimation for bivariate extreme-value distributions. Bernoulli **6**, 835–844 (2000)

Joe, H.: Multivariate Models and Dependence Concepts. Chapman & Hall/CRC, New York (1997)

Longin, F., Solnik, B.: Extreme correlation of international equity markets. J. Financ. **56**, 649–676 (2001)

Nelsen, R.B.: An Introduction to Copulas. Springer, New York (2006)
Peng, L.: Estimation of the coefficient of tail dependence in bivariate extremes. Statist. Probab. Lett. **43**, 349–409 (1999)
Pickands, J.: Multivariate extreme value distributions. Proceedings of the 43rd Session ISI (Buoneos Aires), 399–409 (1981)
Schmidt, R.: Tail Dependence. In Čížek, P., Härdle, W., Weron, R. (eds), *Statistical tools in finance and insurance*, pp. 65-91, Springer Verlag, Heidelberg (2005)
Schmidt, R., Stadmüller, U.: Non-parametric Estimation of Tail Dependence. Scand. J. Stat. **33**,307–335 (2005)
Tse, Y.K., Tsui, A.K.C.: A Multivariate Generalized Autoregressive Conditional Heteroscedasticity model with time-varying correlations. J. Bus. Econ. Stat. **20**, 351–362 (2002).

A Note on Density Estimation for Circular Data

Marco Di Marzio, Agnese Panzera, and Charles C. Taylor

Abstract We discuss kernel density estimation for data lying on the d-dimensional torus ($d \geq 1$). We consider a specific class of product kernels, and formulate exact and asymptotic L_2 properties for the estimators equipped with these kernels. We also obtain the optimal smoothing for the case when the kernel is defined by the product of von Mises densities. A brief simulation study illustrates the main findings.

1 Introduction

A *circular* observation can be regarded as a point on the unit circle, and may be represented by an angle $\theta \in [0, 2\pi)$. Typical examples include flight direction of birds from a point of release, wind and ocean current direction. A circular observation is periodic, i.e. $\theta = \theta + 2m\pi$ for $m \in \mathbb{Z}$. This periodicity sets apart circular statistical analysis from standard real-line methods. Recent accounts are given by Jammalamadaka and SenGupta (2001) and Mardia and Jupp (1999). Concerning circular density estimation we observe that almost all of the related methods appear to have a parametric nature, with the exception of a few contributions on kernel density estimation for data lying on the circle or on the sphere (Bai et al. 1988; Beran 1979; Hall et al. 1987; Klemelä 2000; Taylor 2008). In this paper we consider kernel density estimation when a support point $\boldsymbol{\theta}$ is a point on the d-dimensional torus $\mathbb{T}^d := [-\pi, \pi]^d$, $d \geq 1$. To this end, we define estimators equipped with kernels belonging to a suitable class, and derive their exact and asymptotic L_2 properties.

M. Di Marzio (✉) · A. Panzera
DMQTE, G. d'Annunzio University, Viale Pindaro 42, 65127 Pescara, Italy
e-mail: mdimarzio@unich.it; panzera@yahoo.it

C. C. Taylor
Department of Statistics, University of Leeds, Leeds West Yorkshire LS2 9JT, UK
e-mail: charles@maths.leeds.ac.uk

A. Di Ciaccio et al. (eds.), *Advanced Statistical Methods for the Analysis of Large Data-Sets*, Studies in Theoretical and Applied Statistics,
DOI 10.1007/978-3-642-21037-2_27, © Springer-Verlag Berlin Heidelberg 2012

We also obtain the optimal smoothing degree for the case in which the kernel is a d-fold product of von Mises densities. In particular, in Sect. 2 we discuss the class of toroidal kernels, and in Sect. 3 we obtain the exact and asymptotic integrated mean squared error along with the optimal smoothing for our estimators. Finally, in Sect. 4, we give some numerical evidence which confirms our asymptotic results.

2 Toroidal Kernels

By *toroidal density* we mean a continuous probability density function whose support is $\mathbb{T}^d$ and such that $f(-\boldsymbol{\pi}) \equiv f(\boldsymbol{\pi})$ where $\boldsymbol{\pi} = (\pi, \cdots, \pi)$. For estimating a smooth toroidal density, we consider the following class of kernels introduced by Di Marzio et al. (2011).

Definition 1 (Toroidal kernels). A d-dimensional toroidal kernel with concentration (smoothing) parameters $\boldsymbol{C} := (\kappa_s \in \mathbb{R}_+, s = 1, \cdots, d)$, is the d-fold product $K_{\boldsymbol{C}} := \prod_{s=1}^{d} K_{\kappa_s}$, where $K_\kappa : \mathbb{T} \to \mathbb{R}$ is such that

(i) It admits an uniformly convergent Fourier series $\{1 + 2\sum_{j=1}^{\infty} \gamma_j(\kappa)\cos(j\theta)\}/(2\pi)$, $\theta \in \mathbb{T}$, where $\gamma_j(\kappa)$ is a strictly monotonic function of κ.
(ii) $\int_{\mathbb{T}} K_\kappa = 1$, and, if K_κ takes negative values, there exists $0 < M < \infty$ such that, for all $\kappa > 0$

$$\int_{\mathbb{T}} |K_\kappa(\theta)|\, d\theta \leq M.$$

(iii) For all $0 < \delta < \pi$,

$$\lim_{\kappa \to \infty} \int_{\delta \leq |\theta| \leq \pi} |K_\kappa(\theta)|\, d\theta = 0.$$

These kernels are continuous and symmetric about the origin, so the d-fold product of von Mises, wrapped normal and the wrapped Cauchy distributions are included. As more general examples, we now list families of densities whose d-fold products are candidates as toroidal kernels:

1. Wrapped symmetric stable family of Mardia (1972, p. 72).
2. The extensions of the von Mises distribution due to Batschelet (1981, p. 288, equation (15.7.3)).
3. The unimodal symmetric distributions in the family of Kato and Jones (2009).
4. The family of unimodal symmetric distributions of Jones and Pewsey (2005).
5. The wrapped t family of Pewsey et al. (2007).

Definition 2 (Sin-order). Given the one-dimensional toroidal kernel K_κ, let $\eta_j(K_\kappa) = \int_{\mathbb{T}} \sin^j(\theta) K_\kappa(\theta) d\theta$. We say that K_κ has *sin-order* q if and only if

$$\eta_j(K_\kappa) = 0, \quad \text{for} \quad 0 < j < q, \quad \text{and} \quad \eta_q(K_\kappa) \neq 0.$$

Note that $K_C := \prod_{s=1}^d K_{\kappa_s}$ has sin-order q if and only if K_{κ_s} has sin-order q for $s \in \{1, \cdots, d\}$. Moreover, observe that the quantity $\eta_j(K_\kappa)$ plays a similar rôle as the jth moment of a kernel in the euclidean theory and its expression is given by the following

Lemma 1. *Given a positive even q, if K_κ has sin-order q, then*

$$\eta_q(K_\kappa) = \frac{1}{2^{q-1}} \left\{ \binom{q-1}{q/2} + \sum_{s=1}^{q/2} (-1)^{q+s} \binom{q}{q/2+s} \gamma_{2s}(\kappa) \right\}.$$

Proof. See Appendix.

Remark 1. Notice that if K_κ has sin-order q, then $\gamma_j(\kappa) = 1$ for each $j < q$, and since $\lim_{\kappa\to\infty} \gamma_q(\kappa) = 1$, it results $\eta_q(K_\kappa) = O\left(\{1 - \gamma_q(\kappa)\}2^{1-q}\right)$.

3 Toroidal Density Estimation

Definition 3 (Kernel estimator of toroidal density). Let $\{\boldsymbol{\Theta}_\ell, \ell = 1, \cdots, n\}$, with $\boldsymbol{\Theta}_\ell := (\Theta_{\ell 1}, \cdots, \Theta_{\ell d}) \in \mathbb{T}^d$, be a random sample from a continuous toroidal density f. The kernel estimator of f at $\boldsymbol{\theta} \in \mathbb{T}^d$ is defined as

$$\hat{f}(\boldsymbol{\theta}; \boldsymbol{C}) := \frac{1}{n} \sum_{i=1}^{n} K_{\boldsymbol{C}}(\boldsymbol{\theta} - \boldsymbol{\Theta}_i). \tag{1}$$

From now on we always assume that $\kappa_s = \kappa$ for each $s \in \{1, \cdots, d\}$, *i.e.* we assume that $\boldsymbol{C}$ is a multiset with element κ and multiplicity d.

Letting $\hat{g}$ be a nonparametric estimator of a square-integrable curve g, the mean integrated squared error (MISE) for $\hat{g}$ is defined by $\mathsf{MISE}[\hat{g}] := \int \mathsf{E}[\{\hat{g}(\boldsymbol{\theta}) - g(\boldsymbol{\theta})\}^2] d\boldsymbol{\theta} = \int \{\mathsf{E}[\hat{g}(\boldsymbol{\theta})] - g(\boldsymbol{\theta})\}^2 d\boldsymbol{\theta} + \int \mathsf{Var}[\hat{g}(\boldsymbol{\theta})] d\boldsymbol{\theta}$. In what follows we will derive a Fourier expansion of the exact MISE for the estimator (1). Before stating the main result, we need to introduce a little notation.

Given $\boldsymbol{j} = (j_1, \cdots, j_d) \in \mathbb{Z}^d$, for a function f defined on $\mathbb{T}^d$ we have

$$f(\boldsymbol{\theta}) = \frac{1}{(2\pi)^d} \sum_{\boldsymbol{j} \in \mathbb{Z}^d} c_{\boldsymbol{j}} e^{i\boldsymbol{j}\cdot\boldsymbol{\theta}}, \tag{2}$$

where $i^2 = -1$, $c_{\boldsymbol{j}} := \int_{\mathbb{T}^d} f(\boldsymbol{\theta}) e^{-i\boldsymbol{j}\cdot\boldsymbol{\theta}} d\boldsymbol{\theta}$, and $\boldsymbol{j} \cdot \boldsymbol{\theta}$ is the inner product of $\boldsymbol{j}$ and $\boldsymbol{\theta}$.

Given the d-dimensional toroidal kernel $K_{\boldsymbol{C}}(\boldsymbol{\theta}) = \prod_{s=1}^d K_\kappa(\theta_s)$, and letting $\gamma_{\boldsymbol{j}}(\boldsymbol{C}) := \int_{\mathbb{T}^d} K_{\boldsymbol{C}}(\boldsymbol{\theta}) e^{-i\boldsymbol{j}\cdot\boldsymbol{\theta}} d\boldsymbol{\theta} = \prod_{s=1}^d \gamma_{j_s}(\kappa)$, the estimator in (1), being the convolution between the empirical version of f and $K_{\boldsymbol{C}}$, can be expressed as

$$\hat{f}(\boldsymbol{\theta};\boldsymbol{C}) = \frac{1}{(2\pi)^d}\sum_{\boldsymbol{j}\in\mathbb{Z}^d}\tilde{c}_{\boldsymbol{j}}\gamma_{\boldsymbol{j}}(\boldsymbol{C})e^{i\boldsymbol{j}\cdot\boldsymbol{\theta}}, \tag{3}$$

where $\tilde{c}_{\boldsymbol{j}} := n^{-1}\sum_{\ell=1}^{n} e^{-i\boldsymbol{j}\cdot\boldsymbol{\theta}_\ell}$.

Theorem 1. *Suppose that both f and K_C belong to $L_2(\mathbb{T}^d)$, then*

$$\mathsf{MISE}\left[\hat{f}(\cdot;\boldsymbol{C})\right] = \frac{1}{n(2\pi)^d}\sum_{\boldsymbol{j}\in\mathbb{Z}^d}\left(1-\|c_{\boldsymbol{j}}\|^2\right)\gamma_{\boldsymbol{j}}^2(\boldsymbol{C})$$
$$+\frac{1}{(2\pi)^d}\sum_{\boldsymbol{j}\in\mathbb{Z}^d}\{1-\gamma_{\boldsymbol{j}}(\boldsymbol{C})\}^2\|c_{\boldsymbol{j}}\|^2.$$

Proof. See Appendix.

Now we derive the asymptotic MISE (AMISE) for the estimator in (1) equipped with a kernel given by the d- fold product of univariate second sin-order toroidal kernels.

Theorem 2. *Given the random sample $\{\boldsymbol{\Theta}_\ell, \ell = 1,\cdots,n\}$, consider the estimator $\hat{f}(\cdot;\boldsymbol{C})$ having as the kernel $K_C := \prod_{s=1}^{d} K_\kappa$, with K_κ being a univariate second sin-order toroidal kernel. If:*

(i) K_κ *is such that* $\lim_{\kappa\to\infty}(1-\gamma_2(\kappa))/(1-\gamma_j(\kappa)) = j^2/4$.
(ii) $\lim_{\kappa\to\infty}\eta_2(K_\kappa) = 0$.
(iii) $\lim_{n\to\infty} n^{-1}(2\pi)^{-d}\{1+2\sum_{i=1}^{\infty}\gamma_j^2(\kappa)\}^d = 0$.
(iv) *the hessian matrix of f at $\boldsymbol{\theta}$, $\boldsymbol{H}_f(\boldsymbol{\theta})$, has entries piecewise continuous and square integrable.*

then

$$\mathsf{AMISE}[\hat{f}(\cdot;\boldsymbol{C})] = \frac{1}{16}\{1-\gamma_2(\kappa)\}^2\int \mathrm{tr}^2\{\boldsymbol{H}_f(\boldsymbol{\theta})\}d\boldsymbol{\theta} + \frac{1}{n}\left\{\frac{1+2\sum_{i=1}^{\infty}\gamma_j^2(\kappa)}{2\pi}\right\}^d$$

Proof. See Appendix.

Remark 2. For circular kernels the expansion of the convolution behaves differently from its euclidean counterpart. In fact, here higher order terms do not necessarily vanish faster for whatever kernel, assumption (i) being required to this end. Such an assumption is satisfied by many symmetric, unimodal densities, even though, surely, not by the wrapped Cauchy. Important cases are given by the wrapped normal and von Mises. But also the class introduced by Batschelet (1981, p. 288, equation (15.7.3)) matches the condition, along with the unimodal symmetric densities in the family introduced by Kato and Jones (2009).

The above result can be easily extended to estimators equipped with higher sin-order toroidal kernels. These latter can be constructed from second-sin-order ones as a direct consequence of the result in Lemma 1. For a discussion on higher sin-order and smaller bias, see Di Marzio et al. (2011).

Now we derive the AMISE-optimal concentration parameter for the estimator having as the kernel $V_C(\boldsymbol{\theta}) := \prod_{s=1}^{d} V_\kappa(\theta_s)$, the d-fold product of von Mises density $V_\kappa(\theta) := \{2\pi \mathscr{I}_0(\kappa)\}^{-1} e^{\kappa \cos(\theta)}$, with $\mathscr{I}_j(\kappa)$ denoting the modified Bessel function of the first kind and order j.

Theorem 3. *Given the random sample $\{\boldsymbol{\Theta}_\ell, \ell = 1, \cdots, n\}$, consider the estimator $\hat{f}(\cdot; \boldsymbol{C})$ having as the kernel $V_C(\boldsymbol{\theta})$. Assume that condition (iv) of Theorem 2 holds, and:*

(i) $\lim_{n\to\infty} \kappa^{-1} = 0$.
(ii) $\lim_{n\to\infty} n^{-1}\kappa^{d/2} = 0$.

then the AMISE optimal concentration parameter for $\hat{f}(\cdot; \boldsymbol{C})$ is

$$\left[\frac{2^d \pi^{d/2} n \int \operatorname{tr}^2\{\boldsymbol{H}_f(\boldsymbol{\theta})\} d\boldsymbol{\theta}}{d}\right]^{2/(4+d)}.$$

Proof. See Appendix.

4 Numerical Evidence

In order to illustrate the methodology and results a small simulation was carried out. Taking f to be a bivariate von Mises distribution (with independent components) with concentration parameters 2 and 4, we simulated 50 datasets for each size of $n = 100, 500, 2, 500$. For each dataset a variety of smoothing parameters (κ) were used in the toroidal density estimate, with kernel V_C, and for each value of κ we compute the average integrated squared error (ISE) (over the 50 simulations) and MISE using (4) in the appendix. In addition, for each value of n we compute the AMISE optimal concentration parameter for $\hat{f}(\cdot; \boldsymbol{C})$ given by Theorem 3.

The results are shown in Fig. 1. It can be seen that ISE decreases with n, and that the approximation of MISE improves as n increases. Finally we note that the optimal smoothing parameter given by Theorem 3 also improves with n.

Appendix

Proof of Lemma 1. First observe that for odd j we have that $\sin^j(\theta)$ is orthogonal in $L_1(\mathbb{T})$ to each function in the set $\{1/2, \cos(\theta), \cos(2\theta), \cdots\}$, which implies that $\eta_j(K_\kappa) = 0$. When j is even, $\sin^j(\theta)$ is not orthogonal

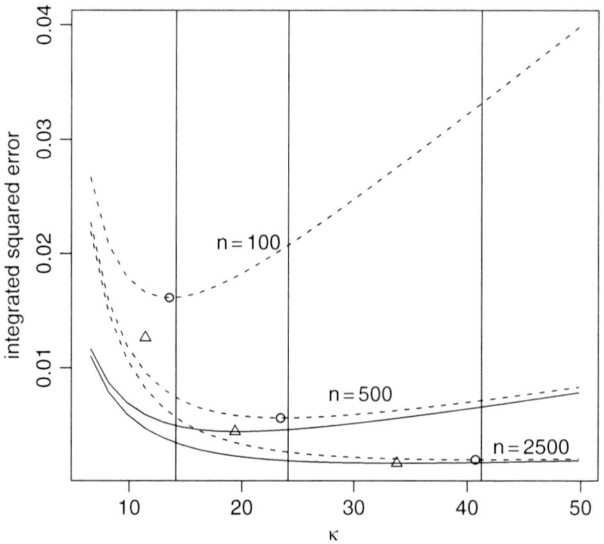

Fig. 1 Asymptotic mean integrated squared error (*dashed lines*) and average integrated squared error (*continuous lines*) over 50 simulations of size n from a bivariate von Mises distribution. The minimum values are shown by *symbols*, and the *vertical lines* represent the AMISE optimal values given by Theorem 3

in $L_1(\mathbb{T})$ to $1/2$ and to the set $\{\cos(2s),\ \ 0 < s \leq j/2\}$, and in particular one has

$$\int_{\mathbb{T}} \frac{\sin^j(\theta)}{2} d\theta = \binom{j-1}{j/2} \frac{\pi}{2^{j-1}} \quad \text{and}$$

$$\int_{\mathbb{T}} \sin^j(\theta)\cos(2s\theta) d\theta = \binom{j}{j/2+s} \frac{(-1)^{j+s}\pi}{2^{j-1}},$$

which gives the result. □

Proof of Theorem 1. First observe that by Parseval's identity $\int_{\mathbb{T}^d}\{f(\boldsymbol{\theta})\}^2 d\boldsymbol{\theta} = (2\pi)^{-d}\sum_{j\in\mathbb{Z}^d}||c_j||^2$, where $||g||$ stands for the L_2 norm of g. Then use the results in (2) and in (3), the identities $\mathsf{E}[\tilde{c}_j] = c_j$, $\mathsf{E}[||\tilde{c}_j - c_j||^2] = n^{-1}(1 - ||c_j||^2)$, and some algebraic manipulations, to get

$$\mathsf{E}\left[\int_{\mathbb{T}^d} \left\{\hat{f}(\boldsymbol{\theta};\boldsymbol{C}) - f(\boldsymbol{\theta})\right\}^2 d\boldsymbol{\theta}\right] = \mathsf{E}\left[\frac{1}{(2\pi)^d}\sum_{j\in\mathbb{Z}^d}\left|\left|\tilde{c}_j\gamma_j(\boldsymbol{C}) - c_j\right|\right|^2\right]$$

$$= \frac{1}{(2\pi)^d} \sum_{j \in \mathbb{Z}^d} \left(\mathsf{E}\left[\|\tilde{c}_j - c_j\|^2 \right] \gamma_j^2(\boldsymbol{C}) + \{1 - \gamma_j(\boldsymbol{C})\}^2 \|c_j\|^2 \right)$$

$$= \frac{1}{n(2\pi)^d} \sum_{j \in \mathbb{Z}^d} \left(1 - \|c_j\|^2\right) \gamma_j^2(\boldsymbol{C}) + \frac{1}{(2\pi)^d} \sum_{j \in \mathbb{Z}^d} \{1 - \gamma_j(\boldsymbol{C})\}^2 \|c_j\|^2.$$

□

Proof of Theorem 2. Put $\boldsymbol{S}_{\boldsymbol{u}} := \{\sin(u_1), \cdots, \sin(u_d)\}^{\mathsf{T}}$, and use $\boldsymbol{D}_f(\boldsymbol{\theta})$ to denote the first-order partial derivatives vector of the function f at $\boldsymbol{\theta}$. Using the expansion $f(\boldsymbol{u}+\boldsymbol{\theta}) = f(\boldsymbol{\theta}) + \boldsymbol{S}_u^{\mathsf{T}} \boldsymbol{D}_f(\boldsymbol{\theta}) + \frac{1}{2} \boldsymbol{S}_u^{\mathsf{T}} \boldsymbol{H}_f(\boldsymbol{\theta}) \boldsymbol{S}_u + O(\boldsymbol{S}_u^{\mathsf{T}} \boldsymbol{S}_u)$, and recalling assumptions (i) and (ii), a change of variables leads to

$$\mathsf{E}[\hat{f}(\boldsymbol{\theta}; \boldsymbol{C})] = \int_{\mathbb{T}^d} K_{\boldsymbol{C}}(\boldsymbol{\psi} - \boldsymbol{\theta}) f(\boldsymbol{\psi}) d\boldsymbol{\psi}$$
$$= f(\boldsymbol{\theta}) + \frac{1}{4}\{1 - \gamma_2(\kappa)\} \mathrm{tr}\{\boldsymbol{H}_f(\boldsymbol{\theta})\} + o(1).$$

Now, recalling assumption (iii), we obtain

$$\mathrm{Var}[\hat{f}(\boldsymbol{\theta}; \boldsymbol{C})] = \frac{1}{n} \int_{\mathbb{T}^d} \{K_{\boldsymbol{C}}(\boldsymbol{\psi} - \boldsymbol{\theta})\}^2 f(\boldsymbol{\psi}) d\boldsymbol{\psi} - \frac{1}{n} \left\{\mathsf{E}[\hat{f}(\boldsymbol{\theta}; \boldsymbol{C})]\right\}^2$$
$$= \frac{1}{n} \int_{\mathbb{T}^d} \{K_{\boldsymbol{C}}(\boldsymbol{u})\}^2 \{f(\boldsymbol{\theta}) + o(1)\} d\boldsymbol{u} - \frac{1}{n}\{f(\boldsymbol{\theta}) + o(1)\}^2$$
$$= \frac{f(\boldsymbol{\theta})}{n} \left\{ \frac{1 + 2\sum_{j=1}^{\infty} \gamma_j^2(\kappa)}{2\pi} \right\}^d + o(1).$$

□

Proof of Theorem 3. First observe that for the von Mises kernel $\gamma_j(\kappa) = \mathscr{I}_j(\kappa)/\mathscr{I}_0(\kappa)$, then follow the proof of Theorem 2 to get

$$\mathrm{MISE}[\hat{f}(\cdot; \boldsymbol{C})] = \frac{1}{4} \left\{ \frac{\mathscr{I}_1(\kappa)}{\kappa \mathscr{I}_0(\kappa)} \right\}^2 \int \mathrm{tr}^2\{\boldsymbol{H}_f(\boldsymbol{\theta})\} d\boldsymbol{\theta} + \frac{1}{n} \left[\frac{\mathscr{I}_0(2\kappa)}{2\pi\{\mathscr{I}_0(\kappa)\}^2} \right]^d + o_p\left(\kappa^{-2} + n^{-1}\kappa^{d/2}\right). \quad (4)$$

Now, replace $\mathscr{I}_1(\kappa)/\mathscr{I}_0(\kappa)$ by 1 with an error of magnitude $O(\kappa^{-1})$, use

$$\lim_{\kappa\to\infty}\left[\frac{\mathscr{I}_0(2\kappa)}{2\pi\{\mathscr{I}_0(\kappa)\}^2}\right]^d = \left(\frac{\kappa}{4\pi}\right)^{d/2},$$

then minimize the leading term of (4). □

References

Z. D. Bai, R. C. Rao, and L. C. Zhao. Kernel estimators of density function of directional data. *Journal of Multivariate Analysis*, 27: 24–39, 1988.

E. Batschelet. *Circular Statistics in Biology*. Academic Press, London, 1981.

R. Beran. Exponential models for directional data. *The Annals of Statistics*, 7: 1162–1178, 1979.

M. Di Marzio, A. Panzera, and C.C. Taylor. Density estimation on the torus. *Journal of Statistical Planning & Inference*, 141: 2156–2173, 2011.

P. Hall, G.S. Watson, and J. Cabrera. Kernel density estimation with spherical data. *Biometrika*, 74: 751–762, 1987.

S. R. Jammalamadaka and A SenGupta. *Topics in Circular Statistics*. World Scientific, Singapore, 2001.

M.C. Jones and A. Pewsey. A family of symmetric distributions on the circle. *Journal of the American Statistical Association*, 100: 1422–1428, 2005.

S. Kato and M.C. Jones. A family of distributions on the circle with links to, and applications arising from, möbius transformation. *Journal of the American Statistical Association*, to appear, 2009.

J. Klemelä. Estimation of densities and derivatives of densities with directional data. *Journal of Multivariate Analysis*, 73: 18–40, 2000.

K. V. Mardia. *Statistics of Directional Data*. Academic Press, London, 1972.

K. V. Mardia and P. E. Jupp. *Directional Statistics*. John Wiley, New York, NY, 1999.

A. Pewsey, T. Lewis, and M. C. Jones. The wrapped *t* family of circular distributions. *Australian & New Zealand Journal of Statistics*, 49: 79–91, 2007.

C. C. Taylor. Automatic bandwidth selection for circular density estimation. *Computational Statistics & Data Analysis*, 52: 3493–3500, 2008.

Markov Bases for Sudoku Grids

Roberto Fontana, Fabio Rapallo, and Maria Piera Rogantin

Abstract In this paper we show how to describe sudoku games under the language of design of experiments, and to translate sudoku grids into contingency tables. Then, we present the application of some techniques from Algebraic Statistics to describe the structure of the sudoku grids, at least for the 4×4 grids. We also show that this approach has interesting applications to both complete grids and partially filled grids.

1 Introduction and Preliminary Material

In recent years, sudoku has become a very popular game. In its most common form, the objective of the game is to complete a 9×9 grid with the digits from 1 to 9. Each digit must appear once only in each column, each row and each of the nine 3×3 boxes. It is known that sudoku grids are special cases of Latin squares in the class of *gerechte designs*, see Bailey et al. (2008). In Fontana and Rogantin (2009) the connections between sudoku grids and experimental designs are extensively studied in the framework of Algebraic Statistics. In this paper we show how to represent sudoku games in terms of $0 - 1$ contingency tables. The connections between contingency tables, design of experiments, and the use of some techniques from Algebraic Statistics allows us to study and describe the set of all sudoku grids.

R. Fontana (✉)
DIMAT Politecnico di Torino, Corso Duca degli Abruzzi 24, 10129 Torino Italy
e-mail: roberto.fontana@polito.it

F. Rapallo
DISTA University of Eastern Piedmont, Viale T. Michel 11, 15121 Alessandria Italy
e-mail: fabio.rapallo@mfn.unipmn.it

M. Piera Rogantin
DIMA Università di Genova, Via Dodecaneso 35, 16146 Genova Italy
e-mail: rogantin@dima.unige.it

A. Di Ciaccio et al. (eds.), *Advanced Statistical Methods for the Analysis of Large Data-Sets*, Studies in Theoretical and Applied Statistics,
DOI 10.1007/978-3-642-21037-2_28,

Although the methodology is simple and can be easily stated for general $p^2 \times p^2$ ($p \geq 2$) sudoku grids, the computations are very intensive and then limited to the 4×4 case. Simply to show the increase of computational complexity, we notice that there are 288 sudoku grids in the 4×4 case, while there are about 6.67×10^{21} grids in the 9×9 case. However, we expect that our work will form a prototype to understand the connections between designed experiments and contingency tables for the general case.

From the point of view of design of experiments, a sudoku can be considered as a fraction of a full factorial design with four factors R, C, B, S, corresponding to rows, columns, boxes and symbols, with p^2 levels each. The three position factors R, C and B are dependent; in fact a row and a column specify a box, but the polynomial relation between these three factors is fairly complicated. A simpler approach consists in splitting row factor R into two pseudo-factors R_1 and R_2, each with p levels, and, analogously, column factor C into C_1 and C_2. Then box factor B corresponds to R_1 and C_1. Factors R_1 and C_1 are named "the band" and "the stack" respectively, and factors R_2 and C_2 are named "the row within a band" and "the column within a stack". Finally, given a digit k between 1 and p^2, factors S_1 and S_2 provide the base-p representation of $k - 1$. It should be noted that two factors for symbols are not essential and they are introduced here because symmetrical designs are easier to handle.

Hence, the full factorial has six factors R_1, R_2, C_1, C_2, S_1 and S_2, each with p levels . In this work we keep our exposition simple by using the integer coding $0, \ldots, p - 1$ for the p levels .

As an example, in a 4×4 sudoku, if the symbol 3 (coded with 10, the binary representation of 2) is in the second row within the first band ($R_1 = 0$, $R_2 = 1$) and in the first column of the second stack ($C_1 = 1, C_2 = 0$), the corresponding point of the design is $(0, 1, 1, 0, 1, 0)$.

As introduced in Fontana and Rogantin (2009), a sudoku grid, as a fraction of a full factorial design, is specified through its indicator polynomial function, and a move between two grids is also a polynomial. With such a methodology, three classes of moves are described. The first class, denoted by $\mathcal{M}_1$, contains permutations of symbols, bands, rows within a band, stacks and columns within a stack. The second class, denoted by $\mathcal{M}_2$, contains the transposition between rows and columns. The last class, denoted by $\mathcal{M}_3$, is formed by all the other moves. The moves in $\mathcal{M}_3$ have a more complex behavior and we refer to (Fontana and Rogantin, 2009, Sect. 12.4.2) for a formal definition. We mention that the moves in $\mathcal{M}_1$ and in $\mathcal{M}_2$ can be applied to all sudoku grids, while the moves in $\mathcal{M}_3$ can be applied only to sudoku grids having a special pattern. Here we give only an example of a move in $\mathcal{M}_3$, see Fig. 1. This move acts on two parts of the grid defined by the intersection of a stack with two rows belonging to different boxes, and it exchanges the two symbols contained in it. But such a move is possible only when the symbols in each selected part are the same. Similar moves are defined for bands and columns, respectively. The relevance of the moves in $\mathcal{M}_3$ will be discussed in the next sections.

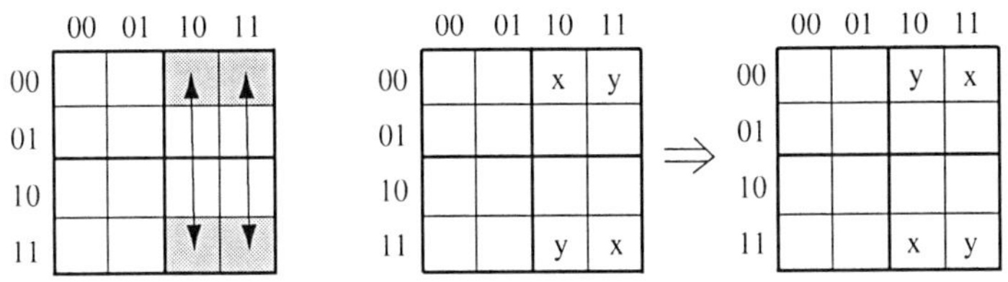

Fig. 1 A move in the class $\mathcal{M}_3$

2 Moves and Markov Bases for Sudoku Grids

The application of statistical techniques for contingency tables in the framework of the design of experiments, and in particular to sudoku grids, is not straightforward. In fact, a sudoku grid is not a contingency table as it contains labels instead of counts. To map a sudoku grid into a contingency table, we need to consider a $p \times p \times p \times p \times p \times p$ table $\mathbf{n}$, with 6 indices r_1, r_2, c_1, c_2, s_1 and s_2, each ranging between 0 and $p-1$. The table $\mathbf{n}$ is a $0-1$ table with $n_{r_1 r_2 c_1 c_2 s_1 s_2} = 1$ if and only if the (r_1, r_2, c_1, c_2) cell of the grid contains the symbol (s_1, s_2) and otherwise it is 0. Notice that $n_{r_1 r_2 c_1 c_2 s_1 s_2}$ is the value of the indicator function of the fraction computed in the design point $(r_1, r_2, c_1, c_2, s_1, s_2)$. This approach has already been sketched in Fontana and Rogantin (2009). A similar approach is also described in Aoki and Takemura (2008) for different applications.

In a sudoku grid each row, each column and each box must contain different symbols. Such constraints translate into the following linear conditions on the corresponding $0-1$ contingency table $\mathbf{n}$:

$$\sum_{s_1, s_2=0}^{p-1} n_{r_1 r_2 c_1 c_2 s_1 s_2} = 1 \ \forall r_1, r_2, c_1, c_2 \qquad \sum_{c_1, c_2=0}^{p-1} n_{r_1 r_2 c_1 c_2 s_1 s_2} = 1 \ \forall\ r_1, r_2, s_1, s_2$$

$$\sum_{r_1, r_2=0}^{p-1} n_{r_1 r_2 c_1 c_2 s_1 s_2} = 1 \ \forall\ c_1, c_2, s_1, s_2 \qquad \sum_{r_2, c_2=0}^{p-1} n_{r_1 r_2 c_1 c_2 s_1 s_2} = 1 \ \forall\ r_1, c_1, s_1, s_2 \, .$$

This system with $4p^4$ linear conditions can be written in the form $A\mathbf{n} = \mathbf{1}$, where A is the appropriate matrix of coefficients. The valid sudoku grids are just its integer non-negative solutions.

Given an integer linear system of equations of the form $A\mathbf{n} = \mathbf{b}$, a move is an integer (possibly negative) table $\mathbf{m}$ such that $A\mathbf{m} = \mathbf{0}$. By linearity, it follows that $A(\mathbf{n} \pm \mathbf{m}) = A\mathbf{n} = \mathbf{b}$. Thus, if $\mathbf{m}$ is a move, and $\mathbf{n}$ is a non-negative solution of the system, then $\mathbf{n}+\mathbf{m}$ and $\mathbf{n}-\mathbf{m}$ are again solutions of the system, when non-negative. A move allows us to move from one solution to another one.

As introduced in Diaconis and Sturmfels (1998), a Markov basis is a finite set of moves $\mathscr{B} = \{\mathbf{m}_1, \dots, \mathbf{m}_k\}$ which connects all the non-negative integer solutions of the system $A\mathbf{n} = \mathbf{b}$. A key result in Algebraic Statistics states that a finite Markov basis always exists, see (Diaconis and Sturmfels, 1998, p. 376). In our framework, given any two sudoku $\mathbf{n}$ and $\mathbf{n}'$, there exists a sequence of moves $\mathbf{m}_{i_1}, \ldots, \mathbf{m}_{i_H}$ in $\mathscr{B}$ and a sequence of signs $\epsilon_{i_1}, \ldots, \epsilon_{i_H}$ ($\epsilon_{i_j} = \pm 1$ for all j) such that

$$\mathbf{n}' = \mathbf{n} + \sum_{j=1}^{H} \epsilon_{i_j} \mathbf{m}_{i_j}$$

and all the intermediate steps

$$\mathbf{n} + \sum_{j=1}^{h} \epsilon_{i_j} \mathbf{m}_{i_j} \quad \text{for all } h = 1, \ldots, H$$

are again non-negative integer solutions of the linear system.

While all the linear constraints in our problem have constant terms equal to 1, it is known that the computation of a Markov basis is independent on the constant terms of the linear system, see e.g. Diaconis and Sturmfels (1998). Therefore, we can select the subset $\mathscr{B}_+^f$ of $\mathscr{B}$ of the moves $\mathbf{m}$ of $\mathscr{B}$ that can be added at least to one sudoku grid, $\mathbf{n}$, in order to obtain another valid sudoku grid, $\mathbf{n}+\mathbf{m}$. In the same way, we denote with $\mathscr{B}_-^f$ the subset of $\mathscr{B}$ formed by the elements that can be subtracted by at least one sudoku grid to obtain yet another valid sudoku grid. It is easy to see that $\mathscr{B}_+^f = \mathscr{B}_-^f$. Thus, we denote this set simply with $\mathscr{B}^f$ and we refer to the moves in $\mathscr{B}^f$ as *feasible moves*.

The approach with Markov bases enables us to connect each pair of sudoku grids and to generate all the sudoku grids starting from a given one.

The actual computation of a Markov basis needs polynomial algorithms and symbolic computations. We refer to Drton et al. (2009) for more details on Markov bases and how to compute them. Although in many problems involving contingency tables Markov bases are small sets of moves, easy to compute and to handle, in the case of sudoku grids we have a large number of moves. Currently, the problem is computationally not feasible already in the case of classical 9×9 grids.

3 The 4 x 4 Sudoku

In this section we consider the case of 4×4 sudoku, i. e., $p = 2$. Using `4ti2` (2007), we obtain the Markov basis $\mathscr{B}$. It contains 34, 920 elements while it is known that there are only 288 4×4 sudoku grids, listed in Fontana and Rogantin (2009). Using such list and some ad-hoc modules written in `SAS-IML` (2004), we have explored this Markov basis finding some interesting facts.

- *Feasible moves.* Among the 34, 920 moves of the Markov basis $\mathscr{B}$ there are only 2, 160 feasible moves. To provide a term of comparison, we note that the cardinality of the set of all the differences between two valid sudoku is $288 \cdot 287/2 = 41,328$; we have checked that $39,548$ of them are different.

- *Classification of moves.* If we classify each of the $2,160$ moves generated by $\mathcal{B}^f$ according to both the number of sudoku that can use it and the number of points of the grids that are changed by the move itself, we obtain the following table.

# of Sudoku that	# of Points moved				Total
can use the move	4	8	10	12	
1	0	0	1,536	192	1,728
2	0	336	0	0	336
8	96	0	0	0	96
Total	96	336	1,536	192	2,160

  - The 96 moves that change 4 points and can be used 8 times are all of type $\mathcal{M}_3$, like the one reported in Fig. 1. We have also verified that these moves are enough to connect all the sudoku grids.
  - The 336 moves that changes 8 points and can be used by 2 tables correspond to an exchange of two symbols, rows and columns. For example, two of them are:

	4	3	
	3		4
4			3
3		4	

$\Rightarrow$

	3	4	
	4		3
3			4
4		3	

4	2	3	1
3	1	2	4

$\Rightarrow$

3	1	2	4
4	2	3	1

  - The remaining $1,728$ moves are suitable compositions of the previous ones. For instance, the move

3		4	1
	1	3	2
2	3		4
1	4	2	

$\Rightarrow$

1		3	4
	3	2	1
3	4		2
2	1	4	

is the composition of a permutation of two symbols, a permutation of two rows and two moves of type $\mathcal{M}_3$.

4 Partially Filled 4 x 4 Grids

Usually, a classical sudoku game is given as a partially filled grid that must be completed placing the right symbols in the empty cells. In the frame of the design of experiments, this procedure corresponds to augment an existing design under suitable properties, in this case those of the *gerechte designs*. For the person who prepares the sudoku grids, it is important to know where to place the givens and which symbols to put in them in order to obtain a grid with a unique completion. The symbols in the non-empty cells in a sudoku game are referred to as *givens*. In this section, we use the Markov bases to study all possible givens for the 4×4 grids.

Notice that the definitions of Markov bases and feasible moves lead us to the following immediate result. When the Markov basis corresponding to a pattern of givens has no moves, then the partially filled grid can be completed in a unique way.

When some cells of the grid contain givens, we have to determine a Markov basis which does not act on such cells. This problem is then reduced to the computation of Markov bases for tables with structural zeros, as each given fixes 4 cells of the table $\mathbf{n}$.

The computation of Markov bases for tables with structural zeros is a known problem, see e.g. Rapallo (2006). As suggested in Rapallo and Rogantin (2007), a way to solve this problem is based on the notion of *Universal Markov basis*, a set of moves with special properties. Unfortunately, the dimensions of the problem make the computation of the Universal Markov bases currently unfeasible.

A different approach is to compute and analyze the Markov basis for all the possible choices C of the cells of the grid that should not be modified. For 4×4 sudoku it means that we should run `4ti2` over 2^{16} configurations corresponding to all the possible subsets of the cells in the grid. To reduce the computational effort we have exploited some symmetries of the sudoku grids, creating a partition of all the possible Cs and computing the Markov basis $\mathcal{B}_C$ only for one representative of each class. An approach to sudoku based on symmetries is also presented in Dahl (2009).

The considered symmetries correspond to moves in $\mathcal{M}_1$ (permutations of bands, of rows within a band, of stacks and of columns within a stack) and in $\mathcal{M}_2$ (transposition) described in Sect. 1, moves that can be applied to all the sudoku grids. Here we describe the construction of the classes of equivalence.

Let π_{e_1,e_2,e_3,e_4} be the transformation acting on the position of a cell:

$$\pi_{e_1,e_2,e_3,e_4}(r_1, r_2, c_1, c_2) = (r_1, r_2, c_1, c_2) + (e_1, e_2, e_3, e_4) \bmod 2 \quad \text{with } e_i \in \{0, 1\} .$$

The permutation of a band corresponds to $(e_1, e_2, e_3, e_4) = (1, 0, 0, 0)$, the permutation of the rows within both the bands corresponds to $(e_1, e_2, e_3, e_4) = (0, 1, 0, 0)$, and the composition of both the permutations corresponds to $(e_1, e_2, e_3, e_4) = (1, 1, 0, 0)$, analogously for stacks and columns.

Let γ_e, $e \in \{0, 1\}$, be the transposition of the position of a cell, if $e = 1$, or the identity, if $e = 0$:

$$\gamma_e(r_1, r_2, c_1, c_2) = \begin{cases} (r_1, r_2, c_1, c_2) & \text{if } e = 0 \\ (c_1, c_2, r_1, r_2) & \text{if } e = 1 . \end{cases}$$

Given $\mathbf{e} = (e_0, e_1, e_2, e_3, e_4, e_5)$, let $\tau_{\mathbf{e}}$ be the composition:

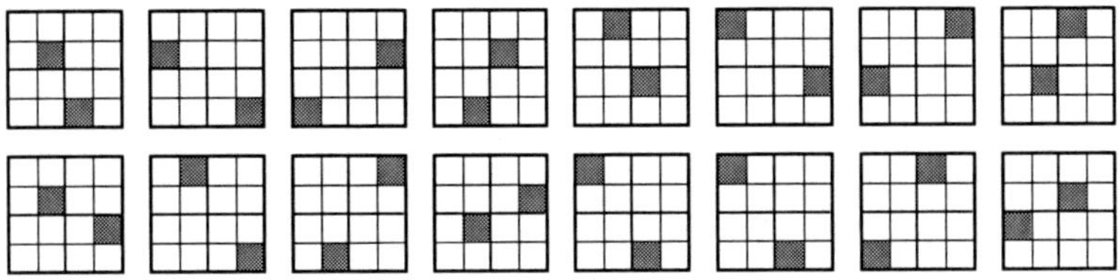

Fig. 2 The 16 grids of an equivalence class with 2 fixed cells, containing the givens

$$\tau_{\mathbf{e}} = \tau_{e_0,e_1,e_2,e_3,e_4,e_5} = \gamma_{e_0} \circ \pi_{e_1,e_2,e_3,e_4} \circ \gamma_{e_5} \quad \text{with } e_i \in \{0,1\} .$$

We notice that the transformation $\tau_{0,0,0,0,0,0}$ is the identity and that the transposition γ_e is considered both the first and the last term in $\tau_{\mathbf{e}}$ because, in general, it does not commute with π_{e_1,e_2,e_3,e_4}. We also point out that the 64 transformations $\tau_{\mathbf{e}}$ do not necessarily cover all the possible transformations but they lead to significant reduction of the problem.

Given a subset of cells C, we denote by $\tau_{\mathbf{e}}(C)$ the transformation of all the cells of C. We say that two choices C_k and D_k, with k fixed cells, are equivalent if there exists a vector $\mathbf{e}$ such that:

$$D_k = \tau_{\mathbf{e}}(C_k)$$

and we write $C_k \sim D_k$. In Fig. 2 a class of equivalence of grids is shown.

Here we show that it is enough to compute the Markov basis only for one representative of each class.

Given a sudoku table $\mathbf{n}$ we denote by $\tilde{\tau}_{\mathbf{e}}(\mathbf{n})$ the transformation $\tau_{\mathbf{e}}$ applied to all the cells of the sudoku:

$$\tilde{\tau}_{\mathbf{e}}(n_{r_1,r_2,c_1,c_2,s_1,s_2}) = n_{\tau_{\mathbf{e}}(r_1,r_2,c_1,c_2),s_1,s_2} \, \forall \; r_1, r_2, c_1, c_2 .$$

In the same way $\tilde{\tau}_{\mathbf{e}}$ can be applied to a move $\mathbf{m}$.

Let C and D be in the same class of equivalence, $D = \tau_{\mathbf{e}}(C)$, and $\mathscr{B}_C^f$ and $\mathscr{B}_D^f$ be the corresponding sets of feasible moves in the Markov bases obtained from C and D. Then:

$$\mathscr{B}_D^f = \tilde{\tau}_{\mathbf{e}}\left(\mathscr{B}_C^f\right) \quad \text{with } \tilde{\tau}_{\mathbf{e}}\left(\mathscr{B}_C^f\right) = \left\{\tilde{\tau}_{\mathbf{e}}(\mathbf{m}),\ \mathbf{m} \in \mathscr{B}_C^f\right\} .$$

In fact, given $\mathbf{m} \in \mathscr{B}_C^f$, it follows that there exist a sudoku $\mathbf{n}$ and a sign ϵ such that $\mathbf{n} + \epsilon\ \mathbf{m}$ is still a sudoku and:

$$\begin{aligned}
\tilde{\tau}_{\mathbf{e}}(\mathbf{n} + \epsilon\ \mathbf{m}) &= \tilde{\tau}_{\mathbf{e}}\left((n + \epsilon\ m)_{r_1,r_2,c_1,c_2,s_1,s_2}\right) \\
&= (n + \epsilon\ m)_{\tau_{\mathbf{e}}(r_1,r_2,c_1,c_2),s_1,s_2} \\
&= n_{\tau_{\mathbf{e}}(r_1,r_2,c_1,c_2),s_1,s_2} + \epsilon\ m_{\tau_{\mathbf{e}}(r_1,r_2,c_1,c_2),s_1,s_2} \\
&= \tilde{\tau}_{\mathbf{e}}(\mathbf{n}) + \epsilon\ \tilde{\tau}_{\mathbf{e}}(\mathbf{m}) .
\end{aligned}$$

Therefore $\tilde{\tau}_{\mathbf{e}}(\mathbf{m})$ is a feasible move for the sudoku $\tilde{\tau}_{\mathbf{e}}(\mathbf{n})$. Moreover as $\mathbf{m}$ does not act on the cells C, $\tilde{\tau}_{\mathbf{e}}(\mathbf{m})$ does not act on the cells $\tau_{\mathbf{e}}(C)$. It follows that $\tilde{\tau}_{\mathbf{e}}(\mathbf{m})$ is in $\mathscr{B}^f_{\tau_{\mathbf{e}}(C)}$.

This methodology allows us to significantly reduce the computation, approximately of 96%, as summarized in the following table, where k is the number of fixed cells and $n_{\text{eq.cl.}}$ is the number of equivalence classes.

k	1–15	2–14	3–13	4–12	5–11	6–10	7–9	8	Total
$\binom{16}{k}$	16	120	560	1,820	4,368	8,008	11,440	12,870	65,534
$n_{\text{eq.cl.}}$	1	9	21	78	147	291	375	456	2,300

In view of this reduction, first, using `4ti2`, we have computed a Markov basis $\mathscr{B}_C$ for one representative C of each of the 2,300 equivalence classes. Then, using some ad hoc modules in SAS-IML, we have selected the feasible moves and obtained $\#\mathscr{S}_C$, the number of sudoku that can use at least one of them. As discussed above, if for a partially filled grid $\mathscr{S}$ the cardinality of the subset of the Markov basis made by the feasible moves is equal to zero, then $\mathscr{S}$ can be completed in a unique way.

The following table displays the results of all the 2,300 runs. It cross-classifies each pattern C with respect to the number k of givens (column) and the number of sudoku $\#\mathscr{S}_C$ that can use at least one move (row).

$\#\mathscr{S}_C \setminus k$	1	2	3	4	5	6	7	8	9	10	11	12	13	14	15
0	0	0	0	0	2	21	92	221	271	250	141	76	21	9	1
24	0	0	0	0	0	8	7	5	0	0	0	0	0	0	0
72	0	0	0	0	3	3	0	0	0	0	0	0	0	0	0
96	0	0	0	0	0	50	158	186	96	40	6	2	0	0	0
120	0	0	0	0	4	22	4	0	0	0	0	0	0	0	0
168	0	0	0	7	29	86	56	18	3	0	0	0	0	0	0
192	0	0	0	2	18	61	50	24	5	1	0	0	0	0	0
216	0	0	0	0	7	0	0	0	0	0	0	0	0	0	0
240	0	0	0	4	43	16	2	0	0	0	0	0	0	0	0
264	0	0	0	16	8	2	0	0	0	0	0	0	0	0	0
288	1	9	21	49	33	22	6	2	0	0	0	0	0	0	0
Total	1	9	21	78	147	291	375	456	375	291	147	78	21	9	1

We can highlight some interesting facts.

- With 1, 2 or 3 fixed cells all the 288 sudoku can use at least one move, and therefore no choice of givens determines the completion of the grid univocally.
- With 4 fixed cells there are 78 patterns (or more precisely equivalent classes of patterns):

- For 49 patterns each of 288 sudoku can use at least one move, so that the completion of the grid is not unique.
- For 16 patterns there are 264 sudoku that can use at least one move of $\mathcal{B}_C$. The other $288 - 264 = 24$ choices of givens determine the completion of the grid univocally.
- Analogously there are 4 patterns (2 and 7 respectively) for which there are $288 - 240 = 48$ choices of givens that determine the completion of the grid univocally ($288 - 192 = 96$ and $288 - 168 = 120$ respectively).

Here we have the verification that the minimum number of givens for the uniqueness of the completion is 4.

- With 5 fixed cells there are 2 patterns for which $\#\mathcal{S}_C = 0$ that means that any choice of givens determines the completion of the grid univocally.
- With 8 fixed cells there are 2 patterns for which $\#\mathcal{S}_C = 288$ that means that any choice of givens do not determine the completion of the grid univocally. Nevertheless for each pattern with 9 fixed cells there is a choice of givens which makes unique the completion of the grids. Then, the maximum number of fixed cells for which any choice of givens do not determine the completion of the grid univocally is 8.
- With 12 fixed cells there are 2 patterns for which 96 choices of givens do not determine the completion of the grid univocally.
- With 13, 14 and 15 fixed cells any choice of givens determines the completion of the grid univocally.

Figure 3a shows that the same pattern of 4 cells (the shades ones) leads to a unique solution, if the givens are chosen like in the left part of the figure, or to more than one solution, if the givens are chosen like in the right part of the figure. Figure 3b is analogous to Fig. 3a but considers a pattern of 12 cells.

Figure 4 shows a pattern of 5 cells for which any choice of the givens corresponds to a unique solution.

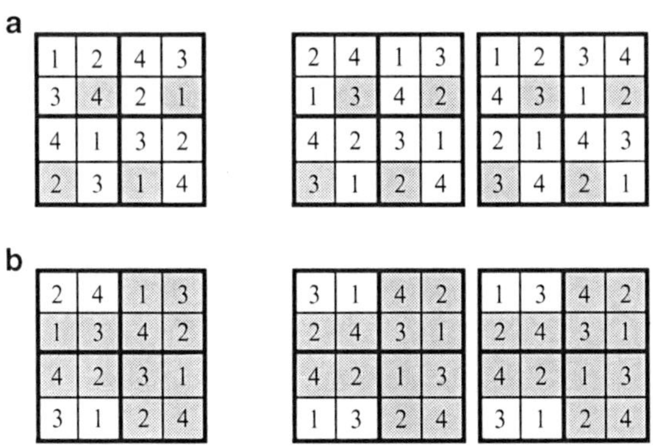

Fig. 3 Fixed a pattern, different choices of givens produce or not the uniqueness. Patterns with 4 and 12 fixed cells

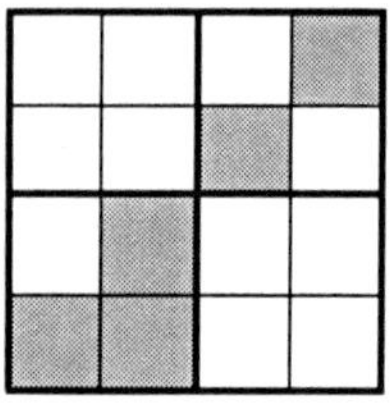

Fig. 4 A pattern of 5 cells for which any choice of the givens produces a unique solution

5 Further Developments

The use of Markov bases has allowed to study the moves between 4×4 sudoku grids, with nice properties for the study of partially filled grids. However, our theory has computational limitations when applied to standard 9×9 grids. Therefore, further work is needed to make our methodology and algorithms actually feasible for 9×9 grids. In particular, we will investigate the following points:

(a) To simplify the computation of Markov bases, using the special properties of the sudoku grids. For instance, some results in this direction is already known for design matrices with symmetries, see Aoki and Takemura (2008), and for contingency tables with strictly positive margins, see Chen et al. (2010).
(b) To characterize the feasible moves theoretically. In fact, in our computations the selection of the feasible moves and the results in Sect. 4 are based on the knowledge of the complete list of sudoku grids. This approach is then unfeasible in the 9×9 case.
(c) To make easy the computation of the Universal Markov basis for our problem, in order to avoid explicit computations for the study of the sudoku grids with givens.

Acknowledgements We thank both the anonymous Referees for their helpful suggestions. The work has been partially funded by an INdAM-GNAMPA project.

References

Aoki, S., Takemura, A.: The largest group of invariance for Markov bases and toric ideals. J. Symb. Comput. **43**(5), 342–358 (2008)
Bailey, R.A., Cameron, P.J., Connelly, R.: Sudoku, Gerechte Designs, Resolutions, Affine Space, Spreads, Reguli, and Hamming Codes. Amer. Math. Monthly **115**(5), 383–404 (2008)
Chen, Y., Dinwoodie, I.H., Yoshida, R.: Markov chains, quotient ideals, and connectivity with positive margins. In: P. Gibilisco, E. Riccomagno, M.P. Rogantin, H.P. Wynn (eds.) Algebraic and Geometric Methods in Statistics, pp. 99–110. Cambridge University Press (2010)
Dahl, G.: Permutation matrices related to Sudoku. Linear Algebra Appl. **430**(8-9), 2457–2463 (2009)
Diaconis, P., Sturmfels, B.: Algebraic algorithms for sampling from conditional distributions. Ann. Statist. **26**(1), 363–397 (1998)
Drton, M., Sturmfels, B., Sullivant, S.: Lectures on Algebraic Statistics. Birkhauser, Basel (2009)

Fontana, R., Rogantin, M.P.: Indicator function and sudoku designs. In: P. Gibilisco, E. Riccomagno, M.P. Rogantin, H.P. Wynn (eds.) Algebraic and Geometric Methods in Statistics, pp. 203–224. Cambridge University Press (2010)

Rapallo, F.: Markov bases and structural zeros. J. Symb. Comput. **41**(2), 164–172 (2006)

Rapallo, F., Rogantin, M.P.: Markov chains on the reference set of contingency tables with upper bounds. Metron **65**(1), 35–51 (2007)

SAS Institute Inc.: SAS/IML®9.1 User's Guide. Cary, NC: SAS Institute Inc. (2004)

4ti2 team: 4ti2–a software package for algebraic, geometric and combinatorial problems on linear spaces. Available at www.4ti2.de (2007)

Part VIII
Application in Economics

Estimating the Probability of Moonlighting in Italian Building Industry

Maria Felice Arezzo and Giorgio Alleva

Abstract It's well known that black market economy and especially undeclared work undermines the financing of national social security programs and hinders efforts to boost economic growth. This paper goes in the direction of shedding light on the phenomenon by using statistical models to detect which companies are more likely to hire off the books workers. We used database from different administrative sources and link them together in order to have an informative system able to capture all aspects of firms activity. Afterward we used both parametric and non parametric models to estimate the probability of a firm to use moonlighters. We have chosen to study building industry both because of its importance in the economy of a country and because its a wide spread problem in that sector

1 Introduction

Moonlighting could be referred either as a multiple-job worker or as a single-job worker in an irregular position because s/he is unknown to the Government since s/he was not regularly registered as an employee. In this paper we focus our attention on the latter and in particular we try to estimate the probability that a firm hires moonlighters. The main idea behind our job is that the probability of interest could be estimated from inspections results of the National Social Security Institute (INPS in the following) and of the Department of Labor. The inspections output are expressed in terms of presence/absence of irregular work. To understand which factors have a major impact on moonlighting, we relate it to a wide variety of firm's characteristics (such as firm localization, labor productivity, labor cost, economic

M.F. Arezzo (✉) · G. Alleva
Methods and Models for Economics, Territory and Finance (MEMOTEF),
University of Rome, Italy
e-mail: mariafelice.arezzo@uniroma1.it; giorgio.alleva@uniroma1.it

A. Di Ciaccio et al. (eds.), *Advanced Statistical Methods for the Analysis of Large Data-Sets*, Studies in Theoretical and Applied Statistics,
DOI 10.1007/978-3-642-21037-2_29,

activity of the firm and so forth). The first researches on moonlighting in Italy go back at the early 1970s. In one of the most important, due to Censis in 1976, an estimate of about three million moonlighters was found. Censis pointed out on one side at the off shoring of manufacturing firms and on the other side at the increased labor cost and to the improved labor conditions to explain the phenomenon. In the 1980s and 1990s moonlighting has slowly but constantly increased until a turning point in 2002, probably due to Bossi-Fini's law, when it has decreased to the same level of 1992. More details on the subject can be found in Alleva et al. (2009b). The entity of moonlighting and its persistency in Italian economy leads us to think that it cannot be regarded as temporary and it highlights the importance of having tools able to track down firms with higher probability to hire irregular workers. Agriculture and Tertiary are traditionally the most exposed sectors (in 2006 the percentage of moonlighters was 18.9% and 13.7% respectively). Within tertiary, commerce, transportation and tourism are far more affected than other sectors. In Industry moonlighting is marginal (3.7%) except for building and construction companies where the average irregularity percentage in 2006 was 11%.

The paper is organized as follows. In Sect. 2 we illustrate the main characteristics of the original datasets and we will explain which ones we used and how we link them. We also report the main results obtained with exploratory data analysis and the evaluation of the quality of the integrated database. In Sect. 3 we will show the main results we had using logistic models and classification and regression trees (CART in the following). A broad description of CART algorithm can be found in Breiman et al. (1983). Conclusion and some points to be developed are discussed in Sect. 4.

2 Creation of the Data Set and the Variables Used

To detect moonlighting, we need to pool firms through common behavioral patterns. In other words we have to find which characteristics construction companies which use irregular employee have in common. We needed two kind of information: the first one is weather a firm has or hasn't moonlighters, the second are all its economic, financial and structural characteristics. It does not exist at the moment an unique informative system that gathers all the information needed; therefore we had first of all to identify useful informative sources and then to combine them in a single dataset. The first kind of information (presence/absence of moonlighters) could be found in inspections archives of INPS and Department of Labor whereas the second ones (firms features) are available in different archives. Table 1 illustrates their main characteristics.

After a thorough data quality assessment, Department of Labor archives was discarded for its information to be improved (data are absent for whole regions and a high percentage of firms doesn't have any identification code). A future use of this data might be possible only if the identification of those firms without VAT code is carried out using other information like firm address and city of location.

Table 1 Datasets characteristics

Data owner	Content	Individual	Dimension
INPS	Inspections outputs (2005–2006)	Inspection	31658 inspections on 28731 firms
Department of Labor	Inspections outputs (2005–2006)	Inspection	62047 inspections
Revenue Agency	Studi di settore (2005–2006).Models: TG69U, TG75U (SG75U),TG50U (SG50U and SG71U), TG70U	Firm	Universe of firms with at most 5 million euros of income
ISTAT	Asia Archives (2005–2006)	Firm	Universe of firms
ISTAT	Balance sheets of Stock and Ltd companies (2005–2006)	Firm	Universe of company/corporation

Information on balance sheets were not used either because they are not available for sole proprietorship enterprises. Data assessment showed that satisfactory territorial and sector coverage are guaranteed in the archives used i.e INPS, ASIA and Revenue Agency. Record linkage was done using VAT numbers and/or tax codes. The original variables were used to build economic indicators which can be grouped in the following different firm's facets: (a) 9 indicators for economic dimension, (b) 13 for organization, (c) 6 for structure, (d) 6 for management, (e) 11 for performance (f) 38 for labor productivity and profitability (g) 3 for contracts award mode (h) 7 variables for location and type. The final dataset had 93 independent variables observed on 14651 building companies with a match rate of 51%. The variable to be predicted is the inspection result which take value 1 if there is at least one moonlighter and 0 otherwise. In the following we will refer to the final dataset as the integrated db because it gathers and integrate information from different sources.

After an in-depth exploratory data analysis (Alleva et al. 2009a,b) we found the following results to be the most interesting:

1. The probability of finding a firm with irregular workers is 22.0%; this probability is defined as the ratio between the number of firms which use moonlighters and the number of inspected firms. Moonlighting is spread out over the whole country, with some peaks in certain regions. The territory with the highest rate of irregularity are Benevento (65.2%), Avellino (48.2%), Rimini (46.6%) and Trapani (45.5%) whereas the lowest are in Catanzaro (3.8%), Imperia (3.8%), Matera (4.7%), Varese (5.6%) and Taranto (5.8%).
2. Firms specialized in duties which requires low skilled labor have a higher probability of using irregular work. In more details Building completion and Civil engineering have respectively 24.5% and 23.4% of moonlighting probability whereas minimum values are reached by Building installation (12.9%) and Site preparation (12.4%).
3. Sole proprietorship and cooperative firms have the highest risk of using irregular labor.

4. Small firms (turnover lower than 50000 euros per year) have the highest risk of using irregular labor.
5. There is a high positive correlation between inspection coverage (number of inspected firms out of active firms) and probability of moonlighting revealing a very efficient inspection activity.

2.1 The Assessment of the Integrated Dataset

As we said we built up our database linking information from INPS, ASIA and Revenue Agency datasets, obtaining a matching rate of 51% which means that we had information on features of interest only for (roughly) half of the firms in original INPS database. In order to evaluate how this loss of information might affect the validity of our models, we conducted a comparative study on the behavior of feature of interest for matched and unmatched firms. In particular we studied inspection coverage and probability of moonlighting for different turnover class and corporate designation typologies and over the territory. We wanted to understand if matched and unmatched data were similar in terms of inspection coverage and probability of moonlighting. If the answer is positive, it means that we didn't exclude firms systematically different from those we built models on. To verify datasets similarity we tested for equality of proportions from two independent samples. In more detail, lets p_i^u and p_i^m be, respectively, the probabilities of moonlighting in unmatched and matched data for level i of a variable of interest. For example we wanted to know if in *any* of the 20 Italian regions the matched and unmatched probability of moonlighting are the same. In this case $i = 1, 2....20$. The significance test is the usual two independent sample proportion test:

$$\begin{array}{l} H_0 : p_i^u = p_i^m \\ H_1 : p_i^u > p_i^m \end{array} \quad i = 1, 2... \tag{1}$$

We checked for:

1. Regions (20 levels)
2. Number of employee (9 classes)
3. Legal structure (5 levels)
4. Turnover (11 classes)

Globally we conducted 45 significance tests on moonlighting probability. We have done the same for inspection coverage. Our results indicates that the major part of these tests are not significant which means that matched and unmatched firms are equivalent in terms of inspection coverage and moonlighting probability. We therefore felt that we could use the sub-sample of matched firms to fit models without obtaining biased results.

3 The Estimation of Probability of Moonlighting

The aim is to associate to any firm a measure of irregularity risk, i.e. the probability of using undeclared workers. This probability can be estimated through statistical models that relate the inspection result (presence/absence of irregular workers) to firm's facets. We used both parametric (logistic) and nonparametric (CART) models. After the information system was built, we wanted three tasks to be achieved before fitting any model: (1) investigate the contribute of each firm's facet on moonlighting, (2) verify the predicting power of the information system (3) reduce the number of independent variables still maintaining as much predicting ability as possible. Goal one was achieved performing a principal component analysis on every single firm's facet considered and then building a tree using only the main factors extracted. Goal two was achieved building a tree with all 93 independent variables. The misclassification rate was used as a measure of informative system predictive power. Variables importance (in terms of impurity reduction capability) and some logical requirements, were used to select a subset of independent variables and therefore to achieve goal three. On the selected variables subset, both logistic and CART models were fitted. The reason why we decided to use both parametric and non parametric models are essentially to exploit their respective advantages. In more details, CART easily allows for many independent variables to be included in the model, even highly correlated ones, and it generally gives better predictive performances than parametric models like logit or probit (Berry and Linoff 2004; Breiman et al. 1983). The main disadvantage of CART is that it doesn't measure the impact of each independent variable on the probability of moonlighting. To obtain this important information we used the logistic model.

3.1 Non Parametric Models: CART

In order to reduce the number of independent variables to use both in parametric and non parametric models and to better understand which are the facets that more than other have an impact on moonlighting, we built two kind of trees. The first one, named extended model, considers all 93 variables and the second one is build using as independent variables the main factors extracted from the principal component analysis (PCA) we run over the six firm aspects we considered. The factors considered were only those with associated eigenvalue greater than one. In Table 2 we summarize the most important results obtained from PCA.

We built several trees using the 21 factors listed in Table 2 and the location and type variables. We found that location and type (i.e. region, province and firm legal structure) are the most relevant to predict moonlighting followed by, in order of decreasing importance, factor 1 of labor productivity and profitability, factors 3 and 1 of performance. The other variables have much less importance and are not reported. Also for the extended model we found the location variables to be the most important followed by variables in the labor productivity and profitability and

Table 2 Principal components analysis

Firm facet	No. original variables	Factor	Interpretation	Explained variability %
Economic dimension	9	1	Market dimension	58.023
		2	Value added and costs dimension	12.007
Structure	6	1	Costs structure	39.221
		2	Value added per firm asset	33.330
Management	6	1	Firm economy wrt[a] managing dept	35.711
		2	Firm economy wrt total earnings	16.892
		3	Firm economy wrt value added	16.671
Performance	11	1	Total performance wrt market reached	27.271
		2	Performance wrt value added	24.153
		3	Total performance excluding personnel costs	21.187
		4	Capital performance (amortization)	9.101
Organization	13	1	Incidence of non-full time employee	21.631
		2	Incidence of part-time employee	20.725
		3	Incidence of senior and middle management	20.108
Labor productivity and profitability	38	1	Productivity of total employee	23.737
		2	Productivity wrt value added	14.278
		3	Profitability of internal employee	11.162
		4	Profitability of external employee	7.980
		5	Productivity wrt value added of external consultant	6.480
		6	Incidence of labor cost on overall firm costs	4.884
		7	Incidence of labor cost on firm earnings	4.816

[a] wrt: with respect to

Table 3 Characteristics and dimension of final nodes (Training sample)

Number	Final node number	Predicted value	Probability associated to predicted value	Number of firms in the final node
1	5	No	0.900	54
2	7	Yes	0.333	521
3	8	No	0.795	110
4	4	Yes	0.592	326
5	2	No	0.833	3366

Table 4 Model predictive ability: the classification matrix

Sample	Observed	Predicted		Correct Percentage
		No	Yes	
Training	No	6927	1102	86.3%
	Yes	1382	863	38.4%
	Overall percentage	80.9%	19.1%	75.8%
Test	No	2936	484	85.8%
	Yes	594	363	37.9%
	Overall percentage	80.6%	19.4%	75.4%

in performance groups. As we said, we used the results obtained in these models and selected 34 independent variables taking into account both variables importance and logical requirement. We therefore run CART on the selected variables subset and the key points we observed on all trees grown are the following:

1. Territorial variables (region and province) are the most important to predict moonlighting.
2. Legal structure is also a crucial variable; in particular sole proprietorships have a higher moonlighting probability.
3. Labor cost systematically have an impact on moonlighting probability.
4. Variables importance ranking appear to be stable over all trees grown. That means that the main risk factors can be identified.
5. It's much easier to classify correctly firms which don't use irregular workers rather than companies that do. To reduce misclassification rate on the latter, we had to increase prior probability on the presence of irregular work value.

Out of more than fifty trees grown, we selected the one with a satisfactory balance between parsimoniousness (measured in terms of final nodes) and predictive ability (measures as correct classification rate on the test sample). Tables 3 and 4 show model characteristics and predictive power respectively.

3.2 *Parametric Models*

As mentioned, for logistic regression we started with the same independent variables subset used for CART. Backward selection procedure was used to identify candidate models. The main results obtained are pretty much in agreement with those found using CART and could be summarized in the following points:

1. Territorial variables (region and province) are the most important to predict irregularities.
2. Companies organization variables and contracts awards mode play a key role in moonlighting prediction.
3. It's much easier to classify correctly companies which don't have irregular employee. We had to reduce to 0.3 the cut off point. In other word we forced the model to classify as irregular all firms which had an estimated probability of using irregular workers greater than or equal to 0.3.
4. All models have a predictive performance greater than 76%.

Since we saw that location variables were very important, we fitted three kind of models including: (a) both province and region (b) region only (c) no location variables. For any of the three types of models, backward selection procedure gave many candidates; we chose those with the best predictive power and goodness of fit. We performed residuals analysis for any models and no critical elements appeared. Scatter plot of predicted probabilities vs Cook's influential statistics showed some points that have to be investigated more in depth because they could be influential. Table 5 shows parameter estimates for the model chosen out of those containing only region as location variable. Table 6 is its classification matrix.

Due to lack of space we cannot report parameters estimates for the other 3 models selected (one including province and two with no location variables at all), but in Table 7 we report the ROC curve test to compare their ability to correctly classify firms. In a Receiver Operating Characteristic (ROC) curve the true positive rate (Sensitivity) is plotted in function of the false positive rate (100-Specificity) for different cut-off points. Each point on the ROC plot represents a sensitivity/specificity pair corresponding to a particular decision threshold. A test with perfect discrimination has a ROC plot that passes through the upper left corner (100% sensitivity, 100% specificity). Therefore the closer the ROC plot is to the upper left corner, the higher the overall accuracy of the test. A measure of accuracy is therefore represented by the area under curve: the closest to one the better the model separates firms with moonlighters from firms without Azzalini and Scarpa (2004).

4 Conclusions and Further Work

Submerged economy has developed features that make it more difficult to detect: the time has come for a careful reflection on a comprehensive definition of irregularities and for more sophisticated methods to aid inspectors in their duties. In this frame, the importance to have a model that identifies the key elements for a firm to use irregular work is self-evident. Our research has just begun, still we have learnt something important. As all models point out clearly, location, labor cost and legal structure are the key elements to identify irregular firms.

Table 5 Parameter estimates

Independent variable	B	p-value	ExpB
Sicilia [a]	1.3754	0.0032	3.9566
Campania	1.0955	0.0183	2.9906
Sardegna	0.6733	0.1699	1.9608
Abruzzo	0.2394	0.6185	1.2706
Calabria	0.1721	0.7171	1.1878
Piemonte	0.0184	0.9687	1.0186
Emilia Romagna	0.0050	0.9914	1.0050
Molise	−0.0356	0.9544	0.9650
Puglia	−0.0534	0.9097	0.9479
Toscana	−0.0601	0.8975	0.9415
Veneto	−0.1440	0.7611	0.8658
Marche	−0.1764	0.7136	0.8382
Trentino Alto Adige	−0.3661	0.4646	0.6933
Lazio	−0.4469	0.3488	0.6395
Friuli Venezia Giulia	−0.7086	0.1787	0.4923
Umbria	−0.7138	0.1710	0.4897
Lombardia	−0.7225	0.1233	0.4855
Liguria	−0.7567	0.1243	0.4691
Basilicata	−1.3865	0.0125	0.2499
Up to 5000 resident [b]	0.2800	0.0023	1.3232
5001–10000	0.2769	0.0030	1.3190
10001–50000	0.1059	0.2151	1.1117
50001–100000	0.0161	0.8788	1.0162
% of works took on a subcontract basis	0.0028	0.0000	1.0028
% of working-days of low skilled workers	−0.1894	0.0241	0.8274
% of working-days of high skilled workers	−0.8175	0.0000	0.4415
% of working-days of employee	−1.7399	0.0000	0.1755
Constant	−1.4146	0.0028	0.2430

[a] Ref: Valle d'Aosta [b] Ref: More than 100000 resident

Table 6 Model predictive ability: the classification matrix

Observed	Predicted		Correct percentage
	No	Yes	
No	7426	1263	85.50%
Yes	1369	930	40.50%
Overall percentage			76.00%

The results obtained are interesting but further investigation is needed. The main points to develop in the next future are:

1. To improve the matching rate of the original datasets.
2. Enrich the information system with other data sources.

Table 7 ROC test to compare logistic models predictive performance

Predicted probability for a firm to use irregular work	AUC[a]	Standard Error	p-value	95% confidence interval	
				Lower	Upper
Model with Province	0.7470	0.0060	0.0000	0.7360	0.7590
Model without Province	0.7000	0.0060	0.0000	0.6880	0.7120
Model 1 without territory	0.5890	0.0070	0.0000	0.5760	0.6020
Model 2 without territory	0.5740	0.0070	0.0000	0.5610	0.5870

[a] Area Under Curve

3. Include socio-economical variables which are able capture the characteristics of the territory where the firms are located; these variables have to be both at region and province level. Multilevel logistical models might be particularly useful to fully interpret territorial influence.
4. Fit non parametric models which generally have higher predictive performance and are more stable than CART, i.e. Random Forests.
5. Calibrate the data in order to properly extend the results to the population.

Further comments on some of the points listed above are due. On behalf of the second remark, an important and new source of information to use is the so called Co-eService which bind firms to communicate beginning, ending and prorogation of any work contract. The system started on march 1st 2008 assuring a real time communication to all agencies which manage information on employment status (besides INPS, the Italian National Insurance at Work Service, the National Welfare Service, and so forth).

Concerning the fifth point, the problem is that the inspected firms we examined are not a random sample since inspections occur upon signalling. Therefore to correctly extend results to the population of building firms, we have to find the most appropriate calibration model.

References

Alleva, G., Arezzo. M.F., Nisi, A.:Il lavoro sommerso nel settore delle costruzioni: analisi esplorativa e rappresentazione territoriale del fenomeno. In Proceedings on Cartographic Studies *From Map to GIS*. Dpt AGEMUS, Sapienza Universitá di Roma, (2009a).

Alleva, G., Arezzo. M.F., Proganó, T.: Definizione e implementazione di parametri di stima e valutazione del lavoro sommerso. ISFORT Working Paper. (2009b)

Azzalini, A.,Scarpa, B: Analisi dei dati e Data mining. Springer,(2004)

Berry, M.J., Linoff, G.S.: Data Mining Techniques. Wiley Publishing, (2004)

Breiman L., Friedman J.H., Olshen R.H., Stone C.J.: Classification and Regression Trees, Wadsworth, (1983)

Use of Interactive Plots and Tables for Robust Analysis of International Trade Data

Domenico Perrotta and Francesca Torti

Abstract This contribution is about the analysis of international trade data through a robust approach for the identification of outliers and regression mixtures called Forward Search. The focus is on interactive tools that we have developed to dynamically connect the information which comes from different robust plots and from the trade flows in the input datasets. The work originated from the need to provide the statistician with new robust exploratory data analysis tools and the end-user with an instrument to simplify the production and interpretation of the results. We argue that with the proposed interactive graphical tools the end-user can combine effectively subject matter knowledge with information provided by the statistical method and draw conclusions of relevant operational value.

1 Introduction

The international trade of commodities produces an enormous amount of data which are collected by Customs, national and European statistical offices (e.g. the Italian ISTAT and the European Commission's EUROSTAT) and international statistical authorities (e.g. United Nations Statistics Division, WTO and OECD). The statistical use of trade data for policy purposes includes relevant applications such as anti-fraud, anti-trade-based money laundering and anti-dumping. Significant discrepancies in the data as reported by the trading countries, such as price outliers and other inconsistencies, can be detected with different statistical approaches.

D. Perrotta (✉)
European Commission, Joint Research Centre, Institute for the Protection and Security of the Citizen, Ispra, Italy
e-mail: domenico.perrotta@ec.europa.eu

F. Torti
Università di MIlano Bicocca, Facolta' di Statistica Italy
e-mail: francesca.torti@unimib.it

A. Di Ciaccio et al. (eds.), *Advanced Statistical Methods for the Analysis of Large Data-Sets*, Studies in Theoretical and Applied Statistics,
DOI 10.1007/978-3-642-21037-2_30,

In general trade values are simply regressed on the corresponding volume of trade and it is sufficient to examine the impact of an observation on a given fit with the traditional approach of deletion diagnostics testing. Typically the diagnostics are monitored through the data "backward", by iteratively deleting one observation at a time (see for example Marasinghe (1985)). Appropriate use of this approach must solve severe multiple testing problems due to the huge number of observations and datasets to be checked simultaneously. Corrections for the multiple outlier tests, such as Bonferroni's correction, must be carefully applied for not loosing too many good signals. Unfortunately in general, especially when the number of observations is large, it is likely to find one or more atypical observations that affect the fitted model so strongly to mask other deviating observations, which remain undetected. This well known masking effect can severely affect the backward estimates.

Another difficult situation occurs concretely in international trade. Import duty rates (which are foreseen for specific goods and third countries) are usually proportional to the value of the imported product and this may induce fraudsters to considerable and sometimes systematic under-declaration of the value, producing groups of outliers in trade data. However, the same phenomenon may take place in situations where the quality level and the price of the good traded vary considerably or whenever specific market events impact unexpectedly on the price. In these circumstances the outlier detection problem is complicated by an even more difficult problem of robust regression mixture estimation. This is the case of the example that we discuss in the paper, where the "good" data and the outliers are aligned towards different regression lines.

To address efficiently the above problems we have chosen the Forward Search, a general method for detecting unidentified subsets and masked outliers and for determining their effect on models fitted to the data. The key concepts of the Forward Search were given by Hadi in (1992), while Atkinson et al. in (2000; 2004) introduced the idea of diagnostic monitoring and applied the method to a wide range of regression and multivariate statistical techniques. A natural extension of the Forward Search to the estimation of regression mixtures was proposed more recently by the authors of the Forward Search and by us (Riani et al. 2008).

Unlike other robust approaches, the Forward Search is a dynamic process that produces a sequence of estimates. It starts from a small, robustly chosen, subset of the data and fits subsets of increasing size in such a way that outliers or the presence of homogeneous subsets of data are revealed by monitoring appropriate diagnostics. In regression the Forward Search typically monitors the evolution of residuals, parameter estimates and inferences as the subset size increases. The results are then presented as "forward plots" that show the evolution of these quantities of interest as a function of sample size. The main two forward plots in regression are recalled in Sect. 2. A more complex forward plot in the context of data response transformation is discussed in Sect. 4.

The first time we made use of a basic forward plot was in 2007 (Perrotta and Torti 2010), when we addressed the example of this paper as multivariate problem and we could identify subsets coming from different populations on the basis of the evolution of the scaled Mahalanobis distance trajectories. Lot of visual inspection and ad-hoc programming were needed to identify the units associated to the groups of different Mahalanobis trajectories. It was clear that the Forward Search approach was lacking of an automatic link among the many forward plots which are monitored during the process. We have treated this aspect only recently, in Perrotta et al. (2009), where we dedicated a full section to highlight the relevance of linking plots in concrete applied problems when it is crucial to present and communicate effectively the statistical results to end-users not necessarily familiar with the statistical methods. This takes place in the Joint Research Centre (JRC) of the European Commission when the authors and other statisticians cooperate with the European Anti-Fraud Office (OLAF) to highlight in European trade data potential cases of fraud, data quality issues and other oddities related to specific international trade contexts.

This paper focuses on the operational use of the interactive tools introduced in Perrotta et al. (2009) for robust trade data analysis. We start with applying the tools to traditional forward plots (Sect. 3), to show how simple is now the type of investigation initiated in Perrotta and Torti (2010). Then we show how a more complex forward plot, the so called fan plot, can be easily used to detect the presence of multiple populations (Sect. 4). Finally (Sect. 5) we show how the subject matter expert job, which focuses on operational conclusions, can be simplified by simply extending interactivity to data tables that typically include categorical or numeric variables which are not analysed, but may have important operational meaning. To better appreciate the overall advantages of interactivity and some new features introduced in this paper (more interactive tables, click-able legends, etc.), we suggest also reading Perrotta et al. (2009) and Perrotta and Torti (2010). To simplify comparison, we use here the same trade dataset example of the previous two papers.

2 Main Forward Plots in Regression

We start by formalising the quantities monitored in the two main forward plots, i.e. the minimum deletion residual and a scaled version of the squared regression residuals. We addressed other useful quantities typically monitored along the search in Perrotta et al. (2009). We consider one univariate response Y and v explanatory variables $X_1, \ldots, X_v$ satisfying (under usual assumptions discussed for example by Seber (1977)) the expression

$$E(y_i) = \beta_0 + \beta_1 x_{i1} + \cdots \beta_v + x_{iv}.$$

Let $\hat{\beta}(m)$ be the estimate of the $(v + 1)$-dimensional parameter vector $\beta = (\beta_0, \beta_1, \cdots, \beta_v)^T$ obtained by fitting the regression hyperplane to subset $S(m)$. From this estimate we compute n squared regression residuals

$$e_i^2(m) = \left[y_i - \left\{\hat{\beta}_0(m) + \hat{\beta}_1(m)x_{i1} + \cdots + \hat{\beta}_v(m)x_{iv}\right\}\right]^2 \qquad i = 1, \ldots, n \quad (1)$$

The observations with the first $m + 1$ smallest squared regression residuals form the new subset $S(m + 1)$. The search starts from an outlier-free subset of $v + 1$ observations satisfying the LMS (Rousseeuw 1984). To detect outliers we examine the minimum deletion residual amongst observations not in the subset

$$r_{\min}(m) = \min \frac{|e_i(m)|}{s(m)\sqrt{\left[1 + x_i^T\{X^T(m)X(m)\}^{-1}x_i\right]}} \quad \text{for } i \notin S(m), \tag{2}$$

where $s(m)$ is the square root of the unbiased estimate of the residual variance $\sigma^2 = E\{y_i - E(y_i)\}^2$ computed from the observations in $S(m)$, $x_i = (x_{i1}, \dots, x_{iv})^T$ is the ith row of the design matrix X and $X(m)$ is the block of X with rows indexed by the units in $S(m)$. Inferences about the existence of outliers require envelopes of the distribution of $r_{\min}(m)$ (see for example Atkinson and Riani (2006)). The more complex problem of detecting and estimating a regression mixture, which characterises the example in the paper, can be addressed following the idea of finding a group of homogeneous observations with the Forward Search, trimming the group and repeating the procedure on the remaining data (Riani et al. 2008). If we have two or more regression groups there will be a point where the stable progression of the regression statistics monitored (2) is interrupted. This breaking point, estimated using the envelopes for $r_{\min}(m)$, identifies a group of homogeneous observations that entered in the subset.

3 Use of Dynamic Forward Plots in the Analysis of Trade Data

The standard forward plots just introduced are static graphs. The analyst can be interested in identifying in the scatterplot the units which entered the subset at specific steps of the search and to appreciate the joint effect of such units on the regression fit or on other statistics monitored in different plots. This requires writing ad-hoc programs, which is clearly an obstacle to the adoption of the method, especially for end-users and practitioners. In this section we illustrate the advantages of adopting more flexible interactive plots. We use an example that was discussed under different perspectives in previous works (e.g Perrotta et al. (2009), Perrotta and Torti (2010) and Riani et al. (2008)), taken from the thousands of international trade datasets that we analyse at the JRC.

We recall the key features of the datasets. The variables that characterise a set of comparable trade flows are the codes of the traded product and the countries of origin and destination (POD). We use the values and volumes traded to estimate the slope of the regression line fit on the POD, which is a sort of "fair price" for that particular POD combination, and we detect trade flows of abnormal price, i.e. low and high price outliers. The example is complex, being characterised by different groups of flows each with a different reference price. Abnormal price transactions are made available to OLAF and its anti-fraud partners in the Member States for

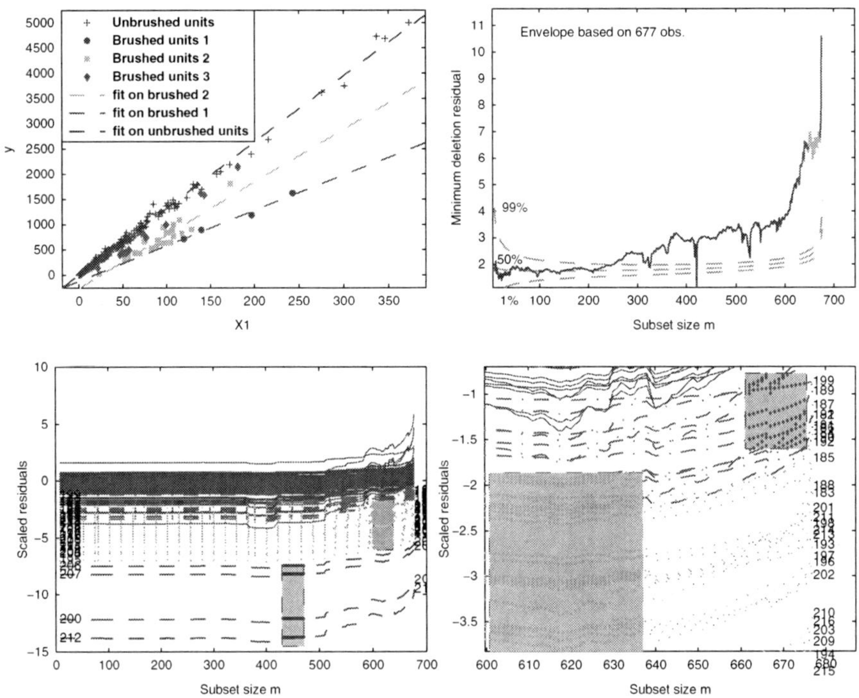

Fig. 1 Fishery trade dataset. On the top panel the scatterplot of trade values and volumes and the minimum deletion residual curve at each step of the Forward Search. On the bottom panel the scaled residuals trajectories throughout the search (*left*) and zoom in an area after step 600 (*right*). The selections are automatically highlighted in all plots. The trajectories labels refer to flows from a specific Member State (records from 181 to 216)

analyses motivated by the identification of potential fraud cases. The complexities and patterns in the dataset are rather frequent in trade data and are also of general interest. Therefore we made the dataset publicly available in our FSDA (Forward Search for Data Analysis) toolbox, freely available at http://www.riani.it/MATLAB.htm. The toolbox can be used with MATLAB to replicate the results of the paper.

Figure 1 gives the main output of the Forward Search applied to such data. The scatterplot (top left) represents the quantity (x axis) and the value (y axis) of the monthly flows of a fishery product imported into the EU from a third country. Almost all the flows represented with circles, squares and diamonds in the lower part of the scatterplot refer to the same EU Member State, that we identify with MS7. Their position in the scatterplot suggests that the prices declared by that Member State are suspiciously low if compared to those of the other Member States: 13 €/Kg for the group of the "fair" import flows, in the scatterplot with symbol '+', and 9.17 and 6.55 €/Kg for the anomalous flows in the groups of circles and squares. In Perrotta and Torti (2010) these groups and the associated price estimates were found by a careful inspection of the scaled squared Mahalanobis distance plot, which in the multivariate context has the same role of the scaled squared residual plot in the lower plots of Fig. 1. The plot that was available in Perrotta and Torti (2010) was unfortunately static and we had to write specific code to identify precisely the units associated to the abnormal Mahalanobis trajectories, while the scaled squared residual plot here is interactive and can be inspected by repeated and natural mouse click operations.

In the lower left side of Fig. 1 we show three consecutive selections of residual trajectories, starting from the most deviating values. The plot is automatically linked to the scatterplot and the minimum deletion residual plot, in the upper part of the figure. Therefore, the units in the two plots that correspond to the selected residual trajectories are simultaneously highlighted with common colours in all plots.

Note that in the scatter plot there is no line fit on the third selection of borderline units (the diamonds). The line was removed automatically by a simple click on the fit legend, that normally becomes greyish and can be clicked again to recover the line fit. In this case it was definitively removed to save space in the plot.

The possibility to zoom on the residuals plot is a conceptually simple but important tool. The zoomed area at the bottom right of the figure shows the residual trajectories that are closer to the dense area of the "fair" trade flows. The labels on the right identify the suspicious flows of MS7, that include records from 181 to 216. The zoom allows to appreciate that almost all selected flows refer to MS7: exceptions are the four dashed upper trajectories that at the last steps of the search (after step 660) leave the selection to join the dense group of the "fair" flows trajectories at the top. These four trajectories correspond to units in the group of the diamonds in the scatterplot, that as expected are border line cases of difficult classification.

In the example the selection started from the residuals plot. However the same can be done from any other forward or traditional plot in the FSDA toolbox.

4 Use of More Complex Forward Plots

This section illustrates a forward plot that would be more difficult to interpret and use without the possibility of linking to other traditional plots. The context is that of data response transformation. Here the null hypothesis is on the Box-Cox transformation parameter (Box and Cox 1964), $\lambda = \lambda_0$, and the added t tests based on constructed variables are known in the statistical literature as "score test for transformation" (Atkinson and Riani 2002). The tool to understand the percentage of observations which are in accordance with the different values of the transformation parameters, is the forward plot of the score test statistic for transformation of the set of constructed variables for different values λ_0, using a separate Forward Search for each λ_0. These trajectories of the score tests can be combined in a single picture named the "fan plot" (Atkinson and Riani 2000).

In general, being the relationship between trade value and quantity supposedly linear, trade data should not require transformation. Different trade scenarios may determine the need of transformation, the most typical being the so called "discount

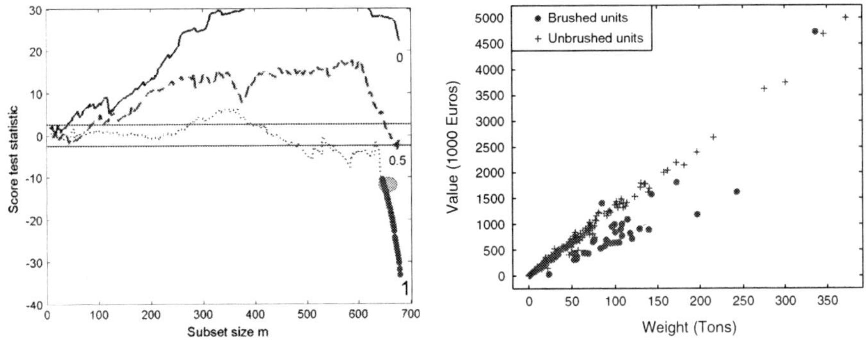

Fig. 2 Fishery Trade Data. Fan plot for $\lambda = (0, 0.5, 1)$ (*left panel*) and the scatter of values (in thousands of euros) against quantities (in tons). The brushed units in the fan plot are automatically highlighted in the scatter plot. The brushed units form a separate cluster from the bulk of the data

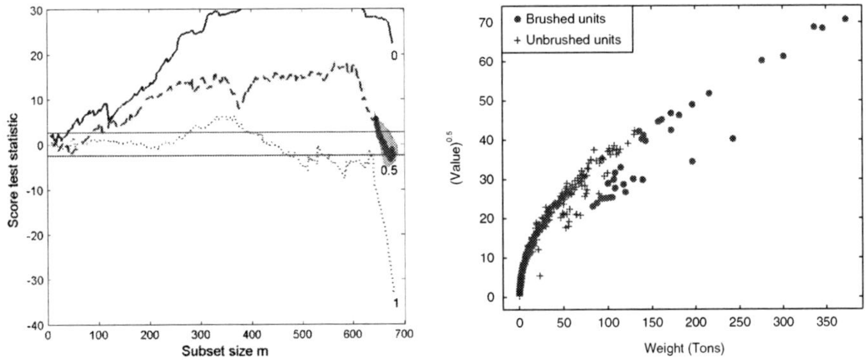

Fig. 3 Fishery Trade Data. Fan plot for $\lambda = (0, 0.5, 1)$ (*left panel*) and the scatter of square root values (in thousands of euros) against quantities (in tons). The brushed units on the $\lambda = 0.5$ trajectory in the fan plot are automatically highlighted in the scatter plot

effect" that happens when prices get lower with the increase of the traded volumes. In this example the need of transformation is due to the presence of multiple populations. The left panel of Fig. 2 shows that the inclusion of the last observations, which enter at some point in the subset, causes strong rejection of the hypothesis of no transformation. After a simple brushing action on the trajectory of the fan plot for $\lambda = 1$ (i.e. no transformation required), it is immediate to see that the brushed units form a well defined cluster below the main cloud in the scatter plot (right panel of Fig. 2). On the other hand, the inclusion of the last observations causes acceptance of the hypothesis of square root transformation (left panel of Fig. 3). However from the scatterplot of the transformed data (right panel of Fig. 3) it is obvious that there is no linear relationship between the transformed response and the independent variable. Thus, the square root transformation is not appropriate.

5 The Perspective of the Trade Analyst

So far we have discussed the possible use of interactive forward plots in the analysis and interpretation of trade data. We have shown how powerful these instruments are for extracting useful information from the data. However, this task still requires a rather profound understanding of the statistical meaning of the various plots. Typically the point of view of the subject matter expert is different. He/she would like to apply the Forward Search automatically on thousands of trade datasets and have the possibility to study in depth the relevant cases with exploratory tools of simple use, integrated in user friendly graphical interfaces.

The FSDA toolbox has been also designed for this purpose. For example, we offer simple interfaces to obtain basic summary statistics and plots such as those of Figs. 4 and 5, which may be of great help for the end-user. In Fig. 4 the user is investigating on trade flows associated to a specific Member State (MS7) and, thus, selects the flows in the data table (left). Summary statistics for the given selection and for the full population are reported automatically in a separate table (top right). The difference in unit price (the variable concerned by the selection) between population and selection is neat, which is the relevant fact of this dataset.

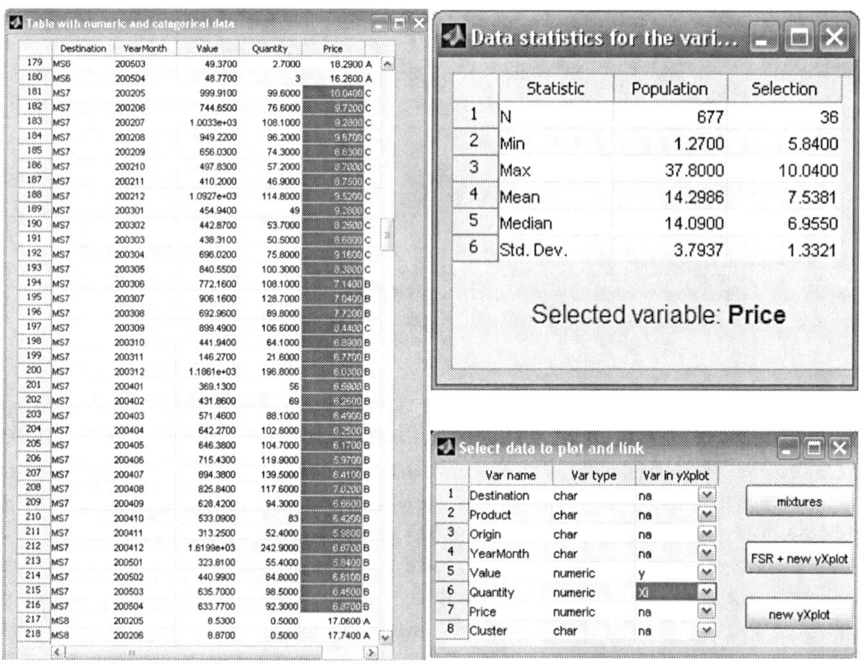

Fig. 4 Example of GUIs available in the FSDA toolbox. When trade flows are selected in the data table (*left*), summary statistics for the given selection and for the full population are reported in a separate table (*top right*). A dialogue box (*bottom right*) is used to generate the plot of value against quantity in the figure below, by selecting such variables and pressing button `new yXplot`

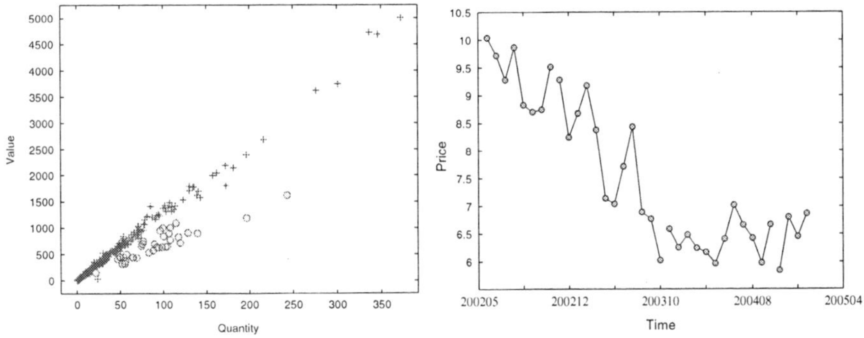

Fig. 5 On the left the scatter plot of value against quantity that results from the selection of cells in the table above. On the right the time evolution in the subset of selected "abnormal" prices. The price variable is obtained as value to quantity ratio

The dialogue table (bottom right) is used to generate the plot of value against quantity in Fig. 5, by selecting such variables and pressing button `new yXplot`. The other two buttons are to apply the Forward Search for the purpose of detecting outliers (button `FSR + new yXplot`) or to estimate regression mixtures (button `mixtures`).

The possibility to inspect the original data table, where all relevant trade variables can be found, at any time and starting from any plot, is an important feature for the end-user. For example, a fact of operational value that has become clear only after inspecting the input table, is that the abnormal flows of MS7 have a clear time structure. In the period analysed the traders in MS7 gradually under-declared their import price, up to half of the price reported by the other MSs, so that to reduce import duty payments. This unexpected trend is more clearly shown in the time series of the price evolution (see Fig. 5), but in general trade prices are not characterised by so neat time evolution and the consultation of the time series is not routine in this application context.

In Perrotta and Torti (2010) such unexpected pattern was found almost incidentally, by off-line inspection of the data table. Clearly this approach is not efficient and the chance of finding unexpected patterns of this type is limited. The GUIs presented in this section, that link tables to plots, are a step towards this type of user need. In Perrotta et al. (2009) we have also discussed the link in the opposite direction, where the points selected in a scatterplot are simultaneously highlighted with common colours in the table.

6 Conclusion

A feature that distinguishes the Forward Search from the other robust approaches is that it is a dynamic process, which produces a sequence of estimates and informative plots. It was therefore natural and even necessary for us to explore and develop new interactive plots. Certainly this work direction was motivated by the needs of our specific anti-fraud customers, but we argue that end-users in many other application domains would equally benefit from coupling the Forward Search with such flexible exploratory analysis tools. We are almost systematically using our interactive plots as an instrument to investigate open methodological issues linked to the Forward Search, such as how the search is affected by the structure of high dense and/or overlapping areas or by observations where the value of one or more variables is repeated. We intend to extend the set of dynamic plots in the FSDA toolbox and to enrich the plots with new interactive functionalities. The implementation of more carefully designed user friendly graphical interfaces to our statistical tools, is our next step. Finally, being the FSDA toolbox based on the commercial product MATLAB (by The Mathworks Inc.), for the main functions we plan to include ports to free statistical platforms such as R and or OCTAVE.

Acknowledgements The work is jointly supported by the University of Parma where the second author is also affiliated as "Cultore della Materia", and the Joint Research Centre of the European Commission, under the institutional work-programme 2007–2013 of the research action "Statistics and Information Technology for Anti-Fraud and Security". We also thank OLAF for encouraging this research.

References

A. C. Atkinson and M. Riani. *Robust Diagnostic Regression Analysis*. Springer–Verlag, New York, 2000.

A. C. Atkinson and M. Riani. Forward search added-variable t-tests and the effect of masked outliers on model selection. *Biometrika*, 89(4):939–946, 2002.

A. C. Atkinson and M. Riani. Distribution theory and simulations for tests of outliers in regression. *Journal of Compuational and Graphical Statistics*, 15:460–476, 2006.

A. C. Atkinson, M. Riani, and A. Cerioli. *Exploring Multivariate Data with the Forward Search*. Springer–Verlag, New York, 2004.

G. E. P. Box and D. R. Cox. An analysis of transformations. *Journal of the Royal Statistical Society. Series B (Methodological)*, 26(2):211–252, 1964.

A. S. Hadi. Identifying multiple outliers in multivariate data. *Journal of the Royal Statistical Society, Series B*, 54:761–771, 1992.

M. G. Marasinghe. A multistage procedure for detecting several outliers in linear regression. *Technometrics*, 27(4):395–399, 1985.

D. Perrotta, M. Riani, and F. Torti. New robust dynamic plots for regression mixture detection. *Advanced Data Analysis and Classification*, pages 263–279, 2009.

D. Perrotta and F. Torti. Detecting price outliers in European trade data with the forward search. In *Data Analysis and Classification: From Exploration to Confirmation*, Studies in Classification, Data Analysis, and Knowledge Organization, pages 415–423. Springer, Berlin-Heidelberg, 2010. Presented at CLADAG 2007, Macerata, Italy.

M. Riani, A. Cerioli, A.C. Atkinson, D. Perrotta, and F. Torti. Fitting mixtures of regression lines with the forward search. In *Mining Massive Data Sets for Security*, pages 271–286. IOS Press, 2008.

P. J. Rousseeuw. Least median of squares regression. *Journal of the American Statistical Association*, 79:871–880, 1984.

G. A. F. Seber. *Linear Regression Analysis*. Wiley, New York, 1977.

Generational Determinants on the Employment Choice in Italy*

Claudio Quintano, Rosalia Castellano, and Gennaro Punzo

Abstract Aim of the paper is to explore some crucial factors playing a significant role in employment decision-making in Italy. In particular, we aim at investigating the influence of family background on the choice to be self-employed rather than salaried; for this end, a series of regression models for categorical data is tested both on sampled workers, taken as a whole, and by gender separation. In this light, to test if the employment choice is context-dependent, environmental attributes are also modeled. In addition to a diversity of determinants, our results shed light on some differences between self-employed workers by first and second generation.

1 Background and Introduction

For a long time, self-employment has consistently been regarded as a residual category of gainful occupation not rewarded by a salary or wage. In the last few years, self-employment has been growing in several developed or developing economies, also due to flourishing of innovative non-standard kinds of work. Since the end of 1970s, when the greatest increase in the proportion of self-employment on the total labour force was occurred (once the crisis of the Taylor-Fordist production model became decisive), Italy has kept a significant incidence of self-employed (20–25%), consistently higher than the UE average. Since 1990s, the ratio between self-employed – those who work for themselves, receiving rewards for their labour

*This paper has been coordinated by C. Quintano and it is the result of the common work of the Authors: R. Castellano has written Sections 1 and 5; G. Punzo has written Sections 2, 3 and 4.

C. Quintano (✉) · R. Castellano · G. Punzo
DSMER – University of Naples "Parthenope", via Medina 40, 80133, Naples, Italy
e-mail: quintcla@hotmail.com; lia.castellano@uniparthenope.it;
gennaro.punzo@uniparthenope.it

A. Di Ciaccio et al. (eds.), *Advanced Statistical Methods for the Analysis of Large Data-Sets*, Studies in Theoretical and Applied Statistics,
DOI 10.1007/978-3-642-21037-2_31, © Springer-Verlag Berlin Heidelberg 2012

and entrepreneurial skills – and wage earners – those who only get returns to their labour and human capital – has substantially been steady over time, although a rising incidence of self-employment is occurred in the construction industry and tertiary sector at the expense of some traditional economic areas.

In literature, several studies argue about the main reasons affecting individual propensities to enter self-employment. Broadly speaking, it may be driven by necessity, by ambition to pursue a business opportunity or by a strive for personal independence (Henrekson 2004). Beyond personal attitudes or requirements, related to personality traits or individual abilities, perceptions, preferences and risk-taking propensity (Verheul et al. 2002; Arenius and Minniti 2005), also the family-specific background as well as the socio-economic environment may play a strong role in occupational choices (Allen 2000; Dunn and Holtz-Eakin 2000). Indeed, a higher possibility of access to material and financial capital – i.e., by transferring stocks, buildings or machineries to offspring, the family, regarded as a substitute banker, might relax capital market constraints (Laferrere and McEntee 1995; Colombier and Masclet 2007) – and a privileged chance to inherit the human knowledge, experience and talents – i.e., education, training and other learning processes (Lentz and Laband 1990; Dunn and Holtz-Eakin 2000) – may enhance the ability to perform specific tasks and encourage to attempt self-employment. In a household context, human capital includes the collection of parental skills acquired in both formal and informal ways which affect children's outcomes (d'Addio 2007).

In this light, the aim of our work is twofold. First, to investigate the influence of family background on the choice to become self-employed rather than employee; thus, crucial differences between self-employed by gender and by first and second generation are also stressed. Second, to test if the occupational choice is also context-dependent; consequently, the impact of some socio-economic regional variations is looked into. At this end, latent variable models for binary outcomes (Long 1997) are estimated on the overall workers as well as by gender separation.

2 Exploring Self-Employment in Italy: Data Sources and Variables

Our analysis draws upon the 2006 Survey on Household's Income and Wealth (SHIW), a biennial split-panel survey carried out by Bank of Italy. With regard to wave 2006, the sample size consists of 7.768 households and 19.551 individuals. It is a valuable data source on a substantial range of socio-economic topics both at household and individual level. Particularly, it provides detailed information on employment status and activity sector as well as on the single income and wealth components over the whole previous calendar year. Most importantly, SHIW also detects a set of retrospective parental information (i.e., education, employment, activity sector), allowing to account for potential generational changes over time.

We focus on all those individuals who are either self-employed or salaried, so as stated by themselves in a question concerning their main employment status, irrespective of activity sector. In such a way, we consider 7.510 individuals, 1.493 (19.88%) of whom are self-employed, defined as anyone working in his/her own business or professional practice (i.e., small employer, owner or member of family business, shareholder/partner, member of profession or arts), any own-account or craft worker and any contingent worker on own account (i.e., freelance regular or occasional collaborator, project worker, etc). So, we leave out all not-employed individuals (i.e., first-job seekers, homemakers, well-off persons, students, pre-school-age children, etc.) as well as all those retired, pensioners or unemployed.

Exploring SHIW data, in 2006, self-employed account for roughly one-fifth the total number of workers, whereas the coexistence ratio between self-employment and wage-employment is close to 25%; these findings are substantially coherent with the evidence from the Continuous Survey on Labour Force, the main Italian data source on work and market labour. A further explorative data analysis stresses that slightly less than one-third of self-employed may be classified as self-employed workers by second generation, since they come from a family where at least one of the two parents is/was self-employed; in this light, the coexistence ratio between the self-employed by first and second generation is close to 50%.

In Table 1, self-employed workers are classified by occupational status and generational stage, i.e., if they are self-employed workers by first or second generation.

First, it is worth stressing how, in Italy, self-employment is quite heterogeneous – it includes small craft workers, whose incidence is rather high, as well as owners of larger enterprises, although the latter stand for just a minority – and how the relative

Table 1 Self-employed workers by occupational status and "generational stage"

Occupational status	All self-employed	First generation	Second generation	
			Same occupation	Not same occupation
Small employer	0.1033	0.1077	0.0263	0.0792
Owner or member of family business	0.1530	0.1501	–	0.1631
Working shareholder/partner	0.0968	0.0820	–	0.1118
Member of profession	0.2046	0.2070	0.0522	0.1503
Own-account worker/craft worker	0.3522	0.3554	0.3352	0.0187
Contingent worker on own account	0.0901	0.0978	–	0.0632

Source: Authors' elaborations on SHIW data (2006)

distribution of self-employed workers by first generation basically reproduces the relative distribution of self-employed taken as a whole. Second, self-employed workers by second generation are classified into the two groups of those who have at least one parent occupied in the same self-employment activity and those self-employed whose parents (or at least one), although self-employed workers, are/were not occupied in the same activity; briefly, it denotes how more than 40% of self-employed workers by second generation has/had at least one parent occupied in the same occupation, essentially own-account or craft worker.

Moreover, by contrasting some personal characteristics, substantial differences between salaried and self-employed workers are highlighted in Table 2. First of all, a higher concentration of self-employment amongst head of households is remarked, whereas head of households' spouses/partners seem to be more "wage-oriented". Similarly, a higher incidence of men in self-employment than their wage-and-salary counterparts as well as a higher concentration of self-employment among Italian citizens are detected.

On average, self-employed tend to be older than salaried; indeed, if the incidence of self-employment is poorer than wage-employment in the lower age-classes, it tends to be higher as we move towards the upper age-classes; in other words, self-employment tends to be more concentrated amongst individuals in mid-career (i.e., between 35 and 44 years of age). At the same time, although there is a higher incidence of self-employed with a low educational level (i.e., pre-primary, primary or lower secondary education) than their wage-and-salary counterparts, it seems to be, on average, a kind of compensating effect of education, in terms of years of schooling, on self-employment. Finally, income levels are surely higher for self-employed as well as the incidence of self-employed owning their main home.

3 A Methodological View: A Latent Variable Model for Binary Outcomes

In order to assess how the employment choice may be affected by some personal attitudes as well as by several components of human and financial capital in a generational perspective, maximum likelihood logit models (Allen 2000), chosen in the sphere of binary response models (BRM), are estimated. This choice is essentially justified by the cross-sectional type of analysis. Indeed, although SHIW design shows a partial overlap of sampling units (Bank of Italy 2008), our study has exclusively been carried out on only one wave (2006), so it need not have modeled correlated data from longitudinal/repeated measures; also, parental information are detected, though in a retrospective way, at the same time of current ones.

Let y_i the outcome variable (*manifest* response) referring to the i_{th} employed individual which is coded 1 if he/she is a self-employed and 0 if a salaried. As we assume that "individuals become and stay self-employed when the relative advantages are higher than in dependent employment" (Arum and Muller 2004), we

Table 2 Main individual characteristics by employment status: summary statistics

Variable	Whole sample		Wage-employed		Self-employed	
*Percent household status:						
- Head of Household (HH)	52.75	–	51.01	–	59.79	–
- HH's spouse/partner	28.28	–	28.94	–	25.65	–
- HH's son/daughter	18.97	–	20.05	–	14.56	–
Gender (1 if *male*)	59.78	(0.5075)	57.68	(0.5108)	68.22	(0.4842)
Citizenship (1 if *Italian*)	95.85	(0.2064)	95.33	(0.218)	97.94	(0.1477)
Age (*years*)	41.16	(10.86)	40.59	(10.67)	43.49	(11.3)
*Percent aged:						
- 16 to 19 years	1.01	–	1.21	–	0.23	–
- 20 to 24 years	4.34	–	4.82	–	2.38	–
- 25 to 34 years	22.13	–	23.19	–	17.85	–
- 35 to 44 years	35.35	–	34.89	–	37.21	–
- 45 to 54 years	25.73	–	26.02	–	24.56	–
- 55 to 64 years	10.53	–	9.48	–	14.74	–
- 65 years and older	0.91	–	0.39	–	3.03	–
Marital status (1 if *married*)	64.92	(0.494)	64.2	(0.4957)	67.8	(0.4858)
Education (*years of schooling*)	11.28	(3.558)	11.27	(3.48)	11.31	(3.861)
*Percent education level:						
- Low (ISCED97: 0; 1; 2A)	38.67	–	38.1	–	41	–
- Medium (ISCED97: 3)	46.75	–	48.14	–	41.13	–
- High (ISCED97: 5; 6)	14.58	–	13.76	–	17.87	–
Annual individual income	17,581	(20,365)	(16,160)	(8,808)	(23,603)	(42,713)
Home ownership (1 if *owner*)	69.96	(0.4854)	67.9	(0.4951)	77.02	(0.4435)

Source: Authors' elaborations on SHIW data(2006) – (Standard deviations in parentheses)

expect that there is an unobservable continuous variable (*latent* variable), y^*, which generates the observed binary y_i's. Thus, we believe that an underlying decisional process, based on comparison between the utilities of the two employment status, leads out to the choice to become a self-employed (j = S) rather than a salaried (j = E). In other words, each individual shows a vector of observed characteristics, X_i, and derives utility from his/her employment status j. At individual level, each utility function, U_{ij}, is composed of an observable utility, $U(X_i; j)$, and an idiosyncratic unobserved utility, u_{ij}. It is assumed that a person prefers to enter self-employment if the utility in this status is higher than the utility in wage-employment. In this light, by following some previous studies (Holtz-Eakin et al. 1994; Dunn and Holtz-Eakin 2000), we assume that utility depends on income as well as a vector of other individual characteristics and family-specific background.

The latent variable, i.e., the relative advantage to self-employment, is supposed to be linearly related to the observed x's through the structural model (Long 1997):

$$y_i^* = U\left(X_i; S\right) - U\left(X_i; E\right) + u_{iS} - u_{iE} = \alpha + x_i\beta + \varepsilon_i \quad y_i = \begin{cases} 1 & if \ y_i^* > \tau \\ 0 & if \ y_i^* \le \tau \end{cases} \tag{1}$$

where β $(=\beta_S - \beta_E)$ is a vector of parameters, ε_i, the error terms, are hypothesized to obey a standard logistic distribution – it is symmetric with mean zero and variance $\pi^2/3 \approx 3.29$ and remarkably similar in shape to the normal distribution with the advantage of a closed-form expression – and τ is the threshold.

Consequently, the positive outcome to become a self-employed ($y_i = 1$) only occurs when the latent response exceeds the threshold; the latter is equal to zero if the intercept is modeled or α if not. The probability of the positive outcome (π_i) is formulated in the cumulative standard logistic probability distribution function (Λ):

$$\pi_i = P(y_i = 1|X) = P\left(y_i^* > \tau\right) = \Lambda(X\beta) = \frac{\exp\left(X\beta\right)}{1 + \exp(X\beta)} \tag{2}$$

BRMs are estimated by maximum likelihood (ML) method, adopting the Fisher's Scoring Algorithm as optimization technique; since a large sample size is adopted in this work, the advantages of the ML asymptotic properties are taken.

3.1 A Set of Predictors for Individual Characteristics and as Proxy of the Different Forms of Capital and Regional Variations

By justifying the logit model as derivation from models of individual behaviour, it allows to define the probability (and to interpret it in utility terms) that the event to be a self-employed occurs. Several explanatory variables are tested according to a stepwise procedure; although some of them are directly available in SHIW data, some others have purposely been constructed on the basis of the same data.

While a first set of individual-level covariates detects some socio-demographic features, a second set is used as proxy for the measurement of workers' capital and both of them help in explaining the decision to enter self-employment. First, we talk about the human and social capital. The former is evaluated through the educational attainment, measured in years of education completed, or even by age, as a potential measure of work experience. The latter, specifically the parental role models and family background, is considered by analysing the impact to have one or both the parents more or less educated or even self-employed themselves; in particular, parents' education is expressed in terms of years spent in schooling by the parent with the higher education level, while parental work status is a derived variable which, in the estimation procedure, has been split into two dummies, using "*neither* of the parents is self-employed" as control group. Second, other two proxies are included to capture some financial and wealth aspects; in particular, the individual income should be cautiously interpreted because of potential endogeneity, since it may also result from the same individual's occupational status.

In addition, to evaluate how the employment choice is also affected by the socio-economic context, the model is enriched by a set of environmental attributes – i.e., unemployment rate, gross domestic product (GDP) and crime rate – related to each Italian region where the worker is located. Area-level variables are selected by the Istat data base of territorial indicators. In particular, GDP, tested to express the wealth level characterizing workers' environment, is computed in an instrumental way as the logarithm (*log*) of the effective value minus the log of the *min* value divided by the log of the *max* value minus the log of the *min* value. Anyway, data structure is not clustered/nested to justify the adoption of multi-level models.

4 Main Empirical Evidence and Effective Interpretation

As extensively illustrated so far, the models aim at sketching a general profile of several individual and family background determinants in employment decision-making. Although the choice of which parameterization to use is arbitrary, it does not affect the estimate of the slopes β's or associated significance tests but only the estimate of the intercept (α); also the probabilities are not affected by the identifying assumption (Long 1997). So, we assume $\tau = 0$; alternatively, we could assume that $\alpha = 0$ and estimate τ.

Broadly, we know the BRMs estimated fit the data well (Table 3) and models' convergence status is consistently satisfied. First, we focus on individual-level determinants of employment choice. Evidence of our analysis, substantially coherent with some empirical studies concerning other countries, highlight how being a *man* with *Italian citizenship* and *older* increases the likelihood to become a self-employed worker. As they say, years of experience in the labor market are often required before starting an activity on own-account; on the other hand, a person, only after a period of salaried work, might discover its own preference for self-employment to acquire autonomy. Moreover, our analysis points to a negative relationship between educational attainment and the probability of being or entering self-employment; as they say, a high level of education may reduce the probability of being self-employed. Lentz and Laband (1990) and Wit (1993) support this evidence arguing that several competences required to be self-employed would depend on the informal transmission of human or social capital and not necessarily through a formal education. Really, we also believe this reasoning may be essentially true for own-account or craft workers, small employers or similar, whose incidence, into the category of self-employment, is particularly high in Italy, but not for members of profession for which a formal education attainment is mandatory. Also, we observe that being currently *not-married* significantly increases the likelihood of self-employment; in other words, as highlighted by Dolton and Makepeace (1990), the family responsibilities seem to have a negative impact on risk-taking and, as consequence, would reduce the probability of being self-employed.

Table 3 Coefficients and standard errors of logit models with latent variable ($\tau = 0$)

Independent variables	Whole sample	Male	Female
Intercept	−3.8600*** (.7486)	−3.2552*** (.7756)	−4.7851** (2.2538)
Individual-level variables Socio-demographic attributes:			
Gender (1 if *male*)	0.4465*** (.1714)	–	–
Citizenship (1 if *Italian*)	0.6944** (.3513)	0.5191(.3632)	2.0672(1.5589)
Age (*years*)	0.0286** (.0133)	0.0192** (.0082)	−0.0052(.0149)
Marital status (1 if *married*)	−0.1591** (.0725)	−0.1454** (.0857)	−0.1734** (.0852)
Human and Social capital:			
Education (*years of schooling*)	−0.0679*** (.0179)	−0.0651*** (.0189)	−0.0874** (.0427)
Parents' educational level	−0.0127** (.0063)	−0.0122** (.0059)	−0.0119** (.0057)
Self-employed parents (1 if *both*)	0.0072** (.0035)	0.0067** (.0029)	0.0039** (.0018)
Self-employed parents (1 if *one*)	0.0051** (.0023)	0.0043** (.0021)	0.0034** (.0016)
Financial capital:			
Annual individual income	0.00002** (1E − 05)	0.00001** (4E − 06)	0.00001(7E − 06)
Home ownership (1 if *owner*)	0.2230(.1520)	0.2831*(.1685)	0.1752(.3758)
Area-level variables			
Unemployment rate	0.0239*(.0127)	0.0202*(.0120)	0.0751** (.0361)
Gross Domestic Product (GDP)	−0.0087** (.0043)	−0.0071** (.0036)	−0.0087** (.0043)
Crime rate	−0.0119** (.0059)	−0.0100*** (.0034)	−0.0092** (.0044)
Log likelihood	1,629.93	1,518.17	1,368.22

Source: Authors' elaborations on SHIW data(2006) – Sign. lev: ***99%; **95%; *90%

Second, as with age and individual education, also the *parents' education level* and the *parental work status*, as proxies for the measurement of workers' human and social capital in a generational perspective, have a significant effect (negative and positive, respectively) on the probability to be self-employed. In particular, the propensity to enter self-employment slightly enhances when both the parents are (or were) self-employed. As they say, self-employment tends to run in families, pointing clearly to strong intergenerational links between parents and children. A joint interpretation of age, generational dimension and financial aspects influencing the propensity to self-employment reminds that older persons are more likely to have received inheritances and to have accumulated capital which can be used to set up a business more cheaply or to overcome borrowing constraints.

Third, in addition to some other significant determinants as proxies of financial capital – i.e., to be home owner may directly influence the likelihood to enter self-employment and, more generally, it denotes a significant positive effect of personal wealth on self-employment propensities – it is worth examining the possible combined effects of individual-level and environmental variables. Briefly,

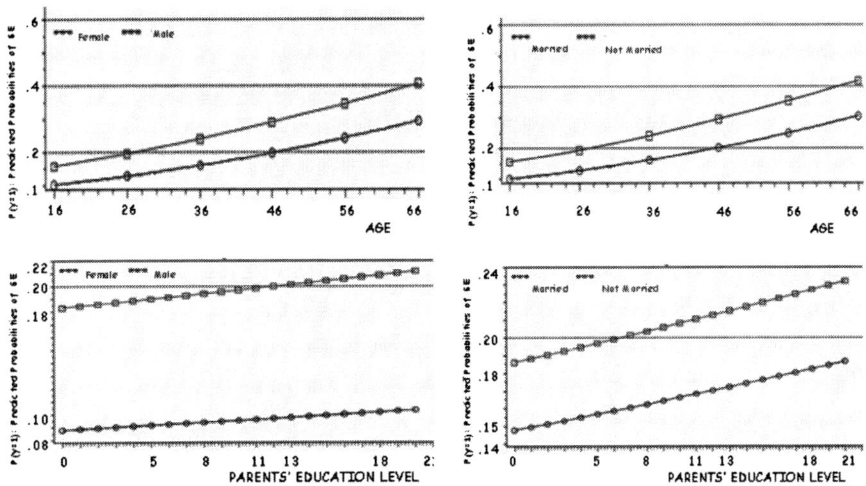

Fig. 1 Predicted probabilities of Self-Employment by gender (*left*) and marital status (*right*)

we note how people living in richer regions, in terms of higher GDP and lower unemployment rates, seems to be less likely to enter self-employment. As they say, self-em-ployment may be considered as a potential way *to wage a war against* the increasing unemployment. Crime levels also appear to negatively affect the propensity to be self-employed. In other words, it emerges a negative relationship between the self-employment choice and the level of "regional socio-economic well-being".

Finally, gender-separation models give further insights into gender differences with regard to reasons may contribute to understanding of why women are usually less likely to become self-employed. In particular, we stress, among other things, how the effect of having self-employed parents is stronger for males and how marriage usually reduces women's chances of self-employment more than men's.

Generally, the non-linearity of BRMs results in difficulty interpreting the effects of the covariates, each of them has a different impact on the dependent variable[1]. In other words, there is no way to directly interpret the coefficients substantively since a coefficient just stands for the expected change in logit, not in employment status, when the covariate increases by one unit. As widely discussed by Long (1997), an effective interpretation method is plotting predicted probabilities over the range of an interval independent variable (Fig. 1). In particular, it is worth stressing how both the age and the parents' education level (if considered alone) have consistently a positive effect on the probability of self-employment. With regard to gender,

[1]In this work, the non-linearity of BRMs estimated is confirmed by predicted probabilities out of the ranges 0.20 and 0.80 (Long 1997). Effective ranges are [0.0423; 0.9813] for the general model, [0.0461; 0.9712] and [0.0750; 0.7684] for the two specific models on male and female workers. Thus, the relationship between the x's and the probability is not approximately linear.

Table 4 Marginal and discrete changes for some explanatory variables (overall model)

Parameter	Discrete changes			Marginal change
	min *to* max	0 *to* 1	−sd/2 *to* +sd/2*	
Age	.2705	.0014	.0360	.0033
Education level	.1597	.0009	.0198	.0016
Parent's education	.0953	.0007	.0258	.0085
Self-Employment parents (*both*)	–	.0392	–	–
Self-Employment parents (*only one*)	–	.0235	–	–
Individual income	.0005	.0001	.0702	.00003
Gender	–	.0744	–	–
Citizenship	–	.0695	–	–
Marital status	–	.0056	–	–

Source: Authors' elaborations on SHIW data and Istat(2006) – **sd* means standard deviation

the gap between the two curves shows that male workers are more likely to enter self-employment than females and the distance slightly increases as workers' age increases, while it is basically steady over all the parents' education levels. Moreover, with regard to the parents' education, the gap between the two curves is lower when the marital status is considered as the variable of interest.

In BRMs the slope of the probability curve is not constant since it depends on the values of the independent variable of interest as well as the other covariates. In this light, the marginal and discrete changes (Long 1997) may be considered as other two effective interpretation methods to summarize the effects of each independent variable on the probability of an event occurring, holding all other variables constant. So, for example, the marginal change highlights that, for one increase in workers' age, the probability to be in self-employment is expected to increase by 0.0033, holding all other variables at the mean value. Most interestingly, the discrete change emphasizes how having both the parents self-employed increases the expected probability to enter self-employment more than having only one parent self-employed (0.0392 *vs* 0.0235) (Table 4).

5 Some Concluding Remarks and Further Developments

In sum, the main evidence of this work highlight, among other things, how family-specific background (and the socio-economic context) may strongly influence the choice to become self-employed (second generation) and how having two self-employed parents has the greatest overall effect. This is essentially due to the intergenerational transmission of methods, experience and knowledge. In other words, self-employed parents might furnish role models and help in accessing to financial capital and business networks. As further developments, we should cre-ate more insights into the heterogeneous category of self-employment; in particular, by testing latent variable models for categorical outcomes, we should aim at evaluating how the generational perspective may differently affect the main sub-groups of own-account or craft workers, entrepreneurs and members of profession.

References

Allen, D.: Social Networks and Self-Employment. Journal of Socio-Economics. **29**, 487–501 (2000)

Arenius, P., Minniti, M.: Perceptual Variables and Nascent Entrepreneurship. Small Business Economics. **24**, 233–247 (2005)

Arum, R., Muller, W.: The Re-emergence of Self-Employment: A Comparative Study of Self-Employment Dynamics and Social Inequality, Princeton University (2004)

Bank of Italy: Supplements to the Statistical Bulletin – Household Income and Wealth in 2006. **XVIII**, 7 (2008)

Colombier, N., Masclet, D.: L'importance de l'environnement familial comme déterminant du travail indépendant. Economie et Statistique. **405–406**, 99–118 (2007)

d'Addio, A.C.: Intergenerational Transmission of Disadvantage: Mobility or Immobility across Generations? A Review of the Evidence for OECD Countries. OECD Social, Employment and Migration Working Papers. **52** (2007)

Dolton, P.J., Makepeace, G.H.: The Earnings of Economics Graduates. Economic Journal. Royal Economic Society. **100 (127)**, 237–250 (1990)

Dunn, T., Holtz-Eakin, D.: Financial Capital, Human Capital and the Transition to Self-Employment: Evidence from Intergenerational Links. Journal of Labor Economics. **18** (2000)

Henrekson, M.: Is More Self-Employment Good or Bad? Comment on David Blanchflower, Conference Self-Employment and Entrepreneurship. Economic Council of Sweden, Stockholm, March 22 (2004)

Holtz-Eakin, D., Joulfaian, D., Rosen, H. Entrepreneurial Decisions and Liquidity Constraints, RAND Journal of Economics. **25 (2)**, 335–347 (1994)

Laferrere, A., McEntee, P.: Self-Employment and Intergenerational Transfers of Physical and Human Capital: An Empirical Analysis of French Data. The Economic and Social Review. **27 (1)**, 43–54 (1995)

Lentz, B., Laband, D.: Entrepreneurial Success and Occupational Inheritance Among Proprietors. Canadian Journal of Economics. **23**, 563–579 (1990)

Long, J.S.: Regression Models for Categorical and Limited Dependent Variables. Advanced Quantitative Techniques in the Social Sciences Series, **7**. Sage (1997)

Verheul, I., Wennekers, A.R.M., Audretsch, D.B., Thurik, A.R.: An Eclectic Theory of Entrepreneurship: Policies, Institutions and Culture, in: Entrepreneurship: Determinants and Policy in a European-US Comparison, Dordrecht: Kluwer Academic Publishers. 11–81 (2002)

Wit, G. de: Determinants of Self-Employment. Physica Verlag (1993)

Route-Based Performance Evaluation Using Data Envelopment Analysis Combined with Principal Component Analysis

Agnese Rapposelli

Abstract Frontier analysis methods, such as Data Envelopment Analysis (DEA), seek to investigate the technical efficiency of productive systems which employ input factors to deliver outcomes. In economic literature one can find extreme opinions about the role of input/output systems in assessing performance. For instance, it has been argued that if all inputs and outputs are included in assessing the efficiency of units under analysis, then they will all be fully efficient. Discrimination can be increased, therefore, by being parsimonious in the number of factors. To deal with this drawback, we suggest to employ Principal Component Analysis (PCA) in order to aggregate input and output data. In this context, the aim of the present paper is to evaluate the performance of an Italian airline for 2004 by applying a model based upon PCA and DEA techniques.

1 Introduction

Modern efficiency measurement begins with Farrell (1957), who introduced the seminal concept of technical efficiency. He drew upon the theory of Pareto optimality stating that a productive system is technically efficient if it either maximises output for a given amount of input or minimises input to achieve a given level of output. Moreover, he introduced the concept of the best practice frontier, also called efficiency frontier: according to him, the measure of technical efficiency is given by the relative distance between the observed production and the nearest benchmark production lying on the frontier. It was the seed for later exploitation, following its rediscovery by Charnes et al. (1978) and subsequent relabelling as CCR-efficiency under the broader heading of Data Envelopment Analysis (DEA) (Stone 2002).

A. Rapposelli (✉)
Dipartimento di Metodi Quantitativi e Teoria Economica, Università "G.D'Annunzio" Chieti-Pescara, Viale Pindaro, 42 Pescara
e-mail: agnese.rapposelli@virgilio.it

A. Di Ciaccio et al. (eds.), *Advanced Statistical Methods for the Analysis of Large Data-Sets*, Studies in Theoretical and Applied Statistics,
DOI 10.1007/978-3-642-21037-2_32,

DEA method measures technical efficiency relative to a deterministic best practice frontier, which is built empirically from observed inputs and outputs using linear programming techniques. Its main advantage is that it allows several inputs and several outputs to be considered at the same time.

The identification of the input and output variables to be used in an assessment of comparative performance is the most important stage in carrying out the assessment: in order to examine relative efficiency of a set of units it is necessary to define a production function which captures the key points of the production process (Coli et al. 2010). Apart of the nature of the inputs and outputs used in assessing efficiency, it must be remembered that questions can also be raised concerning the appropriate number of inputs and outputs for describing an activity process. Introduction of too many, and especially redundant, variables tend to shift the units towards the efficiency frontier, resulting in a large number of units with high efficiency scores (Golany and Roll 1989; Lin 2008). Also in DEA context the first problem in the selection of inputs and outputs is to include factors indiscriminately. As DEA allows flexibility in the choice of inputs and outputs weights, the greater the number of factors included the less discriminatory the method appears to be. In order to individuate a significant number of inefficient organisations, the literature suggests in the case of static analysis that the number of units has to be greater than $3(m+s)$, where $m + s$ is the sum of the number of inputs and number of outputs (Friedman and Sinuany-Stern 1998). Another suggested rule (Dyson et al. 2001) is that, to achieve a reasonable level of discrimination, the number of units has to be at least $2m \times s$. Thus the number of inputs and outputs included in a DEA assessment should be as small as possible in relation to the number of units being assessed. This can be achieved by using Principal Component Analysis (PCA), which is able to reduce the data to a few principal components whilst minimising the loss of information. This process provides therefore a more parsimonious description of a relatively large multivariate data set.

Following on from the above discussion, our objective is to adapt the techniques of efficiency measurement, such as DEA, to airline industry. The production process of air transportation services is characterised by multiple outputs and a large number of categories of costs (inputs) (Banker and Johnston 1994). Hence, this study proposes a modified DEA model that includes PCA results and apply it to measure the technical efficiency of the Italian airline Air One, by comparing its domestic routes for 2004.

The integration of both DEA and PCA techniques have already been proposed in literature. In the last decade, some researchers have made some contributions to combine PCA with DEA in the hope of improving the discriminatory power within DEA and achieving more discerning results. Adler and Golany (2001) tried to apply both DEA and PCA techniques to evaluate West–European airline network configurations, Adler and Berechman (2001) used this methodology to determine the relative quality level of West–European airports, Zhu (1998) and Premachandra (2001) applied the integrated approach to evaluate economic performance of Chinese cities, Liang et al. (2009) applied the PCA–DEA formulation to evaluate the ecological efficiency of Chinese cities. However, it is the first time to see the application of PCA–DEA formulation to route-based performance measurement: hence, this paper enhance the practicability of PCA–DEA.

This paper is organised as follows. Section 2 provides the technical framework for the empirical analysis, Sect. 3 describes inputs and outputs used and lists the results obtained. Finally, Sect. 4 presents conclusions of this study.

2 Methods

This section describes both DEA and PCA techniques and presents the PCA–DEA model to conducting performance measurement in the airline industry.

2.1 *Data Envelopment Analysis (DEA)*

DEA is a linear-programming technique for measuring the relative efficiency of a set of organisational units, also termed Decision Making Units (DMUs). Each DMU represents an observed correspondence of input–output levels.

The basic DEA models measure the technical efficiency of one of the set of n decision making units, DMU j_0, temporarily denoted by the subscript 0, in terms of maximal radial contraction to its input levels (input orientation) or expansion to its output levels feasible under efficient operation (output orientation). Charnes et al. (1978) proposed the following basic linear model, known as CCR, which has an input orientation and assumes constant returns to scale of activities (CRS):

$$e_0 = \min \theta_0 \text{ subject to}$$

$$\theta_0 x_{ij0} - \sum_{j=1}^{n} \lambda_j x_{ij} \geq 0, \quad i = 1, \ldots, m \tag{1}$$

$$\sum_{j=1}^{n} \lambda_j y_{rj} \geq y_{rj0}, \quad r = 1, \ldots, s \tag{2}$$

$$\lambda_j \geq 0, \quad \forall\ j \tag{3}$$

where y_{rj} is the amount of the r-th output to unit j, x_{ij} is the amount of the i-th input to unit j, λ_j are the weights of unit j and θ_0 is the shrinkage factor for DMU j_0 under evaluation. The linear programming problem must be solved n times, once for each unit in the sample, for obtaining a value of θ for each DMU. The efficiency score is bounded between zero and one: a technical efficient DMU will have a score of unity.

Subsequent papers have considered alternative sets of assumptions, such as Banker et al. (1984), who modified the above model to permit the assessment of the productive efficiency of DMUs where efficient production is characterised by variable returns to scale (VRS). The VRS model, known as BCC, differs from the basic CCR model only in that it includes the convexity constraint $\sum_{i=1}^{n} \lambda_j = 1$ in the previous formulation. This constraint reduces the feasible region for DMUs, which results in an increase of efficient units; for the rest, CRS and VRS models work in the same way.

In this study we choose a VRS model, that is also in line with the findings of Pastor (1996) and we use the following output-oriented BCC formulation of DEA method:

$$e_0 = \max \phi_0 \text{ subject to}$$

$$\sum_{j=1}^{n} \lambda_j x_{ij} \leq x_{ij0}, \quad i = 1, \ldots, m \tag{4}$$

$$\phi_0 y_{rj0} - \sum_{j=1}^{n} \lambda_j y_{rj} \leq 0 \quad r = 1, \ldots, s \tag{5}$$

$$\sum_{i=1}^{n} \lambda_j = 1, \tag{6}$$

$$\lambda_j \geq 0, \quad \forall\, j \tag{7}$$

where ϕ_0 is the scalar expansion factor for DMU j_0. DMU j_0 is said to be efficient, according to Farrell's definition, if no other unit or combination of units can produce more than DMU j_0 on at least one output without producing less in some other output or requiring more of at least one input.

2.2 *The PCA–DEA Formulation*

As stated in Sect. 1, Principal Component Analysis (PCA) is a multivariate statistical method devised for dimensionality reduction of multivariate data with correlated variables. This technique accounts for the maximum amount of the variance of a data matrix by using a few linear combinations (termed principal components) of the original variables. The aim is to take p variables X_1, X_2, ..., X_p and find linear combinations of them to produce principal components X_{PC1}, X_{PC2}, ..., X_{PCp} that are uncorrelated. The principal components are also ordered in descending order of their variances so that X_{PC1} accounts for the largest amount of variance, X_{PC2} accounts for the second largest amount of variance, and so on: that is, $\text{var}(X_{PC1}) \geq \text{var}(X_{PC2}) \geq \ldots \geq \text{var}(X_{PCp})$. Often much of the total system variability can be accounted for by a small number k of the principal components, which can then replace the initial p variables without much loss of information (Johnson and Wichern 2002).

We have already highlighted that an excessive number of inputs and outputs will result in an excessive number of efficient units in a basic DEA model: the greater the number of input and output variables, the higher the dimensionality of the linear programming solution space, and the less discerning the analysis. Dyson et al. (2001) argued that omitting even highly correlated variables could have a major influence on the computed efficiency scores. To deal with this drawback, PCA can be combined with DEA to aggregate and then to reduce inputs and outputs. We can use principal component scores instead of original inputs and outputs variables (Adler and Golany 2001): they can be used to replace either all the inputs and/or outputs simultaneously or alternatively groups of variables (Adler and Yazhemsky 2010).

The general DEA formulation has to be modified to incorporate principal components directly into the linear programming problem: hence, the constraints have to be derived from the principal components of original data (Ueda and Hoshiai1997). In particular, constraint (4) is replaced with the following:

$$\sum_{j=1}^{n} \lambda_j x_{PCij} \leq x_{PCij0} \tag{8}$$

$$\sum_{j=1}^{n} \lambda_j x_{Oij} \leq x_{Oij0} \tag{9}$$

and constraint (5) is replaced with the following:

$$\phi_0 y_{PCrj0} - \sum_{j=1}^{n} \lambda_j y_{PCrj} \leq 0 \tag{10}$$

$$\phi_0 y_{Orj0} - \sum_{j=1}^{n} \lambda_j y_{Orj} \leq 0 \tag{11}$$

where x_{Oij} and y_{Orj} denote original input variables and original output variables respectively.

The combination of PCA and DEA techniques enable us to overcome the difficulties that classical DEA models encounter when there is an excessive number of inputs or outputs in relation to the number of DMUs, whilst ensuring very similar results to those achieved under the original DEA method. The advantage of this technique is also that it does not require additional expert opinion (Adler and Berechman 2001), unlike the earliest approach to reducing the number of variables.

3 Case Study

As mentioned in the introduction, this study evaluates the comparative performance of Air One domestic routes for the year 2004.

Set up in 1995, Air One was the leading privately owned domestic operator in Italy. It was a lower cost airline but not low cost (or "no frills" carrier) as it did not fit the low fare model (Lawton 2002). Air One began operating with domestic flights: in addition to the increase in domestic operations (35% of market share and 20 airports served), it expanded its offer by opening international routes. Scheduled passenger air service was the company's core business and was generating approximately 80% of Air One's revenues. In addition to scheduled airline's service, Air One was also operating charter flights and executive flights for passengers and freight, including its postal service. It was also offering maintenance and handling services (Air One 2005). On 13th January 2009, Air One became part of Compagnia Aerea Italiana (CAI), which has taken over Alitalia and Air One as one whole company.

3.1 Data

The sample analysed comprises 30 domestic routes. In order to respect homogeneity assumptions about the units under assessment, we have not included international routes, seasonal destinations and any routes which started during the year 2004.

The domestic airline industry provides a particularly rich setting for this empirical study. In order to assess Air One domestic routes, the inputs and the outputs of the function they perform must be identified. However, there is no definitive study to guide the selection of inputs and outputs in airline applications of efficiency measurement (Nissi and Rapposelli 2008). In the production process under analysis we have identified seven inputs and four outputs to be included in the performance evaluation. The input selected are the number of seat available for sale, block time hours and several airline costs categories such as total variable direct operating costs (DOCs), total fixed direct operating costs (FOCs), commercial expenses, overhead costs and financial costs.

We give a brief overview of inputs used. The number of seats available for sale reflects aircraft capacity. Block time hours is the time for each flight sector, measured from when the aircraft leaves the airport gate to when it arrives on the gate at the destination airport. With regard to the costs categories considered, variable or "flying" costs are costs which are directly escapable in the short run, such as fuel, handling, variable flight and cabin crew expenses, landing charges, passenger meals, variable maintenance costs. These costs are related to the amount of flying airline actually does, hence they could be avoided if a flight was cancelled (Doganis 2002). Fixed or "standing" costs are costs which are not escapable in the short or medium term, such as lease rentals, aircraft insurance, fixed flight and cabin crew salaries, engineering overheads. These costs are unrelated to amount of flying done, hence they do not vary with particular flights in the short run; they may be escapable but only after a year of two, depending on airlines (Doganis 2002). Both DOCs and FOCs are dependent on the type of aircraft being flown (Holloway 1997). Commercial expenses, such as reservations systems, commissions, passengers reprotection, lost and found, and overhead costs, such as certain general and administrative costs which do not vary with output (legal expenses, buildings, office equipment, advertising, etc.), are not directly dependent on aircraft operations (Holloway 1997). Finally, we have included financial costs, such as interests, depreciation and amortisation.

With regard to the output side of the model, the output variables are the number of passengers carried, passenger scheduled revenue, cargo revenue and other revenues. Passenger scheduled revenue is the main output for a passenger focused airline, cargo revenue includes outputs that are not passenger-flight related such as freight and mail services and other revenues includes charter revenue and a wide variety of non-airline businesses (incidental services) such as ground handling, aircraft maintenance for other airlines and advertising and sponsor. Even if incidental services are not airline's core business, they are considered in the production process under evaluation because they utilise part of the inputs included in the analysis (Oum and Yu 1998).

All data have been developed from Financial Statements as at 31st December 2004 and from various internal reports.

3.2 *Empirical Results*

The airline production process defined in this empirical study is characterised by a higher number of inputs than those considered in a previous study (Nissi and Rapposelli 2008), where only two categories of costs have been included as inputs in the DEA model. Unlike the previous paper, in this study all the original input variables are very highly correlated. This is therefore good material for using PCA to produce a reduced number of inputs by removing redundant information.

Table 1 gives the eigenanalysis of the correlation matrix of data set. The first principal component X_{PC1} explains 98.22% of the total variance of the data vector, so the input variables will be included in the DEA model via the first principal component.

It should be noted that principal components used here are computed based on the correlation matrix rather than on covariance, as the variables are quantified in different units of measure. Generally inputs and outputs of DEA models need to be strictly positive, but the results of a PCA can have negative values (Adler and Berechman 2001). It has been argued (Pastor 1996) that the BCC output-

Table 1 Eigenvalues and total variance explained

Component	Eigenvalue	Proportion (%)	Cumulative (%)
1	6.876	98.22	98.22
2	7.420×10^{-2}	1.06	99.28
3	3.509×10^{-2}	0.50	99.78
4	1.410×10^{-2}	0.20	99.98
5	6.151×10^{-4}	8.788×10^{-3}	99.99
6	3.485×10^{-4}	4.978×10^{-3}	100.00
7	3.428×10^{-13}	4.897×10^{-12}	100.00

Table 2 Efficiency ratings of Air One domestic routes

DMU	Score	DMU	Score	DMU	Score	DMU	Score	DMU	Score	DMU	Score
TT	1	Z	1	EE	0.9462	F	0.7912	H	0.7370	JJ	0.6800
A	1	V	1	L	0.8418	U	0.7808	N	0.7159	OO	0.6608
CC	1	J	1	T	0.8275	W	0.7650	QQ	0.7080	SS	0.6431
VV	1	E	0.9866	FF	0.8232	Q	0.7547	G	0.6997	PP	0.6296
DD	1	S	0.9674	X	0.8027	AA	0.7541	B	0.6833	MM	0.6198

oriented model used in the current study is input translation invariant and vice versa. Hence the efficiency classification of DMUs is preserved if the values of principal component X_{PC1} are translated by adding a sufficiently large scalar β (1 in this case) such that the resulting values are positive for each DMU j.

We apply therefore DEA on the translated first component and not on the whole set of the original input variables. In order to incorporate X_{PC1} directly into the linear programming problem the general DEA formulation has to be modified (Sect. 2.2). With regard to the output variables, they are not included in terms of principal components. Hence, only constraint (8) and (11) are used in the DEA model applied.

The DEA analysis has been performed by using DEA-Solver software (Cooper et al. 2000). The efficiency scores of Air One routes, in descending order of efficiency, are shown in Table 2.

Two different remarks can be made. The average level of technical efficiency is 0.8273. Eight routes are fully efficient and three more routes are quite close to the best practice frontier. On the other hand, the remaining DMUs are sub-efficient but they do not show very low ratings. These results suggest that Air One routes are operating at a high level of efficiency, although there is room for improvement in several routes.

4 Conclusions and Future Research

It is well known that the discriminatory power of DEA often fails when there is an excessive number of inputs and outputs in relation to the number of DMUs (Adler and Golany 2002). We have introduced a new model formulation within DEA framework that can be used in efficiency measurement when there is a large number of inputs and outputs variables that can be omitted with least loss of information. This approach has been illustrated with an application to an Italian airline.

However, these results can be improved. This study suggest three main avenue for future research. First of all, further research could include the presence of undesirable outputs (Liang et al. 2009) in the PCA–DEA model proposed, such as the number of delayed flights, which may reflect the service quality of the network. Besides, the usefulness of the method could be explored for large data sets: for example, in further application studies we could add international routes or other air carriers. Finally, we would like to explore the combination between canonical correlation analysis (CCA) and DEA in next future with another data set. We have not used CCA in this paper because generally it is applied when the number of inputs and the number of outputs are very high and when we are not able to distinguish between the input set and the output set, but in this case study we perfectly know which the inputs and the outputs are.

References

Adler, N., Berechman, J.: Measuring airport quality from the airlines' viewpoint: an application of data envelopment analysis. Transp. Policy. **8**, 171–181 (2001)

Adler, N., Golany, B.: Evaluation of deregulated airline networks using data envelopment analysis combined with principal component analysis with an application to Western Europe. Eur. J. Oper. Res. **132**, 260–273 (2001)

Adler, N., Golany, B.: Including principal component weights to improve discrimination in data envelopment analysis. J. Oper. Res. Soc. **53**, 985–991 (2002)

Adler, N., Yazhemsky, E.: Improving discrimination in data envelopment analysis: PCA–DEA or variable reduction. Eur. J. Oper. Res. **202**, 273–284 (2010)

Air One S.p.A.: Annual Report 2004. Rome (2005)

Banker, R.D., Johnston, H.H.: Evaluating the impacts of operating strategies on efficiency in the U.S. airline industry. In: Charnes, A., Cooper, W.W., Lewin, A.Y., Seiford, L.M. (eds.) Data envelopment analysis: theory, methodology and applications, pp. 97–128. Kluwer, Boston (1994)

Banker, R.D., Charnes, A., Cooper, W.W.: Some models for estimating technical and scale inefficiencies in Data Envelopment Analysis. Manag. Sci. **30**, 1078–1092 (1984)

Charnes, A., Cooper, W.W., Rhodes, E.: Measuring the efficiency of decision making units. Eur. J. Oper. Res. **2**, 429–444 (1978)

Coli, M., Nissi, E., Rapposelli, A.: Efficiency evaluation in an airline company: some empirical results. J. Appl. Sci. **11**, 737–742 (2011)

Cooper, W.W., Seiford, L.M., Tone, K.: Data envelopment analysis, a comprehensive test with models, applications, references and DEA-Solver software. Kluwer, Boston (2000)

Doganis, R.: Flying off course: the economics of international airlines. Routledge, London (2002)

Dyson, R.G., Allen, R., Camanho, A.S., Podinovski, V.V., Sarrico, C.S., Shale, E.A.: Pitfalls and protocols in DEA. Eur. J. Oper. Res. **132**, 245–259 (2001)

Farrell, M.J.: The measurement of productive efficiency. J. R. Stat. Soc. Ser. A. **120**, 253–281 (1957)

Friedman, L., Sinuany-Stern, Z.: Combining ranking scales and selecting variables in the DEA context: the case of industrial branches. Comput. Opt. Res. **25**, 781–791 (1998)

Golany, B., Roll, Y.: An application procedure for Data Envelopment Analysis. Manag. Sci. **17**, 237–250 (1989)

Holloway, S.: Straight and level. practical airline economics. Ashgate, Aldershot (1997)

Johnson, R.A., Wichern, D.W.: Applied multivariate statistical analysis. Prentice-Hall, Upper Saddle River (2002)

Lawton, T.C.: Cleared for Take-off. Ashgate, Aldershot (2002)

Liang, L, Li, Y., Li, S.: Increasing the discriminatory power of DEA in the presence of the undesirable outputs and large dimensionality of data sets with PCA. Expert Syst. Appl. **36**, 5895–5899 (2009)

Lin, E.T.: Route-based performance evaluation of Taiwanese domestic airlines using data envelopment analysis: a comment. Transp. Res. E **44**, 894–899 (2008)

Nissi, E., Rapposelli, A.: A data envelopment analysis study of airline efficiency. In: Mantri, J.K. (eds.) Research methodology on data envelopment analysis, pp. 269–280. Universal-Publishers, Boca Raton (2008)

Oum, T.H., Yu, C.: Winning airlines: productivity and cost competitiveness of the world's major airlines. Kluwer, Boston (1998)

Pastor, J.T.: Translation invariance in data envelopment analysis: a generalization. Ann. Op. Res. **66**, 93–102 (1996)

Premachandra, I.M.: A note on DEA vs. principal component analysis: an improvement to Joe Zhu's approach. Eur. J. Oper. Res. **132**, 553–560 (2001)

Stone, M.: How not to measure the efficiency of public services (and how one might). J. R. Stat. Soc. Ser. A. **165**, 405–434 (2002)

Ueda, T., Hoshiai, Y.: Application of principal component analysis for parsimonious summarization of DEA inputs and/or outputs. J. Oper. Res. Soc. Jpn. **40**, 446–478 (1997)

Zhu, J.: Data envelopment analysis vs. principal component analysis: an illustrative study of economic performance of Chinese cities. Eur. J. Oper. Res. **111**, 50–61 (1998)

Part IX
WEB and Text Mining

Web Surveys: Methodological Problems and Research Perspectives*

Silvia Biffignandi and Jelke Bethlehem

Abstract This paper presents a framework of current problems and related literature on Internet/web surveys. It proposes a new classification of research topics on challenging issues. Thereby it takes into account application the role that these surveys play in different research contexts (official statistics, academic research, market research). In addition critical research, open questions and trends are identified. Furthermore a specific section is devoted to bias estimation, which is a critical point in this type of survey. In particular, original bias and variance definitions are proposed.

1 Introduction

There is a growing interest for using web surveys (also called Internet surveys or online surveys) for data collection. This is not surprising as the Internet provides easy access to a large group of potential respondents, and conducting such a survey is relatively cheap and fast. Moreover, costs do not depend too much on the number of interviews. Notwithstanding the appeal of web surveys, there are serious methodological problems. These problems have to be solved in order to obtain reliable survey outcomes. In practice, many surveys are carried out without

---

*The present paper is financially supported from the ex 60% research funds 2007 and 2008 of the University of Bergamo, research responsible Silvia Biffignandi.

S. Biffignandi (✉)
Bergamo University, Via Caniana n. 2, 24127 Bergamo, Italy
e-mail: silvia.biffignandi@unibg.it

J. Bethlehem
Statistics Netherlands, P.O. Box 24500, 2490 HA The Hague, The Netherlands
e-mail: jbtm@cbs.nl

A. Di Ciaccio et al. (eds.), *Advanced Statistical Methods for the Analysis of Large Data-Sets*, Studies in Theoretical and Applied Statistics,
DOI 10.1007/978-3-642-21037-2_33, © Springer-Verlag Berlin Heidelberg 2012

paying much attention to the problems involved in each step of the survey process and on the reliability of the collected data. It is therefore very important: (a) to draw the attention of the practitioners of web surveys to methodological problems and to the need to be cautious when carrying out such surveys and using the results for decision support; and (b) to pay attention to studies investigating methodological aspects that may undermine data quality of web surveys.

WebSm (WebSurvey Methodology) is a European project that has created – since 2003 – a network of partners[1] which have set up a website which publishes in an well organized framework the literature, software references and comments (Lozar Manfreda and Vehovar 2006). The research team formed for this project is currently studying methodological problems in web surveys. This research group periodically (each 2 years) organizes a workshop discussing new frontiers and progress in this research area. Already live workshops have been organized; the last two at the University of Bergamo in 2009 and in the Hague in 2011.

Different studies generally focus on specific aspects or on literature reviews, facing issues from a specific point of view (i.e. from sociological point of view, from marketing or from official statistics on so on). A first book covering every aspect of both theoretical problem and reactional issues in web surveys is Bethlehem and Biffignandi (2012). In our paper, we present an original framework which refers to the different research contexts (official statistics, academic research, market research). In addition, we identify critical research, open questions and trends. Moreover we devote a specific paragraph is to bias estimation; this is a critical and not solved point; in our paper original bias and variance definitions are proposed.

There are two major research areas that are of great importance and on which we focus in our paper:

1. *Design of survey questionnaires (Sect. 2)*. Web surveys can use all kinds of new tools such as sound, video, colours and animation. How such tools will affect response behaviour is not yet completely clear. Therefore, one should be careful when designing a web survey.
2. *Recruitment of participants*. How can a representative sample be selected? If a sample is not representative, can it be corrected such that it becomes representative? And what are the effects of nonresponse? Issues like data collection modes, recruitment, inference and bias are discusses in Sects. 3 and 4.

The impact of these problems and possible solutions depend to a large extent on the type of population which is studied (for example, approaches may be different for household surveys and establishment surveys) and on the type of organization carrying out the web survey (official statistics, universities or market research companies).

It is important to use a proper terminology. Internet surveys are surveys that collect data over the Internet. This includes e-mail surveys, where the respondent

[1] Partners of the European project WebSM (Web Survey Methodology) are: University of Ljubliana, University of Bergamo, Linkoeping University, ZUMA.

returns the questionnaire by email, attaching responses written in a Word or Excel document, or other software which is available to the respondent (for example Blaise). So, the questionnaire is completed off-line. Web surveys are completed on-line. The questionnaire is presented to respondents as a set of web pages. Answers to questions are transmitted immediately after clicking on a submit/next button.

2 Questionnaire Design

Many methodological problems are common to both email and web surveys. Questionnaire design is more critical and relevant for web surveys since they may use special tools such as visual animation, sounds and colours. Therefore, the focus of this paper is on web surveys, but it should be taken into account that many problems are shared with other types of Internet surveys.

Web surveys are self-administered. Therefore, user-friendliness is of the utmost importance. If respondents encounter difficulties when completing the questionnaire, they quit, leaving the researcher with incomplete and often unusable data. Visual elements, pictures, colours and sounds are tools that can make the questionnaire more attractive. Technical limitations (for example low modem speeds), however, may prohibit their use, and an overload of not always relevant information may distract and confuse respondents. The impact of advanced graphical techniques and sound tools on the respondent's behaviour needs more study. It has to be demonstrated that they have a positive impact on the survey participation and data quality. An extensive discussion and review of the features of web questionnaire design is given in Couper (2008). Recent studies have underlined that web questionnaire design problems should be treated in a manner consistent with the way in which sophisticated questionnaire designs problems are addressed (Dillman 2007). Questionnaire length is an important aspect both with respect to the data collection mode and with respect to the target population of a web survey. Actually, in web surveys the questionnaire should be quite short (Czaja and Blair 2005). Nevertheless if panels, closed groups or special target populations are investigated, a longer questionnaire could be effective as well.

3 Data Collection Modes, Recruitment and Inference

Web surveys are substantially affected by coverage problems, because not every member of a target population may have access to the Internet. Moreover, it is often difficult to select a proper probability sample because a proper sampling frame is lacking.

More and more, a web survey is considered as one of the modes in a mixed-mode survey. If the objective is to keep survey costs as low as possible, one could start with a web survey, re-approach the nonrespondents by means of telephone interviewing, and re-contact the remaining nonrespondents in a face-to-face survey. Unfortunately, a mixed-mode survey introduces new problems, like mode effects, e.g. the same question is answered differently in a different data collection mode. Occurrence and treatment of mixed-mode effects need further investigation.

In principle, web surveys can be carried out very quickly. It turns out response times to Internet or web surveys are different from other modes and best procedures have to be identified. For instance, Biffignandi and Pratesi (2000, 2002) have studied response behaviour and the impact of the frequency of reminder procedures. Recent studies have been focusing on the best reminder tools, including sms (which seems to perform well).

Other studies concentrate on keeping respondents focused on the relevant parts of the computer screen, and keeping distraction to a minimum. One of the analysis techniques used is eye-tracking analysis.

Representativity of web surveys is an important issue. Scientifically meaningful results can only be obtained if a proper probability sample is selected and the selection probabilities are known and positive for every member of the population. Unfortunately, many web surveys rely on self-selection of respondents. The survey is simply put on the web. Respondents are those people who happen to have Internet, visit the website and decide to participate in the survey. The survey researcher is not in control of the selection process. These surveys are called self-selection surveys. The bias of different sample selection procedures in web surveys is a key issue. We will focus on this issue on the next section.

Similarly, Internet based panels (so-called access panels) are constructed by wide appeals on well-visited sites and Internet portals. So, they are also based on self-selection. At time of registration, basic demographic variables are recorded. A large database of potential respondents is created in this way. For future surveys, samples are selected from this database. Only panel members can participate in these web panel surveys. The target population is unclear. One can only say that it consists of people who have an Internet connection, who have a non-zero probability of being confronted with the invitation, and decide to participate in the panel. Research in the Netherlands has shown that panel members differ from the general population (Vonk et al. 2006).

Access panels have the advantage that values of basic demographic variables are available for all participants. So the distribution of these variables in the survey can be compared with their distribution in the population. Over- or under-representation of specific groups can be corrected by some kind of weighting adjustment technique. However, there is no guarantee that this leads to unbiased estimates.

To allow for unbiased estimation of the population distribution, a reference survey can be conducted that is based on a true probability sample from the entire target population. Such a reference survey can be small in terms of the number of questions asked. It can be limited to the so-called "webographic" or "psychographic" questions. Preferably, the sample size of the reference survey should be large to allow for precise estimation. A small sample size results in large standard errors of estimates (Bethlehem 2007). Since most (but not all) access panels are based on self-selection, it is impossible to compute unbiased estimates of population characteristics. In order to properly apply statistical inference, probability sampling in combination with a different data collection mode can be applied for panel recruitment. For example, a sample is selected from the general population and respondents are recruited using e.g. CAPI or CATI. Respondents without Internet are provided with Internet access. See Scherpenzeel (2008) for an example. A probabilistic sampling design can be achieved by using some specific methods, such as random digit dialling (RDD). Some probability-based Internet panels have already been constructed in this way (for instance, Huggins and Krotki 2001).

Another approach to correct for lack of representativity is applying propensity scoring methodology. Propensity scores (Rosenbaum and Rubin 1983) have been used to reweight web survey results (Schonlau et al. 2009; Biffignandi et al. 2003; Biffignandi and Pratesi 2005).

4 Bias in Web Surveys

Probability sampling is a crucial prerequisite for making proper inference about the target population of a survey. In web surveys this prerequisite is difficult to achieve. In this section we formalize the bias related to four different approaches.

1. *Selection of a random sample without replacement from the Internet-population.* This is the ideal situation, but it would require a sampling frame listing all elements having access to the Internet. In many practical cases, no such list exists. One empirical approach which allows for a good approximation of the random sample situation is to select a random sample from a larger sampling frame (e.g. a population or address register). The selected sample is contacted by mail or telephone, or another traditional way, then Internet users are asked to complete the questionnaire, With this approach every element in the Internet-population has a positive and known probability ρ_k of being selected for $k = 1, 2, \ldots, NI$. If this is the case, an unbiased estimate of the Internet population mean can be defined (according Horvitz and Thompson 1952). The estimator of the Internet population could be biased with reference to the mean of the total population. The bias is determined by the relative size of non-Internet population and the difference of the mean value of the target variable in the two populations. Summing up, the bias can be written as follows:

$$B(\overline{y}_{HT}) = E(\overline{y}_{HT}) - \overline{Y} = \overline{Y}_I - \overline{Y} = \frac{N_{NI}}{N}(\overline{Y}_I - \overline{Y}_{NI}) \qquad (1)$$

where
N_I = Internet population
$\overline{Y}_I$ = Internet population mean
$\overline{y}_{HT}$ = Horvitz–Thompson mean estimator
$\overline{Y}_{NI}$ = Non internet population.

2. *Self-selection sample.* In many Internet surveys respondents are recruited by means of self-selection. Therefore, each element k in the Internet population has an unknown probability ρ_k of participating in the survey. Since the survey is put on the Web, we might introduce the response propensity of the non-Internet population, which will be 0. Generally, the expected value of the sample mean is not equal to the population mean of the Internet-population. The estimator of the Internet population therefore will be unbiased only if there is no relationship at all between the mechanism causing missingness and target variables of the survey (*Missing Completely At Random* (MCAR)). In general the bias of the sample mean (Bethlehem 2008; Bethlehem and Stoop 2007) can be written as:

$$B(\overline{y}_S) = \frac{C(\rho, Y)}{\overline{\rho}} \tag{2}$$

where

$$C(\rho, Y) = \frac{1}{N_I} \sum_{k=1}^{N} I_k(\rho_k - \overline{\rho})(Y_k - \overline{Y}) \tag{3}$$

is the covariance between the participation probabilities and the values of the survey variable. The bias is determined by the average response propensity and the relationship between the target variable and response behaviour affect the sample mean bias. The bias will be smaller the more likely people are to participate in the survey, i.e. the higher is the average response propensity. The bias will be high in the case the correlation between the values of the target variable and response propensities is high. It has been shown (Bethlehem 2008) that self-selection can cause estimates of population characteristics to be biased, similarly to traditional probability sampling based surveys nonresponse. Since in a web survey the response rate can be very low, the bias in self-selection surveys can be substantially larger than in traditional surveys. If we need to estimate the mean of the total population the bias is computed as follows:

$$B(\overline{y}_S) = E(\overline{y}_S) - \overline{Y} = E(\overline{y}_S) - \overline{Y}_I + \overline{Y}_I - \overline{Y} = \frac{N_{NI}}{N}(\overline{Y}_I - \overline{Y}_{NI}) + \frac{C(\rho, Y)}{\overline{\rho}} \tag{4}$$

i.e. the bias of under coverage (since we have investigated only the Internet population) is added to the self-selection bias due to the Internet population.
3. *Auxiliary variables.* Improved estimation solutions in web surveys are related to the use of auxiliary information, i.e. a set of variables that have been measured in the survey, and for which information on their population distribution is available.

   (a) A possible estimation procedure is based on post-stratification. A difference between the population distribution of a (categorical) auxiliary variable and its sample distribution suggest the need of correcting the sample estimates since the sample is selective. It can be assessed whether or not the sample is representative for the population (with respect to this variable). Adjustment weights are computed on the basis of one or more categorical auxiliary variables. If there is a strong relationship between the target variable Y and the stratification variable X the strata are homogeneous with respect to the target variable. There will be no bias within strata, and post-stratification will remove the bias. If there is an indirect relationship between the mechanism causing missingness and the target variables of the survey, this is called Missing At Random (MAR). The relationship runs through a third variable, and this variable is measured in the survey as an auxiliary variable. In this case estimates are biased, but it is possible to correct for this bias. The bias is due to the differences between Internet and non-Internet population within the strata. The bias of this estimator is equal to

$$B(\overline{y}_{I,PS}) = E(\overline{y}_{I,PS}) - \overline{Y} = \widetilde{Y}_I - \overline{Y} = \sum_{h=1}^{L} W_h \left(\overline{Y}_I^{(h)} - \overline{Y}^{(h)}\right) =$$

$$= \sum_{h=1}^{L} W_h \frac{N_{NI,h}}{N_h} \left(\overline{Y}_I^{(h)} - \overline{Y}_{NI}^{(h)}\right). \qquad (5)$$

(b) A second possible solution, if auxiliary variables are not available, consists of conducting a reference survey. This reference survey is based on a small probability sample. Data collection takes place with a mode different. This means there may be mode effects that have an impact on estimates. Needless to say that such a reference survey will dramatically increase survey costs.

At least one categorical auxiliary variable should be observed both in the web survey and the reference survey. We assume that this variable has a strong correlation with the target variable of the survey. We might apply a kind of post-stratification – like the one described above – using the reference survey data (as a proxy of the population distribution) to estimate the weights and the web survey data to compare distributions. Bethlehem (2008) shows that that, if a reference survey is used, the variance of the post-stratification estimator is equal to

$$V(\overline{y}_{I,RS}) = \frac{1}{m} \sum_{h=1}^{L} W_h \left(\overline{Y}_I^{(h)} - \widetilde{Y}_I\right)^2 + \frac{1}{m} \sum_{h=1}^{L} W_h(1 - W_h) V\left(\overline{y}_I^{(h)}\right)$$

$$+ \sum_{h=1}^{L} W_h^2 V\left(\overline{y}_I^{(h)}\right) \qquad (6)$$

The bias of the estimator is related to the relationship between the target variable and the auxiliary variable used for computing the weights. The bias is small if the target variable has small variability within each stratum.

The bias can be written as

$$B(\overline{y}_{I,RS}) = {}_I E(\overline{y}_{I,RS}) - \overline{Y} = \widetilde{Y}_I - \overline{Y} = \sum_{h=1}^{L} W_h \left(\overline{Y}_I^{(h)} - \overline{Y}^{(h)}\right) =$$

$$= \sum_{h=1}^{L} W_h \frac{N_{NI,h}}{N_h} \left(\overline{Y}_I^{(h)} - \overline{Y}_{NI}^{(h)}\right). \qquad (7)$$

and it is small if the mean of the target variable is very similar in the Internet and in non-Internet population.

The approach based on the use of a reference survey to weight estimation can be successful if such variables can be measured both in the web survey and in the reference survey. One of the advantages of a reference survey is that the best auxiliary variables can be selected for weighting and this will make correction more effective. A disadvantage of a reference survey is that it results in large standard errors. So a reference survey reduces the bias at the cost of a loss in precision.

Empirical analyses have shown that webographics variables seem to work well. Psychographic variables, attitudinal or lifestyle variables often explain at least part of the response behavior. Some authors show, however, that a good explanation of the response behavior is not the rule and sometimes these variables are lacking explanatory power (see for instance Schonlau et al. 2003). In addition, caution is required since attitudinal questions are much less reliable than factual questions. Respondents may be asked about issues they never have thought about, and current circumstances may influence their answers. Therefore, measurement errors could greatly affect the accuracy of answers to attitudinal questions.

The reference survey only works well if it is a real probability sample without nonresponse, or with ignorable nonresponse (MCAR). In practice, reference survey estimates are biased due to nonresponse, therefore the web survey bias is replaced by a reference survey bias. This leaves the problem open.

(c) Another approach related to the use of auxiliary variables is propensity weighting. The idea has been proposed by Rosenbaum and Rubin (1983,1984). Propensity weighting is a often applied by commercial market research agencies, for instance Harris Interactive (Terhanian et al. 2001). Propensity scores are obtained by modelling a variable that indicates whether or not someone participates in the web survey. The *propensity score* $\rho(X)$ is the conditional probability that a person with observed characteristics X participates, i.e.

$$\rho(X) = P(r = 1 | X). \qquad (8)$$

Usually a logistic regression model is used where the indicator variable is the dependent variable and auxiliary variables (mostly attitudinal variables are the explanatory variables). These variables are assumed to explain why someone participates or not. Fitting the logistic regression model comes down estimating the probability (propensity score) of participating, given the values of the explanatory variables. From a theoretical point of view propensity weighting should be sufficient to remove the bias. Assuming that the observed strata have the same propensity score, this can be modelled by a logit model under assumption of missing at random (MAR)

$$\log\left(\frac{\rho(X_k)}{1-\rho(X_k)}\right) = \alpha + \beta' X_k. \tag{9}$$

The estimated propensity scores are used to stratify the population. Each stratum consists of elements with (approximately) the same propensity scores. Five strata are usually sufficient to remove a large part of the bias (Cochran 1968). If indeed all elements within a stratum have the same response propensity, there will be no bias if just the elements in the Internet population are used for estimation purposes. Figure 1 shows the steps of the propensity score procedure.

Shortly speaking the nearest neighbour matching method selects as the match the non-participant with the value of P_j that is closest to P_i. The caliper approach is a variation of nearest neighbor: A match for person i is selected only if the difference is within a pre-specified tolerance (.. The $|P_i - P_j| < \varepsilon,\ j \in I_0$ 1-to-1 nearest neighbor caliper approach is the most common practice. Mahalanobis Metric Matching: (with or without replacement) without p-score consists on randomly ordering subjects, calculate the distance between the first participant and all non-

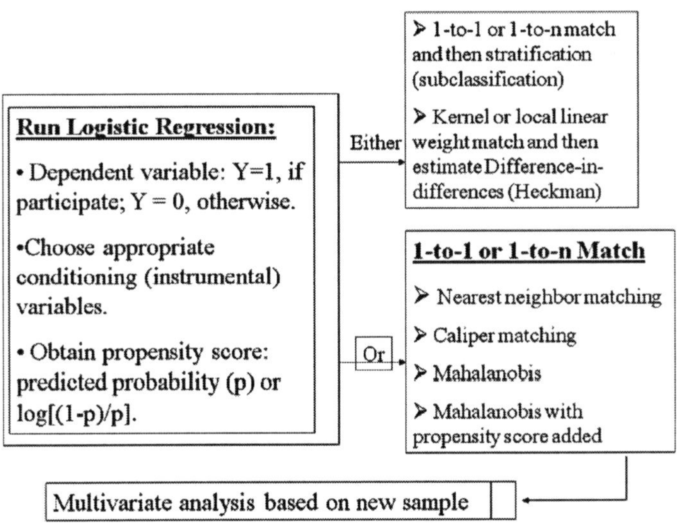

Fig. 1 General procedure for propensity score approach

participants. The distance, $d(i, j)$ can be defined by the Mahalanobis distance:

$$d(i, j) = (u - v)^T C^{-1} (u - v) \tag{10}$$

where u and v are values of the matching variables for participant i and non-participant j, and C is the sample covariance matrix of the matching variables from the full set of nonparticipants. Mahalanobis metric matching with p-score consists simply in adding p-score to u and v. Nearest available Mahalandobis metric matching within calipers defined by the propensity score is computed applying calipers criteria.

Summing up, various approaches using auxiliary variables might be applied in estimating web survey results. Weighting techniques (including propensity weighting) can help to reduce the bias, but only if the sample selection mechanism satisfies the Missing at Random (MAR) condition. This is a strong assumption. It requires weighting variables that show a strong relationship with the target variables of the survey and the response probabilities. Often such variables are not available.

5 Final Comments

The paper reviews main research issues and results of web survey methodology with an original, wide approach that takes into account different field of application (official statistics, academic research and market research) and research viewpoints. It should be underlined that, since some problems are similar, exchanging experiences between university researchers and data collectors working at statistical agencies and at market research firms can be useful in improving methodology of web surveys as well as getting support from empirical evidence. Nevertheless, different challenges that call for different approaches are to be faced in household and business surveys. Household surveys are voluntary (both in private surveys as well as in official surveys), there is no good sampling frame from where one can draw e-mail addresses and, moreover, computer access and competence differ between different populations segments. Business surveys, on the contrary, in official statistics are generally compulsory. Therefore in this case generally we do not face a big nonresponse problem. In such a case the main issues are how the businesses should be convinced that web is better than paper and next to construct web questionnaires that really are better and that makes the respondents feel the response burden is low. As stated in Biffignandi (2010) new research topics are becoming more and more important. Challenges for the near future the use of mixed mode surveys and of incentives to increase participation.

References

Bethlehem J., Biffignandi S. (2012) Handbook of wep surveys, Wiley & Sons, New Jersey.

Bethlehem J. (2007) Reducing the bias of web surveys estimates, CBS Discussion paper 07001, Voorbug/Heelen.

Bethlehem J.G. (2008) How accurate are self-selection web surveys?. Discussion Paper 08014, Statistics Netherlands, The Hague/Heerlen, The Netherlands.

Bethlehem, J.G., Stoop, I. (2007) Online panels - a Paradigm Theft? In M. Trotman et al. (Eds.), The challenges of a changing world. Southampton, UK: ASC, 113–131.

Biffignandi S. (2010) Internet Survey Methodology - recent trends and developments, in International Encyclopedia of Statistical Science, Lovric, Miodrag (Ed.), Springer, ISBN: 978–3–642–04897–5.

Biffignandi S., Pratesi M. (2000) Modelling firms response and contact probabilities in Web surveys, ICESII Proceeding Conference, Buffalo, USA, June 2000.

Biffignandi S., Pratesi M. (2002) Modeling The Respondents' Profile In A Web Survey On Firms In Italy, In A. Ferligoj, A. Mrvar (ed.), Developments in Social Science Methodology, Metodoloski zvezki, 18, 171–185.

Biffignandi S., Pratesi M. (2005) Indagini Web: propensity scores matching e inferenza. Un'analisi empirica e uno studio di simulazione, in Integrazione di dati di fonti diverse: Falorsi P., Pallara A., Russo A. (editors), Angeli, Milano, 131–158.

Biffignandi S., Pratesi M., Toninelli D. (2003) Potentiality of propensity scores methods in weighting for Web surveys: a simulation study based on a statistical register, ISI Conference, Berlino, August (CD).

Cochran W.G. (1968) The effectiveness of adjustment by subclassification in removing bias in observational studies. Biometrics 24.

Couper M.P. (2008) Designing Effective Web Surveys, Cambridge Univ. Press, N.Y..

Czaja R., Blair J. (2005) Designing surveys: A guide to decisions and procedures (2, Sage, Oaks CA.

Dillman D. (2007) Mail and Internet Surveys, The Tailored DsiMethod,Wiley.

Huggins V., Krotki K. (2001) Implementation of nationally representative web-based surveys, Proceedings of the Annual Meeting of the ASA, August 5–9.

Lozar Manfreda K., Vehovar V. (2006) Web survey methodology (WebSM) portal, in B. Aaron, D. Aiken, R.A. Reynolds, R. Woods, J.D. Baker (Eds), Handbook on Research on Electronic Surveys and Measurement, Hershey PA, 248–252.

Rosenbaum, P.R. & Rubin, D.B. (1983), The central role of the propensity score in observational studies for causal effects. Biometrika 70, pp. 41–55.

Rosenbaum, P.R. & Rubin. D.B. (1984), Reducing bias in observational studies using subclassification on the propensity score. JASA, 79, pp. 516–524.

Scherpenzeel, A. (2008), An online panel as a platform for multi-disciplinary research. In: I. Stoop & M. Wittenberg (eds.), Access Panels and Online Research, Panacea or Pitfall?, Aksant, Amsterdam, 101–106.

Schonlau, M., Fricker, R.D. & Elliott, M.N. (2003), Conducting Research Surveys via E-mail and the Web. Rand Corporation, Santa Monica, CA.

Schonlau M., van Soest A., Kapteyn A., Couper M. (2009) Selection Bias in Web Surveys and the Use of Propensity Scores, Sociological Methods & Research, Vol. 37, No. 3, 291–318.

Terhanian, Marcus, Bremer, Smith (2001)Reducing error associated with non-probability sampling through propensity scores, JSM, Atlanta, Georgia, USA.

Vonk, T., Van Ossenbruggen, R. and Willems, P. (2006) The effects of panel recruitment and management on research results, a study among 19 online panels. Panel Research 2006, ESOMAR World Research, ESOMAR Publ.Serv. 317, 79–99.

Semantic Based DCM Models for Text Classification

Paola Cerchiello

Abstract This contribution deals with the problem of documents classification. The proposed approach is probabilistic and it is based on a mixture of a Dirichlet and Multinomial distribution. Our aim is to build a classifier able, not only to take into account the words frequency, but also the latent topics contained within the available corpora. This new model, called $sbDCM$, allows us to insert directly the number of topics (known or unknown) that compound the document, without losing the "burstiness" phenomenon and the classification performance. The distribution is implemented and tested according to two different contexts: on one hand, the number of latent topics is defined by experts in advance, on the other hand, such number is unknown.

1 Introduction

Text categorization can be considered one of the key techniques for handling and organizing data in textual format. Text categorization approaches play a fundamental role in the text mining field and are typically used to classify new documents. Since building text classifiers by hand is difficult, time-consuming and often not efficient, it is worthy to learn classifiers from experimental data.

The literature on text classification problems is quite vast and prolific. Among the proposed models we can mention: the latent semantic analysis (*LSA*) (Deerwester et al., 1990), the probabilistic latent semantic analysis (*pLSA*) (Hofmann, 1999), the latent Dirichlet allocation (*LDA*) (Blei et al., 2003), the correlated topic model (*CTM*) (Blei and Lafferty, 2006) and finally the Independent Factor Topic Models

P. Cerchiello (✉)
Department of Statistics and Applied Economics 'L.Lenti', University of Pavia, Corso Strada Nuova 65, 27100 Pavia Italy
e-mail: paola.cerchiello@unipv.it

A. Di Ciaccio et al. (eds.), *Advanced Statistical Methods for the Analysis of Large Data-Sets*, Studies in Theoretical and Applied Statistics,
DOI 10.1007/978-3-642-21037-2_34,

(*IFTM*) (Putthividhya et al., 2009). All those models are considered generative approaches since they try to represent the word generation process by introducing suitable distributions, in particular the multinomial and the Dirichlet random variables. The more complicated version of those generative models introduces the concept of topics and the relative correlation among them.

Similarly, another interesting research path focuses on the burstiness phenomenon, that is the tendency of rare words, mostly, to appear in burst: if a rare word appears once along a text, it is much more likely to appear again. The above mentioned generative models are not able to capture such peculiarity, that instead is very well modelled by the Dirichlet compound multinomial model (*DCM*). Such distribution was introduced by statisticians (Mosimann, 1962) and has been widely employed by other sectors like bioinformatics (Sjolander et al., 1996) and language engineering (Mackay and Peto, 1994). An important contribution in the context of text classification was brought by Minka (2003) and Madsen et al. (2005) that profitably used *DCM* as a bag-of-bags-of-words generative process. Similarly to *LDA*, we have a Dirichlet random variable that generates a multinomial random variable for each document from which words are drawn. By the way, *DCM* cannot be considered a topic model in a way, since each document derives specifically by one topic. That is the main reason why Doyle and Elkan (2009) proposed in 2009 a natural extension of the classical topic model *LDA* by plugging into it the *DCM* distribution obtaining the so called *DCMLDA*.

Following that research path, we move from *DCM* approach and we propose an extension of the *DCM*, called "semantic-based Dirichlet Compound Multinomial" (*sbDCM*), that permits to take directly topics into account.

2 Dirichlet Compound Multinomial

The Dirichelet Compound Multinomial (or *DCM*) distribution (see Minka (2003), Madsen et al. (2005)) is a hierarchical model where the Dirichlet random variable is devoted to model the Multinomial word parameters θ (the whole set of words as bag-of-bags) and the Multinomial variable models the word count vectors ($\bar{x}$ as bag-of-words) for each document. The distribution function of the (*DCM*) mixture model is:

$$p(\bar{x}|\alpha) = \int_{\theta} p(\bar{x}|\theta)p(\theta|\alpha)d\theta, \tag{1}$$

where the $p(\bar{x}|\theta)$ is the Multinomial distribution and $p(\theta|\alpha)$ is the Dirichlet distribution.

From another point of view, each Multinomial is linked to specific sub-topics and makes, for a given document, the emission of some words more likely than others. Instead the Dirichlet represents a general topic that compounds the set of documents and thus the (*DCM*) could be also described as "bag-of-scaled-documents".

The added value of the (DCM) approach consists in the ability to handle the "burstiness" of a rare word without introducing heuristics. Burstiness is the phenomenon according to which, if a rare word appears once along a text, it is much more likely to appear again.

3 A sbDCM Model

In the (DCM) we have a coefficient (α_w) for each word compounding the vocabulary of the set of documents which is called "corpus". The (DCM) model can be seen as a "bag-of-scaled-documents" where the Dirichlet takes into account a general topic and the Multinomial some specific sub-topics.

Our aim in this contribution is to build a framework that allows us to insert specifically the topics (known or unknown) that compound the document, without losing the "burstiness" phenomenon and the classification performance. Thus we introduce a method to link the α coefficients to the hypothetic topics, indicated with $\beta = \{\beta_t\}$, by means of a function $\alpha = F(\beta)$ which must be positive in β since the Dirichlet coefficients are positive. Note that usually $dim(\beta) << dim(\alpha)$ and, therefore, our proposed approach is parsimonious.

Substituting the new function into the integral in (1), the new model is:

$$p(\bar{x}|\beta) = \int_{\theta} p(\bar{x}|\theta)p(\theta|F(\beta))d\theta. \tag{2}$$

We have considered as function $F(\beta)$ a linear combination based on a matrix $\mathbf{D}$ and the vector $\bar{\beta}$. $\mathbf{D}$ contains information about the way of splitting among topics the observed count vectors of the words contained in a diagonal matrix $\mathbf{A}$ and $\bar{\beta}$ is a vector of coefficient (weights) for the topics. More specifically we assume that:

$$\mathbf{A} = \begin{pmatrix} w_1 & & & \\ & \cdot & & \\ & & w_w & \\ & & & \cdot \\ & & & & w_W \end{pmatrix}, \mathbf{D} = \begin{pmatrix} d_{11} & \cdot & \cdot & \cdot & d_{1T} \\ \cdot & \cdot & & & \cdot \\ \cdot & & d_{wt} & & \cdot \\ \cdot & & & \cdot & \cdot \\ d_{W1} & \cdot & \cdot & \cdot & d_{WT} \end{pmatrix}, \quad \bar{\beta} = \begin{pmatrix} \beta_1 \\ \cdot \\ \beta_t \\ \cdot \\ \beta_T \end{pmatrix} \mathbf{D}^* = \mathbf{A} \times \mathbf{D}$$

$$F(\beta)\text{=}\mathbf{D}^* \times \bar{\beta} = \bar{\alpha} = \begin{pmatrix} \alpha_1 \\ \cdot \\ \alpha_w \\ \cdot \\ \alpha_W \end{pmatrix} \tag{3}$$

Note that:

- $\alpha_w = \sum_t^T d_{wt}^* \beta_t$, with T the number of Topics;
- d_{wt} is the coefficient for word w-th used to define the degree of belonging to topic t-th and by which a portion of the count of word w-th is assigned to that particular topic t-th;
- $d_{wt}^* = w_w \times d_{wt}$.

By substituting this linear combination into integral (2), we obtain the same distribution but with the above mentioned linear combination for each α:

$$p(\bar{x}|\beta) = \frac{n!}{\prod_w^W x_w} \frac{\Gamma\left(\sum_w^W \sum_t^T d_{wt}^* \beta_t\right)}{\Gamma\left(\sum_w^W (x_w + \sum_t^T d_{wt}^* \beta_t)\right)} \prod_w^W \frac{\Gamma\left(x_w + \sum_t^T d_{wt}^* \beta_t\right)}{\Gamma\left(\sum_t^T d_{wt}^* \beta_t\right)}; \quad (4)$$

This model is a modified version of the (*DCM*), henceforth semantic-based (*DCM*) (*sbDCM*) and the new log-likelihood for the set of documents (C) becomes:

$$log(p(C|\beta)) = \sum_d^N \left(log\Gamma\left(\sum_w^W \sum_t^T d_{wt}^* \beta_t\right) - log\Gamma\left(x_d + \sum_w^W \sum_t^T d_{wt}^* \beta_t\right)\right) +$$

$$+ \sum_d^N \sum_w^W \left(log\Gamma\left(x_{dw} + \sum_t^T d_{wt}^* \beta_t\right) - log\Gamma\left(\sum_t^T d_{wt}^* \beta_t\right)\right) \quad (5)$$

In Fig. 1 we report the graphical representation of the new model where the α's are substituted by a function of the β's.

An important aspect of the proposed approach is represented by the number T of topics to be inserted into the *semantic-based DCM*.

T can be:

1. a priori known (i.e. fixed by field experts);
2. unknown, thus to be estimated on the basis of the available data.

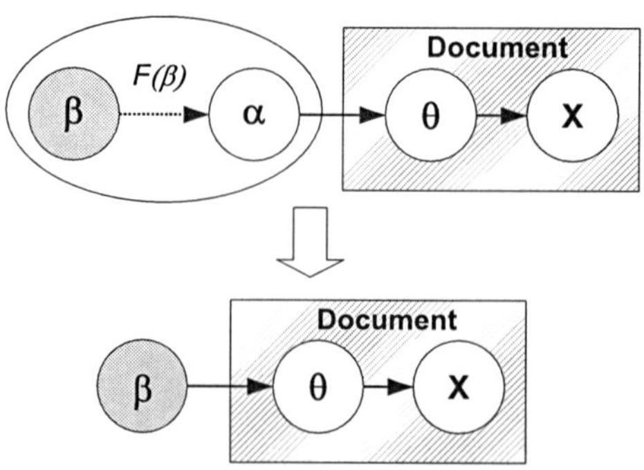

Fig. 1 Hierarchical model sbDCM representation

We treat both the cases, reporting empirical results produced on the basis of real data.

4 sbDCM with T Unknown

Since it is not always possible to known in advance the number of topics contained in a corpora, it becomes very useful to arrange a statistical methodology for discovering efficiently and consistently a suitable number of topics.

In our context we tackle the problem as follows: in order to generate the coefficients contained within the matrix **D** we have used a clustering procedure to group the words (Giudici 2003). The idea is to create groups of words sharing common characteristics that can be interpreted as conceptual topics. Such objective can be accomplished by applying a hierarchical cluster analysis on the association matrix calculated on the complete set of words count. Later on, once completed the analysis by choosing the best number of groups, the cluster distance matrix is used to set the membership percentage (d_{wt}) of each word to each topic. The best number of groups has been chosen by using a combined strategy based on statistical indexes that measure the quota of variability between and within groups. Moreover dendograms and anova test have been investigated to strengthen the analysis.

The matrix (G), employed to obtain words clusters, has been produced by applying the Kruskal–Wallis index (g) to the words count matrix given the qualitative nature of the data. We recall that index g is defined as follows:

$$g = \frac{\frac{12}{N(N+1)} \sum_{i=1}^{s} n_i \left(\bar{r}_i - \frac{N+1}{2}\right)^2}{1 - \frac{\sum_{i=1}^{p}(c_i^3 - c_i)}{N^3 - N}} \tag{6}$$

where n_i is the number of sample data, (N) the total observation number in the s samples, s the number of samples to be compared and $\bar{r}_i$ the average rank of i-th group. The denominator of the index g is a correction factor needed when tied data is present in the data set, where p is the number of recurring ranks and c_i is the recurring frequency of i-th rank.

The training dataset contains 2,051 documents with a vocabulary of 4,096 and the evaluation dataset contains 1,686 documents which are distributed over 46 classes. In Table 1 we list the results obtained by varying the number of clusters (i.e. topics) and we report the most significant ones (5, 11, 17, 23, 46). We indicate with LL_{in} and LL_{out} the log-likelihood before and after the iteration procedure for the parameters updating which is stopped when the predefined error ϵ (10^{-10}) is reached. $AICc_{in}$ and $AICc_{out}$ correspond to the corrected Akaike Information Criterion (AICc) before and after the parameters uploading.

As we can see in Table 1, the percentages of correct classification ($Ind1$) are very close to the original ones with a parameter for each word (4,096 parameters). Of course they depend on the type of classifier employed during the classification

Table 1 Classification results by varying the number of clusters and using matrix (**G**)

Classifier	Measures	sbDCM_5	sbDCM_11	sbDCM_17	sbDCM_23	sbDCM_46	DCM
	LL_{in}	−291,257	−283,294	−270,360	−266,453	−258,061	−222,385
	LL_{out}	−205,912	−204,647	−204,600	−204,604	−204,362	−205,286
	$AICc_{in}$	582,524	566,610	540,754	532,952	516,214	454,264
	$AICc_{out}$	411,834	409,316	409,234	409,254	408,816	420,066
NORMAL	$Ind1$	67.83%	67.71%	67.47%	67.42%	67.65%	67.66%
COMP.	$Ind1$	67.95%	68.66%	68.55%	68.72%	68.60%	68.78%
MIXED	$Ind1$	68.07%	68.13%	67.83%	67.71%	67.89%	68.07%

step. Considering both $sbDCM$ and *DCM*, the differences produced by varying the number of groups are small. Moreover the AICc is always better in the new approach then considering each word as a parameter (*DCM* model). Moreover, if we perform an asymptotic chi-squared test (χ^2_{test}) to decide whether the difference among log-likelihoods (LL), with respect to *DCM*, are significant (i.e. the difference is statistically meaningful if the $|LL_1 - LL_2|$ is greater than 6), we get that the approach based on matrix **G** has the best performance.

5 Semantic-based (*DCM*) with T Known in Advance

A different approach needs to be assessed when the number of available topic T is known in advance. In fact a text corpora could be enriched by several descriptions of treated topics according to the knowledge of field experts.

In more details, the analysis could be provided with a priori knowledge based on ontological schemas that describe the relations among concepts with regards to the general topics of the corpora. An ontology (from which an ontological schema is derived) is a formal representation of a set of concepts within a domain and the relationships between those concepts.

For example, if a text set deals with operational risk management, an ontology can be created on the basis of the four categories of problems (internal processes, people, systems and external events) defined by Basel II Accords. Hence we can suppose that some specific sub-topics, such as those possible operational problems, will be almost surely treated along the texts. Thereby, the ontology structure provides the set of relations among the concepts to which can be associated, by a field expert(s), a certain number of key words. In this line of thinking, we want to use the classes of a given ontology and the associated key words to define the number of topics (T) in advance.

5.1 The Reputational Risk

The ontological schema which we refer to, deals with the so called reputational risk. It is not simple to define and consequently to measure and to monitor the reputation concept since it involves intangible assets such as: honor, public opinion, perception, reliability, merit. By the way, it is a matter of fact, that a bad reputation can seriously affect and condition the performance of a company. Moreover companies tend to act once the adverse event has occurred. According to such approach we can say that there is not a risk management activity, but only a crisis management one.

Reputational risk is generically acknowledged as the ensemble of the economic consequences directly correlated to an alteration of the public evaluation of an entity, such as a public company or a person with a public relevance. Regulators, industry groups, consultants and individual companies have developed elaborate guidelines over the years for assessing and managing risks in a wide range of areas. However, in the absence of agreement on how to define and measure reputational risk, it has been often ignored. An interesting aspect embedded in reputational risk is the possible gap between the perception and the reality: if the reputation of a company is more positive than its underlying reality, this gap leads to a substantial risk.

On the other hand media coverage plays a key role in determining a company reputation. This often occurs when a company reputation has been significantly damaged by unfair attacks from special interest groups or inaccurate reporting by the media. A detailed and structured analysis of what media actors are saying is especially important because the media shape the perceptions and expectations of all the involved actors. Natural language processing technologies enable these services to scan a wide range of outlets, including newspapers, magazines, TV, radio, and blogs. In order to enable the application of the classification textual model $sbDCM$ we have collaborated with the Italian market leader company in financial and economic communication "IlSole24ORE".

5.2 Data Analysis

"IlSole24ORE" team provided us with a set of 870 articles about Alitalia, an Italian flight company, covering a period of one year (sept '07-sept '08).

The 80% of the articles is used to train the model and the remaining 20% to validate the process. The objective is to classify the articles on the basis of the reputation ontology in order to understand the argument treated in the articles. The ontology classes used for the classification are:

- *Identity*: The "persona" of the organization.
- *Corporate Image*: It can be considered the immediate picture of the institution as perceived by stakeholder groups and it of course involves brand value.
- *Integrity*: Personal inner sense of "wholeness" deriving from honesty and consistent uprightness of character.
- *Quality*: The achievement of excellence of an entity. Quality is sometimes certificated by a third part.
- *Reliability*: Ability of a system to perform/maintain its functions in routine and also in different hostile or/and unexpected circumstances. It involve customer satisfaction and customer fidelization.

- *Social Responsibility*: It is a doctrine claiming that an organization or individual has a responsibility to society. It involves foundation campaign and sustainability.
- *Technical Innovation*: The Introduction of new technical products or services. It measures the "RD orientation" of an organization (only for companies).
- *Value For Money*: The extent to which the utility of a product or service justifies its price.

Those classes define the concept of reputation of a company. To link the ontology classes to the textual analysis we use a set of key words for each class of the reputation schema. The available articles are in Italian thus the key words are in Italian as well. For example, since the concept of Reliability involves customer satisfaction and customer fidelization, we employ the following set of key words: *affidabilitá, fiducia, consumatori, risorsa, organizzazione, commerciale, dinamicitá, valore, mercato*. On the basis of these key words, we perform a cluster analysis considering the 9 classes. From the clustering, we derive the matrix **D**.

For classification purposes, we consider topics as document classes to predict and the available news paper articles are 1,321. We compare the classification performance of (*DCM*) and $sbDCM$ by dividing the dataset into training (with 80% of documents) and validation set (with 20% of documents) and with a vocabulary of 2,434 words.

We evaluate the performance by using three kind of discriminant rules, developed in (Rennie et al., 2003) and in addition we propose other four rules through their combination. All the classifiers select the document class (i.e. topic T) with the highest posterior probability:

$$l(d) = argmax_t \left[log\ p(\grave{}_t) + \sum_{w=1}^{W} f_w\ log\ \theta_{tw} \right] \quad (7)$$

where f_w is the frequency count of word w in a document, $p(\grave{}_t)$ is a prior distribution over the set of topics (that we consider uniformly distributed) and $log(\theta_{tw})$ is the weight for word w with regards to topic t.

The weight for each topic is estimated as a function of the α coefficients:

$$\hat{\theta_{tw}} = \frac{N_{tw} + \alpha_w}{N_t + \sum_{w=1}^{N_t} \alpha_w} \quad (8)$$

where N_{tw} is the number of times word w appears in the documents of topic t, N_t the total number of words occurrences in topic c.

We then use comparatively DCM and sbDCM introducing the relative α's parameters into the "Normal", "Mixed" and "Complement" classifier (see Rennie et al. (2003)) and evaluate them on the basis of two appropriate indexes (Table 2):

1. *(Ind1)* The proportion of true positive over the total number of test-documents.
2. *(Ind2)* The proportion of true positive within each class over the number of test documents present in the class.

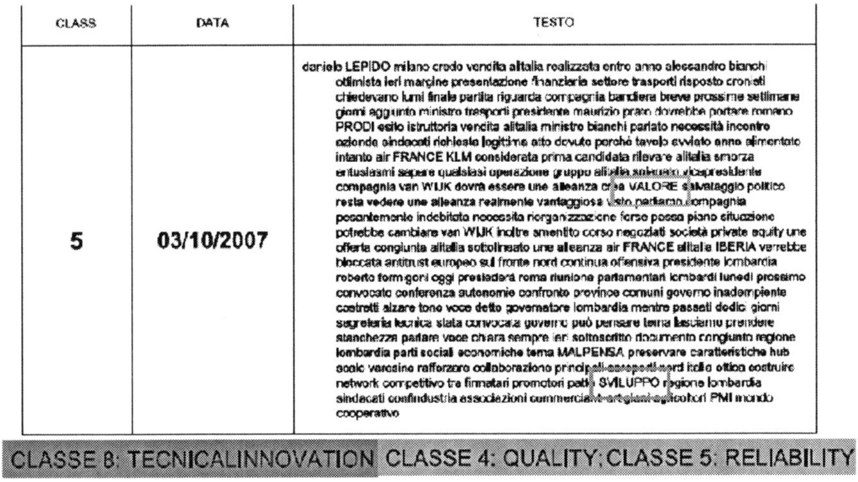

CLASS	DATA	TESTO
5	03/10/2007	daniele LEPIDO milano credo vendita alitalia realizzata entro anno alessandro bianchi ottimista ieri margine presentazione finanziaria settore trasporti risposto cronisti chiedevano lumi finale partita riguarda compagnia bandiera breve prossime settimane giorni aggiunto ministro trasporti presidente maurizio prato dovrebbe portare romano PRODI esito istruttoria vendita alitalia ministro bianchi parlato necessità incontro aziende sindacati richiesta legittima atto dovuto perché tavolo avviato anno alimentato intanto air FRANCE KLM considerata prima candidata rilevare alitalia smorza entusiasmi sapere qualsiasi operazione gruppo alitalia ... vicepresidente compagnia van WIJK dovrà essere une alleanza crea VALORE salvataggio politico resta vedere une alleanza realmente vantaggiosa visto parliamo compagnia pesantemente indebitata necessita riorganizzazione forse possa piano situazione potrebbe cambiare van WIJK inoltre smentito corso negoziati società private equity une offerta congiunta alitalia sottolineato une alleanza air FRANCE alitalia IBERIA verrebbe bloccata antitrust europeo sul fronte nord continua offensiva presidente lombardia roberto formigoni oggi presiederà roma riunione parlamentari lombardi lunedì prossimo convocato conferenza autonomie confronto province comuni governo inadempiente costretti alzare tono voce detto governatore lombardia mentre passati dodici giorni segreteria tecnica stata convocata governo può pensare terna lasciamo prendere stanchezza parlare voce chiara sempre ieri sottoscritto documento congiunto regione lombardia parti sociali economiche tema MALPENSA preservare caratteristiche hub scalo varesino rafforzare collaborazione principali aeroporti nord italia ottica costruire network competitivo tra firmatari promotori patto SVILUPPO regione lombardia sindacati confindustria associazioni commercianti artigiani agricoltori PMI mondo cooperativo

CLASSE 8: TECNICALINNOVATION CLASSE 4: QUALITY; CLASSE 5: RELIABILITY

Fig. 2 Example of classification via $sbDCM$ and a Naive Bayes Classifier. The key words belong to three different classes: technical innovation; quality; reliability

Table 2 Classification results with T known in advance

Classifier	Measures	sbDCM_8	DCM
	LL_{in}	−105.226	−105.385
	LL_{out}	−98.412	−98.286
	$AICc_{in}$	264.462	254.264
	$AICc_{out}$	110.834	112.066
NORMAL	$Ind1$	68.13%	65.66%
//	$Ind2$	62.32%	61.61%
COMP.	$Ind1$	67.1%	68.78%
//	$Ind2$	66%	67.89%
MIXED	$Ind1$	68.07%	66.07%
//	$Ind2$	63.87%	63.87%

The empirical results show good performance in terms of correct classification rate. In Fig. 2 we report an example of output of classification with the coefficients α calculated via $sbDCM$ and the Naive Bayes Classifier.

6 Conclusion

In this contribution we study a methodology useful to classify textual data. Several approaches have been proposed in literature and among them we pay our attention to the (*DCM*) distribution. It is a hierarchical model where the Dirichlet random variable is devoted to model the Multinomial word parameters θ (the whole set of words as bag-of-bags) and the Multinomial variable models the word count vectors ($\bar{x}$ as bag-of-words) for each document. Moving from (*DCM*), we propose a new distribution called $sbDCM$ in which we insert directly the topics treated along the corpora. The topics T can be known in advance (i.e. fixed by field experts) or unknown (i.e. to be estimated on the data).

Two different analysis have been carried out based on the knowledge on T. Both the approaches show comparable results with the standard model ((*DCM*)). Moreover an original contribution is presented for the case T fixed in advance: in order to allow the evaluation of reputational risk, the $sbDCM$ model is trained to classify articles dealing with the reputation of a given company.

Acknowledgements The Author acknowledges financial support from European project MUSING (2006–2010). The Author finally acknowledges Elvio Concetto Bonafede for having supervised the computational aspects.

References

Blei, D. M. & Lafferty, J. D. Correlated topic models, Advances in Neural Information Processing Systems, 18, (2006).

Blei, D. M., Ng, A. Y. & Jordan, M. I., Latent Dirichlet allocation, Journal of Machine Learning Research, 3, pp.993-1022, (2003).

Deerwester, S., Dumais, S., Furnas, G. W., Landauer, T. K. & Harshman, R., Indexing by Latent Semantic Analysis, Journal of the American Society for Information Science, 41, (6), pp. 391-407, (1990).

Doyle, G. & Elkan, C., Accounting for burstiness in topic models. Proceeding of International Conference on Machine Learning, 36, (2009).

Giudici P., Applied Data Mining: Statistical Methods for Business and Industry, Wiley John & Sons, England (2003).

Hofmann, T., Probabilistic Latent Semantic Indexing, Proceedings of Special Interest Group on Information Retrieval SIGIR, (1999).

Mackay, D. J. C. & Peto, L., A Hierarchical Dirichlet Language Model, Natural Language Engineering, 1(3), pp. 1-19, (1994).

Madsen, R. E., Kauchak, D. & Elkan, C., Modeling Word Burstiness Using the Dirichlet Distribution, Proceeding of the 22st International Conference on Machine Learning, (2005).

Minka, T., Estimating a Dirichlet distribution, Technical Report available via *http://research.microsoft.com/en-us/um/people/minka/papers/dirichlet/*, (2003).

Mosimann, J. E., On the Compound Multinomial Distribution, the Multivariate B-Distribution, and Correlations Among Proportions, Biometrika, 49(1 and 2), pp. 65-82, (1962).

Putthividhya, D., Attias, H. T., Nagarajan, S. S., Independent factor topic models, Proceeding of International Conference on Machine Learning, (2009).

Rennie, J. D. M., Shih, L., Teevan, J. & Karger, D. R., Tackling the Poor Assumptions of Naive Bayes Text Classifier, Proceeding of the Twentieth International Conference on Machine Learning, (2003).

Sjolander, K., Karplus, K., Brown, M., Hughey, R., Krogh, A., Mian, I. S. & Haussler, D.. Dirichlet Mixtures: A Method for Improving Detection of Weak but Significant Protein Sequence Homology. Computer Applications in the Biosciences, 12(4), pp. 327-345, (1996).

Probabilistic Relational Models for Operational Risk: A New Application Area and an Implementation Using Domain Ontologies

Marcus Spies

Abstract The application of probabilistic relational models (PRM) to the statistical analysis of operational risk is presented. We explain the basic components of PRM, domain theories and dependency models. We discuss two real application scenarios from the IT services domain. Finally, we provide details on an implementation of the PRM approach using semantic web technologies.

1 Introduction

In statistical models for operational risk, we often need to combine information from multiple entities. An example would be to build a model combining:

- Observations related to failures of technical or other service components.
- Observations related to customer contract related data, like customer claims in the scope of a service contract.

Both classes of observations can be described by suitable theoretical distributions whose parameters can be estimated given actual empirical data. Typically, failure probabilities can be assessed using distributions from reliability theory, while customer claims will follow some loss sizes distribution that is typically taken from one of the "fat tail" distribution families like the Weibull distribution.

In operational risk analysis, combination of these distributions is performed using convolution in order to estimate losses as they accumulate over time and compute meaningful scores like value at risk (VaR), see Giudici (2010).

M. Spies
Chair of Knowledge Management, Department of Computer Science,
LMU University of Munich, Germany
e-mail: marcus.spies@ieee.org
This work has been supported by the European Commission under the umbrella of the MUSING project, contract number 027097, 2006-2010.

A. Di Ciaccio et al. (eds.), *Advanced Statistical Methods for the Analysis of Large Data-Sets*, Studies in Theoretical and Applied Statistics,
DOI 10.1007/978-3-642-21037-2_35, © Springer-Verlag Berlin Heidelberg 2012

In the present service provider scenario, however, the combination of the information from both data sets gets more complicated if we want take into account the service providers who are operating the technical components and also appear as contracting party in customer service contracts. This implies that only component failures that fall in the service domain of a given service contract can be the possible causes of certain customer claims. This dependency can be formulated conveniently using database schemes as usual in relational database modelling (see Sect. 2).

As a consequence, for an integrated statistical model of multivariate observations in a multi-entity setting, we must combine the sampling spaces for both observation classes by considering a suitable constraint structure on the product space. A formal language is required in which we can declare the constraint on possible dependencies between observations and perform computations related to parameter estimation in accordance with the factual entity dependencies underlying the real data at hand.

This requirement for a relational modelling enabled formulation of observation based variable dependency has been noted in several application fields, notably genome analysis, web link analysis, and natural language based information extraction. As a consequence, the new field of probabilistic relational modelling (PRM) and learning is being established by a network of cooperating groups, see Getoor and Taskar (2007). PRM are closely related to Markov random fields, but the approach to defining variables and estimating parameters is different. PRMs can be introduced as extensions of Bayesian networks, which are nothing but single-entity PRMs. Therefore, the PRM approach nicely specializes to well known methods for statistical analysis of graphical models, see Cowell et al. (2007).

Several approaches related to PRM with essentially the same basic assumptions have been published, they build on:

- Using an entity relationship model (ERM) for encoding the relational dependencies in Getoor et al. (2007).
- Using a suitable subset of first order logic (FOL) and model-theoretic semantics to describe entities, attributes and dependencies in Kersting and De Raedt (2007).
- Combining ERM and FOL assertions for capturing constraints in the Directed Acyclic Probabilistic Entity Relationship approach (DAPER) (Heckerman 2007).

The presentation in the paper is motivated by applications and presents examples from the context of the EU integrated project Multi-Industry Semantics based Next Generation Business Intelligence (MUSING, EU contract number 027097). In Sect. 2, we present the basic approach to relational modelling of an application domain, in Sect. 3 we explain dependency models in PRMs and their transformation to Bayesian networks, and outline two case studies. In Sect. 4 we discuss an implementation of the model definition and data collection steps using domain ontologies.

2 Relational Domain Models

The first step towards a probabilistic relational model is to define an explicit formal data theory of the given application domain in a *declarative* way. This theory contains non-logical (factual) axioms about classes, their generic relationships and attributes. In a statistical setting, the attributes correspond to random variables. The theory has models in terms of real-world scenarios allowing for experiments. These experiments consist of drawing samples of objects instantiating the classes. These objects preserve the relationships as stated in theory. What we observe is the values of the attributes corresponding to the class attributes in the theory.

More formally, using concepts of the unified modelling language UML (Object Management Group 2010a,b), let $\mathscr{X} = \{X_1, \ldots, X_n\}$ denote a set of *classes*. Each class X_i is further described by *attributes* $\mathscr{A}(X_i)$. These attributes are assumed to take values in discrete domains, i.e. they are considered as nominal scaled variables. In practice, it is often assumed that the attributes are valued in one element of the set of atomic datatypes in XML Schema and are restricted to a suitably enumerated finite subset in case of an infinite value set.

The classes in $\mathscr{X}$ are further allowed to be related by associations contained in a set $\mathscr{R}$. A direct association is a two-place relation $\rho_{i,j} : X_i \rightarrow X_j$ on classes (X_i, X_j). If a direct association is a mapping, it is also called a reference. Thus, in case of a reference, $|\rho(x_i)| = 1 \;\; \forall x_i \in X_i$. References can be seen as generalizations of *has-a* relations in artificial intelligence. An $m : n$ association is modelled by an explicit association class that contains references to associated members in each participating class. Association classes can combine more than two classes.

The definition of a suitable set of classes, their attributes and associations is a theory of the application domain in question. Such a theory can be formulated according to different modelling approaches in different specific formal languages. One common choice is UML class diagrams (Object Management Group 2010a,b).

Another relevant modelling language is entity relationship modelling (ERM) that allows to derive a relational database schema from the theory. Basically, an ER model interprets classes by schemas of relational tables, references by foreign key relationships, and association classes by explicit join tables containing foreign keys to the tables participating in the association. ERM is based on relational database normalization and schema theories (Beeri et al. 1977, 1983). A particular requirement against an ERM is that references in a domain theory form directed acyclic graph (DAG).

Finally, a class based domain data theory may be formulated using description logic by defining a domain ontology, most commonly based on the web ontology language OWL (Motik et al. 2006). In ontology engineering, the attributes of classes are modelled by *datatype properties*, associations by *object properties*, which are declared functional for references.

Possible transformations and incompatibilities between domain theories from these different modelling languages have been studied extensively and formalized in the Object Management Group's Ontology Definition Metamodel specification (Object Management Group 2009).

3 Probabilities in Relational Domain Models

The second step in setting up a PRM is specifying a probability model covering attributes of classes. The most obvious way is to consider the class attributes as (discrete) random variables. This leads to a family of statistical models studied by Heckerman (2007) and Getoor et al. (2007), and also, in a stronger predicate-logic driven approach, by Laskey (2006). As usual in graphical statistical models, the probability model is stated in terms of statistical dependency or independence assertions (Lauritzen and Spiegelhalter 1988; Cowell et al. 2007).

An attribute dependency assertion is of the form $\Pr(A_j(X_i)|Pa(A_j(X_i)))$ with $A_j(X_i) \in \mathscr{A}(X_i)$, where $Pa(A_j(X_i))$ denotes a set of parent attributes. Thus, dependency assertions are stated for subsets of attributes of one or several classes, possibly with constraints in first-order logic added, see Heckerman (2007). A set of non-circular dependency assertions for $j \in J$ is also called a dependency model, denoted by $\mathscr{D}_J$. If a dependency model contains only attributes and parents from one class, the model translates directly into a conventional Bayesian network.

The interpretation $\mathscr{I}$ (in the usual sense of first-order-logic) of a set of classes and associations from a relational domain model is described as $\mathscr{I}(\mathscr{X}) \cup \mathscr{I}(\mathscr{R})$. A generic element of $\mathscr{I}(X_i)$ is written x_i, with attribute value variable $x_i(a_j)$ for the attribute subscripted j in $\mathscr{A}(X_i)$. It is of key importance to observe that different interpretations of a domain theory usually involve different references. Therefore, if a dependency model contains attributes from several classes, different interpretations usually lead to different entities being related. As a consequence, *the sample space for a dependency model restricted set of class attributes depends on the interpretation of the domain theory*. This is the key statistical issue when moving from single-entity Bayesian networks (BN) to PRMs.

Formally, if $Pa(A_j(X_i))$ is the matrix $\{y_1(a_{y1}), \ldots, y_m(a_{ym})\}$, the sample space is the set of possible observations $\{x_i(a_j), y_1(a_{y1}), \ldots, y_m(a_{ym})\}$ $\forall x_i \in \mathscr{I}(X_i), y_1 \in \mathscr{I}(Y_1), \ldots, y_m \in \mathscr{I}(Y_m)$ s.t. $(x_i, y_1, \ldots y_m) \in \mathscr{I}(\mathscr{R})$ – it contains all observable cases of class instances in the sample and their corresponding attribute value combinations *constrained by* the specific references relating x_i to y_j in the given interpretation.

One basic result from Heckerman (2007), Getoor et al. (2007) is that any $\mathscr{D}$ transforms to exactly one BN for each partial interpretation that fixes class instances and associations only, comprising references. Such a partial interpretation is referred to as skeleton. The random variation is then confined to the attribute values. The practical use of this result is in *multi-entity repeated measures analysis*. As a consequence, the usual computations known from Bayesian network algorithms can be applied (Lauritzen and Spiegelhalter 1988; Cowell et al. 2007).

However, in practical terms, a scenario with specific and fixed instances of all classes, references and other associations is often not useful. Therefore, in terms of *statistical learning*, we generalize over possible skeletons by allowing only as many parameters as the generic dependency assertions require and tie parameters from training on different skeletons accordingly. In general, the log-likelihood function for a PRM can be expressed within each partial interpretation or skeleton as sum across all classes, then across all attributes appearing on the LHS (left hand side) of a dependency assertion, of the conditional log-probabilities of the elements in the skeleton. More formally, given a dependency model $\mathscr{D}_J$, then, for one skeleton σ, we have (see Getoor et al. (2007), p. 162, adapted to our notation, and, in particular, letting $\mathscr{S}_\sigma(x_i(a_j)) = \{y_1(a_{y1}), \ldots, y_m(a_{ym})\}$ if $(x_i, y_1, \ldots, y_m) \in \mathscr{I}(\mathscr{R})$, and else $= \emptyset$), and A_j the range of the attribute subscripted j in $\mathscr{A}_j(X_i)$, and n the count of classes in the domain theory)

$$\begin{aligned}\mathscr{L}(\theta_{\mathscr{D}_J} | \mathscr{I}, \mathscr{S}_\sigma) &= \log \Pr(\mathscr{I} | \mathscr{S}_\sigma, \theta_{\mathscr{D}_J}) \\ &= \sum_{j \in J} \sum_{i=1}^{n} \sum_{x_i \in \mathscr{I}(X_i)} \sum_{a_j \in A_j} \log \Pr(x_i(a_j) | \mathscr{S}_\sigma(x_i(a_j)), \theta_{\mathscr{D}_J})\end{aligned}$$

3.1 *Case Studies*

Customer Claims Risk in IT Services A promising application of the approach is based on the MUSING cooperation with the Italian bank MPS (Monte dei Paschi di Siena). The purpose is to evaluate the risk (in terms of probability and cost) of customer claims following failures in financial IT services providers infrastructures. While the exact relationship between IT failures and customer claims is only known in special cases, investments in network infrastructure and software can be prioritized if probable causes of high cost or high frequency claims can be inferred.

Building on the example of the introduction, a simplified domain theory focussing on corporate customers and network servers can be set up as follows:

1. The relevant classes in this application domain are customers, (IT) service providers, services, server operators, servers, service contracts.
2. Each service contract is between one service provider and one customer for one service.
3. Each claim refers to one service contract.
4. Each service has one provider, failures are summarized in an attribute called downtime average.
5. Each service is deployed on some service components, an $m:n$ relationship, since components can provide different services.
6. A service component is managed by a service operator. The operator defines maintenance policies etc.

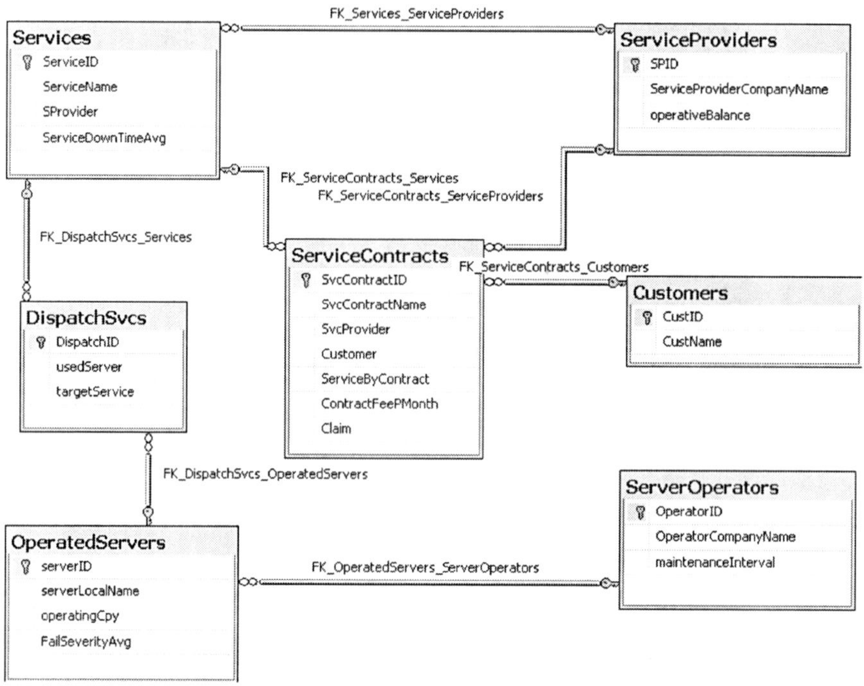

Fig. 1 A database scheme diagram for operational risk analysis in IT services customer relationships, produced with Microsoft (R) Visual Studio 2010

This domain theory (with some extras) is depicted as database scheme diagram in Fig. 1. Note that the overall structure of the diagram is an acyclic directed graph.

In this situation, a PRM based on a fixed partial interpretation for a multi-entity repeated measures analysis is realistic, because entities are linked by contracts for some period of time during which failures, claims, maintenance problems etc can be recorded and analyzed.

To illustrate this, Fig. 2 shows an overlay of a partial interpretation with a suitable ground Bayesian network, simplified (and with a little bit of humour added) from MUSING. In a database setting, the computation of the log-likelihood can be conveniently implemented in SQL (for a simplified example, see Getoor et al. (2007)). Note that the Bayesian network structure would change with variations in the objects' associations, even if the generic dependencies were the same, e.g. those from services downtimes to claims.

Corporate Customers Risk in Services Provisioning A second application case study is the MUSING pilot for assessing operational risks at a virtual network operator (VNO) with a large customer base in Israel. The key practical advantage of the PRM based solution is the ability to integrate corporate customer data with IT operational services protocols into predictions and assessments of likely operational

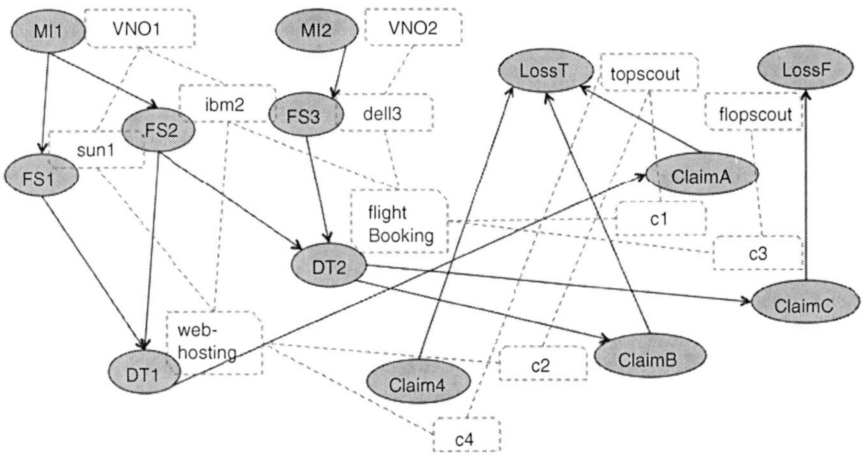

Fig. 2 Overlay of a partial interpretation (dashed items) and the ground Bayesian network (elliptic nodes and directed arcs) for the domain model from Fig. 1. In the partial interpretation, $c1$ to $c4$ denote contract instances linking a service (like *webhosting*), a provider (like *flopscout*) and a customer (like *Tom*), etc. A ground Bayesian network is provided for some attributes of different classes (*MI* for maintenance policy by a *server operator*, *DT* for downtime of a *service*, *FS* for failure severity of a *server*). The Bayesian network exhibits instance specific probabilistic attribute value dependencies

losses and their economic impact on the VNO. This ability has been used to deploy a decision support dashboard application prototype at the VNO. The application integrates indicators from the public part of the corporate balance sheets with operational data for the services the customer requests and computes a qualitative overall riskiness score.

We now turn to the implementation details for both case studies.

4 An Implementation Using a Domain Ontology Development and Runtime Environment

In this section, we introduce and discuss the EU MUSING implementation of the PRM domain modelling and data integration based on domain ontologies using the web ontology language OWL, see Motik et al. (2006).

For the MUSING project, the specific need arose to have a PRM framework available that could easily be combined with existing domain ontologies as they were built for several risk management domains throughout the project using the web ontology language, see Motik et al. (2006). The key motivation for an ontology based representation of data for analytic processing is the combination of statistical analyses with text processing (annotation, text classification etc), see www.musing.eu.

Therefore, a solution was envisioned that would make it very simple to add attribute (datatype property) dependencies to an existing OWL ontology. Moreover, the MUSING ontologies are constantly populated (enriched by new instances) in real application environments. Therefore, simple mechanisms to collect the sufficient statistics of the parametric PRM learning were needed, as well. This could be accomplished using the SPARQL RDF-query language that was extended by SELECT COUNT and UPDATE constructs during the MUSING project lifetime, see Prudhommeaux and Seaborne (2008). This allows to collect counts and store them into appropriate places in the domain ontology itself. The persistent storage of counts is combined with a versioning mechanism in order to handle learning data from various skeletons and from various datasets independently. This was implemented by means of a service oriented architecture offering domain or application specific repositories in addition to the definitory parts (classes, properties, axioms) of the ontologies. In addition, an export utility of the resulting ground Bayesian network an XML representation was implemented. For this, the XML for Bayesian networks format (Microsoft Corp. 2009) was used. Using this utility, applications of MUSING can pass parameter estimates to several commercial or freely available Bayesian network tools for further processing. An import utility for parameter priors could be added easily.

In terms of combining PRMs and ontologies, the approach taken was to define a set of classes suitable for enabling a straightforward *declarative definition* of a dependency model $\mathcal{D}$ in any domain ontology. *This amounts to defining suitable ontology classes and properties such that a dependency assertion can be stated by defining appropriate instances of additional and restrictions on internal classes of a given domain ontology.* These instances shall then be used for collecting counts as sufficient statistics for parameter estimation within a given skeleton.

Two generic classes were needed to back up an implementation of this approach. The first, ObservableAttribute, serves mainly as a source of inheritance relationships for other ontology classes appearing as (object) properties in a dependency model. This means that, in our approach, dependency assertions are expressed in terms of object properties of domain classes. In order to ensure finiteness of the implied discrete random variables, the classes used for any dependency assertion were defined by enumerations. The second class for setting up the construction, CartesianProduct, is instantiated for each dependency assertion. It carries object properties referencing the attributes (object properties) participating in the conditional probability distribution being defined.

Coming back to our running example of the virtual network operator and the assessment of operational risk, a given scenario of services, customers and providers can readily be modelled as a fixed skeleton. For our prototypes, we only used simple dependency models with one element in the LHS and in the RHS of any dependency assertion. At this point, it should be remarked that, due to the triangulation theorem for Bayesian networks (see Lauritzen and Spiegelhalter (1988)), the maximum

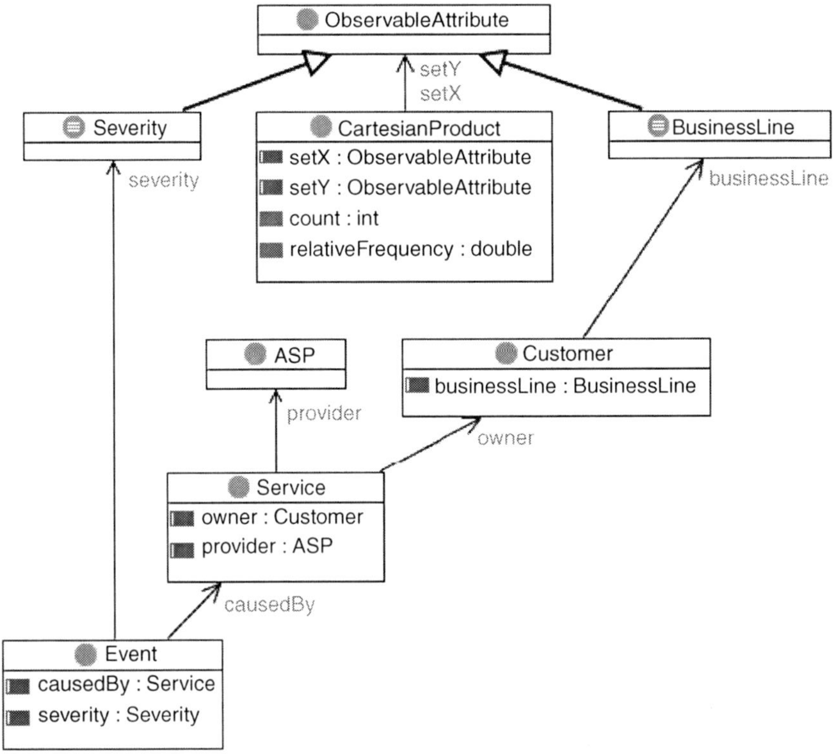

Fig. 3 A domain ontology with two additional entities for probabilistic dependencies. The abstract class ObservableAttribute is used to enable enumerative classes (here severity and business line) for use as discrete random variables. The class CartesianProduct is used to express bivariate dependency assertions. Instances of this class correspond to dependency assertions involving two variables. (Generated with TopBraidComposer of TopQuadrant Inc.)

number of classes involved in a dependency assertion can be limited to three. The approach is illustrated in the subsequent figures, Figs. 3 and 4.

5 Conclusion

This contribution shows a novel application area for probabilistic relational models in risk analysis as implemented in the EU project Multi-Industry Semantics based Next Generation Business Intelligence (MUSING) for operational and credit risks following the Basel II framework (Basel Committee on Banking Supervision 2004). In particular, as MUSING is strongly relying on domain ontologies, we studied an approach to implementing the dependency model definition and data processing

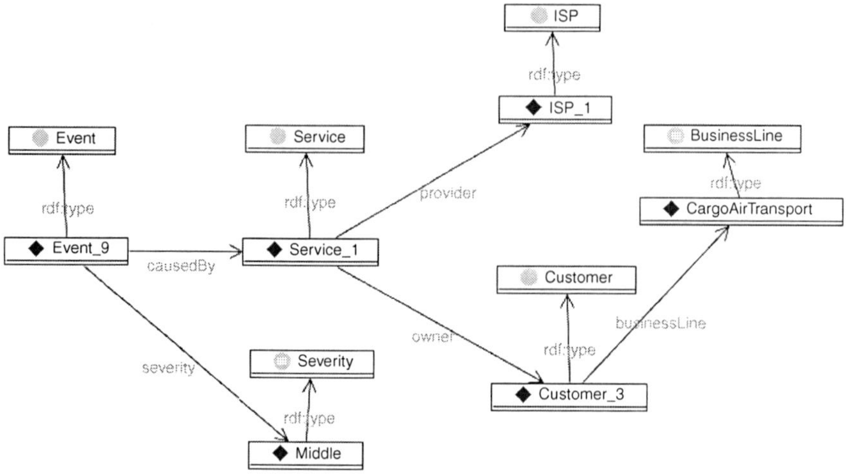

Fig. 4 An example of ontology population (diamond shapes represent instances) affecting two entities involved in a relationship with a dependency assertion (system failure events affecting business lines). The counts used for parameter estimation are stored in the suitable instance of the ObservableAttribute class, the count updating is implemented in SPARQL code that is executed upon committing the population instances to the ontology repository

steps needed for a PRM analysis using ontologies conforming to the web ontology language (OWL) standard. It could be shown that a rather straightforward extension of any domain ontology suffices to enable declarative dependency model definitions and to collect sufficient statistics that are readily exported to standard Bayesian networks tools for further processing.

Acknowledgements The author gratefully acknowledges technical exchanges about practical applications related to IT Operational Risk Management and Credit Risk Management within the MUSING consortium. Special thanks go to Prof. P. Giudici, head of Libero Lenti Laboratory at Pavia University, and Prof. Ron Kenett, Torino University and head of the KPA consultancy (Tel Aviv). The author also acknowledges support by MetaWare S.p.A. of Pisa as overall project coordinator.

References

Basel Committee on Banking Supervision: International convergence of capital measurement and capital standards: A revised framework – comprehensive version (2004). URL http://www.bis.org/publ/bcbs107.htm

Beeri, C., Fagin, R., Howard, J.: A complete axiomatization for functional and multivalued dependencies in database relations. In: Int. Conf. Mgmt of Data, pp. 47–61. ACM (1977)

Beeri, C., Fagin, R., Maier, D., Yannakakis, M.: On the desirability of acyclic database schemes. J. ACM **30**(3), 479–513 (1983)

Cowell, R.G., Dawid, A., Lauritzen, S.L., Spiegelhalter, D.J.: Probabilistic networks and expert systems. Exact computational methods for Bayesian networks. 2nd printing. Information Science and Statistics. New York, NY: Springer. xii, 321 p. (2007)

Getoor, L., Friedman, N., Koller, D., Pfeffer, A., Taskar, B.: Probabilistic relational models. In: L. Getoor, B. Taskar (eds.) Introduction to Statistical Relational Learning, pp. 129–174 (2007)

Getoor, L., Taskar, B. (eds.): Introduction to Statistical Relational Learning. Massachusetts Institute of Technology, MIT Press, Cambridge, MA (2007)

Giudici, P.: Scoring models for operational risk. In: R. Kenett, Y. Raanan (eds.) Operational risk management – a practical approach to intelligent data analysis. Wiley (2010)

Heckerman, D., Meck, C., Koller, D.: Probabilistic entity-relationship models, prms, and plate models. In: Getoor and Taskar (2007), pp. 201–238 (2007)

Kersting, K., De Raedt, L.: Bayesian logic programming: Theory and tool. In: Getoor and Taskar (2007), pp. 291–322

Laskey, K.: MEBN: A logic for open-world probabilistic reasoning. Tech. Rep. GMU C4I Center Technical Report C4I-06-01, George Mason University (2006)

Lauritzen, S., Spiegelhalter, D.: Local computations with probabilities on graphical structures and their application to expert systems. J. R. Statistical Society B **50**(2), 157 – 224 (1988)

Microsoft Corp.: XML for bayesian networks (2009). URL http://research.microsoft.com/dtas/bnformat/xbn_dtd.html

Motik, B., Patel-Schneider, P., Horrocks, I.: OWL 1.1 web ontology language structural specification and functional-style syntax (2006)

Object Management Group: Ontology definition metamodel version 1.0 (2009). URL http://www.omg.org/spec/ODM/1.0/PDF

Object Management Group: Unified modeling language: Infrastructure (2010). URL http://www.omg.org/spec/UML/2.3/Infrastructure/PDF/

Object Management Group: Unified modeling language: Superstructure specification (2010). URL http://www.omg.org/spec/UML/2.3/Superstructure/PDF/

Prudhommeaux, E., Seaborne, A.: SPARQL query language for RDF (2008). URL http://www.w3.org/TR/rdf-sparql-query/

Part X
Advances on Surveys

Efficient Statistical Sample Designs in a GIS for Monitoring the Landscape Changes

Elisabetta Carfagna, Patrizia Tassinari, Maroussa Zagoraiou, Stefano Benni, and Daniele Torreggiani

Abstract The process of land planning, addressed to operate the synthesis between development aims and appropriate policies of preservation and management of territorial resources, requires a detailed analysis of the territory carried out by making use of data stored in Geographic Information Systems (GISs). A detailed analysis of changes in the landscape is time consuming, thus it can be carried out only on a sample of the whole territory and an efficient procedure is needed for selecting a sample of area units. In this paper we apply two recently proposed sample selection procedures to a study area for comparing them in terms of efficiency as well as of operational advantages, in order to set up a methodology which enables an efficient estimate of the change in the main landscape features on wide areas.

1 Introduction

The preservation and management of territorial resources requires detailed knowledge of the landscape and of the evolutionary dynamic that produced its arrangement, as well as the monitoring of the changes occurred in the various landscape features. For these purposes, the analysis of different kinds of geographical data referred to manifold time steps is needed and is currently carried out by making use of databases and maps available for the whole territory. These data are stored and elaborated in Geographic Information Systems (GISs) which have made possible to compile geographical databases with highly detailed content covering large spatial

E. Carfagna (✉) · M. Zagoraiou
Department of Statistical Sciences, University of Bologna, Via Belle Arti 41, 40126 Bologna, Italy
e-mail: elisabetta.carfagna@unibo.it

P. Tassinari · S. Benni · D. Torreggiani
Department of Agricultural Economics and Engineering, Spatial Engineering Division, University of Bologna, v.le Fanin 48, 40127, Bologna, Italy

A. Di Ciaccio et al. (eds.), *Advanced Statistical Methods for the Analysis of Large Data-Sets*, Studies in Theoretical and Applied Statistics,
DOI 10.1007/978-3-642-21037-2_36,

areas. GISs also enable the construction of multi-layer and multi-temporal archives in order to detect changes in land-use/land-cover patterns. However, when a more detailed measurement of the actual variations of main landscape features is required, this detailed analysis is time consuming; thus it can be carried out only on a sample of the whole territory.

In this paper we compare alternative sampling methods for assessing which one is the most appropriate for a detailed measurement of the actual variations of main landscape features in order to provide a statistically sound method for monitoring the landscape transformations in rural areas.

The stratification created in a GIS is described in paragraph 2. Alternative adaptive sample selection procedures are explained in paragraph 3. Then, two adaptive sample selection procedures are applied to the study area and compared. In paragraph 4, the sample sizes needed for reaching a fixed precision of the estimate of the change in building cover density with the two approaches are compared. In paragraph 5, the comparison is made considering a fixed budget and taking into account operational disadvantages quantified with a cost function. Considerations on the most appropriate sample selection procedure for an efficient estimate of the change in the main landscape features on wide areas are given in paragraph 6.

2 Stratification of the Area Frame with a GIS

In order to set up a methodology which enables an efficient estimate of the change on large areas, we have chosen a target study area (787 km^2) situated in the eastern part of the Bologna province, Italy, which comprises the group of ten municipalities known as the "Nuovo Circondario Imolese", analyzed in detail in a previous study (Tassinari et al. 2008). The parameter we have estimated is the change in building cover density, that is the difference between the area covered by buildings in 2005 and in 1975 divided by the land area. This parameter has proven to be useful in analyses supporting the formulation of the strategic and operational planning tools defined by the most recent legislative purviews concerning land government (Benni et al. 2009).

In order to make a detailed analysis on a sample of the study area, we have set up an area frame that subdivides the territory into area units. We have decided to take advantage of the enumeration areas used for the most recent population and housing census (Istat 2001). Thus, the adopted area frame is based on irregular physical boundaries and the size of the area units is not constant. The need to maximize the precision of the estimate of the landscape change suggested the adoption of a stratification which identifies zones characterised by different rates of change in the main landscape features. The stratification we have created is based on a combination of appropriate *land-use/land-cover* classes – referred to the end of the studied period – and of land suitability for agricultural and forestry use.

The stratification has been refined reclassifying the enumeration areas according to the attribution of the majority of their surface to areas already urbanized

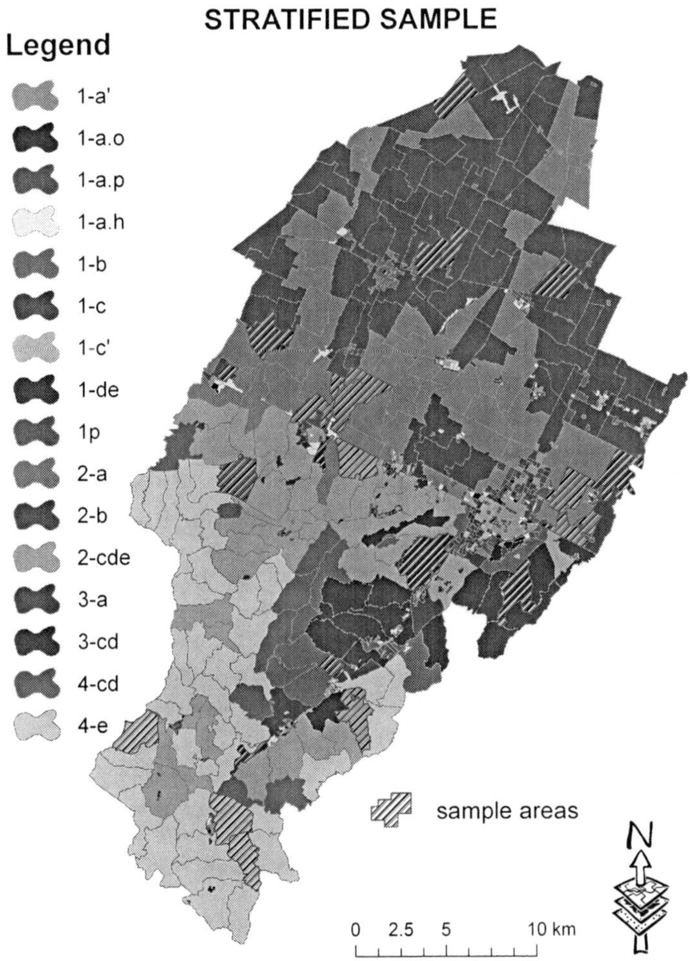

Fig. 1 Map of the stratified sampling frame and spatial distribution of the pilot sample units in the strata

at the beginning of the study period or to areas urbanized during that period. A further subdivision has been made depending on whether the final destination is predominantly residential, productive, or any other type of urban use; for more details see Tassinari et al. (2010). The result is shown in Fig. 1.

The design effect (DEFF, see Kish 1965, 1995) of the stratification is 0.7285, which means that, in case of simple random sampling, a sample size 37% wider is needed to reach the same precision of the estimates.

3 Alternative Adaptive Sample Selection Procedures

Since data acquisition and in depth data analysis are time consuming, it is very important to use very efficient sample designs; however the variability inside the strata is unknown, thus the optimal allocation (Neyman's) cannot be adopted.

Several authors have faced the problem of sample allocation without previous information on the variability inside the strata (see for example Cox 1952; Sandvik et al. 1996; Mukhopadhyay 2005) suggesting various kinds of two steps sampling; but these methods do not allow design unbiased estimates of the parameters.

Thompson and Seber (1996, pp. 189–191) proposed a sequential approach in k phases (phases are sampling steps) which assures unbiased estimates. At each phase, a complete stratified random sample is selected, with sample sizes depending on data from previous phases. Then the conventional stratified estimator, based on the data from the phase, is unbiased for the population total. The weighted average of these estimates is an unbiased estimator of the population total if the weights are fixed in advance (do not depend on observations made during the survey) and each of the strata is sampled at each phase.

Carfagna (2007) proposed a two-steps selection procedure (TSPRN) which involves the permanent random numbers (Ohlsson 1995), in order to allow the allocation of sample units at the second step only in those strata where supplementary selection is needed. In fact, the TSPRN assigns a random number to each unit in each stratum, then, the units are ordered according to the associated number and this order corresponds to the selection order. At the first step, a first subset of the units in a stratum is selected and the other units are included in the sample in the next step according to the same order. This means that only one selection is performed and the set of selected units in each stratum can be considered as a random sample without replacement; thus, there is no need to select at least two units from each stratum at each step. Consequently, the TSPRN is more efficient than the procedure proposed by Thompson and Seber, as shown in Carfagna (2007).

Then, Carfagna and Marzialetti (2009) proposed a kind of adaptive sequential procedure (ASPRN) with permanent random numbers: assign a random number to each unit in each stratum, then order the units according to the associated number; this order corresponds to the selection order. Select a first stratified random sample with probability proportional to stratum size, selecting at least two sample units per stratum and estimate the variability inside each stratum. In case in one stratum the estimated variability is zero, assign to this stratum the variance estimated in the stratum with the lowest positive variance. Then compute Neyman's allocation with sample size $n + 1$ and select one sample unit in the stratum with the maximum difference between actual and Neyman's allocation. Then estimate the parameter of interest and its precision. If the precision is acceptable, the process stops; otherwise, compute Neyman's allocation with the sample size $n + 2$, and so on, until the precision considered acceptable is reached. Due to the use of permanent random numbers, the sample size per stratum depends on the previously selected units but the sample selection does not; thus the ASPRN allows design unbiased and efficient estimates of the parameters of interest.

In principle, the ASPRN should be more efficient than the TSPRN, since it allows updating the estimates of the variances within the strata at each step; on the other side, the ASPRN shows operational drawbacks.

In this work, we have applied the ASPRN and the TSPRN to the above-mentioned study area for comparing them in terms of efficiency as well as of operational drawbacks, in order to set up a methodology which enables an efficient estimate of the change in the main landscape features on large areas.

With both procedures, the usual expansion estimator for stratified random sampling is design unbiased; moreover, although the range of the size of the enumeration areas is wide, the use of the ratio estimator is not necessary, since we have noticed that the change in building cover density is not correlated to the area of the enumeration areas.

Consequently, the estimator of the change in building area D and its variance V can be expressed by the following equations:

$$D = \sum_{h=1}^{L} N_h \sum_{i=1}^{n_h} \frac{B_{hi} - B'_{hi}}{n_h} \tag{1}$$

$$V = \sum_{h=1}^{L} N_h^2 \frac{S_h^2}{n_h} \left(1 - \frac{n_h}{N_h}\right), \tag{2}$$

where N_h is the total number of enumeration areas in stratum h; B_{hi} and B'_{hi} are the surface area covered by buildings in the ith enumeration area belonging to stratum h, respectively at the end and beginning of the considered time span; n_h is the sample size of the hth stratum; L is the total number of strata and S_h^2 is the variance of the built area differences in the hth stratum. The change in built areas cover density Y is estimated dividing D by the area under study.

4 Comparison of the Methods with Fixed Precision of the Estimate

As said before, in principle, the ASPRN should be more efficient than the TSPRN but more cumbersome; thus we have made two kinds of comparison. First, we have compared the number of sample units needed for reaching a chosen precision of the estimate with the two methods in order to assess how much the ASPRN is more efficient than the TSPRN for this kind of application; then we have taken into consideration operational disadvantages through a cost function and we have compared the precision of the estimate obtained with the ASPRN and the TSPRN given the same total budget.

We have adopted the TSPRN for estimating the building cover density and we have selected the pilot (or first step) sample which is constituted of 104 census enumeration areas. We have estimated the change in built areas and the standard

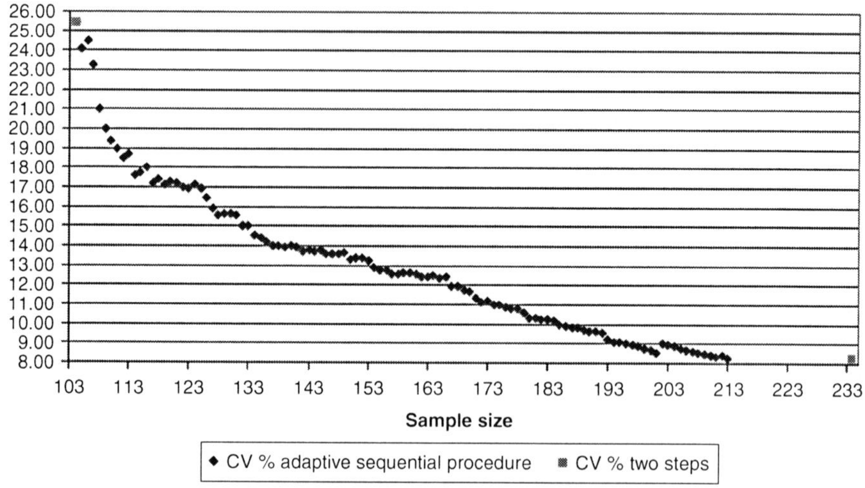

Fig. 2 Coefficient of variation per cent of the estimator of the change in built areas with the TSPRN and the ASPRN with increasing sample size

deviation within each stratum and, using the standard formula for stratified random sampling, we have computed the number of sample units needed for reaching a coefficient of variation of the estimate equal to 10% (234). Thus, we have selected 130 more units with an allocation as near as possible to Neyman's one of 234 sample units. We have obtained a coefficient of variation of the estimate of the change in built areas (CV%) equal to 8.3%, lower than the target CV% (10%) (see the grey square dot on the right side of Fig. 2), this means that the pilot sample has overestimated the standard deviation in some strata. The estimate of the difference between the area covered by buildings in 2005 and in 1975 divided by the land area, with 234 selected units, is 0.00354226 and the variance of the estimator is $8.6 * 10^{-8}$. Then, we have adopted the ASPRN and have noticed that the target CV% (10%) is already reached with a sample size equal to 180 and the same CV% obtained with the TSPRN (8.3%) is reached with 213 sample units, see the trend of the CV% of the estimate with the ASPRN with increasing sample size in Fig. 2 (the black rhomboidal dots).

The outcome of the first kind of comparison is that the same variance and the same CV% of the estimator obtained with the TSPRN (8.3%) is reached with 21 sample units less (– 9%) using the ASPRN. Thus, for estimating the change in building cover density, the adaptive sequential procedure with permanent random numbers is more efficient than the two steps one and more flexible, since it allows reaching exactly the desired precision of the estimate. Moreover, the ASPRN generates estimates of the change in building cover density very similar to the one obtained with the TSPRN, using a smaller sample size.

The efficiency of the ASPRN is due to the improvement, at each step, of the estimates of the variability inside the strata which makes the allocation with ASPRN very similar to Neyman's one, except for the constraint of selecting at least

Table 1 Comparison of proportional, ASPRN and Neyman's allocations

STRATA	N_h	n_h proport. allocation	n_h pilot sample	S_h	n_h ASPRN	n_h Neyman's allocation
1-a.res	142	21	10	1,688	11	11
1-a.prod	72	11	5	24,527	72	83
1-a.dis_ver	18	3	2	3,471	2	3
1-a′	415	61	29	1,448	29	28
1-b	50	7	4	1,323	4	3
1-c	156	23	11	2,366	18	17
1-c′	136	20	9	1,443	9	9
1-de	23	3	2	3	2	0
1p	169	25	12	2,848	23	23
2-a	60	9	4	6,566	18	18
2-b	71	10	5	4,103	14	14
2-cde	36	5	2	996	2	2
3-a	15	2	2	608	2	0
3-cd	18	3	2	1,286	2	1
4-cd	17	3	2	355	2	0
4-e	43	6	3	341	3	1
Total	1,441	213	104	53,372	213	213

two sample units from each stratum and not exceeding the total stratum size, as shown in Table 1. Neyman's and the ASPRN allocations are very different from the proportional one and much more efficient, due to the wide range of the standard deviations in the strata.

5 Comparison of the Methods with Fixed Budget

In order to take into account the operational drawbacks of the ASPRN, we have compared the precision of the estimator of the change in building cover density obtained with the two procedures given the same budget; thus we have specified a cost function and evaluated the budget necessary for estimating the change in building cover density with a CV% of 8.3% with the TSPRN (14,748 €).

The adopted cost function is the following:

$$\begin{aligned} C = {} & C_0 + \text{step cost} + (s \cdot \lceil \phi_1(n_1) \rceil) + (r \cdot n_1) + \\ & + [(\lceil \phi_2(n_1 + n_2) \rceil - \lceil \phi_1(n_1) \rceil) \cdot s)] + (r \cdot n_2) + \\ & + (\text{step cost} \cdot \text{step number}) \end{aligned} \quad (3)$$

where:

- C_0 is the fixed cost. It is given by the preliminary work, corresponding to the cost of two researchers spending two working weeks: 2,226 €. The same fixed cost is considered for the ASPRN and the TSPRN.

- "step cost" is given by the identification of the sampled units in the GIS layer (such operation can be carried out by a researcher or a technician spending one or two hours), plus the statistical elaboration; the estimated cost is about 30 €.
- s is the cost for the map sheets acquisition, their digitization and geo-registration (120 €).
- r is the average survey cost of one areal unit (photo interpretation, digitization etc.) (37 €).
- n_1 is the pilot sample size. The same pilot sample size is used in the adaptive and in the two-steps selection procedures.
- n_2 is the sample size additional to the pilot sample.
- $\phi_1(.)$ and $\phi_2(.)$ are the functions linking the number of selected enumeration areas with the number of map sheets necessary to cover a given number of selected enumeration areas.

We have used two different functions $\phi_1(.)$ and $\phi_2(.)$ because for small sample sizes the number of map sheets increases more rapidly than for large sample sizes. On the basis of the map sheet index and the position of the selected enumeration areas, we have found the trends of these functions, through interpolation. The values assumed by these functions have been rounded to the nearest integer.

The pilot sample selection is considered as the first step, both in the two-steps and in the adaptive selection procedure.

Using the specified cost function, we have obtained that the number of sample units which can be selected and surveyed with the total budget of 14,748 € using the ASPRN is equal to 178 and the corresponding estimate has a CV% equal to 10.8%, much higher than the CV% obtained with the TSPRN (8.3%); mainly due to the fact that, with the ASPRN, the selection of each areal unit is a step of the process, with the consequent cost.

6 Concluding Remarks

The comparison of the two procedures given a fixed budget generates a result opposite to the one obtained comparing the sample sizes needed for reaching the same precision of the estimate. The discrepancy is due to the fact that the second kind of comparison takes into account and quantifies operational disadvantages. We believe that this is an important result because it shows that, in some cases, different procedures have different cost structures and the evaluation of the efficiency without cost considerations can be misleading.

According to our results, we should suggest to use the TSPRN for estimating the change in the main landscape features on wide areas. However, we have to recall that the selection and survey process can be optimised in an operational project, reducing some of the costs specified in the cost function. An interesting future research can be a further analysis of the costs for understanding which costs can be reduced by optimising the selection, survey and estimation procedures when they are applied on large areas and compare the two procedures again. Another area of research is inventing a k steps adaptive procedure, with the optimum number of steps, that is a compromise between the TSPRN and the ASPRN, which reduces the costs to be suffered with the ASPRN and preserves the capability of the ASPRN to generate a sample allocation very close to Neyman's one.

References

Benni, S., Torreggiani, D., Tassinari, P.: Sistemi integrati per l'analisi delle risorse del territorio rurale: alcune note di metodo. Agribusiness Paesaggio & Ambiente **12**, 16–22 (2009)

Carfagna, E.: Crop area estimates with area frames in the presence of measurement errors. In Proceeding of ICAS-IV, Fourth International Conference on Agricultural Statistic. Invited paper, Beijing, 22–24 October 2007

Carfagna, E., Marzialetti, J.: Sequential design in quality control and validation of land cover data bases. J. Appl. Stoch. Models in Business and Industry Volume 25, Issue 2, 195–205 (2009). doi:10.1002/asmb.742

Cox, D.R.: Estimation by double sampling. Biometrika **39**, 217–227 (1952)

Istat, Italian statistic board: 14th Census of Population and Housing. Istat, Rome (2001)

Kish, L.: Survey Sampling. Wiley (1965)

Kish, L.: Methods for design effects. J. Official Stat. **11**, 55–77 (1995)

Mukhopadhyay, N.: A new approach to determine the pilot sample size in two-stage sampling. Commun. Stat. Theory Meth. **34**, 1275–1295 (2005)

Ohlsson, E.: Coordination of samples using permanent random numbers. In: Cox, B.G., Binder, D.A., Chinnapa, B.N., Christianson, A., Colledge, M.J., Kott, P.S. (eds.) Business survey methods, pp. 153–169. Wiley, New York (1995)

Sandvik, L., Erikssen, J., Mowinckel, P., Rodland, E.A.: A method for determining the size of internal pilot studies. Stat. Med. **15**, 1587–1590 (1996)

Tassinari, P., Carfagna, E., Benni, S., Torreggiani, D.: Wide-area spatial analysis: a first methodological contribution for the study of changes in the rural built environment. Biosyst. Engineering **100**, 3, 435–447 (2008)

Tassinari, P., Carfagna, E., Torreggiani, D., Benni, S., Zagoraiou, M.: The study of changes in the rural built environment: focus on calibration and improvement of an areal sampling approach. Biosyst. Eng. **105**, 4, 486–494 (2010)

Thompson, S.K., Seber, G.A.F.: Adaptive sampling. Wiley, New York (1996)

Studying Foreigners' Migration Flows Through a Network Analysis Approach

Cinzia Conti, Domenico Gabrielli, Antonella Guarneri, and Enrico Tucci

Abstract The aim of the paper is to enlighten the advantages of measuring and representing migration flows through the network analysis approach. The data we use, in two different kinds of analysis, regard the changes of residence of foreign population; they are individual data collected by Istat through the Municipal Registers. In the first step, we consider inflows from abroad (average 2005–2006). The countries of origin are identified as "sending nodes" and the local labour market areas are identified as "receiving nodes". In the second step, we examine the internal flows of immigrants between local labour market areas. The analysis is focused on specific citizenships.

1 Aims

The present study exploits network analysis techniques to describe the ties between geographical areas of origin and destination.

In a first step, international mobility is taken into account. Over recent years, the map of migratory flows has radically transformed. In the case of the foreigners' international migration flows the countries of origin are identified as "sending nodes" and the local labour market areas (LLMAs) are identified as "receiving nodes".

In a second step the local labour market areas are considered as nodes to study the internal mobility of the most important foreign communities.

The territorial dimension plays an important role in studying migration trends. For the different communities network analysis allows to show the strongest ties between local labour market areas.

C. Conti (✉) · D. Gabrielli · A. Guarneri · E. Tucci
Istituto Nazionale di statistica (Istat), Direzione Centrale per le Indagini sulle Istituzioni Sociali, V.le Liegi, 13 – 00198 Roma
e-mail: ciconti@istat.it

A. Di Ciaccio et al. (eds.), *Advanced Statistical Methods for the Analysis of Large Data-Sets*, Studies in Theoretical and Applied Statistics,
DOI 10.1007/978-3-642-21037-2_37,

Other studies applied these techniques to migration flows data (Maier and Vyborny 2005); for the Italian case the analyses are carried out only recently (Istat 2007; Istat 2008; Rivellini and Uberti 2009).

2 Source, Data and Territorial Grid

Until 1995 internal migration in Italy has followed a decreasing trend over the years. Since the second half of the last decade, the mobility within Italy has been on the increase again. In a large part this increase is due to the high internal mobility of foreigners in Italy. At the same time, though, the internal migration model underwent another important change. The long-distance movements, that is between different regions, progressively decreased while the percentage of short-distance ones, within the same region, on the total of movements significantly increased; while the short-distance movements accounted for 65% of the total at the beginning of the 1960s', they now exceed and even remain stably over 70% (in 2006, they accounted for 75.3% of the total).

The measurement of internal and international migratory flows is based on the changes of residence between Italian municipalities and between municipalities and other countries; this source is based on individual data and has been extensively restructured in recent years. The quality of the data produced has been considerably improved, its prompt distribution guaranteed and there are a greater number of ad hoc indicators for analysing residential mobility. There are numerous statistical information about the individual features of migrants, in particular their distribution by age, gender and nationality.

To study both international and internal mobility we use local labour market areas. The strong relationship between labour market and foreign people distribution suggests using local labour market areas in order to draw the geography of this population across Italy. The 686 "Local labour market areas" were created on the basis of daily work movements deducted from the last census data. This is the third time that Istat has identified such areas, in occasion of the last three censuses. The number of local labour market areas has dropped by 28.2% between 1981 (when they amounted to 955) and 2001. They correspond to aggregations of local administrative units ("comuni") contiguous one to the other and constitute areas of self-containment, considering the inflows and outflows.

The analysis of the changes of residence, carried out considering the LLMAs as a privileged framework to show the migrants' behaviours, highlights the role of citizenship in shaping the net among the different territories.

3 Methods of Analysis

The method of social network analysis (SNA), a technique for visualizing, describing and analyzing a system of relations, is applied. Both a basic form of SNA displaying the network visually and a statistical approach calculating some

Table 1 Degree centrality for foreign population changes of residence between Italian local labour market areas by main citizenships, Albania, China, Morocco and Romania – average 2005–2006 (absolute values)

	Albania	China	Morocco	Romania
Out-degree (maximum)	15 (Roma)	36 (Milano)	16 (Torino)	58 (Roma)
In-degree (maximum)	11 (Milano)	34 (Milano)	14 (Torino)	17 (Torino)
Out-degree (mean)	3.3	8.0	3.4	13.0
In-degree (mean)	2.4	7.6	3.0	3.7

Source: Istat data

indicators to describe the network and its characteristics are used. The visualization by itself can enhance the understanding of some relationships between the nodes. In the current analyses, different thresholds are used to select the most relevant flows. For all the analysis here presented the techniques of network analysis and graphical network representation allow to provide a summary indication of migration networks in Italy and overcome the "two-by-two" perspective of the origin-destination matrix. Such techniques, therefore, are a particularly useful tool where the objective of the analysis is to identify, by graphic representation, the existence of specific types of networks correlated in part with socio-economic conditions in different geographical areas.

In a first step, flows from abroad are considered. In this case the network is directed but there is no reciprocity of ties.

For internal mobility one-mode networks are adopted. They are directed and it is possible to study the level of attraction and the push role of the nodes considered.

Furthermore, SNA develops a number of indicators that focus on the relationships between the nodes considering for example the degree centrality (see Table 1). Degree centrality measures how many direct connections one node has to other nodes. When ties are directed, it is possible to compute the total number of ties sent (out-degree) and received (in-degree).

4 Main Results

The traditional methods of analysis show that the big cities are the most attractive destinations for the flows from abroad. Using network representations (graphs) it is possible to draw the flows by main citizenships (origin). The local labour market areas are the receiving nodes.

The graphs for the first four largest resident communities indicate (at the centre of the graph) the existence of several destinations common to all or at least to the majority of the communities examined (Fig. 1). As can be expected, these destinations are principally the main cities (Roma, Milano, Torino, etc.); Roma continues to exert a particularly strong force of attraction for Romanians with an average of over 6,600 Municipal Registers enrolments per year.

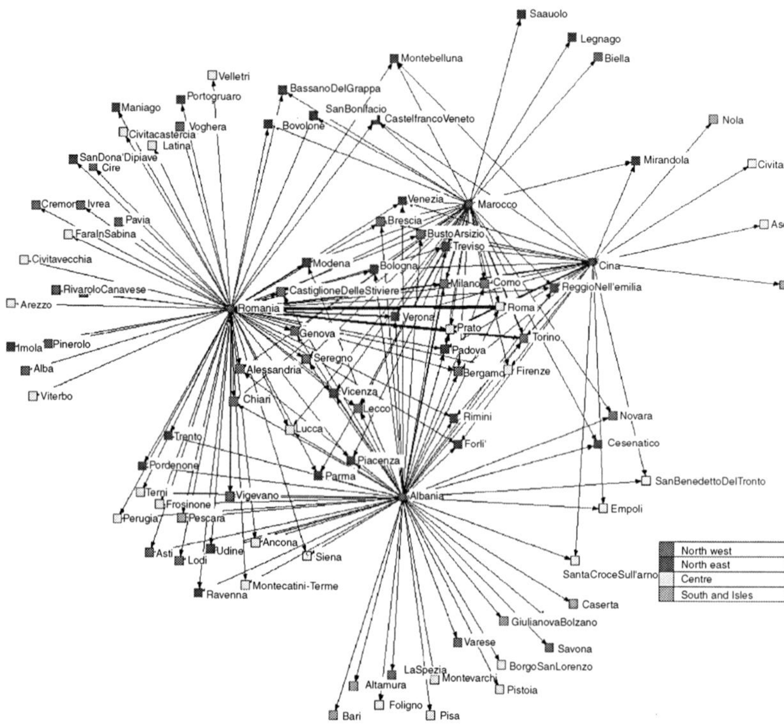

Fig. 1 Networks of foreign population changes of residence from abroad towards Italian Local labour market areas by main citizenships, Albania, China, Morocco and Romania – Average 2005–2006 (absolute values) (*a*).
Source: Istat data. (*a*) Flows above 12 changes of residence per 10 000 residents are considered

However, destinations common to several communities also include LLMAs that are not centred on large cities and which have smaller population size (under 200,000 inhabitants), such as Castiglione delle Stiviere where, on average, 196 Romanians a year enrol at the Municipal Registers.

Beyond these common destinations, each community develops a range of specific target locations where LLMAs consisting of smaller populations assume great importance: each nationality establishes a preferential relationship with specific LLMAs. This phenomenon is certainly related to the different productive specializations of these areas and to the prevalent job opportunities they offer each community. However, it is important not to overlook two aspects that emerge from network analysis: the same community often targets a range of destinations whose LLMAs have considerably different economic and productive characteristics; LLMAs that are similar in terms of type of job opportunities they offer are targeted by different specific communities on a kind of "territorial division" basis. These two elements make it possible to appreciate the action exerted by migration chains that link migrants with specific areas.

It is also clear that some towns with thriving economies, regardless of their main productive vocation, are able to offer different job opportunities and that foreign immigrants are not always employed in the main sectors of local economies but often occupy niches left vacant by Italians.

In 2005–2006 Romanians recorded more than 120 residence enrolments on average in several local labour market areas; none of them, however, is located in the South.

Conversely, Albanians and Chinese immigrants show a preference for a number of southern destinations. In particular, significantly high flows in absolute terms can be observed in Chinese nationals towards several LLMAs in Campania: Nola, Torre del Greco, Avellino.

Considering now the internal mobility the plot of the networks of the foreigners' changes of residence on the Italian territory shows some important aspects. In the North of the country the net seems to be particularly strong and the foreign population very dynamic. The big cities, such as Roma and Milano, seem to play a core role in the network not only as areas of destination but also as sending nodes. Furthermore, many small centres, located in the North of the country, are connected in dynamic networks.

Again in this case, it is appropriate to study migration flows separately for each of the main communities, bearing in mind the results of the research described earlier on foreigners' flows from abroad.

For example, with regard to Romanians, it can first be noted that many of the LLMAs that are destinations for immigrants from abroad are also internal migration destinations: Colleferro, Civita Castellana, Bovolone, to name only several (Fig. 2).[1]

The network, observed for residence enrolments from abroad, does not involve LLMAs in the South and the Islands. Roma occupies a position at the centre of the graph as the point of departure of a large number of flows towards two loosely interconnected fronts: the North-East and the North-West. Moreover, while Roma is linked by outflows to LLMAs in all other areas of the country, Milano and Torino are the main hubs for relocation within the same area. While movements of the Romanian community in Central and North-Western Italy revolve around the local labour market areas of the major cities (in most cases, outflows), no LLMA in the North-East appears to be a main centre of inward or outward-bound relocation.

As regards Moroccan immigrants, the graph indicates that two major separate networks exist. One connects a large number of towns in the North-West with Milano, Bergamo and Brescia at the heart of a network with a high rate of interchange (Fig. 3). The second network connects towns in Emilia-Romagna (with the exception of Firenzuola in Toscana).

Worthy of note is the network of internal movements by Chinese citizens – the only network among those examined that involves towns in Southern Italy (Fig. 4). This aspect had already become known during analysis of movements from abroad to some of the same LLMAs (Avellino, Nola, Torre del Greco). As regards internal

[1]Each node shows a size proportional to its demographic size.

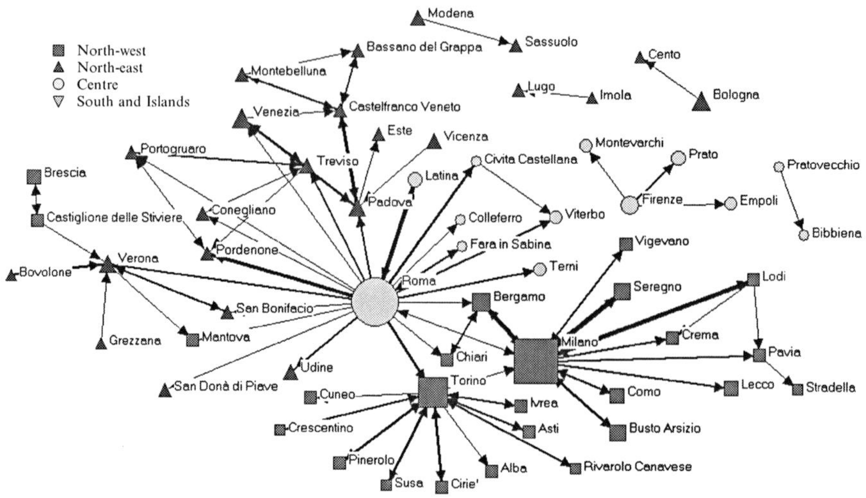

Fig. 2 Networks of Romanian citizens changes of residence between Italian Local labour market areas – average 2005–2006 (absolute values) (*a*).
Source: Istat data. (*a*) Flows above 15 changes of residence are considered

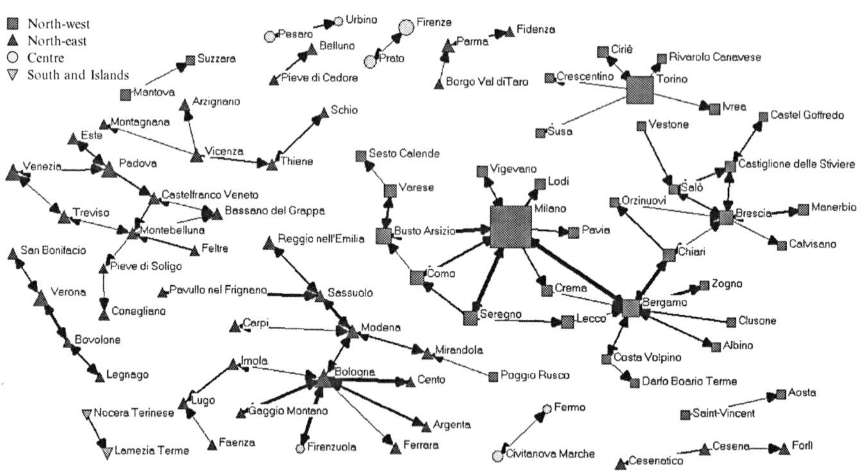

Fig. 3 Networks of Moroccan citizens changes of residence between Italian local labour market areas – average 2005–2006 (absolute values) (*a*).
Source: Istat data. (*a*) Flows above 15 changes of residence are considered

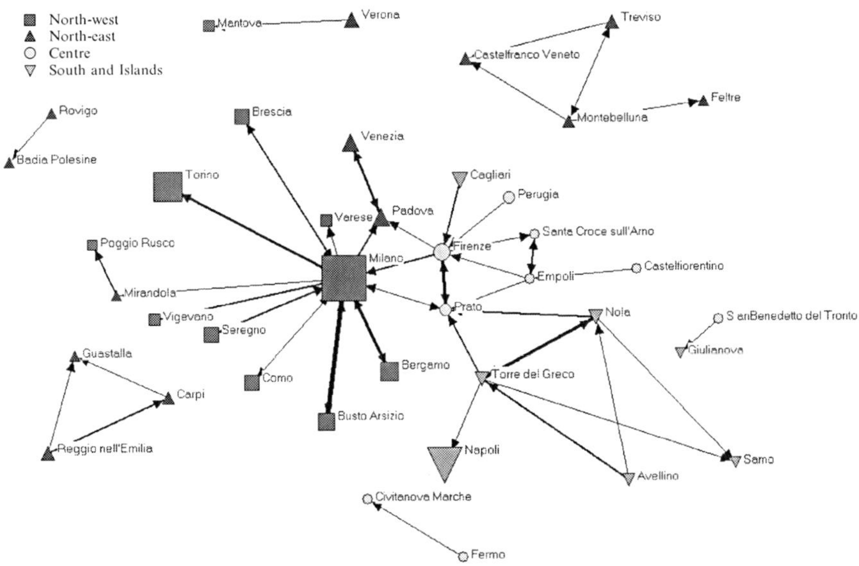

Fig. 4 Networks of Chinese citizens changes of residence between Italian Local labour market areas – average 2005–2006 (absolute values) (*a*).
Source: Istat data. (*a*) Flows above 15 changes of residence are considered

movements, Naples exerts the main force of attraction on the Chinese population from nodes in Central Italy. At the centre of the network is a quadrangle with Milano, Prato, Padova and Firenze at the vertices. Movements radiate out from each vertex of the quadrangle to and from other locations. The areas around Firenze and Prato are peculiar in that they are also connected by migration flows to towns in the South and the Islands.

The main interchange network for Albanians is located entirely within Lombardy, with the exception of Roma which is connected by movements towards Chiari, and it should be remembered that this latter area was found to be particularly important also as regards inflows from abroad. Equally interesting is the Tuscan network. As regards the rest of the country, exchanges emerge between two or three areas and do not constitute authentic networks (Fig. 5).

Among the measures that it is possible to calculate, one of the most interesting for our analysis is degree centrality (Table 1). Citizens from Romania show the highest mean out-degree (largest number of destination nodes). The most important sending node for this community is Roma. The Chinese seem to be the most dynamic. They show the highest mean in-degree (largest number of sending nodes involved in the networks) and, at the same time, a good score in terms of out-degree. In this case, the most relevant node is Milano.

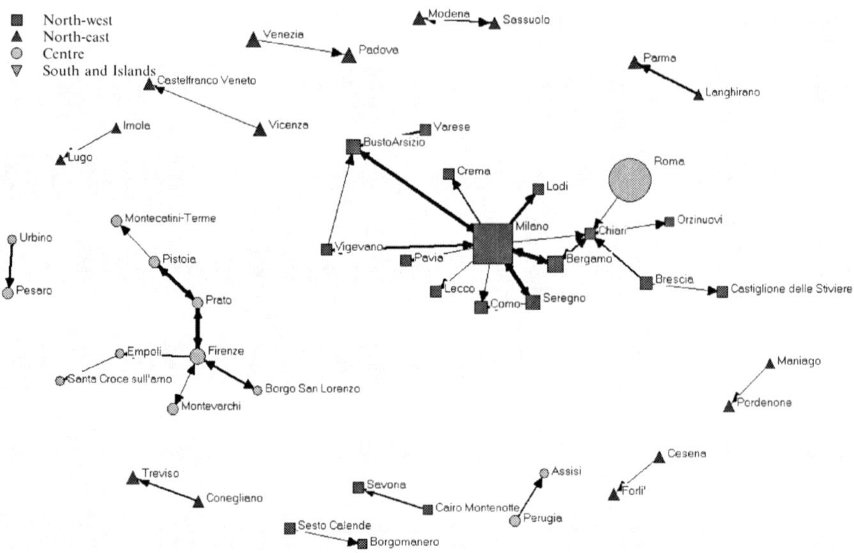

Fig. 5 Networks of Albanian citizens changes of residence between Italian local labour market areas – average 2005–2006 (absolute values) (*a*).
Source: Istat data. (*a*) Flows above 15 changes of residence are considered

5 Conclusions

The advantages of using network analysis techniques in this field are substantial. In this way, smaller LLMAs appear clearly as important points of arrival and settling.

Networks assume peculiar characteristics for each community; in general they include few locations in the South and the Islands.

Areas that attract flows from abroad are equally attractive as regards internal movements. In many cases the large cities act as poles of redistribution of the population towards smaller neighbouring LLMAs. It is certain that LLMAs with a specific productive specialization are particularly attractive to foreigners who again in this case appear to follow routes determined both by demand for specialised labour and, probably, by the pull effect of migration chains or networks.

Migration dynamics, indeed, seem to be deeply determined by the presence and the functioning of a variety of networks at different levels of aggregation. Migration networks can be defined as "groups of social ties formed on the basis of kinship, friendship and common origin" (Massey 1990). This is true both for international migrations and for internal migrations.

References

Cordaz D. (a cura di) (2005) Le misure dell'analisi di rete e le procedure per la loro elaborazione mediante UCINET V, Appendice al volume Salvini, A., *L'analisi delle reti sociali. Risorse e meccanismi*, Ed Plus, Pisa University Press, Pisa.

Istat (2005) I trasferimenti di residenza, Iscrizioni e cancellazioni anagrafiche nel 2002, *Statistiche in breve*, 25 febbraio.

Istat (2007) *Rapporto annuale. La situazione del Paese nel 2006*, Roma.

Istat (2008) *Rapporto annuale. La situazione del Paese nel 2007*, Roma.

Maier G., Vyborny M. (2005) Internal Migration between US States - A Social Network Analysis, *SRE-Discussion 2005/04*, Abteilung für Stadt- und Regionalentwicklung, Department of Urban and Regional Development, Vienna University of Economics and Business Administration (WU-Wien).

Massey, D.S. (1990) "Social structure, household strategies, and the cumulative causation of migration". In: *Population Index*, vol. 56, no.1, pp. 3–26.

Rivellini G., Uberti T. E. (2009) Residential versus educational immigration in Italian provinces: a two-mode network analysis, in M.R. D'Esposito, G. Giordano, M.P. Vitale (a cura di), *Atti del Workshop analisi delle reti sociali: per conoscere uno strumento, uno strumento per conoscere*, Collana Scientifica dell'Università di Salerno, Rubbettino Editore.

SVIMEZ (2007) *Rapporto Svimez 2007 sull'economia del Mezzogiorno*, Il Mulino, Bologna.

Estimation of Income Quantiles at the Small Area Level in Tuscany

Caterina Giusti, Stefano Marchetti and Monica Pratesi

Abstract Available data to measure poverty and living conditions in Italy come mainly from sample surveys, such as the Survey on Income and Living Conditions (EU-SILC). However, these data can be used to produce accurate estimates only at the national or regional level. To obtain estimates referring to smaller unplanned domains small area methodologies can be used. The aim of this paper is to provide a general framework in which the joint use of large sources of data, namely the EU-SILC and the Population Census data, can fulfill poverty and living conditions estimates for Italian Provinces and Municipalities such as the Head Count Ratio and the quantiles of the household equivalised income.

1 Introduction

The principal aim of the EU-SILC survey is to fulfill timely and comparable estimates on income and living conditions in the countries of the European Union, both in a cross-sectional and longitudinal perspective. However, EU-SILC data can be used to produce accurate estimates only at the NUTS 2 level (that is, regional level). Thus, to satisfy the increasing demand from official and private institutions of statistical estimates on poverty and living conditions referring to smaller domains (LAU 1 and LAU 2 levels, that is Provinces and Municipalities), there is the need to resort to small area methodologies.

Small area estimation techniques are employed when sample data are insufficient to produce accurate direct estimates in the domains of interest. The idea of these methods is to use statistical models to link the survey variable of interest with covariate information that is also known for out of sample units. Among these,

C. Giusti (✉) · S. Marchetti · M. Pratesi
Department of Statistics and Mathematics Applied to Economics,
Via Cosimo Ridolfi, 10, Pisa, Italy
e-mail: caterina.giusti@ec.unipi.it; s.marchetti@ds.unifi.it; m.pratesi@ec.unipi.it

A. Di Ciaccio et al. (eds.), *Advanced Statistical Methods for the Analysis of Large Data-Sets*, Studies in Theoretical and Applied Statistics,
DOI 10.1007/978-3-642-21037-2_38, © Springer-Verlag Berlin Heidelberg 2012

a novel approach is represented by M-quantile models. Unlike traditional linear mixed models (Rao, 2003), they do not depend on strong distributional assumptions and they are robust against outlying area values (Chambers and Tzavidis, 2006). Under this approach we focus on the cumulative distribution function of income in each small area and we propose an M-quantile bias adjusted estimator of income quantiles and a bootstrap technique for the estimation of its mean squared error.

The focus is on the estimation of some poverty measures, such as the quantiles of the household equivalised income, at the small area level in Tuscany. For this purpose we combine data coming from the EU-SILC survey 2006 with those from the Population Census 2001. Note that besides the more diffused poverty indicators, the Laeken indicators, the estimation of the quantiles of the household equivalised income, and thus of its cumulative distribution function, can fulfill a more in depth knowledge on the living conditions in the small areas of interest.

2 Small Area Methods for Poverty Estimates

The knowledge of the cumulative distribution function of an income variable in a given area of interest represents an important source of information on the living conditions in that area. From the cumulative distribution function of the household disposable income many quantities (e.g. the median income, the income quantiles) and indicators (e.g. the Head Count Ratio or *at-risk-of-poverty-rate*) can be computed (Giusti et al., 2009).

Let $\mathbf{x}_i$ be a known vector of auxiliary variables for each population unit i in small area j and assume that information for the variable of interest y_{ij}, the household disposable income on an equivalised scale for household i in small area j, is available only for the units in the sample. The target is to use these data to estimate the cumulative distribution function of the household income, which can be defined as:

$$F_j(t) = N_j^{-1}\left(\sum_{i \in s_j} I(y_{ij} \leq t) + \sum_{i \in r_j} I(y_{ij} \leq t)\right) \tag{1}$$

where N_j is the number of population units in small area j, s_j denotes the n_j sampled units in area j and r_j the remaining $N_j - n_j$ not sampled units. The y_{ij} values for the r_j units need to be predicted in each area j under a given small area model. For this purpose we model the quantiles of the conditional distribution of the variable of study y given the covariates. In more detail, the qth M-quantile $Q_q(x;\psi)$ of the conditional distribution of y given x satisfies:

$$Q_q(\mathbf{x}_{ij};\psi) = \mathbf{x}_{ij}^T \boldsymbol{\beta}_\psi(q) \tag{2}$$

where ψ denote the Huber 2 proposal influence function associated with the M-quantile. For specified q and continuous ψ, an estimate $\hat{\boldsymbol{\beta}}_\psi(q)$ of $\boldsymbol{\beta}_\psi(q)$ is obtained via an iterative weighted least squares algorithm.

However, by simply plugging-in the y values predicted under the M-quantile model in (1), using a so called *naïve* estimator, we obtain biased estimates of the cumulative distribution function of y in the small areas. A possible solution is represented by the use of the bias adjusted estimator of the cumulative distribution function proposed by Chambers and Dunstan (1986) (hereafter CD). When (2) holds, this estimator for small area j is:

$$\hat{F}_j^{CD}(t) = N_j^{-1}\left\{\sum_{i\in s_j} I(y_{ij} \le t) + n_j^{-1}\sum_{i\in r_j}\sum_{k\in s_j} I\left\{[\mathbf{x}_{ij}^T\hat{\boldsymbol{\beta}}(\hat{\theta}_j) + (y_{kj} - \hat{y}_{kj})] \le t\right\}\right\} \tag{3}$$

where $\hat{y}_{kj} = \mathbf{x}_{kj}^T\hat{\boldsymbol{\beta}}(\hat{\theta}_j)$ is a linear combination of the auxiliary variables and $\hat{\theta}_j$ is an estimate of the average value of the M-quantile coefficients of the units in area j.

Analytic estimation of the mean squared error of estimator (3) is complex. A linearization-based prediction variance estimator is defined only when the estimator of interest can be written as a weighted sum of sample values, which is not the case with (3). As a consequence, we adapt to the small area problem the bootstrap procedure suggested by Lombardia et al. (2003).

In what follows we describe a semi-parametric bootstrap approach for estimating the mean squared error of small area quantiles.

Having estimated the following model on the sample data

$$y_{ij} = \boldsymbol{x}_{ij}^T\boldsymbol{\beta}_\psi(\theta_j) + \varepsilon_{ij},$$

we compute the estimated M-quantile small area model residuals,

$$e_{ij} = y_{ij} - \boldsymbol{x}_{ij}^T\hat{\boldsymbol{\beta}}_\psi(\hat{\theta}_j).$$

A bootstrap population $\Omega^* = \{y_{ij}^*, \boldsymbol{x}_{ij}\},\ i \in \Omega,\ j = 1,\ldots,d$ can be generated with

$$y_{ij}^* = \boldsymbol{x}_{ij}^T\hat{\boldsymbol{\beta}}_\psi(\hat{\theta}_j) + e_{ij}^*,$$

where the bootstrap residuals e_{ij}^* are obtained by sampling from an estimator of the cumulative distribution function $\hat{G}(t)$ of the model residuals e_{ij}. For defining $\hat{G}(t)$ we consider two approaches:

1. Sampling from the empirical distribution function of the model residuals.
2. Sampling from a smoothed distribution function of these residuals.

For each above mentioned case, sampling of the residuals can be done in two ways:

a. By sampling from the distribution of all residuals without conditioning on the small area – we refer to this as the unconditional approach.
b. By sampling from the conditional distribution of residuals within small area j – we refer to this as the conditional approach.

Hence, there are four possible scenarios by defining e_{ij}^* and consequently the bootstrap population Ω^* using the estimated cumulative distribution function of the model residuals, $\hat{G}(t)$:

- Empirical approach, area unconditioned:

$$\hat{G}(t) = n^{-1} \sum_{j=1}^{d} \sum_{i \in s_j} I(e_{ij} - \bar{e}_s \leq t)$$

- Empirical approach, area conditioned:

$$\hat{G}_j(t) = n_j^{-1} \sum_{i \in s_j} I(e_{ij} - \bar{e}_{sj} \leq t)$$

- Smooth approach, area unconditioned:

$$\hat{G}(t) = n^{-1} \sum_{j=1}^{d} \sum_{i \in s_j} K\left[\frac{t - (e_{ij} - \bar{e}_s)}{h}\right]$$

- Smooth approach, area conditioned:

$$\hat{G}_j(t) = n_j^{-1} \sum_{i \in s_j} K\left[\frac{t - (e_{ij} - \bar{e}_{sj})}{h_j}\right]$$

where h and h_j are smoothing parameters chosen so that they minimise the cross-validation criterion proposed by Bowman et al. (1998), $\bar{e}_s$ and $\bar{e}_{sj}$ are the mean of all the residuals and the mean of the area j residuals respectively, d is the number of small areas, and K is the cumulative distribution function corresponding to a bounded symmetric kernel density k:

$$K(t) = \int_{-\infty}^{t} k(z)dz.$$

The density k is the Epanechnikov kernel density:

$$k(t) = 3/4\left(1 - t^2\right) I(|t| < 1).$$

The bootstrap procedure can be summarised as follows: starting from sample s, selected from a finite population Ω without replacement, we generate B bootstrap populations Ω^{*b} using one of the four above mentioned methods for estimating the distribution of the residuals. From each bootstrap population Ω^{*b}, we select L samples using simple random sampling within the small areas and without replacement in such a way that $n_j^* = n_j$. Bootstrap estimators of the bias and variance of our predictor of the distribution function in area j are defined respectively by:

$$\widehat{\text{Bias}}_j = B^{-1}L^{-1}\sum_{b=1}^{B}\sum_{l=1}^{L}\left(\hat{F}_j^{*bl,CD}(t) - F_{N,j}^{*b}(t)\right),$$

$$\widehat{\text{Var}}_j = B^{-1}L^{-1}\sum_{b=1}^{B}\sum_{l=1}^{L}\left(\hat{F}_j^{*bl,CD}(t) - \bar{\hat{F}}_j^{*bl,CD}(t)\right)^2,$$

where $F_{N,j}^{*b}(t)$ is the distribution function of the bth bootstrap population, $\hat{F}_j^{*bl,CD}(t)$ is the Chambers-Dunstan estimator of $F_{N,j}^{*b}(t)$ computed from the lth sample of the bth bootstrap population and $\bar{\hat{F}}_j^{*bl,CD}(t) = L^{-1}\sum_{l=1}^{L}\hat{F}_j^{*bl,CD}(t)$. The bootstrap mean squared error estimator of the estimated small area quantile is then defined as

$$\widehat{MSE}\left(\hat{F}_j^{CD}(t)\right) = \widehat{\text{Var}}_j + \widehat{\text{Bias}}_j^2.$$

Finally, using a normal approximation we can further compute approximate confidence intervals for the estimated small area quantiles. Alternatively we can obtain a confidence interval picking up appropriate quantiles from the bootstrap distribution of $\hat{F}_j^{CD}(t)$.

Here we show only part of the results of the model based simulations to evaluate the performance of the proposed bootstrap estimators. In particular, we consider the smooth approach area conditioned scenario. We focus on (a) one symmetric and (b) one asymmetric synthetic population to limit the behavior on income distribution. The target parameter is the small area median. The symmetric population has been generated using a linear mixed model with unit errors drawn from $N(0, 16)$ and area effects from $N(0, 1)$ so that we have Gaussian area effects. The asymmetric population has been generated using the χ^2 distribution with 3 and 1 degrees of freedom for the unit errors and area effects respectively. The populations have been generated for every Monte Carlo run.

The results in Table 1 show that the bootstrap estimator works both for symmetric and asymmetric generated populations and that the confidence intervals are very close to the nominal 95% threshold. For further details and simulation results we refer to Tzavidis et al. (2010); we stress here that the method has similar performance also when estimating other population quantiles. Moreover, the technique proposed can be adapted to estimate the Head Count Ratio or other Laeken indicators together with a measure of their variability; however, work on this topic is still in progress.

Table 1 Model based simulations: relative bias (%) and coverage rate of the bootstrap MSE estimators of the median estimator over 30 small areas (500 Monte Carlo runs)

	Min.	1st Qu.	Median	Mean	3rd Qu.	Max.
$RB(\widehat{SE}(\widehat{\text{Median}}))^{(a)}$	−11.90	−6.37	−1.50	−2.58	0.98	4.17
$RB(\widehat{SE}(\widehat{\text{Median}}))^{(b)}$	−15.37	−6.78	−2.05	−2.22	1.68	8.31
$CR(\widehat{SE}(\widehat{\text{Median}}))^{(a)}$	0.92	0.94	0.95	0.95	0.96	0.98
$CR(\widehat{SE}(\widehat{\text{Median}}))^{(b)}$	0.91	0.94	0.95	0.95	0.96	0.98

3 Estimation of Household Income Quantiles in Tuscany

Tuscany Region is a planned domain for which EU-SILC estimates are published, while Tuscany Provinces and Municipalities are unplanned domains. Direct estimates at Provincial level may therefore have large errors and they may not even be computable at Municipality level, thereby requiring resort to small area estimation techniques. Data sources for the present application are the 2006 EU-SILC survey (for the n_j sampled units in area j) and the 2001 Population Census of Italy, with a total of 1388260 households in Tuscany (for the not sampled r_j units in area j). In 2006 the EU-SILC regional sample size in Tuscany was of 1525 households; 54 municipalities were included in the sample. The small areas of interest are the 10 Tuscany Provinces, with sample sizes n_j ranging from 71 (Province of Grosseto) to 457 (Province of Firenze). Due to the large sample size in the Province of Firenze and to the differences characterizing that territory, we consider the Municipality of Florence, with 125 units out of 457, as a stand-alone small area (though not shown as separate area in the maps). The characteristic of interest y is the household disposable income, referring to year 2005, equivalised according to the Eurostat guidelines.

As already underlined, the knowledge of the cumulative distribution function of an income variable in any area of interest represents an important source of information for the living conditions in that area. Other quantities, such as the mean of a given income variable, can be highly influenced by outlying values. From the cumulative distribution function of the household disposable income many quantities (e.g. the median income, the income quantiles) and monetary poverty indicators can be computed. Moreover, it is important to note that the knowledge of the cumulative distribution function of the income allows to also estimate the proportion of population whose income is immediately under or above a given poverty threshold. In this application we are interested in estimating the small area median as well as the first and the third quartiles of the household equivalised income in the small areas.

The following auxiliary variables are known for each unit in the population and have resulted significant in the models for income:

- Size of the household in terms of the number of components of the household i in the small area j (integer value).
- Age of the head of the household i in the small area j (integer value).
- Years in education of the head of the household i in the small area j (integer value).
- Working position of the head of the household i in the small area j (employed/unemployed in the previous week).
- Tenure status of household i in the small area j (owner/tenant).

Table 2 and Figs. 1, 2 and 3 show the income distribution by Province in Tuscany, as represented by respectively the 1st quartile, the median and the 3rd quartile of the household equivalised income in each area estimated using the CD M-quantile predictor. In each map the Provinces are grouped in four different classes of colors,

Table 2 Sample (n_j) and population (N_j) size, estimated quartiles and percentage coefficient of variation (CV%) of the equivalised household income, Tuscany Provinces

	n_j	N_j	1st Qu. (CV%)	Median (CV%)	3rd Qu. (CV%)
Massa-Carrara (MS)	110	80810	8633,9 (7,1)	12591,5 (5,4)	17093,2 (5,7)
Lucca (LU)	109	146117	10265,1 (6,1)	14924,1 (4,9)	20317,5 (5,3)
Pistoia (PT)	124	104466	11158,3 (5,2)	15791,8 (4,3)	22315,8 (4,5)
Prov. Firenze	332	216531	12076,9 (3,0)	16414,5 (2,4)	21500,8 (2,8)
Livorno (LI)	105	133729	10916,7 (6,0)	15659,8 (4,7)	20696,3 (5,2)
Pisa (PI)	143	150259	11706,5 (4,6)	16738,5 (3,8)	23195,4 (4,3)
Arezzo (AR)	159	123880	11704,7 (4,5)	16415,9 (3,6)	22532,1 (3,7)
Siena (SI)	119	101399	11461,0 (5,2)	16851,9 (4,0)	23399,0 (4,4)
Grosseto (GR)	71	87720	9396,6 (8,5)	13887,8 (6,2)	19695,1 (6,3)
Prato (PO)	128	83617	11885,0 (4,8)	16761,3 (3,9)	23411,3 (4,1)
Mun. Firenze	125	159724	11912,2 (5,0)	16923,3 (4,0)	23380,5 (4,1)

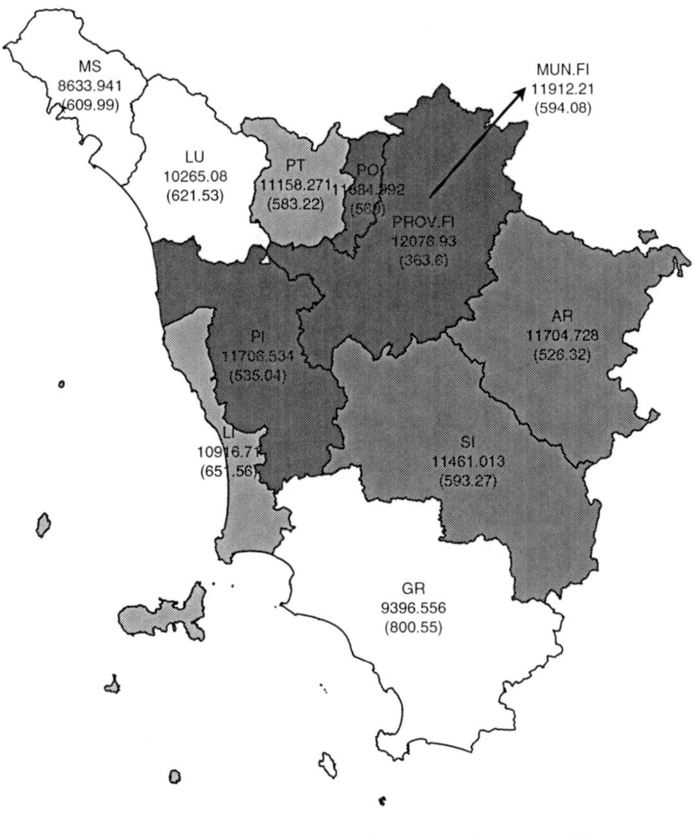

Fig. 1 First quartile of the household equivalised income and corresponding root mean squared error (in parenthesis)

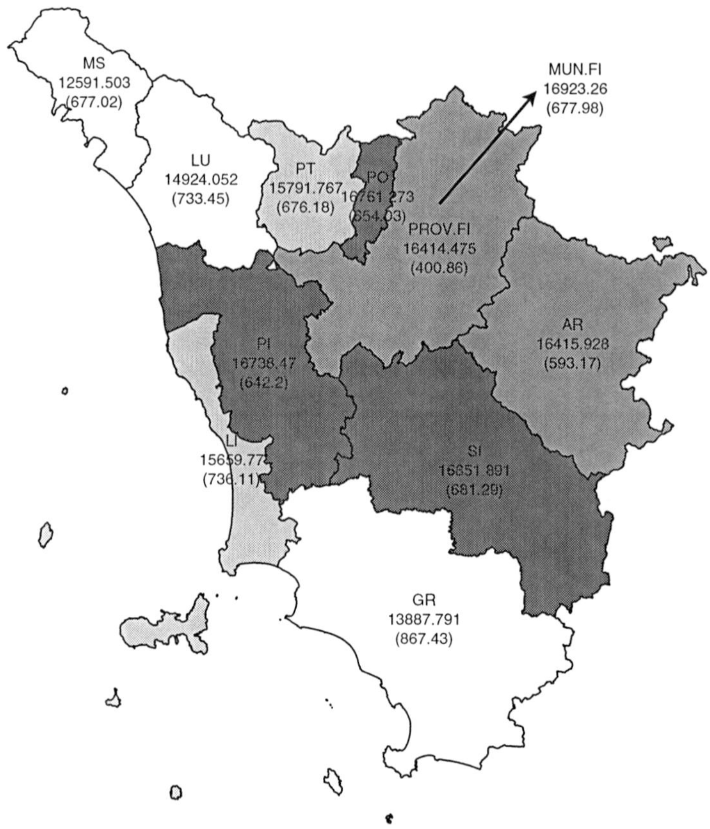

Fig. 2 Median household equivalised income and corresponding root mean squared error (in parenthesis)

where the darkest color corresponds to higher values of the estimates of the target variable.

As we can see, there are differences between the estimates: the Provinces of Massa-Carrara (MS), Grosseto (GR) and Lucca (LU) have the smaller values for each quartile (lighter color); at the opposite the Provinces of Prato (PO), Pisa (PI) and the Municipality of Florence (MUN. FI) are always in the higher class of estimates (darkest color). Moreover, from the maps we can appreciate the changes in the ranking of the remaining areas going from a quartile to another, as it happens for the Provinces of Siena (SI) and Pistoia (PT). Indeed, each area is characterised by a different income distribution function; thus, a given area can have an higher proportion of households below low income values when compared with the other areas, while having at the same time a relative lower proportion of households below high income values. This depends on the behavior of the income distribution function, and in particular on its steepness for ranging income values. Next steps of our analysis will regard the estimation of a higher number of quantiles of the household income in each area, to better track the differences in the income distributions.

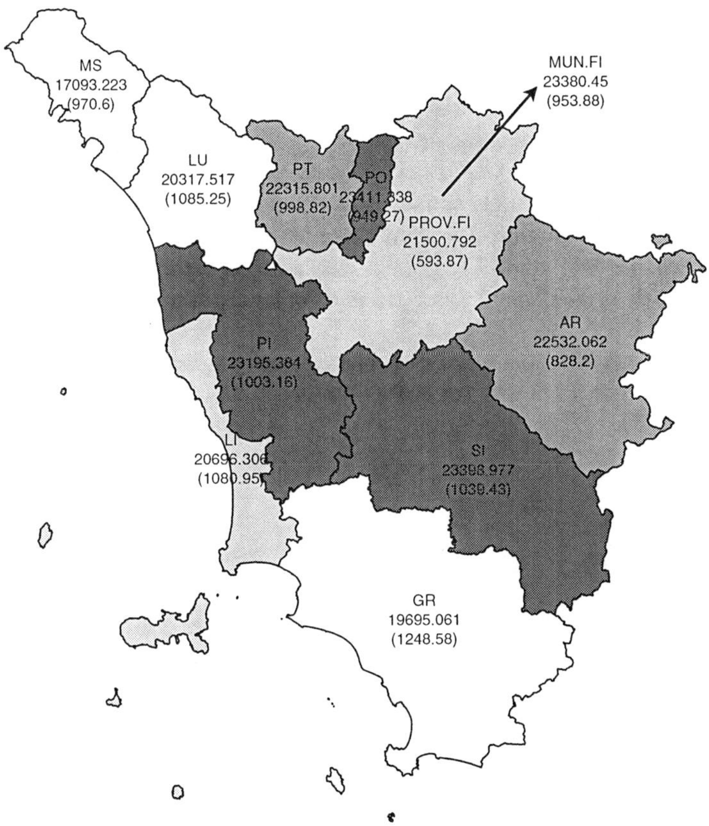

Fig. 3 Third quartile of the household equivalised income and corresponding root mean squared error (in parenthesis)

Note that another important finding is the estimation of a measure of variability for the estimated income quantiles using the proposed bootstrap approach. Indeed, point estimates alone do not suffice to get a complete picture of the income distribution in the different areas and also to make comparisons between the areas. The estimation of the root mean squared error and of the percentage coefficient of variation (see Table 2 and Figs. 1, 2 and 3) allows to conduct these comparisons on the ground of a more accurate statistical information, for example by building confidence intervals for the estimates.

4 Conclusions

In this work we propose an M-quantile bias adjusted estimator for income quantiles at a small area level. The estimation of the quantiles of the equivalised household income is an important information to understand the living conditions and poverty in a given area of interest. Detailed information at local level can be used efficiently by the policy makers to develop "ad-hoc" interventions against poverty.

It is important to accompany the small area estimates with a measure of their variability. We suggest a bootstrap technique to estimate the mean squared error of the proposed small area quantiles estimator.

The application to EU-SILC data in Tuscany shows the potentialities of these methods in fulfilling accurate estimates at Provincial level. Future developments will focus on the estimation at a smaller area level, such as the Municipality level. Furthermore, we will consider the point and interval estimation of other poverty measures, such as the quantile share ratio and other Laeken indicators.

Acknowledgements This work has been financially supported by the European Project SAMPLE "Small Area Methods for Poverty and Living Condition Estimates", European Commission 7th FP - www.sample-project.eu.

References

Bowman A, Hall P, Prvan T (1998) Bandwidth selection for the smoothing of distribution functions. Biometrika 85:799–808

Chambers R, Dunstan R (1986) Estimating distribution functions from survey data. Biometrika 73:597–604

Chambers R, Tzavidis N (2006) M-quantile models for small area estimation. Biometrika 93:255–268

Giusti C, Pratesi M, Salvati S (2009) Small area methods in the estimation of poverty indicators: the case of Tuscany. Politica Economica 3:369–380

Lombardia M, Gonzalez-Manteiga W, Prada-Sanchez J (2003) Bootstrapping the Chambers-Dunstan estimate of finite population distribution function. Journal of Statistical Planning and Inference 116:367–388

Rao J (2003) Small Area Estimation. Wiley, New York

Tzavidis N, Marchetti S, Chambers R (2010) Robust estimation of small area means and quantiles. Australian and New Zealand Journal of Statistics 52:167–186

The Effects of Socioeconomic Background and Test-taking Motivation on Italian Students' Achievement

Claudio Quintano, Rosalia Castellano, and Sergio Longobardi

Abstract The aim of this work is to analyze the educational outcomes of Italian students and to explain the differences across Italian macro regions. In addition to the "classic" determinants of student achievement (e.g. family socioeconomic background) we investigated the extent to which the test-taking motivation may contribute to influence the results from assessment test and to explain, partially, the Italian territorial disparities. Therefore, a two stage approach is provided. Firstly, the data envelopment analysis (DEA) is applied to obtain a synthetic measure of the test-taking motivation. Secondly, a multilevel regression model is employed to investigate the effect of this measure of test-taking motivation on student performance after controlling for school and student factors.

1 Introduction

Many studies focus on the role played by human capital on the economic growth. A widespread literature recognizes the positive relationship between human capital and growth (Mankiw et al. 1992) investigating various education-related determinants of economic growth. Empirically, Romer (1989), incorporating a human capital proxy variable (adult literacy) as one of the regressors of the growth rate of GDP, found a positive effect of adult literacy rates on economic growth. Hanushek and Woessmann (2007) highlighted that education could raise income levels mainly by speeding up technological progress. Barro and Sala-i-Martin (2004) reveal a positive association between male secondary and higher schooling and economic growth. A similar result is obtained by Gemmell (1996), who utilizes an especially constructed measure of school attainment. In any case, there is no doubt that

C. Quintano (✉) · R. Castellano · S. Longobardi
University of Naples "Parthenope" Via Medina 40, Naples
e-mail: claudio.quintano@uniparthenope.it

A. Di Ciaccio et al. (eds.), *Advanced Statistical Methods for the Analysis of Large Data-Sets*, Studies in Theoretical and Applied Statistics,
DOI 10.1007/978-3-642-21037-2_39, © Springer-Verlag Berlin Heidelberg 2012

education quality is one of the factors that enter into the determination of growth (Montanaro 2006). Therefore, we try to analyze the determinants of Italian human capital on the basis of the results from an international students assessment. Indeed, one of the alternatives in measuring the quality of human capital is assessing skills directly. Even if the tests cannot completely measure attitudes and motivation, which are obviously to be included in the human capital definition, their results are able to capture a large part of labour force quality and to enrich research on the causal relationship between education and economic outcomes. Thus, average reading literacy, math literacy, and science literacy of the high school students are thought to be good predictors of the economic growth of a nation (OECD/UNESCO-UIS 2003). Our aim is to analyze the educational outcomes (as a proxy of the human capital) of Italian students focusing on the territorial differences within the country. The first objective is to detect the school and student variables which influence the students' performance. The second objective is to examine whether and to what extent the students' test-taking motivation can influence the result of standardized tests and how much of the Italian territorial disparities is attributable to differences in the test-taking motivation. Consequently, we propose a two stage approach: firstly, the data envelopment analysis (DEA) is applied to obtain a synthetic measure which expresses the test-taking motivation, secondly, a multilevel regression is performed to analyze the effect of this indicator on student performance after controlling for school and students' variables. The research was carried out regarding Italian students and schools by analyzing the data from the last edition (2006) of the OECD' PISA (Programme for International Student Assessment) survey. The paper is structured as follows: in Sect. 2, on the basis of the PISA data, the results of Italian students and the territorial differences are briefly described. In Sect. 3, the relationship between student effort and test outcomes is discussed. Section 4 illustrates the proposed methodology to determinate an effort indicator. Section 5 contains some methodological aspects related to the data envelopment analysis and to the multilevel approach. Section 6 focuses on the results of the multilevel analysis. Finally, some concluding remarks are made.

2 The Italian Literacy Divide

The PISA (Programme for International Student Assessment) survey, which takes place every three years, is carried out by the Organization of Economic Cooperation and Development (OECD)[1] and it has been designed to assess the skills and knowledge of 15-year-old students in three main areas – reading, mathematics and science. Fifty-seven countries participated in the last edition of PISA (2006),

[1]The OECD initiated the first cycle of the PISA in 2000, the second cycle occurred in 2003 and the third in 2006. 43 countries participated in PISA 2000, 41 in 2003 and 57 in 2006.

including all 30 OECD countries. In Italy, approximately 22,000 15-year-olds from about 800 schools participated. In 2006, performance of Italian 15-year olds was lower than the performance of their counterparts from most of the countries that participated in PISA. The Italian students in PISA 2006 reached an average test score of 462 points in mathematics, 469 in reading and 475 in science, being under the OECD average of 500. The gap between Italian students and top performing countries like Korea and Finland is extremely high and Italy performs significantly worse than all OECD countries, excepting the Republic of Slovak, Turkey, Spain and Greece. The very poor performance of Italian students is due to significant territorial differences within the country[2]. Indeed, 15 year-old students in the Italian Southern regions performed very low in each assessment area which contributed to Italy's standing in international comparisons. For each PISA cycle (2000, 2003, and 2006) the average score differs strongly among the Northern and the Southern regions and these marked differences originate a wide North–South divide which is called *literacy divide*. These disparities are confirmed by several surveys, at national and international level, such as the National System of Assessment conducted by the Italian Evaluation Institute of the Ministry of Education (INVALSI), the Progress in International Reading Literacy Study (PIRLS) and the Trends in International Mathematics and Science Study (TIMMS)[3].

3 The Student Test-Taking Motivation

The issue of test-taking motivation is highly relevant in the context of international comparative studies. In a summary of 12 studies investigating the effects of student test-taking motivation on test performance, Wise and DeMars (2005) found that well-motivated students outperformed less-motivated students with an average effect size exceeding half a standard deviation. Wise and DeMars were also able to show a near zero correlation between self-reports of test-taking motivation and measures of ability, a finding that suggests there was no confounding of motivation and ability. The PISA test is considered as a *low stake test* since the students perceive an absence of personal consequences associated with their test performance. There is a widespread consensus (Chan et al. 1997; Wolf and Smith 1995) that assessments which have no direct consequences for students, teachers or schools underestimate student ability. Because no established measures of test-taking motivation were available in the PISA context, we developed a composite indicator considering

[2]This topic is also discussed in Quintano et al. (2009).

[3]TIMMS and PIRLS are conducted by the International Association for the Evaluation of Educational Achievement (IEA). The IEA (www.iea.nl) is an independent, international cooperative of national research institutions and governmental research agencies.

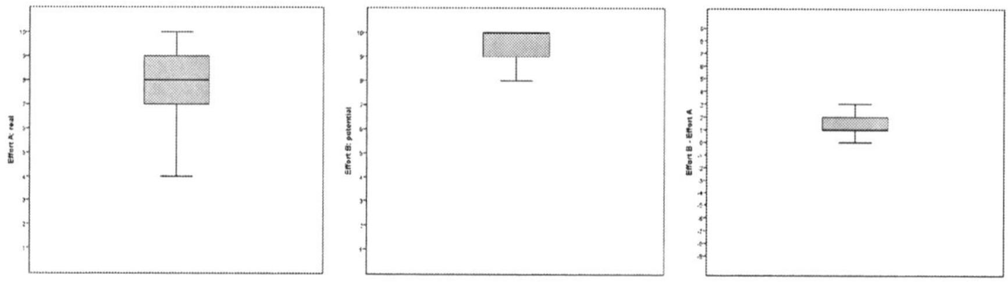

Fig. 1 Distribution of the real effort (*effort a*), the potential effort (*effort b*) and the difference between real and potential effort.
Source: Authors' elaborations on OECD-PISA 2006 database.

three proxies of the students' motivation: (a) students'self-report effort, (b) test non-response rate, (c) questionnaire non-response rate.

In the PISA cognitive test there are two items asking for how the students are motivated to do their best on the PISA test, the first one asks for the real student effort (effort a) and the second the potential effort (effort b), i.e., the effort which the students would have used if the results of PISA test would have influenced their semester grade in school (Fig. 1).

The low variability of the self reported effort and the low difference between the real and the potential effort lead to integrate the self report effort with two proxies of the student motivation: the test non-response rate and the questionnaire non-response rate. The "Test non-response rate" is computed on the basis of the number of missing or invalid answers in the PISA cognitive test:

$$NR_i^T = \left(\frac{m_i^T + i_i^T}{{}_a n_i^T} \right) \tag{1}$$

where m_i^T and i_i^T denote, respectively, the number of item non-responses and of invalid responses given to the cognitive test, while ${}_a n_i^T$ represents the number of applicable questions for the ith student computed by the difference between the total number of questions and the number of the not applicable questions. This ratio could be influenced by the competence of the student[4] since if a student does not know the correct answer he could skip the question (missing response) or invalidate the answer, then as second variable we consider the "Questionnaire non-response rate" computed on the basis of the number of missing or invalid answers in the PISA student questionnaire which are not related with the proficiency level:

[4]The correlation, at student level, between the science literacy score and the three proxies of student motivation (questionnaire non-response rate, students' self-report effort and test non-response rate) is equal to: −0.157, 0.240 and −0.591 respectively. Although the literacy score and the test non-response rate show an high correlation, the lack of direct measures of test taking motivation and the hypothesis that less motivated students tend to give many missing responses suggest to include this variable in order to better describe the students' response behaviour.

$$NR_i^Q = \left(\frac{m_i^Q + i_i^Q}{{}_a n_i^Q} \right) \tag{2}$$

where m_i^Q and i_i^Q denote, respectively, the number of item non-responses and of invalid responses given to the student questionnaire, while ${}_a n_i^Q$ represents the number of applicable questions for the ith student computed by the difference between the total number of questions and the not applicable questions. The Fig. 2 confirms the territorial connotation of the students' response behaviour.

To obtain a measure which expresses the collaboration of the student with a positive scale, the complement to one of the two non-response rates are computed:

$$I_i^T = 1 - NR_i^T \quad I_i^Q = 1 - NR_i^Q \tag{3}$$

In this way, we obtain two indicators (I_i^T and I_i^Q) which increase in correspondence of a high degree of students' motivation (low values of non-response rates) and decrease if low effort is invested in the test (high values of non- response rates). Before aggregating the three measures of student motivation, each variable has been rescaled by a linear scaling technique (LST) to avoid that the composite indicator will be implicitly weighted towards the variable with larger range:

$$Y_{resc} = \frac{Y_i - Y_{\min}}{Y_{\max} - Y_{\min}} \tag{4}$$

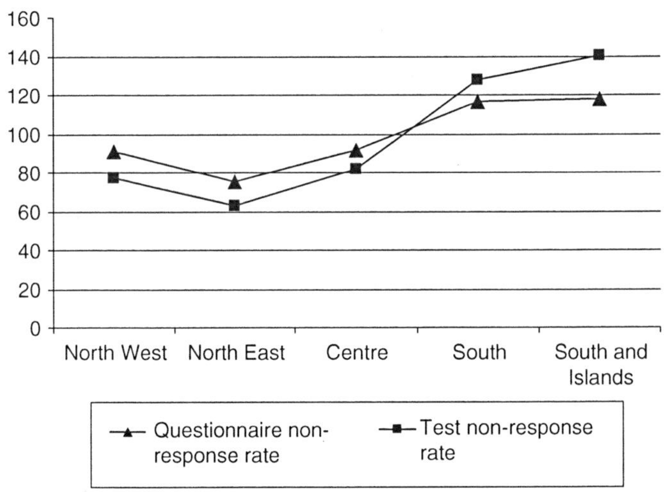

Fig. 2 Questionnaire non-response rate and test non-response rate (Italy = 100)
Source: Authors' elaborations on OECD-PISA 2006 database.

where: Y_i denotes the value of a generic indicator for the i student, Y_{max} and Y_{min} are, respectively, the lower and the higher value of the indicator for all units (student). After the rescaling procedure, each indicator is aggregated at school level (by the mean of student measure) and then they are summarized, by a data envelopment analysis, into one synthetic indicator called "Effort Indicator" (EI) which expresses the test-taking motivation of students.

4 Some Methodological Aspects

4.1 Data Envelopment Analysis (DEA)

In order to aggregate the proxies of students' motivation, we propose a DEA model to obtain an *Effort Indicator* (EI), at school level, as a weighted average of the three indicators, y_k, that is:

$$EI_j = \sum_{k=1}^{h} y_{kj} w_k \tag{5}$$

with y_{kj} as the value of indicator k for school j.

The weights (w_k) that maximize the effort indicator for each school are endogenous computed by solving the following linear programming problem:

$$EI_j = \max_{w_k} \sum_{k=1}^{h} y_{kj} w_k \qquad \textit{subject to:} \tag{6}$$

$$\sum_{k=1}^{h} y_{kj} w_k \leq 1 \forall j = 1, ..., n \quad (\textit{normalisation constr.})$$

$$w_k \geq \varepsilon \forall k = 1, ..., h \quad (\textit{non-negativity constr.})$$

This is a slight modification of the multiplier form of the DEA linear programming problem (Charnes et al. 1978). Indeed in this problem each school produces h output, the h indicators of test-taking motivation, using only one input with value equal to one. As a result, the highest relative weights are accorded to those dimensions for which the school j achieves the best performance (in relative terms) when compared to the other schools. As pointed by Despotis (2005), this model is formally equivalent to an input-oriented constant return to scale (CRS) DEA model. The weights are not fixed a priori and the only restriction in the formulation above is that they should be higher of an infinitesimal ε to assure that none of the weights will take a value lower or equal to zero. For each school we obtain $0 \leq EI_j \leq 1$, with higher values indicating a higher level of students' motivation. The chart (Fig. 3) shows the variation of effort indicator in correspondence of the average performance

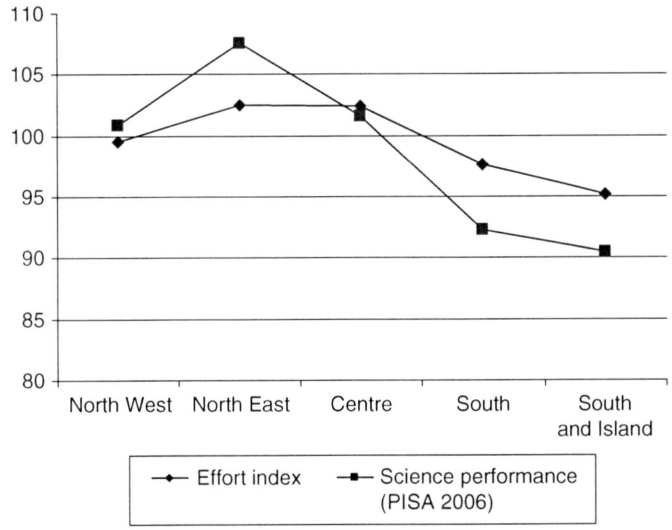

Fig. 3 Effort indicator and Science literacy scores (Italy = 100)
Source: Authors' elaborations on OECD-PISA 2006 database.

at macro region level. The correlation between this indicator and the science literacy scores is equal to 0.47 at national level.

These empirical evidences lead us to suppose that the students of Southern schools have a low awareness of the PISA test importance; consequently it seems that their weak motivation plays an important role to determine low test scores. In the next section, a multilevel regression framework is employed to highlight the role of test-taking motivation on students' performance and on territorial differences with controlling for several student and school variables.

4.2 The Multilevel Model

To identify the possible determinants of the Italian student achievement a multilevel regression model with random intercept is applied[5]. The choice of the multilevel approach is suggested by the hierarchical structure of the PISA data where students (level-one units) are nested in schools (level-two units). The two-level random intercept regression model for the ith student in the jth school is written as:

[5]See Raudenbush and Bryk (2002) and Snijders and Bosker (1999), *inter alia*, for a relevant discussion on multilevel models.

$$Y_{ij} = \gamma_0 + \sum_{s=1}^{f} \beta_s x_{sij} + \sum_{t=1}^{g} \beta_t z_{tj} + \varepsilon_{ij} + U_{oj} \quad (7)$$

where x_s are f variables at student level and z_t are the g variables at school level while ε_{ij} and U_{oj} denote the error components respectively at students and school level, these components are supposed to be normally distributed and uncorrelated:

$$\varepsilon_{ij} \sim \mathrm{IID} - \mathrm{N}(0, \sigma^2) \quad U_{0j} \sim \mathrm{IID} - \mathrm{N}(0, \tau^2) \quad \mathrm{cov}(\mathrm{U_{oj}}, \varepsilon_{ij}) = 0 \quad (8)$$

The model requires a preliminary estimate of the empty model (model with no independent variables) to split the total variation on the dependent variable (student's achievement in Science) into within and between variance. This partition allows to estimate the proportion of the total variation attributable to differences in schools and how much of the total variation as well as the within and between variation are explained by student and school characteristics. Then a *block entry* approach is adopted (Choen and Choen 1983) which consists to the gradual addition of the first and second level covariates.

5 Main Results

The proposed multilevel analysis has required seven models (Table 1) to compare the impacts of student and school characteristics, including in the last model the effort indicator. The dependent variable is the student science achievement measured by five scaled scores called *plausible values*.[6] The set of independent variables is composed by six student-level and seven school-level characteristics derived from the PISA questionnaires. After the empty model (model 1), the second model contains some exogenous variables (gender, immigrant status, and the index of Economic, Social and Cultural Status) that influence other variables but are not affected by other variables. This model provides the opportunity to analyze the absolute effects of individual variables on science achievement. It shows a gender gap of 11.37 points in favour of males and a gap of 43.75 points in favour of non immigrants. The effect of student socioeconomic status is small even if it is statistically significant. In the third model, a set of endogenous student covariates (time spent on homework, enjoyment of science study, awareness of environmental issues) which expresses the students' behaviors and the attitudes towards learning science is introduced. The awareness of science issues might be considered an important predictor of science achievement, furthermore, the

[6]The multilevel analyses proposed in this paper are developed by the Hierarchical Linear Model (HLM) software (Raudenbush, Bryk, Cheong and Congdon 2000) in order to handle plausible values as the dependent variable. The three continuous variables at student level are centred on the school mean while the five continuous variables at school are centred on the grand mean.

students who enjoy science more are 11 points ahead of students who enjoy science less. The multilevel modelling proceeded by including variables describing school characteristics. These factors have a stronger influence on student's outcomes. The school socioeconomic composition shows statistically significant effect on student's achievement, this result is substantially coherent with several studies (Marks 2006; Korupp et al. 2002), moreover the public-school students score 50 points higher than their private-school classmates confirming a "remedial" role for the Italian private school sector offering relatively low rewards to talent and, therefore, attracting relatively low-talent students (Bertola et al. 2007). Another Italian peculiarity is the wide variation of results between different types of upper secondary school, the gap between the schools focusing on humanities or on sciences (Lyceum) and the schools with other study programmes (technical institutes, professional institutes or vocational training) is very high, in favour of Lyceum, and it reflects the social segregation due to family choices (OECD 2009). Once the macro area dummies are introduced (fifth model), a comparison with the previous model shows that the gap related to the school socioeconomic composition is decreased. These large differences in pupils' performance between macro regions could reflect different socioeconomic conditions, not explained by the index of economic and social status, rather than regional differences in school efficiency (OECD 2009). The model sixth shows a positive relation between the computer with web (as a proxy for the quantitative level of the school's facilities) and the students performance while the quality of school's educational resources might not influence the students' achievement. The last multilevel model brings in the effort indicator as control variable. This variable shows a high and significant impact on student performance and this factor involves the reduction of the macro area coefficients. Indeed, after controlling for the effort indicator, the gap of the Italian Southern area is narrowed down from 30 points to 21 points (29%). This might confirm that the Southern students have spent less effort than the students of the North and Centre of Italy. Furthermore, the effort indicator allows to explain a larger amount of variance, indeed, after controlling for test-taking motivation, the accounted total variance among schools (compared to the sixth model) increases from 74% to 80%.

6 Concluding Remarks

This paper provided an analysis of Italian students' achievement on the basis of OECD-PISA 2006 data. The final aim was the identification of the main determinants of the poor performance of Italian students focusing on the geographical disparities within Italy. The multilevel analysis has confirmed the role of the socioeconomic context to influence the student achievement. This result supports the Coleman's conceptualization of social capital. According to Coleman (1990), families with higher socio economic status have larger social capital (entailing social relations and networks) that helps children develop an identity that allows them to better understand and value cognitive development. However, the differences in

Table 1 Results of multilevel regression

	[1]	[2]	[3]	[4]	[5]	[6]	[7]
Intercept	460.09***	467.78***	466.31***	489.85***	501.29***	509.04***	502.94**
Student level variables							
Exogenous variables							
Gender (ref.= male)		−11.37***	−7.34***	−8.08***	−8.31***	−8.22***	−8.14***
Immigrate (ref. = native)		−43.75***	−38.57***	−36.51***	−38.36***	−38.36***	−38.20***
Index of socioeconomic status		5.60***	2.70***	2.69***	2.64***	2.65***	2.67**
Endogenous variables							
Hours spent on homework (ref. = 0-2)							
0			−5.96***	−6.38***	−6.96***	−6.88***	−5.91***
>4			0.37	0.07	0.60	0.53	0.40
Awareness of environmental issues			22.20***	22.20***	22.13***	22.15***	22.17***
Enjoyment of science			11.51***	11.47***	11.43***	11.44***	11.50***
School level variables				70.82***	52.39***	50.07***	49.55***
Index of socioeconomic status							
Private school (ref. = public school)				−49.33***	−64.30***	−57.38***	−55.33***
Study programme (ref. = Lyceum)							
Other study programmes				−21.59***	−38.88***	−46.82***	−43.63***
Lower secondary school				−88.64***	−103.88***	−100.87***	−89.23***
Macro area (ref.= Center)							
North					34.44***	29.64***	30.02***
South					−27.39***	−29.65***	−21.49***
Computers with web						0.19***	0.14****
Index of quality of educational resources						0.55	−0.66
Effort indicator							1.83***
Variance components							
Variance between schools	5,347.98	5,275.51	5,280.67	2,172.40	1,439.68	1,361.89	1,071.50
Variance within schools	4,674.31	4,555.00	3,999.11	4,000.72	4,003.26	4,003.57	4,005.11

Significance level: *** 99%;** 95%;* 90%
Source: Authors' elaborations on OECD-PISA 2006 database

the socioeconomic and cultural status of students and schools are not sufficient to explain the literacy divide between the Centre-North and the South of Italy (Checchi 2004). Therefore, the analysis takes into account the role of test-taking motivation to "boost" the North-South differences. The effort indicator, computed by a data envelopment analysis, has allowed to highlight that the score differences are also influenced by the lower effort and engagement of the Southern students. All of this confirms the complexity of the issue of differences between Italian regions and it suggests the needs for long terms policies to improve the equity of Italian educational system. Furthermore, an information and awareness campaign concerning the importance of PISA survey, focusing on Southern students and teachers, could be useful to improve their test-taking motivation.

References

Barro, R.J., Sala-i-Martin, X.: Economic growth. Second Edition. Cambridge, MA: MIT (2004)

Bertola, G., Checchi, D., Oppedisano, V.: Private school quality in Italy. IZA Discussion pap. ser. **3222** (2007)

Chan, D., Schmitt, N., DeShon, R. P., Clause, C. S., Delbridge, K.: Reactions to cognitive ability tests: the relationships between race, test performance, face validity perceptions, and test-taking motivation. J. Appl. Psychol. **82**, 300–310 (1997)

Charnes, A., Cooper, W.W., Rhodes, E.: Measuring the efficiency of decision making units. Eur. J. Oper. Res. **2**, 429–444 (1978)

Checchi, D.: Da dove vengono le competenze scolastiche. Stato e Mercato.**3,** 413–454 (2004)

Choen, J., Choen, P.: Applied multiple regression correlation analysis for the behavioural sciences. Hillsdale, Lawrence Erlbaum (1983)

Coleman, J. Foundations of social theory. Cambridge: Harvard University Press. (1990).

Despotis, D.K.: A reassessment of the human development index via data envelopment analysis. J. Oper. Res. Soc.**56**, 969–980 (2005)

Gemmell, N.: Evaluating the impacts of human capital stocks and accumulation on economic growth: some new evidence. Oxford Bulletin of Economics and Statistics. **58 (1)**, 9–28 (1996)

Hanushek, Eric A., Woessmann, L.: The role of education quality for economic growth. World bank Policy Res. Work. Pap. Ser.**4122**, (2007)

Korupp, S.E., Ganzeboom, H.B.G., Van Der Lippe, T.: Do Mothers Matter? A Comparison of models of the influence of mothers' and fathers' educational and occupational status on children's educational attainment. Qual. & Quant. **36 (1)**, 17–42 (2002)

Mankiw, G.N., Romer, D., Weil, D.N.: A contribution to the empirics of economic growth. Q. J. Econ. **107**, 407–437 (1992)

Marks, G. N.: Are between- and within-school differences in student performance largely due to socio-economic background? Evidence from 30 countries. Educ. Res. **48 (1)**, 21–40 (2006)

Montanaro, P.: Learning divides across the Italian regions: some evidence from national and international surveys. Bank of Italy Occasional Paper. **14** (2006)

OECD, OECD Economic Surveys: Italy 2009. **8** (2009)

OECD, UNESCO Institute for Statistics: Literacy skills for the world of tomorrow: Further Results from PISA 2000. Paris (2003)

Quintano, C., Castellano, R., Longobardi, S.: L'influenza dei fattori socio economici sulle competenze degli studenti italiani. Un'analisi multilevel dei dati PISA 2006. Riv. di Economia e Statistica del Territorio. **2**, 109–150 (2009)

Raudenbush, S.W., Bryk, A.S.: Hierarchical linear models: Applications and data analysis methods. Thousand Oaks: Sage (2002)

Romer, P.: Human Capital and growth: theory and evidence. NBER Working P.**3137** (1989)
Snijders, T.A.B., Bosker, R.J.: Multilevel analysis. Thousand Oaks: Sage (1999)
Wise, S.L., DeMars, C.E.: Low examinee effort in low-stakes assessment: Problems and potential solutions. Educ. Assess. **10 (1)**, 1–17 (2005)
Wolf, L.F., Smith, J.K.: The consequence of consequence: motivation, anxiety, and test performance. Appl. Meas. Educ. **8 (3)**, 227–242 (1995)

Part XI
Multivariate Analysis

Firm Size Dynamics in an Industrial District: The Mover-Stayer Model in Action

F. Cipollini, C. Ferretti, and P. Ganugi

Abstract In the last decade, the District of Prato (an important industrial area in the neighborhood of Florence, Italy) suffered a deep shrinkage of exports and added value of the textile industry, the core of its economy. In this paper we aim to investigate if such a crisis entailed a firm downsizing (evaluated as number of employees) of the same industry and, possibly, of the overall economy of the District. For this purpose we use the Mover Stayer Model. Data are represented by two panels from ASIA-ISTAT data. The main results of the analysis are that: (1) the textile industry is affected by a relevant downsizing of the firm size; (2) such a process takes place through a slightly changed level of concentration; (3) the mentioned changes does not seem to spread to the overall economy.

1 Introduction

The deep crisis which, in terms of exports and added value, has affected the textile District of Prato, has involved a downsizing of firms belonging to its typical industry. On the basis of Coeweb-ISTAT data, during the period 2001–2009 the annual average rate of total exports was −5.2%.

Because of this, it can be interesting to investigate the firm size dynamics of both the textile industry (TI) and the overall economy (OE) of the area. The analysis can shed some light on important aspects of the industrial organization of the District:

F. Cipollini
Department of Statistics, Università di Firenze, 50134 Italy
e-mail: cipollini@ds.unifi.it

C. Ferretti (✉) · P. Ganugi
Department of Economics and Social Sciences, Università Cattolica del Sacro Cuore, 29100 Piacenza, Italy
e-mail: camilla.ferretti@unicatt.it; piero.ganugi@unicatt.it

A. Di Ciaccio et al. (eds.), *Advanced Statistical Methods for the Analysis of Large Data-Sets*, Studies in Theoretical and Applied Statistics,
DOI 10.1007/978-3-642-21037-2_40,

- The intensity of the downsizing of the typical industry and the resulting equilibrium configuration.
- The possible spreading to the whole economy of the area.
- The possible change of firm's concentration.

From the methodological point of view, for modeling the dynamics of the firms size we use the Mover Stayer model which, in comparison to the classical Markov Chain model, takes into account some firm heterogeneity. The dynamics and the equilibrium configuration resulting from the model allow to analyze the mentioned points concerning the evolution of the District. The size of each firm is evaluated by means of the variable "total workers" (*dipendenti* plus *indipendenti*) as stored in the ASIA-ISTAT data for firms in the Provincia of Prato.

The work is organized as follows: Sect. 2 shortly summarizes the theory of the Mover-Stayer model and the estimation techniques for the model parameters proposed by Goodman (1961) and Frydman (1984); in Sect. 3 the model is applied to our data with the aim to estimate the equilibrium distributions, needed to evaluate possible changes in concentration. A comparison with a classical Markov Chain model is also given. The last section is devoted to conclusions and remarks.

2 The Model

The Mover-Stayer Model (MS) can be placed in the framework of latent class models, described for the first time in Blumen et al. (1955) with the aim to create a more flexible model than a simple Markov Chain (MC).[1] A *Latent Class Model* relates a discrete random variable X to a set of latent classes $\{a_k\}$, so that

$$Pr(X = x) = \sum_k Pr(X = x|a_k)Pr(a_k).$$

In the case considered in the work, $X(t, j) \in \{1, \dots, k\}$ denotes the value of the variable of interest at time t for the unit j of the population considered. The units are grouped in two latent classes: the *Movers*, for which $X(t, j)$ evolves according to a Markov Chain with transition matrix M, and the *Stayers*, for which $Pr(X(t, j) \equiv X(0, j), \forall t) = 1$. Let $S = \text{diag}(\{s_i\})$, where s_i denotes the probability that a unit starting from the i-th state is a Stayer. Then the global one-time transition matrix is given by the mixture

$$P = S + (I - S)M. \tag{1}$$

Denoting as $P^{(t)}$ the transition matrix at time t, then

[1] In what follows we will adopt the following conventions: if $x_1, \dots, x_K$ are real numbers then $\text{diag}(\{x_i\})$ indicates the $k \times k$ diagonal matrix with diagonal elements equal to $x_1, \dots, x_K$.

$$P^{(t)} = S + (I - S)M^t \tag{2}$$

that differs from the Markov property $P^{(t)} = P^t$.

In the following we will compare both the MC and the MS, but we choose the latter mainly for two reasons: (1) the MS adjusts the tendency of the MC to underestimate the diagonal elements of M (see Blumen et al. (1955) and Spilerman (1972)); (2) the parameter s_i represents an additional information which allows us to gain a deeper understanding about the mobility of firms among the states.

2.1 Estimation

According to the model definition, the couple (s, M) has to be estimated, where $s = (s_1, \ldots, s_k) \in [0, 1]^k$ and M is a transition matrix, so that it has to satisfy $M_{ij} \geq 0$ and $\sum_{j=1}^{k} M_{ij} = 1$. The estimation cannot be as simple as in the MC because a unit observed to remain in its starting state might be a Stayer, but also a Mover which never moves during the observed time, and this last event occurs with non-zero probability. Estimation's techniques for this model are in Blumen et al. (1955), Goodman (1961), Spilerman (1972) and Frydman (1984).

From now on we will use the following notation: $n_{ij}(t)$ is the number of units in state i at time t and in state j at time $t + 1$; $n_i(t)$ is the number of units in state i at time t; c_i is the number of units remaining in the starting state i during the whole period taken into account.

In order to compare the MC and the MS we first remind the maximum likelihood estimator proposed in Anderson and Goodman (1957) of the transition matrix M of a simple Markov chain:

$$\hat{m}_{ij} = \frac{\sum_t n_{ij}(t)}{\sum_t n_i(t)}. \tag{3}$$

2.1.1 Goodman (1961)

Goodman (1961) describes some different estimation methods, differing on the basis of the available data. If c_i, $n_i(t)$ and $n_{ij}(t)$ are observed for every $i, j = 1, \ldots, k$ and $t = 1, \ldots, T$, the maximum likelihood estimator for the matrix M has elements:

$$\hat{m}_{ij} = \begin{cases} \left[\sum_{t=1}^{T} n_{ii}(t) - Tc_i\right] / \left[\sum_{t=1}^{T} n_i(t) - Tc_i\right] & \text{if } i = j \\ \left[\sum_{t=1}^{T} n_{ij}(t)\right] / \left[\sum_{t=1}^{T} n_i(t) - Tc_i\right] & \text{if } i \neq j \end{cases} \tag{4}$$

We note that this estimator coincides with (3) applied only on the group of individuals which are observed to move at least once. The estimator of s_i is $\hat{s}_i = c_i / n_i(0)$ (Goodman, 1961, p.854).

2.1.2 Frydman (1984)

Defining $n_{ij}^* = \sum_{t=1}^{T} n_{ij}(t)$ and $n_i^* = \sum_{t=0}^{T-1} n_i(t)$, Frydman (1984) prove that the Maximum Likelihood estimator of the diagonal elements of the M matrix is the solution in x of

$$[n_i^* - Tn_i(0)] \cdot x^{T+1} + [Tn_i(0) - n_{ii}] \cdot x^T + [Tc_i - n_i^*] \cdot x + n_{ii} - Tc_i = 0$$

and then recursively calculates the ML estimates of the remaining parameters $\hat{s}_i$ and $\hat{m}_{ij}$ when $i \neq j$ (Frydman, 1984, p.634).

3 Empirical Application

In this Section we apply the MS to the ASIA-textile industry (3,465 firms) and ASIA-overall economy (19,559 firms) for the period 2000–2006. Following the official UE classification, we grouped the population of firms in eight states labeled by roman numbers, as in Table 1.

3.1 Exploratory Analysis

In Table 2 we show the observed relative frequencies of the eight states for both the textile industry and the overall economy (for the sake of space, we show only the first, middle and last observed years). The mass of the first state tends to grow up for both groups of firms. The tendency to become smaller and smaller is evident for the textile industry, and the cause is mostly the fall in the demand of textile goods. This is not the case for the overall economy which, including heterogeneous industries, is influenced by several and compensating levels of demand. Since not-textile industry looks stable as the overall economy, changes in the distribution of the latter are mostly ascribable to textile industry.

Tables 3 and 4 show the observed firm transitions for both the textile industry and the overall economy: the number corresponding to row i and column j is the conditional relative frequency of firms in the j-th state at 2006 given that they are in the i-th state at 2000. The main differences concerns the last row, showing that the biggest firms in textile industry have a greater tendency to move than in the overall economy.

Table 1 Firm size expressed as number of total workers

State	I	II	III	IV	V	VI	VII	VIII
Size=n. of workers	1	2	[3, 5]	[6, 9]	[10, 20]	[21, 49]	[50, 100]	> 100

Table 2 Empirical distributions of ASIA-textile, not-textile and overall economy (percentage frequencies)

Group	Year	I	II	III	IV	V	VI	VII	VIII
TI	2000	16.08	19.25	23.67	14.75	16.94	7.16	1.65	0.52
	2003	17.37	19.60	23.52	15.24	16.02	6.09	1.82	0.35
	2006	21.76	21.10	21.36	14.31	14.55	5.17	1.44	0.32
Not-TI	2000	46.36	20.99	19.84	6.84	4.42	1.12	0.31	0.11
	2003	45.64	19.79	20.39	7.75	4.65	1.26	0.32	0.19
	2006	46.23	20.11	20.24	7.31	4.36	1.25	0.32	0.17
OE	2000	40.99	20.68	20.52	8.24	6.64	2.19	0.55	0.18
	2003	40.64	19.76	20.94	9.08	6.67	2.12	0.59	0.21
	2006	41.90	20.29	20.44	8.55	6.17	1.94	0.52	0.20

Table 3 Empirical transition matrix for the ASIA-textile industry (conditional probabilities – year 2006 given year 2000 – expressed as percentages)

2000–2006	I	II	III	IV	V	VI	VII	VIII
I	81.69	11.31	4.85	1.08	0.72	0.36	0.00	0.00
II	21.89	65.37	10.94	1.35	0.30	0.15	0.00	0.00
II	9.27	20.85	54.15	12.80	2.68	0.24	0.00	0.00
IV	5.09	6.65	27.79	45.40	14.48	0.59	0.00	0.00
V	6.64	3.07	7.84	23.34	55.71	3.24	0.00	0.17
VI	3.23	3.63	2.82	2.42	29.84	52.82	5.24	0.00
VII	3.51	0.00	1.75	1.75	1.75	35.09	52.63	3.51
VIII	11.11	0.00	0.00	0.00	0.00	5.56	38.89	44.44

Table 4 Empirical transition matrix for the ASIA-overall economy (conditional probabilities – year 2006 given year 2000 – expressed as percentages)

2000–2006	I	II	III	IV	V	VI	VII	VIII
I	82.84	11.47	4.93	0.60	0.11	0.05	0.00	0.00
II	25.32	54.86	17.95	1.46	0.40	0.02	0.00	0.00
II	9.00	17.87	58.68	12.38	1.92	0.15	0.00	0.00
IV	4.16	4.16	26.24	49.07	15.63	0.68	0.06	0.00
V	6.08	2.54	6.24	20.32	57.04	7.47	0.23	0.08
VI	3.50	2.56	3.03	2.10	25.41	54.31	8.62	0.47
VII	2.80	0.93	3.74	3.74	1.87	25.23	49.53	12.15
VIII	11.11	0.00	0.00	2.78	0.00	2.78	19.44	63.89

3.2 *The Mover-Stayer*

We employed both the estimation techniques mentioned in Sect. 2.1 for estimating the parameters of the MC and the MS on the data described in Sect. 3.1.

In order to evaluate the goodness-of-fit of the estimated model we cannot use the transition matrices, because of the presence of empty cells, then we calculated the yearly expected marginal distributions since 2001 up to 2006, comparing them with

Table 5 χ^2-goodness-of-fit test on yearly marginal probabilities for the MS and MC models (0.05-critical value = 14.07)

Model	Group	2001	2002	2003	2004	2005	2006
MS (Frydman)	TI	6.97	13.62	8.60	7.46	1.10	1.75
	OE	4.97	14.49	17.38	10.06	4.08	1.47
MS (Goodman)	TI	7.13	13.96	8.32	6.28	0.76	3.44
	OE	4.97	13.63	16.39	8.40	3.42	2.99
MC	TI	7.11	13.24	8.73	7.87	1.30	0.68
	OE	4.92	15.09	18.80	11.49	5.02	0.81

Table 6 Distances between empirical and expected transition matrices

Step	Group	Frydman	Goodman	MC
t=1	TI	0.099	0.092	0.102
	OE	0.107	0.105	0.109
t=6	TI	0.218	0.297	0.198
	OE	0.139	0.216	0.239

the corresponding observed distributions by means of a χ^2-goodness-of-fit-test. Let $f_i^{(t)}$ be the probability, given by the model, of being in the state i at time t, and let $n_i^{obs}(t)$ be the observed number of firms in the same state at the same time. Under the null hypothesis $n_i^{obs}(t) = nf_i^{(t)}$, where n is the total number of firms, it holds:

$$\sum_i \frac{(n_i^{obs}(t) - nf_i^{(t)})^2}{nf_i^{(t)}} \sim \chi_7^2$$

(see D'Agostino and Stephens (1986)). The results of the test (Table 5) substantially do not reject the null hypothesis both in the MS and MC.

In order to choose the best one between the MS (with the two estimation methods) and the MC, following Frydman et al. (1985) we compare the 2-norm of the residual matrices $P_{obs}^{(t)} - P_{exp}^{(t)}$, that is the distance between the observed and the expected t-steps transition matrix, under the considered model (Table 6, for the norm definition see Golub and Van Loan (1996)).

The MS is then a better choice in terms of distance with respect of MC. Among Goodman's and Frydman's estimation methods for the MS we choose the latter which is known to have better statistical properties (Frydman (1984, Sect. 4)).

Tables 7 and 8 show the estimated conditional transition probabilities $\hat{p}_{ij}$ and the percentage of Stayers in every state for the two group of firms. As mentioned before, $\hat{s}_i$ can be considered as an index for the *persistence* of firms in the states: we note for example the difference between the percentage of Stayers in the VIII state. On the other hand the estimated matrix $\hat{P}$ represents an important tool to evaluate the expected dynamics of the population under study. Comparing the textile

Table 7 Expected transition matrix $\hat{P}$ and $\hat{s}_i$ for the textile industry (percentage)

	I	II	III	IV	V	VI	VII	VIII	Stayers
I	94.93	3.99	0.87	0.11	0.05	0.03	0.03	0.00	44.95
II	6.70	87.72	5.17	0.26	0.12	0.02	0.00	0.00	42.81
III	1.38	8.05	83.21	7.07	0.25	0.04	0.00	0.00	23.33
IV	0.76	0.82	13.61	77.22	7.50	0.09	0.00	0.00	20.39
V	0.38	0.38	1.03	10.17	85.78	2.26	0.00	0.00	23.14
VI	0.16	0.32	0.48	0.63	11.42	84.78	2.14	0.08	32.83
VII	0.28	0.00	0.00	0.00	0.28	10.81	87.53	1.11	29.90
VIII	2.62	0.00	0.00	0.00	0.00	0.00	13.10	84.28	11.51

Table 8 Expected transition matrix $\hat{P}$ and $\hat{s}_i$ for the overall economy (percentage)

	I	II	III	IV	V	VI	VII	VIII	Stayers
I	95.29	4.01	0.63	0.04	0.02	0.01	0.01	0.00	54.56
II	8.36	82.64	8.74	0.21	0.04	0.01	0.00	0.00	32.70
III	1.40	8.20	84.13	6.14	0.11	0.02	0.00	0.00	26.96
IV	0.54	0.73	12.01	80.57	6.07	0.09	0.00	0.00	31.93
V	0.33	0.31	1.03	12.21	82.35	3.77	0.00	0.00	12.41
VII	0.28	0.32	0.32	0.84	10.45	84.92	2.79	0.08	30.65
VII	0.62	0.15	0.31	0.00	0.15	10.84	84.51	3.41	16.61
VIII	0.83	0.41	0.41	0.41	0.00	0.41	6.20	91.32	26.72

Table 9 Equilibrium distributions (percentage) for the textile industry and the overall economy

	I	II	III	IV	V	VI	VII	VIII
TI	33.22	25.30	21.32	12.01	3.74	3.64	0.68	0.09
OE	42.32	20.49	20.60	9.62	4.72	1.70	0.39	0.16

industry and overall economy in terms of transition probabilities, we note that the main differences among the two groups are attributable to the last row of $\hat{P}$, that is to the tendency of the largest firms toward downsizing. For example we note that in textile industry the probability to move from state VIII to state I after one step is about three times as much as in the overall economy.

Further research about an index describing the dynamics, with the aim of improving the comparison among groups could be relevant. As auxiliary tool we use the expected equilibrium distributions for both the overall economy and the textile industry (Table 9), evaluated calculating the limit for $t \to \infty$ in (2).

The equilibrium distribution represents the expected trend in absence of shocks, and allows us to better appreciate the presence of downsizing. With this aim, and in order to evaluate the different behavior of textile industry and overall economy, we compare the equilibrium with the first and the last observed distribution (year 2000 and 2006, both in Table 2). Furthermore we compare 2000 and 2006 distributions together. Dependence among data could invalidate the result of a statistical test (D'Agostino and Stephens, 1986, pp. 88-89), then we evaluate dissimilarities among distributions using the *Hellinger Distance*, defined

Table 10 Hellinger distances between 2000, 2006 and Equilibrium distributions

Group	2000-equilibrium	2006-equilibrium	2000–2006
TI	0.2041	0.1639	0.0675
OE	0.0206	0.0106	0.0125

by $d(p,q) = \sqrt{1 - \sum_i \sqrt{p(i)q(i)}}$, where p and q are two discrete probability distributions. Results are displayed in Table 10.

The textile industry shows significant changes in the distribution, overall economy is more stable, confirming remarks of Sect. 3.1.

Finally we remark that downsizing for textile industry, joined with the stability of overall economy of the district is not surprising: according to Baumol et al. (2003) we know that during the 1980s and 1990s American economy is characterized by downsizing in manufacturing and upsizing in the services.

3.3 *Concentration*

The equilibrium distribution provides information about possible changes in the industrial organization and, in particular, in firm concentration. The tool for evaluating it is not trivial because states represent grouped data. Lorenz curve might be a possible non parametric approach but grouped data tend to severely underestimate the true value of the Gini index (see Arnold and Press (1989)). A grouping correction of Lorenz curve is feasible but represents a specific topic that we cannot face in the remaining of the paper. We then choose to model the equilibrium with Pareto 1 distribution, whose shape parameter α is directly related to Gini index: the Paretian cumulative distribution function is

$$F_\alpha(x) = 1 - \left(\frac{x_0}{x}\right)^\alpha, \; x \geq x_0, \; \alpha > 0$$

and the Gini index related to this distribution is $G(\alpha) = (2\alpha - 1)^{-1}$ (concentration increases when α decreases).

We recognize some drawbacks of this approach, since Pareto 1 is a zero modal distribution and, by consequence, it is non suitable for modeling the whole distribution of textile industry which during years 2000–2005 has mode in the third state (see also Kleiber and Kotz (2003) and Bonini and Simon (1958) on this point).

Possible changes of concentration related to higher or smaller values of α, thus, have to be interpreted as regarding the units with size greater than a fixed value x_0, taken as starting value of the tail, and not the whole set of firms. On the other hand, such units are the most relevant in terms of added value and employment, and hence provide an important information about the industrial organization of the whole district. The estimator we use is OBRE (Optimal Bounded Robust Estimator),

Table 11 Estimated $\hat{\alpha}$ parameter for the Pareto distribution for textile industry and overall economy (last three states, 0.05-critical value = 5.99)

Group	Year	$\hat{\alpha}$	SE	χ^2
TI	2000	1.75	0.051	0.84
	2006	1.67	0.057	5.62
	Equilibrium	2.07	0.081	2.11
OE	2000	1.67	0.037	2.15
	2006	1.61	0.038	0.56
	Equilibrium	1.64	0.042	0.33

introduced and implemented on grouped data of income distribution by Feser and Ronchetti (1997). OBRE has the advantage to be a robust estimator against small perturbations in the model. Assumed a cumulative distribution function F_θ for the random variable X, the estimator is the solution in θ of

$$\sum_{i=1}^{I} \psi_i(\theta) p_i^{\gamma} = 0 \tag{5}$$

where ψ_i is an suitable function of the observed frequency in the i-th state, γ is an arbitrary parameter (for the functional form of $\psi_i(\theta)$ and the best choice of γ see Feser and Ronchetti (1997)) and p_i is the theoretical relative frequency of the i-th state calculated using F_θ. We established the tail threshold as the number of the states on whom the Pareto 1 is fitted with a p-value smaller than 5%. Using this rule, this distribution is fitted to the last three states of 2000, 2006 and equilibrium distribution resulting from the MS, which include 2.92%, 2.66% and 2.25% units of the whole set of overall economy, and 9.33%, 6.93% and 4.42% of the textile industry. The α estimates obtained through OBRE, and the corresponding standard errors, are reported in Table 11, for the choice $\gamma = 2$ (which is the best choice in terms of χ^2-value though, varying γ, the estimated $\hat{\alpha}$ is very stable). Compared with tail of the textile 2006 distribution, the estimates calculated on the tail of the equilibrium distribution of the same industry reveal a slight decrease of concentration.

4 Conclusions

In this work we studied the dynamics (in terms of number of workers) of firms belonging to the industrial District of Prato. According to its ability to take into account some firms heterogeneity, we investigated such dynamics by means of the Mover Stayer model applied to two ASIA-ISTAT panel data concerning the overall economy and the textile industry of the Provincia of Prato. The analysis shows that:

- The textile industry is affected by a relevant downsizing of the firm size.
- Such a downsizing takes place through a slight decrease of concentration.
- The mentioned changes do not seem to spread to the overall economy.

Further research will follow two different directions: from the methodological point of view we are interested in the continuous time version of MS, and in the development of more robust statistical tools, based on the transition matrices, to check the goodness of fit of the models and to compare the dynamics of different populations. On the other hand, from the economical point of view, it could be relevant to verify if the presence of downsizing remarked by Baumol et al. (2003) for United States is confirmed in the Italian industry, for example comparing several industrial Districts. The separate analysis of the service industry can be a further topic to be investigated in order to verify this issue.

References

Anderson, T. and Goodman, L. (1957). Statistical Inference about Markov Chains. *Ann. Math. Statist.*, **28**(1), 89–110.

Arnold, B. and Press, J. (1989). Bayesian estimation and prediction for Pareto data. *J. Amer. Statist. Assoc.*, **84**(408), 1079–1084.

Baumol, W. J., Blinder, A., and Wolff, E. (2003). *Downsizing in America*. Russell Sage Foundation, New York.

Blumen, I., Kogan, M., and McCarthy, P. (1955). *The Industrial Mobility of Labor as a Probability Process*. Cornell University Press.

Bonini, C. and Simon, H. (1958). The size distribution of business firms. *Am. Econ. Rev.*, **48**(4), 607–617.

D'Agostino, R. and Stephens, M. (1986). *Goodness of Fit Tecniques*. CRC Press.

Feser, M. and Ronchetti, E. (1997). Robust Estimation for Grouped Data. *J. Amer. Statist. Assoc.*, **92**(437), 333–340.

Frydman, H. (1984). Maximum likelihood estimation in the mover-stayer model. *J. Amer. Statist. Assoc.*, **79**, 632–638.

Frydman, H., Kallberg, J., and Kao, D. (1985). Testing the adequacy of Markov Chain and Mover-Stayer Models as representation of Credit Behavior. *Oper. Res.*, **33**(6), 1203–1214.

Golub, G. and Van Loan, C. (1996). *Matrix Computation*. Johns Hopkins University Press, 3rd edition.

Goodman, L. (1961). Statistical Methods for the Mover-Stayer Model. *J. Amer. Statist. Assoc.*, **56**, 841–868.

Kleiber, C. and Kotz, S. (2003). *Statistical size Distribution in Economics and Actuarial Sciences*. Wiley Ser. Prob. Stat.

Spilerman, S. (1972). Extensions of the Mover-Stayer Model. *Am. J. Sociol.*, **78**, 599–626.

Multiple Correspondence Analysis for the Quantification and Visualization of Large Categorical Data Sets

Alfonso Iodice D'Enza and Michael Greenacre

Abstract The applicability of a dimension-reduction technique on very large categorical data sets or on categorical data streams is limited due to the required singular value decomposition (SVD) of properly transformed data. The application of SVD to large and high-dimensional data is unfeasible because of the very large computational time and because it requires the whole data to be stored in memory (no data flows can be analysed). The aim of the present paper is to integrate an incremental SVD procedure in a multiple correspondence analysis (MCA)-like procedure in order to obtain a dimensionality reduction technique feasible for the application on very large categorical data or even on categorical data streams.

1 Introduction

In social, behavioural, environmental sciences as well as in marketing, a large amount of information is gathered and coded in several attributes. In most cases the aim is to identify pattern of associations among the attribute levels. A data-mart selected from a categorical data base can be represented as a binary data matrix whose rows correspond to records and whose columns correspond to the levels of the attributes.

A well-known exploratory method to describe and visualize this type of data is multiple correspondence analysis (MCA) (Greenacre 2007). MCA is widely

A.I. D'Enza (✉)
Università di Cassino, Cassino, Italy
e-mail: iodicede@gmail.com

M. Greenacre
Universitat Pompeu Fabra, Barcelona, Spain
e-mail: michael@upf.es

A. Di Ciaccio et al. (eds.), *Advanced Statistical Methods for the Analysis of Large Data-Sets*, Studies in Theoretical and Applied Statistics,
DOI 10.1007/978-3-642-21037-2_41,

applied in different fields: from marketing to social sciences, behavioural and environmental sciences. MCA is the generalization of correspondence analysis (CA) to more than two categorical variables. CA and MCA can be viewed as an adaptation to categorical data of principal component analysis (PCA, Jolliffe (2002)). As PCA, MCA aims to identify a reduced set of synthetic dimensions maximizing the explained variability of the categorical data set in question. The advantages in using MCA to study associations of categorical data are then to obtain a simplified representation of the multiple associations characterizing attributes as well as to remove noise and redundancies in data. The exploratory and visualization-based approach characterizing MCA provides immediate interpretation of the results.

Roughly speaking the MCA implementation consists of a singular value decomposition (SVD) – or the related eigenvalue decomposition (EVD) – of properly transformed data. The applicability of MCA on very large data sets or on categorical data streams is limited due to the required SVD. The application of the SVD to large and high-dimensional data is unfeasible since it requires a computational time that is quadratic in the data size; furthermore the SVD input matrix must be complete (no missings) and stored in memory (no data flows can be analysed) (Brand 2003). Since the SVD characterizes many techniques aiming at dimension reduction, noise suppression and clustering, many contributions in the literature aim to overcome the SVD computational problem by updating the SVD incrementally. For example Zhao et al. (2006) propose a scalable dimension-reduction technique for continuous data such as incremental PCA that exploits SVD updating procedures.

In some applications the data set to be analysed can be stratified in different subgroups. The need for splitting data in chunks is two-fold: (a) if the amount of data to analyse is very large, or data are produced at a high rate (data flows), it can be convenient or necessary to process it in different "pieces"; (b) if the data in question refer to different occasions or positions in time or space, a comparative analysis of data stratified in chunks can be suitable.

The aim of the present contribution is to propose an MCA-like procedure that can be updated incrementally as new chunks of data are processed. The proposed procedure is obtained by integrating an incremental SVD with a properly modified MCA procedure. The low-dimensional quantification and visualization of categorical attributes via this MCA-like procedure is a promising approach to investigate the association structures and for fast clustering purposes.

The present paper is structured as follows: in Sect. 2 the motivation of stratifying data into chunks is described; Sect. 2.1 briefly recalls the computations of MCA and in Sect. 2.2 some options are recalled to study the association structures along the different chunks. Section 3 contains a description of the basic procedure for incremental SVD. Section 4 described the modification to the MCA procedure necessary to embed the incremental SVD procedure. An application is proposed in Sect. 5 and a last section is dedicated to future work.

2 Multiple Correspondence Analysis of Data Chunks

The stratification of a data set into different chunks can be performed for computational reasons, when the data set is too large to be analysed as a whole. Depending on the context, chunks can be defined according to an external criterion, related to time or another characteristic. For example, consider the market basket analysis framework. Since the aim is to study the buying behaviours of customers, then a suitable task is to monitor customer choices along weeks or months, as well as to appreciate customer reactions to promotion policies. Chunks are determined in this case according to time or to different promotions on products.

Furthermore, in evaluating higher-education systems, CA or MCA can be suitably applied to analyse student careers (with attribute levels indicating whether or not an examination is passed): the stratification in this case could be used to compare the overall behaviour of students from different universities or at different academic years. Further applications involving large amounts of categorical data are in text mining and web-clickstream analysis among others.

2.1 *Computations of Multiple Correspondence Analysis*

Let us consider a data-mart resulting from the selection of n records and Q attributes from a categorical data base. Let $J_q, q = 1, \ldots, Q$, be the levels of each categorical attribute. The indicator matrix $\mathbf{Z}$ has n rows and J columns, where $J = \sum_{q=1}^{Q} J_q$. The general element is $z_{ij} = 1$ if the ith record assumes the jth attribute level, $z_{ij} = 0$ otherwise.

Two possible approaches to MCA consist of the application of CA algorithm to the indicator matrix $\mathbf{Z}$ and to the Burt matrix $\mathbf{C} = \mathbf{Z}^{\mathsf{T}}\mathbf{Z}$ (see appendix of Greenacre and Blasius (2006) for details). Let us briefly recall the computations based on the Burt matrix: let the correspondence matrix be $\mathbf{P} = \frac{\mathbf{C}}{(n \times Q^2)}$, with $(n \times Q^2)$ being the grand total of $\mathbf{C}$. Let the vector $\mathbf{r}$ contain the row sums of $\mathbf{P}$, which are also column sums since $\mathbf{P}$ is symmetric. In MCA, as well as in CA, the association structure is revealed by performing a SVD (or equivalently EVD, in this particular case) of the standardized residuals matrix

$$\mathbf{S} = \mathbf{D}_r^{-1/2} \left(\mathbf{P} - \mathbf{r}\mathbf{r}^{\mathsf{T}} \right) \mathbf{D}_r^{-1/2} \tag{1}$$

where $\mathbf{D}_r$ is the diagonal matrix of the elements of $\mathbf{r}$. The general element of the $\mathbf{S}$ matrix is then

$$s_{ij} = \frac{(p_{ij} - r_i r_j)}{\sqrt{r_i r_j}}. \tag{2}$$

The SVD of $\mathbf{S}$ is

$$\mathbf{S} = \mathbf{U}\Sigma\mathbf{U}^{\mathsf{T}} \tag{3}$$

where Σ is the diagonal matrix of singular values in descending order, and $\mathbf{U}$ is the matrix of singular vectors, respectively. Note that only the $J - Q$ positive singular values are retained. The reduced space representation of row and column profiles is obtained via the SVD results. In particular, the principal coordinates of the rows, or columns since $\mathbf{S}$ is symmetric, are

$$\mathbf{F} = \mathbf{D}_r^{-1/2}\mathbf{U}\Sigma. \tag{4}$$

2.2 *Evolutionary MCA Solutions*

The exploratory study of association of multiple categorical data chunks can be approached in different ways. The most basic and straightforward way is to perform a MCA on each data chunk. In this case, however, each MCA solution refers to a single chunk and a comparison among the solutions is difficult, since the displays refer to different subspaces.

An important option is the three-way correspondence analysis (Carlier and Kroonenberg 1996) for the analysis of three-way contingency tables. Although this approach is particularly useful in studying association structures at different occasions, chunks are supposed to have same size in order to consider cubes of data. A further approach exploiting some of the properties of CA (and MCA) aims at a comparative study of association structures. Once a chunk (or a set of chunks) is selected as a reference analysis, further chunks are projected as supplementary information, as proposed by Iodice D'Enza and Palumbo (2007, 2009) and Palumbo and Iodice D'Enza (2009). This approach allows a direct comparison among the displays of the chunks since they all refer to the same reference subspace. The weakness of this approach is in the prominence of the associations characterizing the reference chunks: the resulting subspace is constant and it conditions all the further visualizations. Then a quite natural step forward is to update the reference MCA solution incrementally as new chunks are processed in order to have a comparable and updated subspace of projection.

3 Enhanced Methods for SVD

The SVD is computed via a batch time algorithm with a computational complexity $O\left(nQ^2 + n^2Q + Q^3\right)$, (Golub and van Loan 1996): then SVD becomes unfeasible as the values of n and Q increase. Furthermore, because all of available data are necessary to obtain the decomposition, it cannot be suitably applied to data flows.

The contribution can be roughly categorized in batch methods and incremental methods. The batch methods aim to reduce the computational effort of the SVD extending its applicability: the Lanczos bilinear diagonalization methods (Baglama and Reichel 2007) represent a relevant example of such a contribution.

The incremental approach aims to update an existing SVD solution when new data comes in. These methods have the advantage over the batch methods as they can be applied to subsequent chunks of data without the need to store the previously analysed data in memory. This has motivated the description of numerous SVD updating methods, e.g., the contribution by Businger (1970) and by Bunch and Nielsen (1978). The SVD update process, although of high complexity, presents several advantages. Most updating procedures rely on the *dominant* SVD which is a decomposition retaining the r largest singular values and the related singular vectors: this is done in order to reduce the overall computational complexity. Examples of such procedures are in the proposals by Chandrasekaran et al. (1997), Levy and Lindenbaum (1998), Chahlaoui et al. (2001), Kwok and Zhao (2003) and Brand (2003, 2006). These methods approximate the dominant SVD after a single scan of the matrix, and they maintain only a low-rank factorization: that is, these methods are able to approximate the SVD using less memory and computation than direct full-rank SVD methods (Baker et al. 2008).The reference procedure is the online SVD proposed by Brand (2003; 2006) for updating the decomposition when additive modifications occur on the starting data matrix: this kind of modification is particularly useful in embedding the SVD updates in a MCA procedure. The main steps of the procedure are described in Sect. 3.1.

3.1 *Incremental SVD Procedure*

In the present paper we refer to the online SVD procedure proposed by Brand (2003) in the context of recommender systems. Consider a $n \times p$ continuous data matrix $\mathbf{X}$ and its *rank-r* SVD ($\mathbf{U}\Sigma\mathbf{V}^{\mathsf{T}}$) representing an approximation of $\mathbf{X}$ according to the r highest singular values. Let $\mathbf{A}$ and $\mathbf{B}$ be two modification matrices with c columns each and n and p rows, respectively. The aim is to obtain the SVD of $\mathbf{X} + \mathbf{A}\mathbf{B}^{\mathsf{T}}$ by updating $\mathbf{U}$, Σ and $\mathbf{V}$. The detailed explanation of the procedure is in the contributions by Brand, in the appendix of Brand (2003) and in Brand (2006). Here follow the most relevant steps:

(a) Perform a modified Gram-Schmidt procedure for the QR decomposition as follows:

$$[\mathbf{U}\ \mathbf{A}] \overset{QR}{\rightarrow} [\mathbf{U}\ \mathbf{P}_{\mathrm{A}}] \begin{bmatrix} \mathbf{I} & \mathbf{U}^{\mathsf{T}}\mathbf{A} \\ \mathbf{0} & \mathbf{R}_{A} \end{bmatrix},$$

$$[\mathbf{V}\ \mathbf{B}] \overset{QR}{\rightarrow} [\mathbf{V}\ \mathbf{P}_{\mathrm{B}}] \begin{bmatrix} \mathbf{I} & \mathbf{V}^{\mathsf{T}}\mathbf{B} \\ \mathbf{0} & \mathbf{R}_{B} \end{bmatrix}; \qquad (5)$$

note that the QR decomposition factorize a matrix into an orthogonal matrix and an upper triangular matrix. The matrices $\mathbf{P}_A, \mathbf{Q}_A, \mathbf{P}_B, \mathbf{Q}_B$ are blocks of the matrix resulting from the QR decompositions.

(b) Consider the following matrix

$$\mathbf{K} = \begin{bmatrix} \Sigma & \mathbf{0} \\ \mathbf{0} & \mathbf{0} \end{bmatrix} + \begin{bmatrix} \mathbf{U}^{\mathsf{T}}\mathbf{A} \\ \mathbf{R}_A \end{bmatrix} \begin{bmatrix} \mathbf{V}^{\mathsf{T}}\mathbf{B} \\ \mathbf{R}_B \end{bmatrix}^{\mathsf{T}} \tag{6}$$

and compute the *rank-*$(r + c)$ SVD of $\mathbf{K}$

$$\mathbf{K} = \mathbf{U}^{'}\Sigma^{'}\mathbf{V}^{'\mathsf{T}}. \tag{7}$$

(c) The SVD of $\mathbf{X} + \mathbf{A}\mathbf{B}^{\mathsf{T}}$ is obtained as

$$\mathbf{X} + \mathbf{A}\mathbf{B}^{\mathsf{T}} = \left([\mathbf{U}\ \mathbf{P}_{\mathrm{A}}]\,\mathbf{U}^{'}\right)\Sigma^{'}\left([\mathbf{V}\ \mathbf{P}_{\mathrm{A}}]\,\mathbf{V}^{'}\right)^{\mathsf{T}} = \mathbf{U}^{''}\Sigma^{'}\mathbf{V}^{''\mathsf{T}}. \tag{8}$$

The matrix $\mathbf{X}$ has the same singular values as $\mathbf{K}$ whereas the left and right singular vectors depend on both the corresponding singular vectors of the matrix $\mathbf{X}$ ($\mathbf{U}$ and $\mathbf{V}$) and those of the matrix $\mathbf{K}$ ($\mathbf{U}'$ and $\mathbf{V}'$). Note that the procedure requires a rank-$(r + c)$ SVD of $\mathbf{K}$ instead of a rank-r SVD of a pre-updated matrix $\mathbf{X}$. However, as pointed out by Brand (2006), the matrix $\mathbf{K}$ is sparse and highly structured, so it is easily diagonalized. Furthermore, the SVD update does not require the starting matrix $\mathbf{X}$ to be kept in memory, only the starting SVD and $\mathbf{A}$ and $\mathbf{B}$ matrices are needed.

4 Update of MCA-Like Results

In order to integrate the SVD update procedure in MCA it is necessary to define both $\mathbf{A}$ and $\mathbf{B}$ matrices to reflect new upcoming data. In particular, let the starting matrix be $\mathbf{X} = \mathbf{S}$, the original matrix of standardized residuals. $\mathbf{A}$ and $\mathbf{B}$ must be such that $\mathbf{A}\mathbf{B}^{\mathsf{T}}$ contains the additive modifications of $\mathbf{S}$ when new rows of $\mathbf{Z}$ are added. Let $\mathbf{Z}^+$ be the indicator matrix of upcoming data and let $\mathbf{C}^+ = \mathbf{Z}^{+\mathsf{T}}\mathbf{Z}^+$ the corresponding Burt matrix. Since

$$\begin{bmatrix} \mathbf{Z} \\ \mathbf{Z}^+ \end{bmatrix}^{\mathsf{T}} \begin{bmatrix} \mathbf{Z} \\ \mathbf{Z}^+ \end{bmatrix} = \mathbf{C} + \mathbf{C}^+$$

the aim is to properly transform $\mathbf{C}^+$ to update the starting matrix $\mathbf{S}$. Let n^+ be the number of added rows, with $\left[(n + n^+) \times Q^2\right]$ being the updated *grand total*. Define the correspondence matrix update as

$$\mathbf{P}^{+} = \frac{\mathbf{C}^{+}}{(n + n^{+}) \times Q^{2}}. \tag{9}$$

Let $\mathbf{r}^{+}$ be the vector of rows (columns) margins of $\mathbf{P}^{+}$. The standardized residuals updating matrix is

$$\mathbf{S}^{+} = \mathbf{D}_{r}^{-1/2} \left(\mathbf{P}^{+} - \mathbf{r}^{+}\mathbf{r}^{+\mathsf{T}} \right) \mathbf{D}_{r}^{-1/2}. \tag{10}$$

Note that the centring operators are the update of the margins, while the residuals are divided by the original independence condition: this is done in order to keep the updating quantity in the same scale as the original matrix **S**. This update does not reconstruct exactly the standardized residual matrix. Then the modification matrices to input in the SVD update are defined as

$$\mathbf{A} = \mathbf{D}_{r}^{-1/2} \left(\mathbf{P}^{+} - \mathbf{r}^{+}\mathbf{r}^{+\mathsf{T}} \right) \text{ and } \mathbf{B} = \mathbf{D}_{r}^{-1/2}.$$

The actual residuals are divided by the starting independent condition instead of the updated one, this leads to some difference in the usual interpretation of the attribute points on the map. In particular, the centring operator is updated, then the centre of the map still represents the independence condition. The distance of the points from the centre is still a weighed Euclidean metric, but it is not a chi-square distance, since the weights are not updated. However, this affects the scale of the configuration of points but not the relative position of the points.

5 Example of Application

The data set is taken from the multinational ISSP survey on environment in 1993 and it is provided with the R package ca (Nenadić and Greenacre 2007). The number of considered attributes is $Q = 7$, the number of respondents is $n = 871$. In particular, attributes correspond to four substantive questions from the ISSP survey (see Table 1), then there are three demographic variables such as gender, age and education. The total number of attribute levels is $J = 34$. The possible answers to each question are: (1) strongly agree, (2) somewhat agree, (3) neither agree nor disagree, (4) somewhat disagree, (5) strongly disagree (notice that it is usual practice to not mix substantive and demographic variables as active variables, but we do so here merely to illustrate the technical functioning of our procedure). Although the data dimensionality is quite small, the aim is to show the procedure on real data and compare the results with those of a regular MCA on the whole data matrix. In order to update incrementally a starting solution we randomly selected $n_0 = 100$ respondents. The remaining 771 respondents were treated as further information and split into $k = 10$ chunks of approximately equal sizes.

Table 1 The ISSP survey data set

Attribute	n. of levels
A: We believe too often in science, and not enough in feelings and faith	5
B: Overall, modern science does more harm than good	5
C: Any change humans cause in nature, no matter how scientific, is likely to make things worse	5
D: Modern science will solve our environmental problems with little change to our way of life	5
Gender	2
Age	6
Education	6

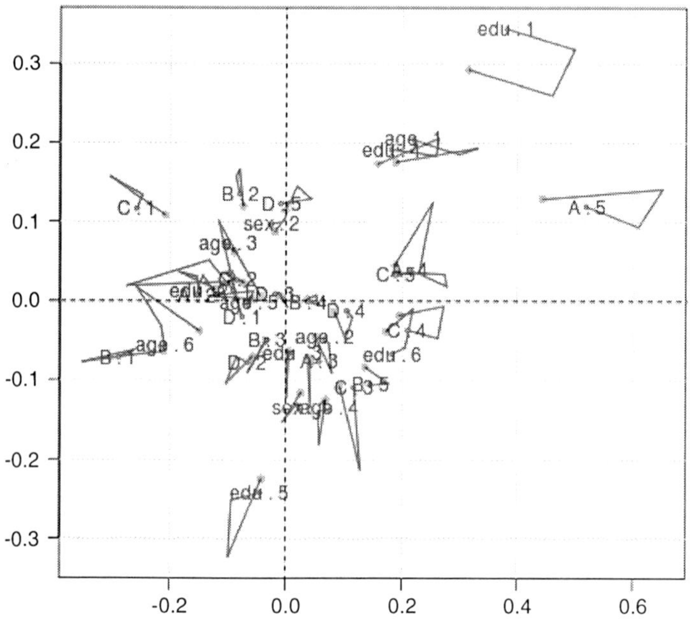

Fig. 1 Trajectories of the attribute points: starting, two intermediates and final positions

Figure 1 shows the trajectories of the attribute points: each broken line goes from the starting position (corresponding to the first n_0 respondents), then passes through the third and the sixth intermediate frames and ends at the final position of the corresponding attribute level. The attribute labels are placed at the end of the corresponding trajectory. What Fig. 1 shows is that the amount of change in the attribute level positions is proportional to their distance from the centre of the map. Since the centre of the map corresponds to the independence condition, attributes placed close to the centre of the map are characterized by a lower level of association with one other. Then highly associated attribute levels are more "sensitive" to the new data

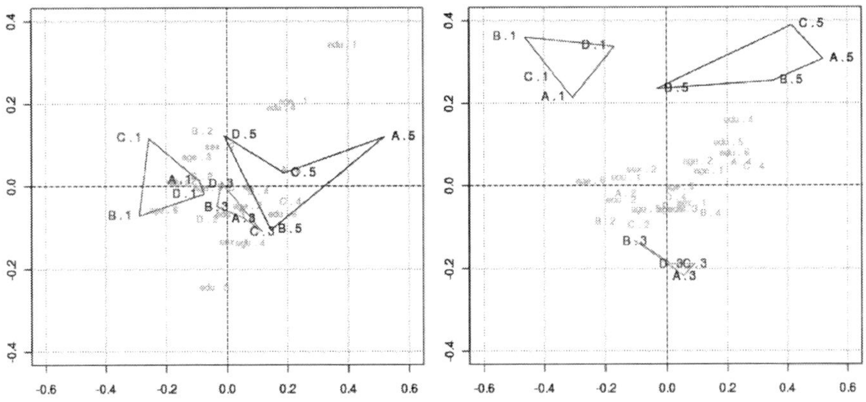

Fig. 2 Solution of the incremental MCA-like procedure (*left*) and of the MCA (*right*)

being incorporated as their level of association is more likely to change. The evolutionary structure of association could be better detected by a dynamic visualization, showing the frame-by-frame configuration of points. As the trajectory shows, there are some changes from one frame to another, but the overall structure of association of the whole data set does not change in a relevant way. This is due to the fact that there is a well-defined association structure underlying the data set in question.

Some further comments arise when comparing the incrementally updated solution with the MCA solution resulting for the analysis of the whole data set. In order to ease the comparison, as Fig. 2 shows, we linked together the points corresponding to the response "strongly agree" for each of the four substantive questions. We did the same for the responses "neither agree nor disagree" and "strongly disagree". Looking at the three resulting polygons in the left- and right-hand side of Fig. 2 it can be seen that the relative positions of the polygons are fairly similar and they present the "horse shoe" effect which is typical of CA and MCA. Although the comparability of the obtained result with MCA is a desirable feature, it has to be noted that, in the case of the incremental MCA-like solution, the polygons are less spread over the map compared to the MCA map. This is due to the update of the Burt matrix (see formula (10)): in particular, for sake of comparability, the updating residuals are standardized by the starting marginals, and this condition keeps the axis scale artificially fixed. This is not a desirable feature, so more enhanced updates are needed.

6 Future Work

The proposed approach is mainly aimed to extend the usability of a powerful exploratory and visualization-based technique like MCA to a wider range of application. Future developments are needed and further enhancements are possible, for example to enhance reconstruction of the updated standardized residual matrix. Another development of the procedure will be to focus on the computations based on the indicator matrix rather than the Burt matrix: this makes it possible to use lower rank modifications of the SVD that require a lower computational effort. In order to make more comparable the frame-by-frame visualizations the idea is to adopt a Procrustean rotation matrix that compensates for the "rotation effect" due to the subspace update. One last step is to exploit the SVD update to take missing values into account.

References

J. Baglama and L. Reichel (2007). Augmented implicitly restarted Lanczos bidiagonalization methods. *SIAM J. Sci. Comput.*, 27, 19–42.

Baker C., Gallivan K. and Van Dooren P. (2008) Low-rank incremental methods for computing dominant singular subspaces. Computer Science and Mathematics Oak Ridge National Laboratory (**url:** http://www.csm.ornl.gov/~cbaker/Publi/IncSVD/copper08.pdf).

Brand M. (2003). Fast online svd revision for lightweigth recommender systems. *SIAM International Conference on Data Mining*.

Brand M. (2006). Fast low-rank modifications of the thin singular value decomposition. *Linear Algebra and its Applications*, 415, 20–30.

Bunch J. R. and Nielsen C. P. (1978). Updating the singular value decomposition. *Numerische Mathematik*, 31(2):111–129.

Businger P (1970). Updating a singular value decomposition. *BIT*, 10(3), 376–385.

Carlier A. and Kroonenberg P.M. (1996). Decompositions and biplots in threeway correspondence analysis. *Psychometrika*, 61, 355–373.

Chahlaoui Y., Gallivan K. and Van Dooren P. (2001). An incremental method for computing dominant singular spaces. *Computational Information Retrieval*, SIAM., 53–62.

Chandrasekaran S., Manjunth B. S., Wang Y.F., Winkeler J. and Zhang H. (1997). An eigenspace update algorithm for image analysis, *Graphical Models and Image Processing*, 59(5), 321–332.

Golub G. and van Loan A. (1996). *Matrix Computations*, John Hopkins U. Press.

Greenacre M.J. (2007). *Correspondence Analysis in Practice*, second edition, Chapman and Hall/CRC.

Greenacre M.J. and Blasius J. (2006). *Multiple Correspondence Analysis and Related Methods*, Chapman & Hall/CRC, first edition.

Jolliffe I.T. (2002). *Principal Component Analysis*, Springer-Verlag, second edition.

Kwok J. and Zhao H. (2003). Incremental eigen decomposition. *Proceedings of the International Conference on Artificial Neural Networks (ICANN)*, 270–273, Istanbul, Turkey.

Iodice D'Enza A. and Palumbo F. (2009). A regulatory impact analysis (RIA) approach based on evolutionary association patterns. *Statistica Applicata*, accepted, in press.

Iodice D'Enza A. and Palumbo F. (2007). Binary data flow visualization on factorial maps. *Revue Modulad*, n.36.

Levy A. and Lindenbaum, M. (1998). Sequential Karhunen-Loeve basis extraction and its application to images. *Technical Report CIS9809*, Technion.

Nenadić O. and Greenacre M.J. (2007). Correspondence analysis in R, with two- and three-dimensional graphics: The ca package. *Journal of Statistical Software*, 20(3), 1–13.

Palumbo F. and Iodice D'Enza A. (2009). Clustering and dimensionality reduction to discover interesting patterns in binary data. *Advances in Data Analysis, Data Handling and Business Intelligence*. Springer, 45–55.

R Development Core Team (2009). R: A language and environment for statistical computing. R Foundation for Statistical Computing, Vienna, Austria. ISBN 3-900051-07-0, URL: http://www.R-project.org.

Zhao H., Chi P. and Kwok J. (2006). A novel incremental principal component analysis and its application for face recognition. *Systems, Man and Cybernetics, Part B: Cybernetics, IEEE Transactions*, 35, 873–886.

Multivariate Ranks-Based Concordance Indexes

Emanuela Raffinetti and Paolo Giudici

Abstract The theoretical contributions to a "good" taxation have put the attention on the relations between the efficiency and the vertical equity without considering the "horizontal equity" notion: only recently, measures connected to equity (iniquity) of a taxation have been introduced in literature. The taxation problem is limited to the study of two quantitative characters: however the concordance problem can be extended in a more general context as we present in the following sections. In particular, the aim of this contribution consists in defining concordance indexes, as dependence measures, in a multivariate context. For this reason a k-variate ($k > 2$) concordance index is provided recurring to statistical tools such as ranks-based approach and multiple linear regression function. All the theoretical topics involved are shown through a practical example.

1 An Introduction to Concordance Index Problem

The issue of defining a concordance index often recurs in the statistical and economical literature. Although the presentation is general we will refer, for sake of clarity, to the taxation example throughout: in particular, the concordance index is strictly connected to the "horizontal equity" topic according to which people who own the same income level have to be taxed for the same amount (see e.g. Musgrave 1959).

The analysis is focused on considering n pairs of ordered real values, (x_i, y_i), $i = 1, 2, \ldots, n$, whose components describe measures of two quantitative variables referred to each element of a statistical population: let us denote by X and Y the income amount before taxation and the income amount after taxation. Our interest is in defining the i-th individual rank with respect to variable X (denoted by $r(x_i)$)

E. Raffinetti (✉) · P. Giudici
University of Pavia, Via Strada Nuova 65, Italy
e-mail: emanuela.raffinetti@unipv.it; giudici@unipv.it

A. Di Ciaccio et al. (eds.), *Advanced Statistical Methods for the Analysis of Large Data-Sets*, Studies in Theoretical and Applied Statistics,
DOI 10.1007/978-3-642-21037-2_42,

and to variable Y (denoted by $r(y_i)$). Furthermore, suppose that $x_i \neq x_j$, $y_i \neq y_j$, $i \neq j$.

In a situation of perfect horizontal equity one can show that

$$r(x_i) = r(y_i), \qquad i = 1, 2, \ldots, n \tag{1}$$

whereas, in a situation of perfect horizontal iniquity, one gets

$$r(y_i) = n + 1 - r(x_i), \qquad i = 1, 2, \ldots, n. \tag{2}$$

Obviously the definition of the "horizontal equity" requires the existence of an *ordering* among individuals before taxation and the knowledge of each individual income amount after taxation. Furthermore, getting an equity index requires that the available data are referred to the single considered units and not to grouped data because these ones do not allow the identification of individuals reordering after the taxation process. The purpose is then identifying an index able to stress potential *functional monotone relations* between variables leading to study the degree of concordance or discordance among the involved quantitative variables.

Statistical literature provides a wide set of association indicators such as the Kendall-τ, the Spearmann-ρ and the Gini index: as well known these indexes assume values between -1 and $+1$ and, in particular one can show that they are equal to

$$-1 \Leftrightarrow (\forall i) \text{ if } r(y_i) = n + 1 - r(x_i) \tag{3}$$

$$+1 \Leftrightarrow (\forall i) \text{ if } r(x_i) = r(y_i). \tag{4}$$

These indexes, however, even if they are invariant with respect to monotone transformations, are unfortunately based on observations ranks and the same ranks can remain unchanged also after the redistribution process in spite of each individual income extent is substantially changed.

For this reason one has to define equity measures based also on the considered extent character: a possible solution to this problem can be identified in resorting to the Lorenz curve and the dual Lorenz curve. In the taxation context the analysis is limited to the bivariate case (see e.g. Muliere 1986): in the following sections we consider the extension of this problem in a more general case when one considers more than two variables.

Furthermore, R^2 is not suited in this context as it looks for linear relationships.

2 Concordance Problem Analysis in a Multivariate Context

The objective of this analysis concerns the definition of concordance measures in a multidimensional context: the study is then oriented to the achievement of a concordance index in presence of a random vector $(Y, X_1, X_2, \ldots, X_k)$.

The followed procedure is very simple and consists in applying a model able to describe the relation among the target variable Y and the explanatory variables $X_1, X_2, \ldots, X_k$: in order to define a concordance index in the hypothesis that the dependent variable Y is conditioned by more than one explanatory variable, one can recur to the multiple linear regression model (see e.g. Leti 1983). Thus the estimated variable Y values can be obtained recurring to the following relation

$$E(Y_i|X_1, X_2, \ldots, X_k) = \beta_0 + \beta_1 x_{1i} + \beta_2 x_{2i} + \ldots + \beta_k x_{ki} = \hat{y}_i; \quad (5)$$

obviously the gap between the observed value y_i and the estimated value $\hat{y}_i$, where $i = 1, 2, \ldots, n$, on the basis of the regression represents the residual deviance of the model that has to be minimized (see e.g. Leti 1983).

2.1 *Proposal: Multivariate Ranks-Based Approach*

The starting point is based on building the response variable Lorenz curve, L_Y (characterized by the set of ordered pairs $(i/n, 1/(nM_Y)\sum_{j=1}^{i} y_{(j)})$, where $y_{(i)}$ denotes the y_i ordered in an increasing sense and M_Y is the Y mean) and the so called dual Lorenz curve of the variable Y, L'_Y, (characterized by the set of ordered pairs $(i/n, 1/(nM_Y)\sum_{j=1}^{i} y_{(n+1-j)})$, where $y_{(n+1-j)}$ denotes the y_i ordered in a decreasing sense) (see e.g. Petrone and Muliere 1992). The analysis proceeds in estimating the variable Y values according to the multiple linear model application. First of all we estimate the regression coefficients using the usual ordinary least square method: the purpose is getting the estimated Y values, $\hat{y}_i$, for each $i = 1, 2, \ldots, n$.

Once computed the $\hat{y}_i$, one can proceed to the construction of the concordance function based on ordering the Y values with respect to the ranks assigned to the $\hat{y}_i$. Let us denote this ordering with $(y_i|r(\hat{y}_i))$ and, more specifically, by y_i^*: the set of pairs $\left(i/n, 1/(nM_Y)\sum_{j=1}^{i} y_j^*\right)$ defines the concordance curve denoted with $C(Y|r(\hat{y}_i))$.

Through a direct comparison between the set of points that represent the Lorenz curve, L_Y, and the set of points that represent the concordance curve, $C(Y|r(\hat{y}_i))$, one can show that a perfect "overlap" is provided only if

$$\sum_{j=1}^{i} y_{(j)} = \sum_{j=1}^{i} y_j^* \text{ for every } i = 1, 2, \ldots, n, \quad (6)$$

that is if and only if $r(y_i) = r(\hat{y}_i)$: obviously, it implies that if the residual deviance of the model decreases the concordance setting is attained due to the fact that the y_i preserve their original ordering also with respect to $r(\hat{y}_i)$.

The further comparison between the set of points that represent the Y dual Lorenz curve, L'_Y, and the set of points that represent the concordance curve, $C(Y|r(\hat{y}_i))$, allows to conclude that there is a perfect "overlap" if and only if

$$\sum_{j=1}^{i} y_{(n+1-j)} = \sum_{j=1}^{i} y_j^* \text{ for every } i = 1, 2, \ldots, n. \tag{7}$$

Recalling the following inequalities

$$\begin{cases} \sum_{j=1}^{i} y_j^* \geq \sum_{j=1}^{i} y_{(j)} \\ \sum_{j=1}^{n} y_j^* = \sum_{j=1}^{n} y_{(j)} \end{cases}$$

and

$$\begin{cases} \sum_{j=1}^{i} y_j^* \leq \sum_{j=1}^{i} y_{(n+1-j)} \\ \sum_{j=1}^{n} y_j^* = \sum_{j=1}^{n} y_{(n+1-j)} \end{cases}$$

provided that $\sum_{j=1}^{i} y_{(j)} \leq \sum_{j=1}^{i} y_j^* \leq \sum_{j=1}^{i} y_{(n+1-j)}$ we have that $L_Y \leq C(Y|r(\hat{y}_i)) \leq L'_Y$, as also shown in Fig. 1.

A multivariate concordance index can be then provided: its expression is the following

$$C_{Y,X_1,X_2,\ldots,X_k} = \frac{\sum_{i=1}^{n-1} \{i/n - (1/(nM_Y)) \sum_{j=1}^{i} y_j^*\}}{\sum_{i=1}^{n-1} \{i/n - (1/(nM_Y)) \sum_{j=1}^{i} y_{(j)}\}}: \tag{8}$$

this index represents the ratio of Y and $(Y|r(\hat{y}_i))$ concentration areas (Gini indexes): the concordance index enable to express the contribution of the k explanatory variables to the variable concentration. In particular the numerator of (8) describes

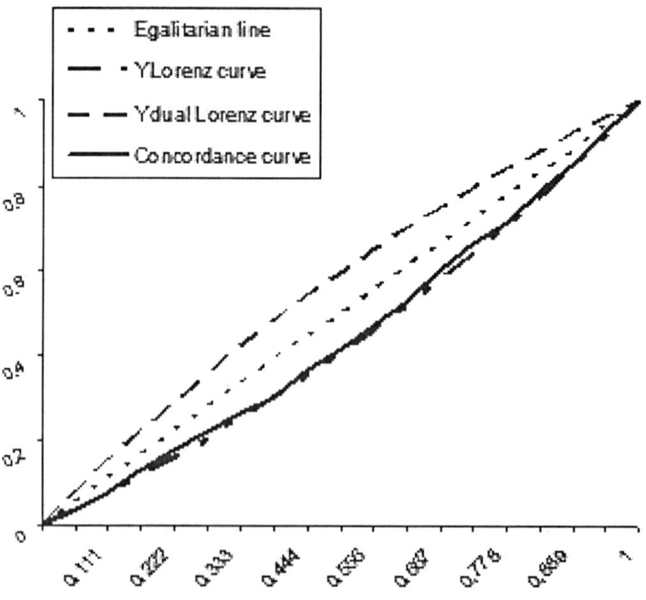

Fig. 1 Y Lorenz curve, Y dual Lorenz curve and concordance function

the "gap" between the ordinates of points that lie on the egalitarian line and the ordinates of points that lie on the concordance curve, provided that these points have the same x-axis values: in the same manner the denominator of (8) defines the "gap" between the ordinates of points that lie on the egalitarian line and the ordinates of points that lie on the Y Lorenz curve.

Through some mathematical steps one can provide an alternative concordance index expression:

$$C_{Y,X_1,X_2,\dots,X_k} = \frac{2\sum_{i=1}^{n} i y_i^* - n(n+1)M_Y}{2\sum_{i=1}^{n} i y_{(i)} - n(n+1)M_Y}. \qquad (9)$$

Proof. Let's try to simplify (8) by operating both in the numerator and in the denominator in the same manner.

After multiplying both the numerator and the denominator for nM_Y, and by applying the products in (8) we get

$$C_{Y,X_1,X_2,\dots,X_k} = \frac{M_Y \sum_{i=1}^{n} i - \sum_{i=1}^{n}\sum_{j=1}^{i} y_j^*}{M_Y \sum_{i=1}^{n} i - \sum_{i=1}^{n}\sum_{j=1}^{i} y_{(j)}}.$$

Since $\sum_{i=1}^{n} i = \frac{n(n+1)}{2}$, one obtains

$$C_{Y,X_1,X_2,\dots,X_k} = \frac{n(n+1)M_Y - 2\sum_{i=1}^{n}\sum_{j=1}^{i} y_j^*}{n(n+1)M_Y - 2\sum_{i=1}^{n}\sum_{j=1}^{i} y_{(j)}}. \qquad (10)$$

Finally, verified that $\sum_{i=1}^{n}\sum_{j=1}^{i} y_j^* = \sum_{i=1}^{n}(n+1-i)y_j^*$ and $\sum_{i=1}^{n}\sum_{j=1}^{i} y_{(j)} = \sum_{i=1}^{n}(n+1-i)y_{(j)}$ are jointly true, we have

$$\sum_{i=1}^{n}\sum_{j=1}^{i} y_j^* = n(n+1)M_Y - \sum_{i=1}^{n} i y_i^* \qquad (11)$$

which substituted in (10) gives directly

$$C_{Y,X_1,X_2,\dots,X_{k-1}} = \frac{2\sum_{i=1}^{n} i y_i^* - n(n+1)M_Y}{2\sum_{i=1}^{n} i y_{(i)} - n(n+1)M_Y}. \qquad \square$$

Now $\sum_{i=1}^{n} i y_i$ is an arrangement increasing function. By arrangement we mean a real valued function of a vector arguments in $\mathbb{R}^n$ that increases in value if the components of the vector arguments become more similarly arranged (see e.g. Muliere 1986). We can conclude that:

Remark 1. $-1 \leq C_{Y,X_1,X_2,\dots,X_{k-1}} \leq +1$.

Proof. It is sufficient to prove that $\sum_{i=1}^{n} i y_{(i)} \geq \sum_{i=1}^{n} i y_i^*$. This can be proved, for instance, directly by looking at the systems of equations of page 4: since

$$\sum_{j=1}^{i} y_j^* \geq \sum_{j=1}^{i} y_{(j)}$$

is intuitively true for all i, then we also have that

$$\sum_{i=1}^{n}\sum_{j=1}^{i} y_j^* \geq \sum_{i=1}^{n}\sum_{j=1}^{i} y_{(j)};$$

now, because of the aforementioned relationship (11) we have

$$n(n+1)M_Y - \sum_{i=1}^{n} i y_i^* \geq n(n+1)M_Y - \sum_{i=1}^{n} i y_{(i)},$$

which gives $\sum_{i=1}^{n} i y_{(i)} \geq \sum_{i=1}^{n} i y_i^*$. □

Remark 2. $C_{Y.X_1,X_2,\ldots,X_{k-1}} = +1$ if and only if concordance function overlaps with the Lorenz curve.

Proof. Concordance function overlaps with the Lorenz curve if and only if $\sum_{j=1}^{i} y_{(j)} = \sum_{j=1}^{i} y_j^* \Rightarrow r(y_i) = r(y_i^*)$ for every $i = 1, 2, \ldots, n$. □

Remark 3. $C_{Y.X_1,X_2,\ldots,X_{k-1}} = -1$ if and only if concordance function overlaps with the dual Lorenz curve.

Proof. This remark can be proved, similarly to Remark 1, from the second system of page 4 by first noticing that:

$$\sum_{i=1}^{n}(n+1-i)y_{(i)} = \sum_{i=1}^{n} y_{(n+1-i)} i$$

so

$$\sum_{i=1}^{n} i y_{(i)} = n(n+1)M_Y - \sum_{i=1}^{n} y_{(n+1-i)} i$$

and therefore by applying this equivalence in the denominator of (8) we get an equivalent formulation of the concordance index based on L_Y':

$$\begin{aligned} C_{Y.X_1,\ldots,X_k} &= \frac{2\sum_{i=1}^{n} i y_i^* - n(n+1)M_Y}{n(n+1)M_Y - 2\sum_{i=1}^{n} i y_{(n+1-i)}} \\ &= -\frac{2\sum_{i=1}^{n} i y_i^* - n(n+1)M_Y}{2\sum_{i=1}^{n} i y_{(n+1-i)} - n(n+1)M_Y}. \end{aligned}$$

Finally, since from the second system of equations of page 4 we have $\sum_{j=1}^{i} y_j^* \leq \sum_{j=1}^{i} y_{(n+1-i)}$, $\forall i$, then the result follows similarly to Remark 1 proof. □

An alternative concordance measure, which provides a measure of distance between concordance function and the Y Lorenz curve, is the Plotnick indicator (see e.g. Plotnick 1981) expressed by

$$I^*_{Y.X_1,X_2,\dots,X_k} = \frac{\sum_{i=1}^n iy_{(i)} - \sum_{i=1}^n iy_i^*}{2\sum_{i=1}^n iy_{(i)} - (n+1)\sum_{i=1}^n y_{(i)}}. \tag{12}$$

Furthermore, one can verify that:

$$I^*_{Y.X_1,X_2,\dots,X_k} = 0 \Leftrightarrow r(\hat{y}_i) = r(y_i) \Rightarrow \sum_{i=1}^n iy_{(i)} = \sum_{i=1}^n iy_i^*, \tag{13}$$

$$I^*_{Y.X_1,X_2,\dots,X_k} = 1 \Leftrightarrow r(\hat{y}_i) = n+1-r(y_i) \Rightarrow \sum_{i=1}^n iy_i^* = \sum_{i=1}^n (n+1-i)y_{(i)}. \tag{14}$$

2.2 Some Practical Results

Suppose to have data concerning 18 business companies three characters: Sales revenues (Y) (expressed in thousands of Euros), Selling price (X_1) (expressed in Euros) and Advertising investments (X_2) (expressed in thousand of Euros). These data are shown in Table 1.

Table 1 Data describing Sales revenues, Selling price and Advertising investments expressed in Euros

ID Business company	Sales revenues	Selling price	Advertising investments
01	350	84	45
02	202	73	19
03	404	64	53
04	263	68	31
05	451	76	58
06	304	67	23
07	275	62	25
08	385	72	36
09	244	63	29
10	302	54	39
11	274	83	35
12	346	65	49
13	253	56	22
14	395	58	61
15	430	69	48
16	216	60	34
17	374	79	51
18	308	74	50

Table 2 Results

Ordered y_i	$r(y_i)$	$\widehat{y}_i$	Ordered $\widehat{y}_i$	$r(\widehat{y}_i)$	y_i ordered by $r(\widehat{y}_i)$
202	1	231.07	231.07	1	202
216	2	291.41	234.10	2	253
244	3	270.46	245.57	3	304
253	4	234.10	251.56	4	275
263	5	282.73	270.46	5	244
274	6	310.41	282.73	6	263
275	7	251.56	291.41	7	216
302	8	310.47	308.07	8	385
304	9	245.57	310.41	9	274
308	10	373.26	310.47	10	302
346	11	363.04	356.71	11	350
350	12	356.71	360.99	12	430
374	13	380.97	363.04	13	346
385	14	308.07	373.26	14	308
395	15	413.45	380.68	15	404
404	16	380.68	380.97	16	374
430	17	360.99	411.05	17	451
451	18	411.05	413.45	18	395

The model used to describe relations among the involved variables is based on linear regression. The application of ordinary least square method leads to the following estimated regression coefficients $\beta_0 \cong 98.48$, $\beta_1 \cong 0.63$, $\beta_2 \cong 4.57$ so the regression line is

$$\hat{y}_i = 98.48 + 0.63x_{1i} + 4.57x_{2i}$$

Once getting the estimated Y values, we assign their ranks and finally order Y values according to $\hat{y}_i$ ranks. All the results are summarized in Table 2: through all these information we can compute concordance index in a multivariate context using (8) recalling that y_i^* represent the Y variable values ordered with respect to $\hat{y}_i$ ranks. Concordance index assumes value 0.801 proving that there is a strong concordance relation among the response variable Y and the explanatory variables X_1, X_2: this conclusion is well clear in Fig. 1 where concordance curve (denoted with the continuous black line), is very close to Y variable Lorenz curve (denoted by the dash dot line). A further verification of this result is provided by the Plotnick indicator (12), whose numerical value is very close to 0, meaning that the distance between concordance function and Lorenz curve is minimum.

3 Conclusion

Through this analysis it has been proved that dependence study can be led in terms of concordance and discordance topics: the choice of a linear regression model is limited when one considers only quantitative variable. In the described context we referred to quantitative variables because we started from the source of the concordance problem involving the income amount before and after taxation intended as a quantitative character.

A future extension can regard the application of the concordance index analysis in cases when one of the considered variable is binary and the adopted model is a logistic regression.

Another important development is establishing if there exists a relation between the determination coefficient, intended as a dependence measure in a linear regression model, and the concordance index: our further research, focused on this topic, is in progress.

Acknowledgements The authors acknowledge financial support from the European grant EU-IP MUSING (contract number 027097). The paper is the result of the close collaboration among the authors, however, it has been written by Emanuela Raffinetti with the supervision of Prof. Paolo Giudici.

References

Leti, G.: Statistica descrittiva. Il Mulino (1983)

Muliere, P.: Alcune osservazioni sull'equità orizzontale di una tassazione. Scritti in onore di Francesco Brambilla. Ed. by Bocconi Comunicazione **2**, (Milano, 1986)

Musgrave, R.A.: The Theory of Public Finance. New York, Mc Graw Hill (1959)

Petrone, S., Muliere, P.: Generalized Lorenz curve and monotone dependence orderings. Metron **Vol L**, No. 3–4 (1992)

Plotnick, R.: A Measure of Horizontal Inequity. The review of Economics and Statistics, **2**, 283–288 (1981)

Methods for Reconciling the Micro and the Macro in Family Demography Research: A Systematisation

Anna Matysiak and Daniele Vignoli

Abstract In the second half of the twentieth century, the scientific study of population changed its paradigm from the macro to the micro, so that attention focused mainly on individuals as the agents of demographic action. However, for accurate handling of all the complexities of human behaviours, the interactions between individuals and the context they belong to cannot be ignored. Therefore, in order to explain (or, at least, to understand) contemporary fertility and family dynamics, the gap between the micro and the macro should be bridged. In this contribution, we highlight two possible directions for bridging the gap: (1) integrating life-course analyses with the study of contextual characteristics, which is made possible by the emergence of the theory and tools of multi-level modelling; and (2) bringing the micro-level findings back to macro outcomes via meta-analytic techniques and agent-based computational models.

1 The Need to Bridge the Gap Between the Micro and the Macro Perspectives in Family Demography Research

After mid-twentieth century the scientific study of population changed its paradigm from the macro to the micro so that the main focus of attention has been devoted to individuals as the agents of demographic action. Event-history analysis was born from the need to develop a comprehensive theoretical framework for studying events

A. Matysiak (✉)
Institute of Statistics and Demography, Warsaw School of Economics, ul. Madalińskiego 6/8, 02-513 Warsaw, Poland
e-mail: amatys@sgh.waw.pl

D. Vignoli
Department of Statistics "G. Parenti", University of Florence, Viale Morgagni 59, 50134, Italia
e-mail: vignoli@ds.unifi.it

A. Di Ciaccio et al. (eds.), *Advanced Statistical Methods for the Analysis of Large Data-Sets*, Studies in Theoretical and Applied Statistics,
DOI 10.1007/978-3-642-21037-2_43, © Springer-Verlag Berlin Heidelberg 2012

that occur within the life-course (Courgeau and Lelievre 1997). This new approach led to a much wider set of research into human behaviours than classical macro-demographic analysis. It also allowed to shift the research from the mere description of phenomena to its interpretation, avoiding the risk of ecological fallacy (Salvini and Santini 1999).

Apart from numerous benefits this shift from the macro to the micro brought also some disadvantages. First, for many years the importance of the social and economic context in which individuals live was disregarded and its potential effect on fertility and family behaviours was ignored. Second, the improvement in the access to the individual-level data and development of the techniques of event-history analysis led to an explosion in the number of micro-level studies. These micro-level studies are generally fragmented, however, and often provide contradictory results. Third, more progress is needed as regards the inference about the macro-level outcomes from the micro-level studies. Drawing conclusions from the micro-level studies on macro-level phenomena risks atomistic fallacy as micro-level studies focus often on a specific situation, constituting only a piece in the overall puzzle of understanding contemporary fertility and family dynamics. Additionally, inference can be complicated by possible interactions of micro-level processes.

Recently, a renewed interest in linking macro- and micro-level research has been recorded in many disciplines of social science (e.g. Voss 2007). Scientists now emphasize that bridging the gap between micro- and macro-approaches in family demography research is a prerequisite for a deeper understanding of contemporary fertility and family dynamics. This new trend is reflected in two international demographic research projects conducted within the EU Framework Programmes: Mic-Mac (Willekens et al. 2005) and Repro (Philipov et al. 2009).

Sharing this view, in this contribution we outline the directions for research and the analytical methods which will facilitate successful reconciliation of the micro and the macro in family demography research. In what follows we propose to bridge the macro-to-micro gap by: (1) integrating life-course analyses with contextual characteristics, feasible owing to the emergence of the theory and tools of multi-level modelling; and (2) bringing the micro-level findings back to macro-outcomes via meta-analytic techniques and agent-based computational models. Before we proceed with our analytical suggestions, we briefly present the concept of methodological individualism which initially drove the shift from the macro to the micro level in family demography research.

2 Methodological Individualism

The major inference of methodological individualism is that understanding individual behaviour is crucial for explaining the social phenomena observed at the macro level. Various versions of this doctrine have developed across disciplines. They range from the more extreme, which suggest that social outcomes are created exclusively by individual behaviours, to the less absolute, which additionally assign an important role to social institutions and social structure (Udehn 2002). Such a moderate version of methodological individualism was proposed by Coleman (1990) and adopted in demography (De Bruijn 1999: 19–22).

According to Coleman, the relation between an individual and society runs both from the macro to the micro level and from the micro to the macro level. There are three mechanisms corresponding to this process are: (1) the situational mechanism in which context influences individual background; (2) the action formation mechanism within which individual background affects individual behaviour; and (3) the transformational mechanism which transforms individual actions into a social outcome (see also Hedström and Swedberg 1999; Billari 2006).

Individual life choices are at the centre of this theoretical model. Individuals do not live in a vacuum, however, but are embedded in a social environment – i.e., in a macro-context. This context is a multi-level and multidimensional "structure of institutions that embody information about opportunities and restrictions, consequences and expectations, rights and duties, incentives and sanctions, models, guidelines, and definitions of the world" (De Bruijn 1999: 21). Such information is continuously being transmitted to individuals who acquire, process, interpret, and evaluate it. In this way, the context influences people's life choices, reflected in occurrence or non-occurrence of demographic events, which are subsequently transformed into a social outcome that is observed at the macro level.

An improvement in the availability of longitudinal data as well as the development of event-history analysis tools allowed social researchers to achieve a deeper insight into the action-formation mechanism, or at least into the manner in which the individual background influences people's behaviours. Much less attention has so far been paid to exploring the situational and transformational mechanisms. Below we elaborate on ways these macro-to-micro and micro-to-macro gaps can be closed in empirical research by using the most suitable analytical methods available. Alongside the presentation of these methods, we document a series of examples from literature. For consistency in the general reasoning of this paper, all illustrations refer to the field of family demography.

3 Bridging the Macro-to-Micro Gap: Multi-Level Event-History Analyses

Life-course theory and event-history techniques, which aim to explore people's life choices, have become standard practice in family and fertility research. However, these approaches ignore the fact that individuals are by their very nature nested in households, census tracts, regions, countries, etc., and that these situational contexts affect people's decisions. In light of the conceptual framework proposed by Coleman (1990), this significantly limits our ability to understand human behaviours (Pinnelli 1995; De Rose 1995; Blossfeld 1996; Santini 2000; Rosina and Zaccarin 2000).

Furthermore, such approaches also cause technical problems, as applying single-level models to hierarchically structured data leads to a bias in the model estimates. The reason for this is that single-level models assume the independence of observations which are in fact dependent, as they are nested within one unit. For instance, households residing within the same neighbourhood are likely to have similar characteristics.

The most influential approach that has been created to account for the hierarchical structure of the data is multi-level modelling. Multi-level models see individuals as behavioural agents, embedded in social units (tracts, regions, countries, etc.). They allow the analyst to detect the effect of the context on individual behaviour as well as to identify the macro-characteristics which are mainly responsible for the contextual effect (Borra and Racioppi 1995; Micheli and Rivellini 2000; Zaccarin and Rivellini 2002). The natural implication of these methods is that they blur the artificial boundaries between micro and macro analyses (Voss 2007). Multi-level event-history analysis in particular represents a challenging and so far not much explored opportunity for bridging the gap between analysis of events unfolding over the life-course (the micro approach) and the contextual (macro) approach in family demography research. However, while the methods (and corresponding software packages) are relatively well-established, data availability is a critical point.

In order to conduct a multi-level event-history analysis, longitudinal individual data should be linked with the time-series of contextual indicators. This requires data on the migration histories of the individuals, together with all their other life-course careers, as well as time-series data for contextual indicators. Consequently, this method has so far mainly been employed on cross-sectional data.

Only recently have some researchers started to investigate the influence of macro-level factors on family-related behaviours from a longitudinal perspective. Still fewer have allowed for a hierarchical structure by taking into account the unobserved community-level factors or even by introducing some contextual indicators into models in order to explicitly study their impact on family-related behaviours. As an example we refer to the study by Adserà (2005), who used a multi-level event-history model in order to explore the impact of regional unemployment on childbearing, employing data from the European Community Household Panel (ECHP 1994–2001). The study was conducted on a pooled dataset for thirteen European countries and included information on the country-level gender unemployment gap and the long-term unemployment rate, which was introduced into the model on a higher level than the individual one. Adserà's results clearly indicate that a higher gender gap in unemployment and a higher long-term unemployment rate slow down the transition to motherhood and higher order births.

To summarise, the existing macro-to-micro studies generally make use of data from a national, a regional, or even a municipal level. The available literature not only indicates the differences between countries or regions in the timing of fertility or in fertility intentions, but also demonstrates that a proper accounting for context may change the influence of individual-level factors (Philipov et al. 2009). Consequently, future research should give better recognition to multi-level event-history approaches.

4 Bridging the Micro-to-Macro Gap: Meta-Analyses and Agent-Based Computational Models

Despite the problems with data availability, the contextual influence on action formation is already quite well understood. By contrast, the transformational mechanism (the transfer from the micro to the macro level) is as yet largely unexplored. At the same time, the rapid development of micro-level studies increases the need to summarize the existing individual-level empirical evidence and to relate them to the macro-level outcomes. In this section, we elaborate on two possible ways of bridging the micro-macro gap from the bottom up, namely meta-analysis and agent-based computational models.

4.1 Meta-Analytic Techniques

Meta-analysis, also referred to as a quantitative literature review, can facilitate drawing general conclusions from micro-level findings. This methodology, relatively new in the social sciences, was developed in order to synthesise, combine and interpret a large body of empirical evidence on a given topic. It offers a clear and systematic way of comparing results of different studies, standardised for the country analysed, the method applied, the control variables employed, the sample selected, etc.

In order to conduct a meta-analysis, papers researching a topic of interest are collected in a systematic manner. Estimated coefficients are selected across studies and recalculated in a standardised way into comparable indicators (i.e. effect sizes). The effect sizes constitute the units of statistical analysis, and can be combined into single summary indicators or analysed using regression techniques. The quintessence of this approach is quantifying the effect of interest on the basis of the available micro-level empirical studies.

Meta-analysis has only recently been adopted in family demography research. The very few such studies in this field include meta-analyses of: the aggregate relationship between a population's age structure and its fertility as hypothesised by Easterlin (Waldorf and Byun 2005), the impact of modernisation and strength of marriage norms on divorce risks in Europe (Wagner and Weiss 2006), and the micro-level relationship between fertility and women's employment in industrialised economies (Matysiak and Vignoli 2008). In order to give a better insight into the meta-analysis method, we elaborate shortly on the meta-study by Matysiak and Vignoli (2008). It aimed to synthesise micro-level findings on the relationship between fertility and women's employment in industrialised economies. Two effects were analysed: that of women's work on fertility (90 studies) and that of having young children on women's employment entry (55 studies). The authors found that the micro-level relationship between the two variables is still negative, but its magnitude varies across countries, differing in their welfare policies, the labour market structures and the social acceptance of women's work. This variation in the magnitude of the micro-level relationship explains the existence of the positive cross-country correlation between fertility and women's labour supply, which has been observed in OECD countries since the mid-1980s (Engelhardt et al. 2004).

Meta-analysis certainly is a useful tool for summarising and synthesising the abundant micro-level research. Its unquestionable strength is that effect estimates produced within its framework have higher external validity than those obtained in individual studies owing to the generality of results across various research papers (Shadish et al. 2002). Nevertheless, a weakness of this method lies in the assumption that the micro-to-macro transformation can be achieved through a simple summation of individual-level actions into a macro-level outcome. According to Coleman (1990), the complex interactions between and within social groups, as well as the heterogeneity of individuals, preclude such a simple aggregation. Since demographic choices are made by interacting and heterogeneous individuals, this assumption, implicit in meta-analysis, may not be valid.

4.2 Agent-Based Computational Models

Agent-based computational models come as a solution to this problem. They seem to be the most powerful tool which is available for transforming the micro results to the macro-level outcomes and which allows to account for heterogeneity among individuals and for the complexity of individual-level interactions (Billari and Ongaro 2000; Billari 2006). It includes micro-simulation, which models macro processes on the basis of empirical models (i.e. event-history models, or even multilevel event-history models), as well as formal models of demographic behaviours, which operationalise decision-making processes at the micro level and simulate their outcomes in terms of macro-level indicators. The additional advantage of agent-based computational models is that they allow study of the impact of policy interventions on demographic behaviours, taking into account policy side effects as well as the interactions of policy with other elements of the social system (Van Imhoff and Post 1998). Below we give one example of micro-simulation that was run with the goal of, among others, assessing the macro-level consequences of an increase in women's employment on fertility (Aassve et al. 2006).

The first study was conducted in two steps. First, using the British Household Panel Study, the authors estimated a multi-process hazard model of five interdependent processes: childbirth, union formation, union dissolution, employment entry, and employment exit. They found the employment parameter in the fertility equation to be strongly negative. The micro-simulation conducted in the second step showed, however, that increasing the hazard of employment entry by 10% and decreasing the hazard of employment exit by another 10% led to a decline in the proportion of women having their second child before the age of 40 by only 0.2% points. This was much less than one could have expected from the analysis of the parameter estimates in the fertility equation. The underlying reason was that employment affected fertility also in an indirect way: it had a positive impact on the time spent in a union, which in turn facilitated childbearing. In short, the negative direct and the positive indirect effect of employment on fertility cancelled each other out, resulting in very small general effects of employment on fertility. This study clearly demonstrated that interpreting parameters from a hazard model alone is not enough to conclude on the subsequent macro-level developments. The interactions between the processes should also be taken into account.

5 Towards an Empirical Implementation of the Theoretical Model: Implications for Data Collection and an Avenue for Future Research

The concepts and relationships presented in this paper are summarised in Fig. 1, which illustrates the theoretical model of methodological individualism in the context of family demography research (see also Muszyńska 2007: 169; Philipov et al. 2009: 17). The scheme of the theory is supplemented with information on analytical methods that could support formation of a comprehensive explanation of the mechanisms and factors driving change in family-related outcomes, as observed at the macro-level. In short, multi-level event-history models are suggested for operationalising the situational and action formation mechanisms, while meta-analyses and agent-based computational models are viewed to be the most suitable for quantifying the transformational mechanism.

We believe that in the future it will be possible to implement this full theoretical model in a single study in the field of family demography. The major challenge to be faced at that stage will be collection of suitable data. Today, in fact, the gap between the analytical tools available and the proper data seems to be the most important barrier preventing population scientists from following the research framework suggested. Conducting a multi-level event-history analysis requires data on the migration histories of individuals together with all other life-histories,

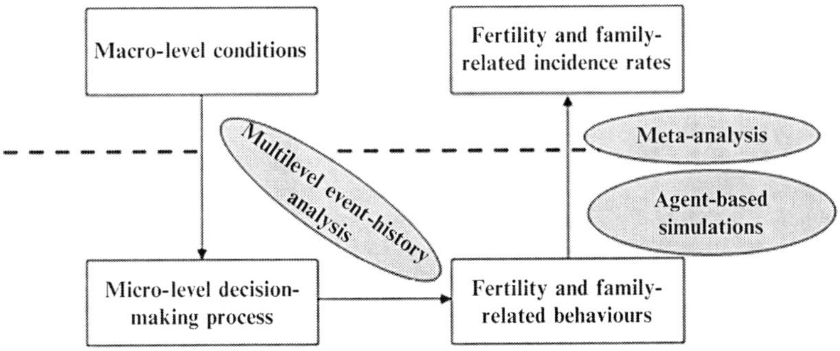

Fig. 1 Theoretical model for the explanation of family and fertility dynamics complemented with the most suitable methods for its implementation

as well as time-series contextual data. Similarly, performing a micro-simulation requires information on several individual life-histories that are often closely connected. To date, such data are not available. It should be noted, however, that substantial advancement in this direction has been made within the Generations and Gender Programme (GGP) (Vikat et al. 2007; Kveder 2009). Its international harmonised database will include individual life-histories of respondents residing in over twenty developed countries. It will additionally be supplemented by the Contextual Database, which contains high quality data at the national or regional level (Spielauer 2006). Furthermore, other contextual indicators can be found in the Family Database developed by the OECD or in the EDACWOWE Portal developed within the RECWOWE (Reconciling work and welfare in Europe) project. A serious drawback of the GGP is its very limited scope of information on migration histories of the respondents, which impedes the possibilities of linking the longitudinal individual data with the time-series of contextual indicators. In future data collection programmes, care should be taken to eliminate this shortcoming.

References

Adserà, A.: Vanishing children: From high unemployment to low fertility in developed countries. American Economic Review, **95**(2), 189–193 (2005).

Aassve, A., Burgess, S., Propper, C., Dickson, M.: Employment, family union and childbearing decisions in Great Britain. Journal of the Royal Statistical Society, **169**(4), 781–804 (2006).

Billari F.C.: Bridging the gap between micro-demography and macro-demography. In: Caselli, G., Vallin, J., Wunsch, G. (Eds.) Demography: analysis and synthesis Vol. 4, pp. 695–707. Academic Press (Elsevier), New York (2006).

Billari, F.C., Ongaro, F.: Quale ruolo per una demografia computazionale? Proceedings of the XL Riunione Scientifica della Società Italiana di Statistica, Firenze, 04 26-28 2000, pp. 245–256 (2000).

Blossfeld, H.P.: Macro-sociology, Rational Choice Theory, and Time A Theoretical Perspective on the Empirical Analysis of Social Processes. European Sociological Review, **12**(2), 181–206 (1996).

Borra, S., Racioppi, F.: Modelli di analisi per dati complessi: lintegrazione tra micro e macro nella ricerca multilevel. In: Sosiet Italiana di Statistica, Continuit e discontinuit nei processi demografici. L'Italia nella transizione demografica, pp. 303–314, April 20-21 1995, Università degli Studi della Calabria, Arcavacata di Rende: Rubbettino, Soveria Mannelli (1995).

Coleman, J. S.: Foundations of social theory, Harvard University Press, Harvard (1990).

Courgeau, D., Lelievre, E.: Changing Paradigm in Demography. Population: An English Selection. **9**(1), 1–10 (1997).

De Bruijn, B. J.: Foundations of demographic theory: choice, process, theory, Thela Thesis, Amsterdam (1999).

De Rose, A.: Uniformit di modelli individuali e divergenze di modelli collettivi nello studio dei comportamenti familiari. In: Sosietà Italiana di Statistica, Continuit e discontinuit nei processi demografici. L'Italia nella transizione demografica, pp. 323–330, April 20-21 1995, Università degli Studi della Calabria, Arcavacata di Rende: Rubbettino, Soveria Mannelli (1995).

Engelhardt, H., Kogel, T., Prskawetz, A.: Fertility and women's employment reconsidered: A macro-level time-series analysis for developed countries, 19602000. Population Studies, **58**(1), 109–120.

Hedström, P., Swedberg, R.: Social mechanisms. An analytical approach to social theory, Cambridge University Press, Cambridge (1999).

Kveder, A.: Generation and Gender Program Micro-Macro Data Source on Generational and Gender Ties, Proceedings of the Conference. In: Italian Statistical Society, Statistical Methods for the Analysis of Large Data-Sets, pp. 35–38, Invited Papers, September 23-25, 2009, Pescara, Italy, (2009).

Matysiak, A., Vignoli, D.: Fertility and Women's Employment: A Meta-Analysis. European Journal of Population, **24**(4), 363–384, (2008).

Micheli, G., Rivellini, G.: Un contesto significativamente influente: appunti per una modellazione multilevel ragionata. Proceedings of the XL Riunione Scientifica della Societ Italiana di Statistica, Firenze, 04 26-28 2000, pp. 257–272, (2000).

Muszyńska M.: Structural and cultural determinants of fertility in Europe. Warsaw School of Economics Publishing, Warsaw (2007).

Philipov, D., Thvenon, O., Klobas, J., Bernardi, L., Liefbroer, A.: Reproductive Decision-Making in a Macro-Micro Perspective (REPRO). State-of-the-Art Review. European Demographic Research Papers 2009(1), Vienna Institute for Demography (2009).

Pinnelli, A.: Introduzione alla sessione "Dimensione micro e macro dei comportamenti demografici: quadri concettuali e modelli di analisi". In: Sosietà Italiana di Statistica, Continuit e discontinuit nei processi demografici. L'Italia nella transizione demografica, pp. 285–290, April 20-21 1995, Università degli Studi della Calabria, Arcavacata di Rende: Rubbettino, Soveria Mannelli (1995).

Rosina, A., Zaccarin, S.: Analisi esplicativa dei comportamenti individuali: una riflessione sul ruolo dei fattori macro. Proceedings of the XL Riunione Scientifica della Società Italiana di Statistica, Firenze, 04 26-28 2000, pp. 273–284 (2000).

Salvini, S., Santini, A.: Dalle biografie alle coorti, dalle coorti alle biografie. In Billari, F., Bonaguidi, A., Rosina, A., Salvini, S., Santini, S. (Eds.). Quadri concettuali per la ricerca in demografia, Serie Ricerche teoriche, Dipartimento di Statistica "G. Parenti", Firenze (1999).

Santini, A.: Introduzione alla sessione specializzata: Analisi micro e macro negli studi demografici. Proceedings of the XL Riunione Scientifica della Società Italiana di Statistica, Firenze, 04 26-28 2000, pp. 241–243 (2000).

Shadish, W. R., Cook, T. D., Campbell D. T.: Experimental and quasi-experimental designs for generalized causal inference, Houghton Mifflin, Boston (2002).

Spielauer, M.: The Contextual Database of the Generations and Gender Programme. MPIDR Working Paper WP-2006-030, Rostock (2006).

Udehn, L.: The changing face of methodological individualism. Annual Review of Sociology, **28**, 479–507 (2002).

Van Imhoff, E., Post, W.: Microsimulation models for population projection, Population (English Edition), **10**(1), 97–136 (1998).

Vikat, A., Spéder, Z., Beets, G., Billari, F. C., Bühler, C., Désesquelles, A., Fokkema, T., Hoem, J. M., MacDonald, A., Neyer, G., Pailhé, A., Pinnelli, A., Solaz, A.: Generations and Gender Survey (GGS): Towards a better understanding of relationships and processes in the life course. Demographic Research, **17**, Article 14, 389–440 (2007).

Voss, P.: Demography as a Spatial Social Science, Population Research and Policy Review, **26**(4), 457–476 (2007).

Wagner, M., Weiss, B.: On the Variation of Divorce Risks in Europe: Findings from a Meta-Analysis of European Longitudinal Studies. European Sociological Review, **22**(5), 483–500 (2006).

Waldorf, B., Byun, P.: Meta-analysis of the impact of age structure on fertility. Journal of Population Economics,**18**, 15–40 (2005).

Willekens, F. J.: Understanding the interdependence between parallel careers. In Siegers, J.J., de Jong-Gierveld, J., van Imhoff, E. (Eds.). Female labour market behaviour and fertility: A rational-choice approach. Berlin: Springer (1991).

Willekens, F. J.: The life-course approach: Models and analysis. In Van Wissen, L. J. G., Dykstra, P. A. (Eds.) Population issues. An interdisciplinary focus, Dordrecht: Kluwer Academic Publishers (1999).

Willekens, F., de Beer, J., van der Gaag, N.: MicMac From demographic to biographic forecasting. Paper prepared for presentation at the Joint Eurostat-ECE Work Session on Demographic Projections, September 21-23, 2005, Vienna (2005).

Zaccarin, S., Rivellini, G., Multilevel analysis in social research: An application of a cross-classified model. Statistical Methods & Applications, **11**, 95–108 (2002).

CPSIA information can be obtained at www.ICGtesting.com
Printed in the USA
LVOW080500150812

294388LV00007B/4/P

9 783642 210365